INTRODUCTORY ALGEBRA

FOURTH EDITION

INTRODUCTORY ALGEBRA

FOURTH EDITION

MERVIN L. KEEDY
PURDUE UNIVERSITY

MARVIN L. BITTINGER
INDIANA UNIVERSITY
PURDUE UNIVERSITY AT INDIANAPOLIS

ADDISON-WESLEY PUBLISHING COMPANY

Reading, Massachusetts ▪ Menlo Park, California
London ▪ Amsterdam ▪ Don Mills, Ontario ▪ Sydney

SPONSORING EDITOR:	*Patricia Mallion*
PRODUCTION EDITORS:	*Robert Hartwell Fiske and Martha K. Morong*
TEXT DESIGNERS:	*Marshall Henrichs and Vanessa Piñeiro*
ILLUSTRATOR:	*VAP Group Limited*
COVER DESIGNER:	*Richard Hannus*
ART COORDINATOR:	*Susanah H. Michener*
PRODUCTION MANAGER:	*Herbert Nolan*
PRODUCTION COORDINATOR:	*Sheryl Stenzel*

The text of this book was composed in Century Schoolbook by Syntax International.

Library of Congress Cataloging in Publication Data

Keedy, Mervin Laverne.
  Introductory algebra.

  Includes index.
  1. Algebra. I. Bittinger, Marvin L. II. Title.
QA152.2.K43 1982   512.9   82–13771
ISBN 0–201–14785–8

Reprinted with corrections, October 1984

ISBN 0–201–14785–8
    HIJ–MU–89876

The success of the first three editions of this text has been most gratifying. We wish to thank the many instructors and students who have both praised and criticized our product. The criticism has helped us tremendously in making significant improvements in the book.

WHAT'S NEW IN THE FOURTH EDITION?

In addition to the computerized testbank and videotapes mentioned in the list of supplements that follows this preface, there are the following new features.

- *Extension Exercises.* At the end of each chapter is a set of extension exercises, keyed to the sections of the chapter. These exercises require the student to go beyond the individual objectives of the sections. They may, for example, ask the student to synthesize the objectives of several sections.

- *Aside Comments.* A new feature of this edition is the use of aside comments to students. Many of them are cautions or suggestions for avoiding common errors. These comments are enclosed in colored octagons.

- *Polishing of Presentation.* We have further improved the development in spots that have been troublesome for some learners. The changes are numerous, and many of them are not minor. We expect that the improvements, based on users' suggestions, will make this edition a good deal more teachable than the previous editions.

FEATURES OF THE BOOK

One of the important features of this book is its format. Most pages have an outer margin, used for material of several types, including developmental exercises and statements of objectives.

- *Objectives.* The objectives for each section are stated in the margin. They have been made easy to identify with a domino symbol ▮. That same symbol is placed in the text and also in the exercises and tests beside the corresponding material, making the text easy to read and usable for self-study.

- *Margin Exercises.*   The exercises in the margins are designed to get the student *involved actively* in the development, an important feature. Those exercises are always very much like the examples in the text and also like the homework exercises.

- *Exercise Sets.*   The exercises for each section are placed at the end of the section on tear-out sheets. The spacing of the answer lines matches that in the *Answer Booklet* to make for quick and easy grading. The exercises are carefully indexed to the objectives, and identified by the domino symbols ⚫⚫. The exercises are also paired, so that each even-numbered exercise is very much like the one that precedes it. They are also carefully graded, the easier exercises coming first. Answers to the odd-numbered exercises are in the back of the book. Special, optional calculator exercises and optional challenge exercises have also been included at the ends of certain exercise sets.

- *Readiness Checks.*   A readiness check is a short set of exercises that can be used to determine whether a student has acquired the skills that are prerequisite to success with a chapter. Readiness checks have been placed at the beginnings of all chapters except the first one. They are indexed to places in the text where the topic concerned is covered. The readiness checks are optional.

- *Tests and Reviews.*   At the end of each chapter there is a set of questions that can be used as a test or a review, and there is a final exam at the end of the book. The items are again indexed to the places in the text where the material concerned is covered. Additional test materials are provided in the *Instructor's Manual with Tests* and the computerized testbank.

TEACHING MODES

This book can be used in many ways. We list a few.

- *As an ordinary textbook.*   To use the book this way, the instructor and students simply ignore the margin exercises.

- *For a modified lecture.*   To bring student-centered activity into the class, the instructor stops lecturing and has the students do margin exercises at appropriate times.

- *For a no-lecture class.*   The instructor makes assignments that students do on their own, including working exercise sets. During the following class period the instructor answers questions, and students have an extra day or two to polish their work before handing it in. In the meantime, they are working on the next assignment. This method provides individualization while keeping a class together. It minimizes the number of instructor hours required and has been found to work well with large classes.

- *In a learning laboratory.*   This book has a low reading level and high understandability, so it is quite suitable for use in a learning laboratory or any kind of learning situation that is essentially self-study.

INTENDED AUDIENCE AND SCOPE

This book is one of a series of basic mathematics texts, which include the following:

> *Arithmetic*
> *Introductory Algebra*
> *Intermediate Algebra*
> *College Algebra* (or *Algebra and Trigonometry*)

The first book in the series, *Arithmetic*, covers arithmetical skills as a prelude to an introductory algebra course. By appropriate choices of topics it can also be used for a terminal course in arithmetic. The present book, *Introductory Algebra*, the second book of the series, is for students who have not studied algebra. It is a beginner's introduction, although it may also be appropriate for those needing a review. The third book, *Intermediate Algebra*, is intended for use by students who have finished a first course in algebra in high school or have taken a comparable course, such as that provided by this book. *Intermediate Algebra* starts "at the beginning," but proceeds faster than an introductory book does, and covers the usual topics of intermediate algebra, including quadratics, exponential functions, and logarithms. It provides adequate preparation for a course in finite mathematics or brief calculus. The last book, *College Algebra*, provides the level of algebraic maturity usually desired as a prerequisite to the mainstream calculus.

A Placement Test is available to help place students in the appropriate book of the series.

ACKNOWLEDGMENTS

We wish to express great thanks to Don Campbell, of Cochise College, for his suggestions regarding the alternate factoring method in Chapter 4. Our thanks go to the following people who reviewed the manuscript: Mary Jean Brod, of the University of Montana; Donald R. Johnson, of Scottsdale Community College; and Barbara Poole, of North Seattle Community College. To the following participants in the Focus Group, we are also grateful: Laura Cameron, of the University of New Mexico; Marianna McClymonds, of Phoenix College; Jacky Peterson, formerly of Arizona State University and now in private industry; Dennis H. Sorge, of Purdue University; James W. Stark Jr., of Lansing Community College; and John D. Whitesitt, of Southern Oregon State College.

M.L.K.
M.L.B.
January, 1983

SUPPLEMENTS

This book is one of a series. The series is accompanied by the following kinds of supplements.

■ *Computerized Testbank.*   Test items can be randomly selected and a test constructed by a computer, using the testbank. The bank can be used with almost all computer hardware currently available. The bank contains approximately ten test items for each objective considered to be a terminal objective. The test question format is multiple choice, but the questions are worded in such a way that an ordinary constructed-response test can also be constructed, merely by leaving off the answer choices. The computerized testbank is available to users in three ways:

1. *On computer tape.* In this mode the bank can be used with most large computing systems. Users may have to specify the kind of tape their system accepts.

2. *On mini-floppy discs for TRS-80 microcomputers.*

3. *On mini-floppy discs for Apple II microcomputers.*

Final copies of tests can be produced from our programs where a letter quality printer is available (such as NEC spinwriter, DIABLO, or CUME) by using the printout and writing in only some special characters and certain graphs. If a printer is not available, the computer will still generate tests. Notation such as fractions and exponents, requiring spacing other than the usual one-line-at-a-time, will be printed in coded form. That notation is then to be added and the tests can then be typed and duplicated.

The authors owe very special thanks to John Whitesitt, of Southern Oregon State College, for his work in designing and producing the computerized testbank.

■ *Printed Testbank.*   The same test items appear in the printed form of the testbank as in the computerized testbank. The printed bank is usable by those who do not wish to use a computer or who do not have access to a computer.

- *Videotape Cassettes.*   Color TV is used to help students over the rough spots. A lecturer (John Jobe, of Oklahoma State University) speaks to students and works out examples, with lucid explanations. These video cassettes can be used to supplement lectures or in individualized learning situations, such as learning laboratories.

- *Audiotape Cassettes.*   In the tape recorder, the student has the voice of a friendly tutor, who helps him or her by talking through examples and problems, using the book itself as a visual device, to make a truly audiovisual presentation. These audiotapes are especially useful in individualized learning situations, such as learning laboratories, and are an excellent help for students who have missed a number of classes.

- *Instructor's Manual with Tests.*   In this booklet there are five alternative forms of each chapter test and five alternative forms of the final examination. They are classroom ready, and all answers are provided.

- *Answer Booklet.*   This booklet contains answers to *all* exercises. Answers to the odd-numbered exercises are in the text. The booklet is a great time-saver for the instructor, and it can also be made available to students, if desired, at low cost.

- *Student's Guide to Exercises.*   Complete worked-out solutions are provided in this booklet. There are solutions to about 70 percent of the odd-numbered exercises in the exercise sets. That represents a change from the third edition, where most of the worked-out exercises were margin exercises, as found in the paperback books of the series. The change was made because it was found that help with exercises from the exercise sets themselves was preferred by students.

- *Placement Test.*   To help place a student properly in one of the algebra books of the series, a test has been constructed. The test is handy because it is short, although its brevity may make it appear less useful than it actually is. It was thoroughly evaluated statistically and is known to give a good screening with respect to placement into one of four categories into which the books of the series fall.

  I ARITHMETIC, Fourth Edition

 II INTRODUCTORY ALGEBRA, Fourth Edition
    ALGEBRA: AN INTRODUCTORY COURSE*

III INTERMEDIATE ALGEBRA, Fourth Edition
    ALGEBRA: AN INTERMEDIATE COURSE*

IV COLLEGE ALGEBRA, Third Edition
    FUNDAMENTAL COLLEGE ALGEBRA, Second Edition*
    ALGEBRA AND TRIGONOMETRY, Third Edition
    FUNDAMENTAL ALGEBRA AND TRIGONOMETRY, Second Edition*
    TRIGONOMETRY: TRIANGLES AND FUNCTIONS, Third Edition

*These texts are hardcover versions of the paperbound texts.

More precise placement within a specific book can be made by using an alternative form of the final examination as a pretest.

CONTENTS

IBM

$127.50
62.47
13.12
108.00
24.37
62.50
1152.75
68.00
47.76
31.22
78.62
94.00
103.00
27.31
16.50
75.00
913.47
702.03
12.08
16.42
19.35
111_

OFF PUSH

THE NUMBERS OF ORDINARY ARITHMETIC AND THEIR PROPERTIES

1

Terminating or repeating decimals represent numbers of arithmetic. They are important in algebra.

OBJECTIVES

After finishing Section 1.1, you should be able to:

◖•◗ Write several numerals for any number of arithmetic by multiplying by 1.

◖••◗ Find simplest fractional notation for a number of arithmetic.

◖•••◗ Add, subtract, and multiply with fractional notation.

1. Write three fractional numerals for $\frac{2}{3}$.

2. Write three fractional numerals for $\frac{1}{2}$.

3. Write three fractional numerals for $\frac{3}{5}$.

1.1 NUMBERS AND ALGEBRA

In this chapter we do two things. We review arithmetic. We also give an introduction to what algebra is and how it works. Later we go into more detail about algebra.

The symbols we use to name numbers are called *numerals*. Thus symbols such as 4, 8, XVI, and $\frac{3}{4}$ are numerals. In algebra we also use symbols such as x, y, a, and b that represent numbers to calculate and solve problems.

◖•◗ NUMBERS OF ORDINARY ARITHMETIC

The numbers used for counting are called *natural numbers*. They are

1, 2, 3, 4, 5, 6, 7, 8, 9, 10, 11,   and so on.

The *whole numbers* consist of the natural numbers and zero. They are

0, 1, 2, 3, 4, 5, 6, 7, 8, 9, 10, 11,   and so on.

The *numbers of ordinary arithmetic* consist of the whole numbers and the fractions such as $\frac{2}{3}$ and $\frac{9}{5}$. These numbers are sometimes called the *numbers of arithmetic*. All the numbers of arithmetic can be named by *fractional notation a/b*.

Example 1 Write three fractional numerals for the number *three fourths*.

$$\frac{3}{4}, \quad \frac{6}{8}, \quad \frac{600}{800}$$

Example 2 Write three fractional numerals for the number *one third*.

$$\frac{1}{3}, \quad \frac{2}{6}, \quad \frac{10}{30}$$

> Numerals are *names* for numbers. A number has many names.

Any fractional numeral with a denominator three times the numerator names the number *one third*.

DO EXERCISES 1–3 (IN THE MARGIN AT THE LEFT).

Whole numbers can also be named with fractional notation.

Examples

3. $0 = \dfrac{0}{1} = \dfrac{0}{4} = \dfrac{0}{56}$, and so on.

4. $2 = \dfrac{2}{1} = \dfrac{6}{3} = \dfrac{128}{64}$, and so on.

5. $1 = \dfrac{2}{2} = \dfrac{5}{5} = \dfrac{642}{642}$, and so on.

DO EXERCISES 4 AND 5 (IN THE MARGIN).

To find different names, we can use the idea of multiplying by 1.

Examples

6. $\dfrac{2}{3} = \dfrac{2}{3} \times \boxed{1}$

$= \dfrac{2}{3} \times \boxed{\dfrac{5}{5}}$ Using $\dfrac{5}{5}$ for 1

$= \dfrac{10}{15}$

> How did we choose $\frac{5}{5}$? We just picked a number—
> in this case, 5—and multiplied by $\frac{5}{5}$.

7. $\dfrac{2}{3} = \dfrac{2}{3} \times \boxed{1}$

$= \dfrac{2}{3} \times \boxed{\dfrac{13}{13}}$ This time we use $\dfrac{13}{13}$ for 1.

$= \dfrac{26}{39}$

DO EXERCISES 6 AND 7.

●● SIMPLEST FRACTIONAL NOTATION

The simplest fractional notation for a number has the smallest possible numerator and denominator. To simplify, we reverse the process used in Examples 6 and 7.

Examples   Simplify.

8. $\dfrac{10}{15} = \dfrac{2 \times \boxed{5}}{3 \times \boxed{5}}$ Factoring numerator and denominator
Factoring means writing as a product.

$= \dfrac{2}{3} \times \boxed{\dfrac{5}{5}}$ Factoring the fraction

$= \dfrac{2}{3}$ We "removed" a factor of 1.

> We have now simplified. We have the smallest possible
> numerator and denominator.

4. Write three fractional numerals for 1.

5. Write three fractional numerals for 4.

6. Multiply by 1 to find three different names for $\dfrac{4}{5}$.

7. Multiply by 1 to find three different names for $\dfrac{8}{7}$.

ANSWERS ON PAGE A–1

Simplify.

8. $\dfrac{18}{27}$

9. $\dfrac{38}{18}$

10. $\dfrac{56}{49}$

Simplify.

11. $\dfrac{27}{54}$

12. $\dfrac{48}{12}$

9. $\dfrac{36}{24} = \dfrac{6 \times 6}{4 \times 6} = \dfrac{3 \times 2 \times 6}{2 \times 2 \times 6} = \dfrac{3}{2} \times \dfrac{2 \times 6}{2 \times 6} = \dfrac{3}{2}$

You may be in the habit of canceling. For example, you might have done Example 9 as follows. This is a shortcut for the procedure used in Examples 8 and 9.

$$\dfrac{\overset{3}{\cancel{18}}}{\underset{2}{\cancel{\cancel{12}}}} \quad \text{or} \quad \dfrac{36}{24} = \dfrac{3 \times \cancel{12}}{2 \times \cancel{12}} = \dfrac{3}{2}$$

> *Caution:* The difficulty with canceling is that it is applied incorrectly in situations such as the following:
>
> $$\dfrac{\cancel{2} + 3}{\cancel{2}} = 3, \qquad \dfrac{\cancel{4} + 1}{\cancel{4} + 2} = \dfrac{1}{2}, \qquad \dfrac{1\cancel{5}}{\cancel{5}4} = \dfrac{1}{4}.$$
>
> Wrong!   Wrong!   Wrong!
>
> In each of these situations the expressions canceled were *not* factors. Factors are parts of products. For example, in $2 \cdot 3$, 2 and 3 are factors, but in $2 + 3$, 2 and 3 are *not* factors. If you can't factor, you can't cancel! If in doubt, don't cancel!

DO EXERCISES 8–10.

We can put a factor of 1 in the numerator or denominator.

Examples   Simplify.

10. $\dfrac{18}{72} = \dfrac{2 \times 9}{8 \times 9}$

$\qquad = \dfrac{1 \times 2 \times 9}{4 \times 2 \times 9}$ Putting a 1 in the numerator

$\qquad = \dfrac{1}{4} \times \dfrac{2 \times 9}{2 \times 9}$

$\qquad = \dfrac{1}{4}$

11. $\dfrac{72}{9} = \dfrac{8 \times 9}{1 \times 9} = \dfrac{8}{1} \times \dfrac{9}{9} = \dfrac{8}{1} = 8$

DO EXERCISES 11 AND 12.

◖◗◗ MULTIPLICATION, ADDITION, AND SUBTRACTION

Example 12   Multiply and simplify: $\dfrac{5}{6} \cdot \dfrac{9}{25}$.

$$\frac{5}{6} \cdot \frac{9}{25} = \frac{5 \cdot 9}{6 \cdot 25} \qquad \text{Multiplying numerators and denominators}$$

$$= \frac{1 \cdot 5 \cdot 3 \cdot 3}{2 \cdot 3 \cdot 5 \cdot 5} \qquad \text{Factoring numerator and denominator}$$

$$= \frac{3 \cdot 5 \cdot 1 \cdot 3}{3 \cdot 5 \cdot 2 \cdot 5}$$

$$= \frac{\boxed{3 \cdot 5}}{\boxed{3 \cdot 5}} \cdot \frac{1 \cdot 3}{2 \cdot 5} \qquad \text{Factoring the fraction}$$

$$= \frac{3}{10} \qquad \text{``Removing'' a factor of 1}$$

In Example 12, we used a dot $\cdot$ instead of $\times$ for a multiplication sign. It means exactly the same thing as $\times$.

DO EXERCISES 13 AND 14.

We can multiply by 1 to find common denominators.

Example 13   Add and simplify: $\dfrac{3}{8} + \dfrac{5}{12}$

Method 1:

$$\frac{3}{8} + \frac{5}{12} = \frac{3}{8} \cdot \frac{\boxed{12}}{\boxed{12}} + \frac{5}{12} \cdot \frac{\boxed{8}}{\boxed{8}} \qquad \text{Multiplying by 1}$$

$$= \frac{36}{96} + \frac{40}{96}$$

$$= \frac{76}{96} \qquad \text{Adding}$$

$$= \frac{19}{24} \cdot \frac{\boxed{4}}{\boxed{4}} \qquad \text{Factoring}$$

$$= \frac{19}{24} \qquad \text{``Removing'' a factor of 1 to simplify}$$

You may know how to add by first finding the least common multiple of the denominators. If you do, the problem can be done as follows.

Method 2:   The least common multiple of the denominators is 24. We multiply by 1 to obtain the least common multiple for each denominator.

$$\frac{3}{8} + \frac{5}{12} = \frac{3}{8} \cdot \frac{\boxed{3}}{\boxed{3}} + \frac{5}{12} \cdot \frac{\boxed{2}}{\boxed{2}} \qquad$$

Multiplying by 1. Since $3 \cdot 8 = 24$, we multiply the first number by $\frac{3}{3}$. Since $2 \cdot 12 = 24$, we multiply the second number by $\frac{2}{2}$.

$$= \frac{9}{24} + \frac{10}{24}$$

$$= \frac{19}{24}$$

Multiply and simplify.

13. $\dfrac{6}{5} \cdot \dfrac{25}{12}$

14. $\dfrac{3}{8} \cdot \dfrac{5}{3} \cdot \dfrac{7}{2}$

ANSWERS ON PAGE A–1

Add and simplify.

15. $\dfrac{5}{6} + \dfrac{7}{10}$

16. $\dfrac{1}{4} + \dfrac{1}{3}$

Subtract and simplify.

17. $\dfrac{4}{5} - \dfrac{4}{6}$

18. $\dfrac{5}{12} - \dfrac{2}{9}$

Using this method we have less simplifying to do at the end. In this example we had none at all.

Use this method if you recall how to find least common multiples. We consider this in detail in Chapter 8.

DO EXERCISES 15 AND 16.

Example 14 Subtract and simplify: $\dfrac{9}{8} - \dfrac{4}{5}$.

$$\frac{9}{8} - \frac{4}{5} = \frac{9}{8} \cdot \frac{5}{5} - \frac{4}{5} \cdot \frac{8}{8} = \frac{45}{40} - \frac{32}{40} = \frac{13}{40}$$

Example 15 Subtract and simplify: $\dfrac{7}{10} - \dfrac{1}{5}$.

$$\frac{7}{10} - \frac{1}{5} = \frac{7}{10} - \frac{1}{5} \cdot \frac{2}{2} = \frac{7}{10} - \frac{2}{10} = \frac{5}{10} = \frac{1}{2} \cdot \frac{5}{5} = \frac{1}{2}$$

DO EXERCISES 17 AND 18.

> In arithmetic you usually write $1\frac{1}{8}$ rather than $\frac{9}{8}$. In algebra you will find that the so-called *improper* symbols such as $\frac{9}{8}$ are more useful.

NAME   CLASS   ANSWERS

EXERCISE SET 1.1

◖•◗ Write four different fractional numerals for each number.

1. $\dfrac{4}{3}$ 2. $\dfrac{5}{9}$ 3. $\dfrac{6}{11}$ 4. $\dfrac{15}{7}$

5. $\dfrac{2}{11}$ 6. 1   7. 5   8. 0

◖••◗ Simplify.

9. $\dfrac{8}{6}$ 10. $\dfrac{15}{25}$ 11. $\dfrac{17}{34}$ 12. $\dfrac{35}{25}$

13. $\dfrac{100}{50}$ 14. $\dfrac{13}{39}$ 15. $\dfrac{250}{75}$ 16. $\dfrac{12}{18}$

◖•••◗ Compute and simplify.

17. $\dfrac{1}{4}\cdot\dfrac{1}{2}$ 18. $\dfrac{11}{10}\cdot\dfrac{8}{5}$ 19. $\dfrac{17}{2}\cdot\dfrac{3}{4}$ 20. $\dfrac{11}{12}\cdot\dfrac{12}{11}$

1. _____

2. _____

3. _____

4. _____

5. _____

6. _____

7. _____

8. _____

9. _____

10. _____

11. _____

12. _____

13. _____

14. _____

15. _____

16. _____

17. _____

18. _____

19. _____

20. _____

ANSWERS

21. $\dfrac{1}{2} + \dfrac{1}{2}$ **22.** $\dfrac{1}{2} + \dfrac{1}{4}$ **23.** $\dfrac{4}{9} + \dfrac{13}{18}$ **24.** $\dfrac{4}{5} + \dfrac{8}{15}$

21. _____

22. _____

23. _____

24. _____

25. $\dfrac{3}{10} + \dfrac{8}{15}$ **26.** $\dfrac{9}{8} + \dfrac{7}{12}$ **27.** $\dfrac{5}{4} - \dfrac{3}{4}$ **28.** $\dfrac{12}{5} - \dfrac{2}{5}$

25. _____

26. _____

27. _____

28. _____

29. $\dfrac{13}{18} - \dfrac{4}{9}$ **30.** $\dfrac{13}{15} - \dfrac{8}{45}$ **31.** $\dfrac{11}{12} - \dfrac{2}{5}$ **32.** $\dfrac{15}{16} - \dfrac{2}{3}$

29. _____

30. _____

31. _____

32. _____

1.2 DECIMAL AND EXPONENTIAL NOTATION

DECIMAL NOTATION

Another notation for the numbers of arithmetic is *decimal notation*. It is based on ten. Standard notation, such as 9345 and 20,867, is decimal notation for whole numbers. Decimal notation for fractions, such as 0.9 and 34.248, contains *decimal points*.

■· FROM DECIMAL TO FRACTIONAL NOTATION

To convert from decimal to fractional notation, we can multiply by 1, as shown in the following examples.

Examples   Write fractional notation.

1. $34.2 = \dfrac{34.2}{1} \cdot \boxed{\dfrac{10}{10}}$ Multiplying by 1

$= \dfrac{342}{10}$ 34.2 is 34 and two tenths, or 342 tenths.

> Why did we multiply by $\frac{10}{10}$? We did not want a decimal point in the answer, and we knew that multiplying 34.2 by 10 would give us a whole number, 342. Thus we multiplied by $\frac{10}{10}$ to rename the number.

2. $16.453 = \dfrac{16.453}{1} \cdot \boxed{\dfrac{1000}{1000}}$ Multiplying by 1

$= \dfrac{16453}{1000}$ 16.453 is 16,453 thousandths.

DO EXERCISES 1 AND 2.

■·· FROM FRACTIONAL TO DECIMAL NOTATION

To convert from fractional to decimal notation we divide.

Example 3 $\frac{3}{8}$ means $3 \div 8$, so we divide:

$$
\begin{array}{r}
0.375 \\
8\overline{)3.000} \\
\underline{2\ 400} \\
600 \\
\underline{560} \\
40 \\
\underline{40} \\
\end{array}
$$

Sometimes we get a repeating decimal when we divide.

OBJECTIVES

After finishing Section 1.2, you should be able to:

■· Convert from decimal to fractional notation.

■·· Convert from fractional to decimal notation.

■··· Tell the meaning of exponential notation such as 5^3.

·■· Tell the meaning of expressions with exponents of 0 and 1.

■·■ For expressions such as nnn, n^3, and xyz, give their meanings when the variables are replaced by numbers.

Write fractional notation. You need not simplify.

1. 1.62

2. 35.431

Write decimal notation. Use a bar for repeating decimals.

3. $\dfrac{7}{8}$

4. $\dfrac{4}{5}$

5. $\dfrac{9}{11}$

6. $\dfrac{23}{9}$

7. Write exponential notation for $10 \cdot 10 \cdot 10 \cdot 10$.

8. What is the meaning of 5^4?

9. What is the meaning of x^5?

Example 4 $\dfrac{4}{11}$ means $4 \div 11$, so we divide:

$$
\begin{array}{r}
0.3636\ldots \\
11\overline{)4.00} \\
3\ 3 \\
\hline
70 \\
66 \\
\hline
4\ 0 \\
3\ 3 \\
\hline
70 \\
66 \\
\hline
4
\end{array}
$$

Thus $\dfrac{4}{11} = 0.363636\ldots$. Such decimals are often abbreviated by putting a bar over the repeating part, as follows.

$$\frac{4}{11} = 0.\overline{36}$$

DO EXERCISES 3–6.

••• EXPONENTIAL NOTATION

Shorthand notation for $10 \cdot 10 \cdot 10$ is called *exponential notation*.

For $\underbrace{10 \cdot 10 \cdot 10}_{3\ \text{factors}}$ we write 10^3.

This is read "ten cubed" or "ten to the third power." We call the number 3 an *exponent* and we say that 10 is the *base*. For $10 \cdot 10$ we write 10^2, read "ten squared," or "ten to the second power."

Example 5   Write exponential notation for $10 \cdot 10 \cdot 10 \cdot 10 \cdot 10$.

$$10 \cdot 10 \cdot 10 \cdot 10 \cdot 10 = 10^5$$

DO EXERCISE 7.

An exponent of 2 or greater tells us how many times the base is used as a factor.

Examples

6. 3^5 means $3 \cdot 3 \cdot 3 \cdot 3 \cdot 3$.

7. 7^4 means $7 \cdot 7 \cdot 7 \cdot 7$.

8. If we use n to stand for a number, n^4 means $n \cdot n \cdot n \cdot n$.

DO EXERCISES 8 AND 9.

1 AND 0 AS EXPONENTS

Look for a pattern.

$$10 \cdot 10 \cdot 10 \cdot 10 = 10^4$$
$$10 \cdot 10 \cdot 10 = 10^3$$
$$10 \cdot 10 = 10^2$$
$$10 = 10^?$$
$$1 = 10^?$$

We are dividing by 10 each time. To continue the pattern we would say that

$$10 = 10^1 \quad \text{and} \quad 1 = 10^0.$$

We make the following agreement. It is a definition.

For any number n,

n^1 **means** n,

and if n is not 0, then

n^0 **means 1.**

If n is 0, then n^0 is meaningless. We explain why later.

Examples What is the meaning of each of the following?

9. 5^1 *Answer:* 5

10. 8^1 *Answer:* 8

11. 3^0 *Answer:* 1

12. 5^0 *Answer:* 1

DO EXERCISES 10 AND 11.

In summary, note that there are many kinds of names, or notations, for a number. For example,

0.64 (decimal notation),

$\frac{16}{25}$ (fractional notation),

and

0.8^2 (exponential notation)

are all names for the *same* number. Later, we study *percent notation*, which gives us another name, 64%.

10. What is 4^1?

11. What is 6^0?

ANSWERS ON PAGE A–1

12. Write exponential notation for *nnnnn*.

13. What is the meaning of y^3? Do not use $\times$ or $\cdot$.

14. What number does *Prt* represent if *P* stands for 2000, *r* stands for 0.07, and *t* stands for 1?

15. What does x^3 represent if *x* stands for 2?

8·8 VARIABLES

A letter that can be replaced by different numbers is called a *variable*. If we write two or more variables together, such as

$$nnn, \quad \text{or} \quad lwh, \quad \text{or} \quad Prt,$$

we agree that the numbers are to be multiplied.

Examples

13. *nnn* means $n \cdot n \cdot n$. It also means n^3.

14. *lwh* means $l \cdot w \cdot h$.

15. If *P* is replaced by 1000, *r* is replaced by 0.08, and *t* is replaced by 3, then *Prt* means $1000 \times 0.08 \times 3$, or 240.

DO EXERCISES 12–15.

NAME

EXERCISE SET 1.2

Review the objectives for this lesson.

▮ • Write fractional notation.

1. 29.1 **2.** 16.33 **3.** 4.67 **4.** 3.1415

5. 3.62 **6.** 13.617 **7.** 18.789 **8.** 309.62

▮ • • Write decimal notation.

9. $\dfrac{1}{4}$ **10.** $\dfrac{1}{2}$ **11.** $\dfrac{3}{5}$ **12.** $\dfrac{6}{5}$

13. $\dfrac{2}{9}$ **14.** $\dfrac{4}{9}$ **15.** $\dfrac{1}{8}$ **16.** $\dfrac{5}{8}$

17. $\dfrac{5}{11}$ **18.** $\dfrac{7}{11}$ **19.** $\dfrac{1}{12}$ **20.** $\dfrac{23}{12}$

▮ • • • What is the meaning of each of the following?

21. 5^2 **22.** 8^3 **23.** m^3 **24.** t^2

ANSWERS

1. _____

2. _____

3. _____

4. _____

5. _____

6. _____

7. _____

8. _____

9. _____

10. _____

11. _____

12. _____

13. _____

14. _____

15. _____

16. _____

17. _____

18. _____

19. _____

20. _____

21. _____

22. _____

23. _____

24. _____

ANSWERS

25. _____

26. _____

27. _____

28. _____

29. _____

30. _____

31. _____

32. _____

33. _____

34. _____

35. _____

36. _____

37. _____

38. _____

39. _____

40. _____

41. _____

42. _____

43. _____

44. _____

45. _____

What is the meaning of each of the following? Do not use $\times$ or $\cdot$.

25. x^3 **26.** m^2 **27.** y^4 **28.** p^5

What is the meaning of each of the following?

29. y^0, $y \neq 0$ **30.** h^0, $h \neq 0$ **31.** p^1 **32.** z^1

33. M^1 **34.** 9.68^0 **35.** $\left(\dfrac{a}{b}\right)^0$, $a \neq 0$, $b \neq 0$ **36.** $(ab)^1$

What does each of the following mean when n is replaced by 10?

37. n^2 **38.** n^3 **39.** n^5 **40.** n^4

What does each of the following mean when P is replaced by 1000, r is replaced by 0.16, and t is replaced by 2?

41. r^2 **42.** r^3 **43.** Prt **44.** Pr^2
[*Hint:* This is $P \cdot r^2$, not $(Pr)^2$.]

All exercises below a line like this are extra and optional.

_____

45. ▦ What powers of 6 are too large for your calculator readout?

1.3   PROPERTIES OF NUMBERS OF ARITHMETIC

Some properties of the numbers of arithmetic are so simple that they may seem unimportant. However, they are important, especially in algebra, because they allow us to manipulate unknown numbers. We develop them now.

● PARENTHESES. SYMBOLS OF GROUPING

What does $5 \times 2 + 4$ mean? If we multiply 5 by 2 and add 4, we get 14. If we add 2 and 4 and multiply by 5, we get 30. To tell which operation to do first, we use parentheses.

Examples

1. $(3 \times 5) + 6$ means $15 + 6$, or 21.
2. $3 \times (5 + 6)$ means 3×11, or 33.

DO EXERCISES 1–5.

●● GROUPING AND ORDER

What does $3 + 5 + 4$ mean? Does it mean $(3 + 5) + 4$ or $3 + (5 + 4)$? Either way the sum is 12, so it doesn't matter. In fact, if we are doing addition only, we can group numbers in any manner. This means that we really don't need parentheses if we are doing only addition. This illustrates another property.

> For any numbers a, b, and c,
>
> $(a + b) + c = a + (b + c)$.   **(The associative law of addition)**

Another basic property of numbers is that they can be added or multiplied in any order. For example, $3 + 2$ and $2 + 3$ are the same. Also, $5 \cdot 7$ and $7 \cdot 5$ are the same.

> For any numbers a and b,
>
> $a + b = b + a$;   **(The commutative law of addition)**
> $a \cdot b = b \cdot a$.   **(The commutative law of multiplication)**

DO EXERCISES 6–9.

The commutative and associative laws together help make addition easier.

Example 3   Add $3 + 4 + 7 + 6 + 8$. Look for combinations that make ten.

Add 3 and 7 to make 10, 4 and 6 to make 10, and then add the two 10's. Finally add the 8. The sum is 28.

DO EXERCISES 10 AND 11.

OBJECTIVES

After finishing Section 1.3, you should be able to:

● Do calculations as shown by parentheses.

●● Use grouping and order in addition and multiplication and tell which laws are illustrated by certain sentences.

●●● Use the distributive law to do calculations like $(4 + 8) \cdot 5$ in two ways.

●●●● Use the distributive law to factor expressions like $3x + 3y$.

Evaluate expressions like $4x + 4y$ and $4(x + y)$ when numbers are given for the letters.

Do these calculations.

1. $(5 \times 4) + 2$

2. $5 \times (4 + 2)$

3. $(4 \times 6) + 2$

4. $5 \times (2 \times 3)$

5. $(6 \times 2) + (3 \times 5)$

Do these calculations.

6. $17 + 10$

7. $10 + 17$

8. 26×70

9. 70×26

Add. Look for combinations that make ten.

10. $5 + 2 + 3 + 5 + 8$

11. $1 + 5 + 6 + 9 + 4$

Which laws are illustrated by these sentences?

12. $61 \times 56 = 56 \times 61$

13. $(3 + 5) + 2 = 3 + (5 + 2)$

14. $4 + (2 + 5) = (4 + 2) + 5$

15. $7 \cdot (9 \cdot 8) = (7 \cdot 9) \cdot 8$

Do the calculations as shown.

16. a) $(2 + 5) \cdot 4$

  b) $(2 \cdot 4) + (5 \cdot 4)$

17. a) $(7 + 4) \cdot 7$

  b) $(7 \cdot 7) + (4 \cdot 7)$

Compute.

18. $3 \cdot 5 + 2 \cdot 4$

19. $4 \cdot 2 + 7 \cdot 1$

For multiplication, does the grouping matter? For example, is $2 \cdot (5 \cdot 3) = (2 \cdot 5) \cdot 3$? Our experience with arithmetic tells us that multiplication is also associative.

> **For any numbers a, b, and c,**
>
> $a \cdot (b \cdot c) = (a \cdot b) \cdot c.$ **(The associative law of multiplication)**

Examples   Which laws are illustrated by these sentences?

  4. $3 + 5 = 5 + 3$ Commutative law of addition (order changed)

  5. $(2 + 3) + 5 = 2 + (3 + 5)$ Associative law of addition (grouping changed)

  6. $(3 \cdot 5) \cdot 2 = 3 \cdot (5 \cdot 2)$ Associative law of multiplication

DO EXERCISES 12–15.

●●● THE DISTRIBUTIVE LAW

If we wish to multiply a number by a sum of several numbers, we can either add and then multiply or multiply and then add.

Example 7   Compute in two ways: $(4 + 8) \cdot 5$.

$$\left.\begin{array}{c} (4 + 8) \cdot \boxed{5} \\ 12 \cdot 5 \\ 60 \end{array}\right\} \text{Adding and then multiplying}$$

$$\left.\begin{array}{c} (4 \cdot \boxed{5}) + (8 \cdot \boxed{5}) \\ 20 + 40 \\ 60 \end{array}\right\} \text{Multiplying and then adding}$$

DO EXERCISES 16 AND 17.

The property we are investigating is the *distributive law of multiplication over addition*. Before we state it formally, we need to make an agreement about parentheses. We agree that in an expression like $(4 \cdot 5) + (3 \cdot 7)$, we can omit the parentheses. Thus $4 \cdot 5 + 3 \cdot 7$ means $(4 \cdot 5) + (3 \cdot 7)$. In other words, we do the multiplications first.

DO EXERCISES 18 AND 19.

> **In an expression such as $ab + cd$, it is understood that parentheses belong around ab and cd. In other words, the multiplications are to be done first.**

Using our agreement about parentheses, we now state the distributive law.

For any numbers a, b, and c,

$$a(b + c) = ab + ac.$$ **(The distributive law)**

> Note that there are *two* operations involved in the distributive law: addition and multiplication.

We cannot omit parentheses on the left above. If we did we would have $ab + c$, which by our agreement means $(ab) + c$.

Note that the distributive law can be extended to more than two numbers inside the parentheses.

$$a(b + c + d) = ab + ac + ad.$$

The distributive law would apply to the following situation. Someone decides to invest \$1000 in one bank at 8%, and \$2000 in another bank at 8%. At the end of one year the total interest from the two investments would be

$$(\ 8\%\ \cdot 1000) + (\ 8\%\ \cdot 2000).$$

The same interest would also have been made by investing the entire \$3000 in just one bank. The interest is

$$8\%\ \cdot (1000 + 2000), \quad \text{or} \quad 8\% \cdot 3000.$$

DO EXERCISE 20.

░░ FACTORING

Any equation can be reversed. Thus for the distributive law we could also write

$$ab + ac = a(b + c).$$

The distributive law is the basis for a process called *factoring*.

To *factor* an expression is to write it as a product.

Example 8   Factor: $3x + 3y$.

By the distributive law,

$$\boxed{3}\, x + \boxed{3}\, y = \boxed{3}\, (x + y).$$

When we write $3(x + y)$, we say we have *factored* $3x + 3y$. That is, we have written it as a product.

Example 9   Factor: $5x + 5y + 5z$.

$$\boxed{5}\, x + \boxed{5}\, y + \boxed{5}\, z = \boxed{5}\, (x + y + z)$$

Note that we manipulated the expression on the left according to the distributive law. We did not have to know what the unknowns x, y, and z were.

DO EXERCISES 21–23.

Do these calculations.

20. a) $(0.08 \times 1000) + (0.08 \times 2000)$

   b) $0.08 \times (1000 + 2000)$

Factor.

21. $4x + 4y$

22. $5a + 5b$

23. $7p + 7q + 7r$

Factor. Then evaluate both the original and factored expressions when $x = 4$ and $y = 3$.

24. $5x + 5y$

⠏⠏ FACTORING AND EVALUATING

It is important to realize that when we factor an expression like $3x + 3y$ the factored expression represents the same number as the original one, no matter what numbers we use for x and y.

Example 10   Factor $4x + 4y$. Then evaluate both the original and factored expressions when x stands for 2 and y stands for 3.

First, factor $4x + 4y$:

$$4 \; x \; + \; 4 \; y \; = \; 4 \; (x + y).$$

Evaluate $4x + 4y$:

$$
\begin{aligned}
4x + 4y &= 4 \cdot 2 + 4 \cdot 3 \qquad \text{Replacing } x \text{ by 2 and } y \text{ by 3} \\
&= 8 + 12 \\
&= 20.
\end{aligned}
$$

Evaluate $4(x + y)$:

$$
\begin{aligned}
4(x + y) &= 4(2 + 3) \qquad \text{Replacing } x \text{ by 2 and } y \text{ by 3} \\
&= 4 \cdot 5 \\
&= 20.
\end{aligned}
$$

To say that x stands for 2, we may also write $x = 2$. That is, we are agreeing to use x and 2 as names of the same number.

25. $7x + 7y$

Example 11   Factor $6x + 6y$. Then evaluate both the original and factored expressions when $x = 3$ and $y = 8$.

a) $6x + 6y = 6(x + y)$ Factoring

b) $6x + 6y = 6 \cdot 3 + 6 \cdot 8$ Evaluating $6x + 6y$

$$
\begin{aligned}
&= 18 + 48 \\
&= 66
\end{aligned}
$$

c) $6(x + y) = 6(3 + 8)$ Evaluating $6(x + y)$

$$
\begin{aligned}
&= 6 \cdot 11 \\
&= 66
\end{aligned}
$$

DO EXERCISES 24 AND 25.

In summary, we list the five properties considered in this section:

The associative laws: For any numbers a, b, and c,

$$a + (b + c) = (a + b) + c, \qquad a(bc) = (ab)c.$$

The commutative laws: For any numbers a and b,

$$a + b = b + a, \qquad ab = ba.$$

The distributive law: For any numbers a, b, and c,

$$a(b + c) = ab + ac.$$

NAME   CLASS

EXERCISE SET 1.3

█ **·** █ Do these calculations.

1. $(10 + 4) + 8$ **2.** $10 \times (9 + 4)$

3. $(10 \cdot 7) + 19$ **4.** $(10 \cdot 7) + (20 \cdot 14)$

█ **· ·** █ Do these calculations. Choose grouping and ordering to make the work easy.

5. $8 + 4 + 5 + 2 + 6 + 15 + 1$ **6.** $9 + 6 + 3 + 4 + 1 + 7 + 11$

7. $14 + 3 + 12 + 7 + 8 + 6 + 9$ **8.** $17 + 7 + 16 + 3 + 4 + 3 + 8$

Which laws are illustrated by these sentences?

9. $67 + 3 = 3 + 67$ **10.** $15 \cdot 44 = 44 \cdot 15$

11. $6 + (9 + 5) = (6 + 9) + 5$ **12.** $8 \cdot (7 \cdot 6) = (8 \cdot 7) \cdot 6$

█ **· · ·** █ Compute in two ways.

13. $(6 + 7) \cdot 4$ **14.** $(8 + 10) \cdot 2$

█ **: :** █ Factor.

15. $9x + 9y$ **16.** $7w + 7u$

17. $\frac{1}{2}a + \frac{1}{2}b$ **18.** $\frac{3}{4}x + \frac{3}{4}y$

19. $1.5x + 1.5z$ **20.** $0.7a + 0.7b$

ANSWERS

21. _____

22. _____

23. _____

24. _____

25. _____

26. _____

27. _____

28. _____

29. _____

30. _____

31. _____

32. _____

33. _____

34. _____

35. _____

36. _____

21. $4x + 4y + 4z$

22. $10a + 10b + 10c$

23. $\frac{4}{7}a + \frac{4}{7}b + \frac{4}{7}c + \frac{4}{7}d$

24. $\frac{3}{5}x + \frac{3}{5}y + \frac{3}{5}z + \frac{3}{5}w$

⚅ Factor. Then evaluate both expressions when $x = 5$ and $y = 10$.

25. $9x + 9y$ **26.** $8x + 8y$ **27.** $10x + 10y$ **28.** $2x + 2y$

Factor. Then evaluate both expressions when $a = 0$ and $b = 9$.

29. $5a + 5b$ **30.** $7a + 7b$ **31.** $20a + 20b$ **32.** $14a + 14b$

The distributive law is a basis for the ordinary procedure we use to multiply. We show this in two parts as follows.

Example 1   Consider $36 \cdot 7$.

Part I:

$$36 \cdot 7 = (30 + 6) \cdot 7$$
$$= 30 \cdot 7 + 6 \cdot 7$$
$$= 210 + 42$$
$$= 252$$

Part II:

$$
\begin{array}{r}
36 \\
\times\ 7 \\
\hline
252
\end{array}
$$

Example 2   Consider $27 \cdot 36$.

Part I:

$$27 \cdot 36 = (20 + 7) \cdot 36$$
$$= 20 \cdot 36 + 7 \cdot 36$$
$$= 720 + 252$$
$$= 972$$

Part II:

$$
\begin{array}{r}
36 \\
\times\ 27 \\
\hline
252 \\
720 \\
\hline
972
\end{array}
$$

Show the methods of Part I and Part II for each product.

33. $25 \cdot 8$ **34.** $67 \cdot 4$ **35.** $12 \cdot 39$ **36.** $45 \cdot 67$

1.4   USING THE DISTRIBUTIVE LAW

The distributive law is the basis of many procedures in both arithmetic and algebra. Below are some examples of other procedures based on this property and some further examples of factoring.

◌   MULTIPLYING

In the expression $x + y + z$, the parts separated by plus signs are called *terms*. Thus x, y, and z are terms in $x + y + z$. The distributive law is the basis of a procedure called "multiplying." For example, consider

$$8(a + b).$$

Using the distributive law we multiply each term of $(a + b)$ by 8:

$$8 \cdot (a + b) = 8 \cdot a + 8 \cdot b.$$

Examples   Multiply.

1. $3(x + 2) = 3 \cdot x + 3 \cdot 2$ Using the distributive law

   $= 3x + 6$

2. $6(s + 2t + 5w) = 6 \cdot s + 6 \cdot 2t + 6 \cdot 5w$ Using the distributive law

   $= 6s + 12t + 30w$

We multiplied each term inside the parentheses by the factor outside.

DO EXERCISES 1–3.

◌◌   FACTORING

In the expression $6x + 3y + 9z$ the terms are $6x$, $3y$, and $9z$. In this case, the terms are products. To factor, look for a factor common to all the terms. Then "remove" it, so to speak, using the distributive law.

Example 3   Factor: $6x + 3y + 9z$.

$6x + 3y + 9z = 3 \cdot 2x + 3 \cdot y + 3 \cdot 3z$ The common factor is 3.

$= 3(2x + y + 3z)$ Using the distributive law

Note that factoring is the reverse of multiplying. In fact, you can check factoring by multiplying.

Example 4   Factor: $7y + 21z + 7$.

$7y + 21z + 7 = 7 \cdot y + 7 \cdot 3z + 7 \cdot 1$ The common factor is 7.

$= 7(y + 3z + 1)$

Caution! Be sure not to omit the 1.

Caution! Be sure not to omit the common factor 7.

DO EXERCISES 4–7.

OBJECTIVES

After finishing Section 1.4, you should be able to:

◌   Use the distributive law to multiply expressions like $5(x + 3)$.

◌◌   Factor expressions like $5x + 10$ by using the distributive law.

◌◌◌   Collect like terms in expressions like $3x + 4y + 5x + 3y$.

Multiply.

1. $5(y + 3)$

2. $4(x + 2y + 5)$

3. $8(m + 3n + 4p)$

Factor.

4. $5x + 10$

5. $12 + 3x$

6. $6x + 12 + 9y$

7. $5x + 10y + 25$

8. Factor: $Q + Qab$.

Collect like terms.

9. $6y + 2y$

10. $4x + x$

11. $x + 0.03x$

12. $10p + 8p + 4q + 5q$

13. $7x + 3y + 4x + 5y$

Collect like terms.

14. $4y + 12y$

15. $3s + 4s + 6w + 7w$

16. $5x + 4y + 4x + 6y$

17. $5a + b + a + 0.07b$

Example 5   Simple interest on a principal of P dollars invested at interest rate r for t years is given by Prt. In t years, principal P will grow to the amount

$$(\text{Principal}) + (\text{Interest}) = P + Prt.$$

Factor this expression.

$$P + Prt = \boxed{P} \cdot 1 + \boxed{P}\, rt = \boxed{P}\,(1 + rt)$$

DO EXERCISE 8.

●●● COLLECTING LIKE TERMS

If two terms have the same letters, they are called *like* terms, or *similar* terms.* We can often simplify expressions by *collecting* or *combining like terms*. (We could also say *collecting* or *combining similar terms*.)

Examples   Collect like terms.

6. $3\,\boxed{x} + 4\,\boxed{x} = (3 + 4)\,\boxed{x}$ Using the distributive law
   $\qquad\qquad\quad = 7x$

7. $2x + 3y + 5x + 8y = 2\,\boxed{x} + 5\,\boxed{x} + 3\,\boxed{y} + 8\,\boxed{y}$ Regrouping and reordering using the associative and commutative laws

   $\qquad\qquad = (2 + 5)\,\boxed{x} + (3 + 8)\,\boxed{y}$ Factoring

   $\qquad\qquad = 7x + 11y$

8. $7x + x = 7 \cdot \boxed{x} + 1 \cdot \boxed{x} = (7 + 1)\,\boxed{x} = 8x$

9. $x + 0.05x = 1 \cdot \boxed{x} + 0.05\,\boxed{x} = (1 + 0.05)\,\boxed{x} = 1.05x$

DO EXERCISES 9–13.

With practice we can leave out some steps, collecting like terms mentally. Note that constants like 4 and 7 are considered like terms.

Examples   Collect like terms.

10. $5y + 2y + 4y = 11y$

11. $3x + 7x + 2y = 10x + 2y$

12. $3a + 5a + 8t + 2t = 8a + 10t$

13. $8p + q + p + 0.3q = 9p + 1.3q$

14. $4 + x + 7 = x + 11$

15. $3x + 25 + 7y + 8x + 11 = 11x + 7y + 36$

DO EXERCISES 14–17.

---

* Later we will amend this definition.

NAME _____   CLASS _____

ANSWERS
1. _____
2. _____
3. _____
4. _____
5. _____
6. _____
7. _____
8. _____
9. _____
10. _____
11. _____
12. _____
13. _____
14. _____
15. _____
16. _____
17. _____
18. _____
19. _____
20. _____
21. _____
22. _____
23. _____
24. _____

EXERCISE SET 1.4

⬤⬤ Multiply.

1. $3(x + 1)$

2. $2(x + 2)$

3. $4(1 + y)$

4. $9(s + 1)$

5. $9(4t + 3z)$

6. $8(5x + 3y)$

7. $7(x + 4 + 6y)$

8. $8(9x + 5y + 8)$

9. $5(3x + 9 + 7y)$

10. $4(5x + 8 + 3z)$

⬤⬤ Factor. Check by multiplying.

11. $2x + 4$

12. $9x + 27$

13. $6x + 24$

14. $5y + 20$

15. $9x + 3y$

16. $15x + 5y$

17. $14x + 21y$

18. $18x + 24y$

19. $5 + 10x + 15y$

20. $7 + 14b + 56w$

21. $8a + 16b + 64$

22. $9x + 27y + 81$

23. $3x + 18y + 15z$

24. $4r + 28s + 16t$

ANSWERS

●●● Collect like terms.

25. _____

26. _____

27. _____

28. _____

29. _____

30. _____

31. _____

32. _____

33. _____

34. _____

35. _____

36. _____

37. _____

38. _____

39. _____

40. _____

41. _____

42. _____

25. $2x + 3 + 3x + 9$

26. $7y + 9 + 8 + 20y$

27. $10a + a$

28. $16x + x$

29. $2x + 9z + 6x$

30. $3a + 5b + 7a$

31. $41a + 90c + 60c + 2a$

32. $42x + 6b + 4x + 2b$

33. $x + 0.09x + 0.2t + t$

34. $0.01a + 0.23b + a + b$

35. $8u + 3t + 10u + 6u + 2t$

36. $5t + 6h + t + 8t + 9h$

37. $23 + 5t + 7y + t + y + 27$

38. $45 + 90d + 87 + 9d + 3 + 7d$

39. $\frac{1}{2}b + \frac{2}{3} + \frac{1}{2}b + \frac{2}{3}$

40. $\frac{2}{3}x + \frac{5}{8} + \frac{1}{3}x + \frac{3}{8}$

41. $2y + \frac{1}{4}y + y$

42. $\frac{1}{2}a + a + 5a$

1.5   THE NUMBER 1 AND RECIPROCALS

The number 1 has some very special properties important in both arithmetic and algebra. When we multiply any number by 1, we get that same number.

> **For any number n,**
> $$n \cdot 1 = n.$$

When we divide a number by 1, we get the same number with which we started. Since a/b means $a \div b$, hence $n/1 = n \div 1 = n$.

> **For any number n,**
> $$\frac{n}{1} = n.$$

When we divide a number by itself, the result is the number 1. This is true for any number except zero. We will see later why we do not divide by zero.

> **For any number n, except zero,**
> $$\frac{n}{n} = 1.$$

Examples   Simplify.

1. $\dfrac{3}{5} \cdot \boxed{\dfrac{7}{7}} = \dfrac{3}{5} \cdot \boxed{1} = \dfrac{3}{5}$

2. $\dfrac{\frac{3}{5}}{1} = \dfrac{3}{5}$

3. $\dfrac{\frac{4}{3}}{\frac{4}{3}} = \boxed{1}$

DO EXERCISES 1–4.

● RECIPROCALS

Two numbers whose product is 1 are called *reciprocals* of each other. All the numbers of arithmetic, except zero, have reciprocals.

Examples

4. The reciprocal of $\dfrac{2}{3}$ is $\boxed{\dfrac{3}{2}}$ because $\dfrac{2}{3} \cdot \boxed{\dfrac{3}{2}} = \dfrac{6}{6} = 1.$

5. The reciprocal of 9 is $\boxed{\dfrac{1}{9}}$ because $9 \cdot \boxed{\dfrac{1}{9}} = \dfrac{9}{9} = 1.$

6. The reciprocal of $\dfrac{1}{4}$ is $\boxed{4}$ because $\dfrac{1}{4} \cdot \boxed{4} = 1.$

DO EXERCISES 5–8.

OBJECTIVES

After finishing Section 1.5, you should be able to:

● Find the reciprocal of any number.

●● Divide by multiplying by a reciprocal.

●●● Graph numbers of arithmetic on a number line.

●● Use the proper symbol >, <, or = between two fractional numerals.

Simplify.

1. $\dfrac{4}{7} \cdot \dfrac{11}{11}$

2. $\dfrac{67}{67}$

3. $\dfrac{\frac{2}{3}}{1}$

4. $\dfrac{\frac{7}{5}}{\frac{7}{5}}$

Find the reciprocal of each number.

5. $\dfrac{4}{11}$

6. $\dfrac{15}{7}$

7. 5

8. $\dfrac{1}{3}$

Divide by multiplying by 1.

9. $\dfrac{\dfrac{3}{5}}{\dfrac{4}{7}}$

10. $\dfrac{\dfrac{5}{4}}{\dfrac{3}{2}}$

11. $\dfrac{\dfrac{9}{7}}{\dfrac{4}{5}}$

Divide by multiplying by the reciprocal of the divisor.

12. $\dfrac{4}{3} \div \dfrac{7}{2}$

13. $\dfrac{3}{5} \div \dfrac{7}{4}$

14. $\dfrac{\dfrac{2}{9}}{\dfrac{5}{7}}$

● ● RECIPROCALS AND DIVISION

The number 1 and reciprocals can be used to explain division of numbers of arithmetic. To divide, we can multiply by 1, choosing carefully the symbol for 1.

Example 7   Divide $\dfrac{2}{3}$ by $\dfrac{7}{5}$.

$$\dfrac{\dfrac{2}{3}}{\dfrac{7}{5}} = \dfrac{\dfrac{2}{3}}{\dfrac{7}{5}} \times \dfrac{\dfrac{5}{7}}{\dfrac{5}{7}} \quad \text{Multiplying by } \dfrac{\frac{5}{7}}{\frac{5}{7}}.$$

We use $\frac{5}{7}$ because it is the reciprocal of $\frac{7}{5}$.

$$= \dfrac{\dfrac{2}{3} \times \dfrac{5}{7}}{\dfrac{7}{5} \times \dfrac{5}{7}} \quad \text{Multiplying numerators and denominators}$$

$$= \dfrac{\dfrac{10}{21}}{1}$$

$$= \dfrac{10}{21}$$

After multiplying we got 1 for a denominator. This was because we used the reciprocal of the divisor, $\frac{7}{5}$, for both the numerator and denominator of the symbol for 1.

DO EXERCISES 9–11.

When multiplying by 1 to divide, we get a denominator of 1. What do we get in the numerator? In Example 7, we got $\frac{2}{3} \times \frac{5}{7}$. This is the product of $\frac{2}{3}$, the dividend, and $\frac{5}{7}$, the reciprocal of the divisor.

> **To divide, multiply by the reciprocal of the divisor:**
>
> $$\dfrac{a}{b} \div \dfrac{c}{d} = \dfrac{a}{b} \cdot \dfrac{d}{c}.$$

Example 8   Divide by multiplying by the reciprocal of the divisor.

$$\dfrac{1}{2} \div \dfrac{3}{5} = \dfrac{1}{2} \cdot \dfrac{5}{3} = \dfrac{5}{6} \qquad \dfrac{5}{3} \text{ is the reciprocal of } \dfrac{3}{5}$$

After dividing, simplification is often possible and should be done.

Example 9   Divide.

$$\dfrac{2}{3} \div \dfrac{4}{9} = \dfrac{2}{3} \cdot \dfrac{9}{4} = \dfrac{18}{12} = \dfrac{3 \cdot 6}{2 \cdot 6} = \underbrace{\dfrac{3}{2} \cdot \dfrac{6}{6}}_{\text{Simplifying}} = \dfrac{3}{2}$$

DO EXERCISES 12–14.

⬤⬤⬤ THE NUMBER LINE AND ORDER

The order of numbers of arithmetic can be shown on a number line.

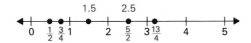

DO EXERCISES 15 AND 16.

Note that $\frac{1}{2}$ is less than 1, and $\frac{1}{2}$ is to the left of 1 on the number line. To say that $\frac{1}{2}$ is less than 1, we write $\frac{1}{2} < 1$.

> **For any numbers a and b,**
>
> $$a < b \qquad \text{(read ``}a\text{ is less than }b\text{'')}$$
>
> **means that a is to the left of b on the number line.**

The number $\frac{5}{2}$ is to the right of $\frac{1}{2}$. This means that $\frac{5}{2}$ is greater than $\frac{1}{2}$. To say that $\frac{5}{2}$ is greater than $\frac{1}{2}$, we write $\frac{5}{2} > \frac{1}{2}$.

> **For any numbers a and b,**
>
> $$a > b \qquad \text{(read ``}a\text{ is greater than }b\text{'')}$$
>
> **means that a is to the right of b on the number line.**

Sentences such as $\frac{5}{2} > \frac{1}{2}$ and $x < 2$ are called *inequalities*.

Examples   Determine whether true or false. Use the number line above.

10. $1.5 > 1$;   true 1.5 is to the right of 1

11. $\frac{5}{2} < \frac{3}{4}$;   false $\frac{5}{2}$ is not to the left of $\frac{3}{4}$

12. $\frac{1}{2} < 6$;   true $\frac{1}{2}$ is to the left of 6

DO EXERCISES 17–19.

⬛⬛ COMPARING NUMBERS

We want to develop a more efficient way of comparing numbers of arithmetic. Consider $\frac{4}{5}$ and $\frac{3}{5}$. The number line shows that $\frac{4}{5} > \frac{3}{5}$.

Note also that $4 > 3$. When denominators are the same, we just compare numerators.

Graph each number on a number line.

15. $\dfrac{6}{5}$

16. $\dfrac{17}{18}$

Determine whether true or false.

17. $\dfrac{1}{2} < \dfrac{3}{4}$

18. $\dfrac{3}{4} > \dfrac{1}{2}$

19. $\dfrac{13}{4} < \dfrac{5}{2}$

ANSWERS ON PAGE A–2

Use the proper symbol, $>$, $<$, or $=$, between each pair of numerals.

20. $\dfrac{3}{4}$ $\dfrac{4}{4}$

21. $\dfrac{1}{2}$ $\dfrac{1}{2}$

22. $\dfrac{22}{19}$ $\dfrac{21}{19}$

Use the proper symbol $>$, $<$, or $=$.

23. $\dfrac{9}{5}$ $\dfrac{11}{7}$

24. $\dfrac{16}{12}$ $\dfrac{13}{8}$

25. $\dfrac{23}{7}$ $\dfrac{29}{9}$

26. $\dfrac{23}{13}$ $\dfrac{18}{11}$

For any numbers of arithmetic $\dfrac{a}{b}$ and $\dfrac{c}{b}$,

$$\frac{a}{b} > \frac{c}{b} \text{ when } a > c,$$

$$\frac{a}{b} < \frac{c}{b} \text{ when } a < c,$$

$$\text{and } \frac{a}{b} = \frac{c}{b} \text{ when } a = c.$$

DO EXERCISES 20–22.

It is not so easy to tell which of $\frac{5}{7}$ and $\frac{2}{3}$ is larger. Let's find a common denominator and compare numerators.

Example 13   Insert the proper symbol $>$, $<$, or $=$ between $\frac{5}{7}$ and $\frac{2}{3}$. We multiply by 1 to find a common denominator:

$$\frac{5}{7} = \frac{5}{7} \cdot \frac{3}{3} = \frac{15}{21} \quad \text{and} \quad \frac{2}{3} = \frac{2}{3} \cdot \frac{7}{7} = \frac{14}{21}.$$

Since $15 > 14$, it follows that $\frac{15}{21} > \frac{14}{21}$, so $\frac{5}{7} > \frac{2}{3}$.

Example 14   Insert the proper symbol $>$, $<$, or $=$ between $\frac{7}{10}$ and $\frac{8}{9}$.

$$\frac{7}{10} = \frac{7}{10} \cdot \frac{9}{9} = \frac{63}{90} \quad \text{and} \quad \frac{8}{9} = \frac{8}{9} \cdot \frac{10}{10} = \frac{80}{90}$$

Since $63 < 80$, it follows that $\frac{63}{90} < \frac{80}{90}$, so $\frac{7}{10} < \frac{8}{9}$.

Example 15   Insert the proper symbol $>$, $<$, or $=$ between $\frac{6}{27}$ and $\frac{2}{9}$.

$$\frac{6}{27} = \frac{6}{27} \cdot \frac{9}{9} = \frac{54}{243} \quad \text{and} \quad \frac{2}{9} = \frac{2}{9} \cdot \frac{27}{27} = \frac{54}{243}$$

Since the numerators (and denominators) are the same, $\frac{6}{27} = \frac{2}{9}$.

We could also do these problems using a calculator. We find decimal notation and compare.

Example 16   Insert the proper symbol between $\frac{5}{7}$ and $\frac{2}{3}$.

$$\frac{5}{7} \approx 0.7142857143 \quad \text{🖩, rounded to ten decimal places}$$

$$\frac{2}{3} \approx 0.6666666667$$

Thus $\frac{5}{7} > \frac{2}{3}$.

DO EXERCISES 23–26.

NAME	CLASS	ANSWERS

EXERCISE SET 1.5

● Find the reciprocal of each number.

1. $\dfrac{3}{4}$ **2.** $\dfrac{5}{8}$ **3.** $\dfrac{1}{8}$ **4.** $\dfrac{1}{10}$

5. 1   **6.** 9

●● Divide by multiplying by the reciprocal of the divisor.

7. $\dfrac{7}{6} \div \dfrac{3}{5}$ **8.** $\dfrac{7}{5} \div \dfrac{3}{4}$ **9.** $\dfrac{8}{9} \div \dfrac{4}{15}$ **10.** $\dfrac{3}{4} \div \dfrac{3}{7}$

11. $\dfrac{1}{4} \div \dfrac{1}{2}$ **12.** $\dfrac{1}{10} \div \dfrac{1}{5}$ **13.** $\dfrac{\frac{13}{12}}{\frac{39}{5}}$ **14.** $\dfrac{\frac{17}{6}}{\frac{3}{8}}$

15. $100 \div \dfrac{1}{5}$ **16.** $78 \div \dfrac{1}{6}$ **17.** $\dfrac{3}{4} \div 10$ **18.** $\dfrac{5}{6} \div 15$

●●● Graph each number on the number line.

19. $\dfrac{5}{4}$ **20.** $\dfrac{7}{6}$ **21.** $\dfrac{15}{16}$ **22.** $\dfrac{11}{12}$

ANSWERS

1. _____
2. _____
3. _____
4. _____
5. _____
6. _____
7. _____
8. _____
9. _____
10. _____
11. _____
12. _____
13. _____
14. _____
15. _____
16. _____
17. _____
18. _____
19. See graph.
20. See graph.
21. See graph.
22. See graph.

⚅ Use the proper symbol $>$, $<$, or $=$ between each pair of numerals.

23. _____

23. $\dfrac{1}{2}$ $\dfrac{2}{4}$ 24. $\dfrac{9}{12}$ $\dfrac{3}{4}$ 25. $\dfrac{11}{15}$ $\dfrac{13}{24}$ 26. $\dfrac{19}{16}$ $\dfrac{5}{4}$

24. _____

25. _____

26. _____

27. $\dfrac{13}{8}$ $\dfrac{8}{5}$ 28. $\dfrac{21}{16}$ $\dfrac{5}{4}$ 29. $\dfrac{4}{5}$ $\dfrac{8}{10}$ 30. $\dfrac{8}{9}$ $\dfrac{16}{18}$

27. _____

28. _____

29. _____

31. $\dfrac{7}{22}$ $\dfrac{1}{3}$ 32. $\dfrac{8}{23}$ $\dfrac{1}{3}$

30. _____

31. _____

32. _____

33. _____

33. ▦ Find decimal notation for the reciprocal of 0.3125.

34. ▦ Find decimal notation for the reciprocal of 0.24.

34. _____

35. ▦ Use the proper symbol $>$, $<$, or $=$.

$\dfrac{1439}{2007}$ $\dfrac{2359}{2876}$

35. _____

36. ▦ Find decimal notation for each. Use this to order from largest to smallest.

$\dfrac{11}{13}, \dfrac{17}{20}, \dfrac{2}{3}, \dfrac{35}{37}, \dfrac{5}{6}, \dfrac{23}{25}$

36. _____

1.6 SOLVING EQUATIONS

● EQUATIONS AND SOLUTIONS

An *equation* is a number sentence with $=$ for its verb. For example,

$$3 + 2 = 5, \qquad 7 - 2 = 8, \qquad \text{and} \qquad x + 6 = 13.$$

Some equations are true. Some are false. Some are neither true nor false.

Examples

1. The equation $3 + 2 = 5$ is true.

2. The equation $7 - 2 = 8$ is false.

3. The equation $x + 6 = 13$ is neither true nor false because we don't know what x stands for.

An equation says that the symbols on either side of the equals sign stand for, or name, the same number. For example, $5 - 4 = 3 + 7$ says that $5 - 4$ names the same number as $3 + 7$. This equation is false.

DO EXERCISES 1–3.

The letter x in $x + 6 = 13$ is called a *variable*. Some replacements for the variable make the equation true. Some make it false.

> **The replacements that make an equation true are called** *solutions*. **To** *solve* **an equation means to find all of its solutions.**

There are several ways to solve equations.

Example 4 Solve $x + 6 = 13$ by trial.

If we replace x by 2 we get a false equation: $2 + 6 = 13$.
If we replace x by 8 we get a false equation: $8 + 6 = 13$.
If we replace x by 7 we get a true equation: $7 + 6 = 13$.

No other replacement makes the equation true, so the only solution is 7.

DO EXERCISES 4–8.

●● SOLVING EQUATIONS $x + a = b$

Solving equations is an important part of algebra. Here we consider enough equation-solving techniques to give you a feel for this part of algebra. Later, we study equation solving in more detail. Consider the true sentence

$$2 + 3 = 5.$$

If we subtract 3 on both sides, we still get a true equation.

$2 + 3 =$	5	Two names of the same number
$- 3$	-3	Subtract 3 on both sides.
2	2	We get a true equation.

OBJECTIVES

After finishing Section 1.6, you should be able to:

● Solve simple equations by trial.

●● Solve equations of the type $x + a = b$ by subtracting a.

●●● Solve equations of the type $ax = b$ by dividing by a.

1. Write three true equations.

2. Write three false equations.

3. Write three equations that are neither true nor false.

4. Find three replacements that make $x + 5 = 12$ false.

5. Find the replacement that makes $x + 5 = 12$ true.

Solve by trial.

6. $x + 4 = 10$

7. $3x = 12$

8. $3y + 1 = 16$

ANSWERS ON PAGE A–2

Solve. Be sure to check.

9. $x + 9 = 17$

10. $x + 234 = 507$

11. $x + 2.78 = 7.65$

Suppose we want to solve $x + 3 = 5$. We subtract 3 from both sides.

$$\begin{array}{c|c} x + 3 = & 5 \\ -3 & -3 \\ \hline x & 2 \end{array}$$

We subtract 3 to "undo" the addition $x + 3$. This gets x by itself. This equation is easy to solve. It is true when x is replaced by 2. We check to see if $x + 3 = 5$ is true when x is replaced by 2:

$$\begin{array}{c|c} x + 3 = 5 \\ \hline 2 + 3 & 5 \\ 5 \end{array}$$

Thus 2 is a solution.

> **To solve $x + a = b$, subtract a from both sides.**

Example 5   Solve: $x + 6 = 13$.

$$x + 6 = 13$$
$$x = 13 - 6 \qquad \text{Subtracting 6. Subtraction "undoes" addition.}$$
$$x = 7$$

The solution to $x = 7$ is obviously the number 7. It is also the solution of the original equation. We check to find out. To do this we replace x by 7. When we simplify, we get 13 on both sides.

$$\textit{Check:} \quad \begin{array}{c|c} x + 6 = 13 \\ \hline 7 + 6 & 13 \\ 13 \end{array}$$

The solution is 7.

In the preceding you may have questioned using the technique when you could "see" the answers. In the next example it is not so easy to "see" the answer.

Example 6   Solve: $x + 1.64 = 9.08$.

$$x + 1.64 = 9.08$$
$$x = 9.08 - 1.64 \qquad \text{Subtracting 1.64}$$
$$x = 7.44$$

$$\textit{Check:} \quad \begin{array}{c|c} x + 1.64 = 9.08 \\ \hline 7.44 + 1.64 & 9.08 \\ 9.08 \end{array}$$

The solution is 7.44.

DO EXERCISES 9–11.

Example 7   Solve: $x + \dfrac{3}{4} = \dfrac{5}{6}$.

$$x = \frac{5}{6} - \frac{3}{4} \qquad \text{Subtracting } \frac{3}{4}$$

$$x = \frac{5}{6} \cdot \boxed{\frac{4}{4}} - \frac{3}{4} \cdot \boxed{\frac{6}{6}} \qquad \begin{array}{l}\text{Multiplying by 1}\\ \text{to get a common}\\ \text{denominator.}\end{array}$$

$$x = \frac{20}{24} - \frac{18}{24} = \frac{2}{24}$$

$$x = \frac{1}{12} \cdot \boxed{\frac{2}{2}} = \frac{1}{12} \qquad \text{Simplifying}$$

Check:

$$\begin{array}{c|c} x + \dfrac{3}{4} = \dfrac{5}{6} \\ \hline \dfrac{1}{12} + \dfrac{3}{4} & \dfrac{5}{6} \\[2mm] \dfrac{1}{12} + \dfrac{9}{12} & \\[2mm] \dfrac{10}{12} & \\[2mm] \dfrac{5}{6} & \end{array}$$

The solution is $\frac{1}{12}$.

DO EXERCISE 12.

◦◦◦   SOLVING EQUATIONS $ax = b$

Consider the true equation

$\quad 5 \cdot 9 = 45$.

If we divide by 5 on both sides, we still get a true equation:

$$\frac{5 \cdot 9}{5} = \frac{45}{5}, \quad \text{or} \quad 9 = 9.$$

Suppose we want to solve $5x = 45$. We divide by 5 on both sides to undo the multiplication $5x$.

$$\frac{5x}{5} = \frac{45}{5} \qquad \boxed{\text{We can also think of this as multiplying by } \tfrac{1}{5}.}$$

or $x - 9$.

The solution of $x = 9$ is 9. We check to see if 9 is a solution of $5x = 45$:

$$\begin{array}{c|c} 5x = 45 \\ \hline 5 \cdot 9 & 45 \\ 45 & \end{array}$$

Thus 9 is a solution.

12. Solve: $x + \dfrac{11}{8} = \dfrac{5}{2}$.

ANSWER ON PAGE A–2

Solve. Be sure to check.

13. $5x = 35$

14. $8x = 36$

15. $1.2x = 6.72$

Which of these divisions are possible?

16. $\dfrac{21}{6}$

17. $\dfrac{13}{0}$

18. $\dfrac{0}{8}$

19. $\dfrac{11}{10 - 10}$

20. $\dfrac{9}{12 - (4 \cdot 3)}$

21. $\dfrac{P}{x - x}$

ANSWERS ON PAGE A–2

To solve $ax = b$ when a is nonzero, divide on both sides by a (or multiply on both sides by $1/a$).

Example 8   Solve: $5x = 16$.

$$5x = 16$$

$$x = \frac{16}{5}, \quad \text{or } 3.2 \qquad \text{Dividing by 5, or multiplying by } \tfrac{1}{5}$$

We divide by 5 to "undo" the multiplication and get x by itself. This equation is easy to solve. It is true when x is replaced by $\frac{16}{5}$.

Check:

$$
\begin{array}{c|c}
\multicolumn{2}{c}{5x = 16} \\
\hline
5(3.2) & 16 \\
16 &
\end{array}
$$

The solution is 3.2. (It could also be left as $\frac{16}{5}$.)

Example 9   Solve: $4.7x = 40.42$.

$$4.7x = 40.42$$

$$x = \frac{40.42}{4.7} \qquad \text{Dividing by 4.7, or multiplying by } 1/4.7$$

$$x = 8.6$$

Check:

$$
\begin{array}{c|c}
\multicolumn{2}{c}{4.7x = 40.42} \\
\hline
4.7(8.6) & 40.42 \\
40.42 &
\end{array}
$$

The solution is 8.6.

DO EXERCISES 13–15.

DIVISION BY ZERO

We cannot use the preceding method to solve an equation $ax = b$ when $a = 0$, because this would result in division by zero. But, why? The division $a \div b$ or a/b is defined to be some number c such that $a = bc$. Thus $a/0$ would be some number c such that $0 \cdot c = a$. But $0 \cdot c = 0$, so the only possible number a that could be divided by 0 is 0. Look for a pattern.

a) $\dfrac{0}{0} = 5$ because $0 = 0 \cdot 5$ **b)** $\dfrac{0}{0} = 789$ because $0 = 0 \cdot 789$

c) $\dfrac{0}{0} = 17$ because $0 = 0 \cdot 17$ **d)** $\dfrac{0}{0} = \dfrac{1}{2}$ because $0 = 0 \cdot \dfrac{1}{2}$

It looks as if $\frac{0}{0}$ could be any number at all. This would be very confusing, getting any answer we want when we divide 0 by 0. Thus we agree to exclude division by zero.

We never divide by zero.

DO EXERCISES 16–21.

| NAME | CLASS | ANSWERS |

EXERCISE SET 1.6

● Solve by trial.

1. $x + 8 = 10$ **2.** $x + 5 = 13$

3. $x - 4 = 5$ **4.** $x - 6 = 12$

5. $5x = 25$ **6.** $7x = 42$

7. $5y + 7 = 107$ **8.** $9x + 5 = 86$

9. $7x - 1 = 48$ **10.** $4y - 2 = 10$

●● Solve. Be sure to check.

11. $x + 17 = 22$ **12.** $x + 18 = 32$

13. $x + 56 = 75$ **14.** $x + 47 = 83$

15. $x + 2.78 = 8.44$ **16.** $x + 3.04 = 4.69$

17. $x + 5064 = 7882$ **18.** $x + 4112 = 8007$

19. $x + \dfrac{1}{4} = \dfrac{2}{3}$ **20.** $x + \dfrac{1}{3} = \dfrac{4}{5}$

ANSWERS
1. ___
2. ___
3. ___
4. ___
5. ___
6. ___
7. ___
8. ___
9. ___
10. ___
11. ___
12. ___
13. ___
14. ___
15. ___
16. ___
17. ___
18. ___
19. ___
20. ___

21. _____

22. _____

23. _____

24. _____

25. _____

26. _____

27. _____

28. _____

29. _____

30. _____

31. _____

32. _____

33. _____

34. _____

35. _____

36. _____

21. $x + \dfrac{2}{3} = \dfrac{5}{6}$ **22.** $x + \dfrac{3}{4} = \dfrac{7}{8}$

●●● Solve. Be sure to check.

23. $6x = 24$ **24.** $4x = 32$ **25.** $4x = 5$ **26.** $6x = 27$

27. $10y = 2.4$ **28.** $9x = 3.6$ **29.** $2.9y = 8.99$ **30.** $5.5y = 34.1$

31. $6.2x = 52.7$ **32.** $9.4x = 23.5$ **33.** $\dfrac{3}{4}x = 35$ **34.** $\dfrac{4}{5}x = 27$

Solve.

35. ▦ $x + 506{,}233 = 976{,}421$ **36.** ▦ $0.1265x = 1065.636$

1.7 SOLVING PROBLEMS

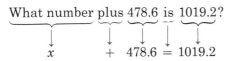 Why learn to solve equations? Because applied problems can be solved using equations. To do this, we first translate the problem situation to an equation and then solve the equation. The problem situation may be explained in words (as in a textbook) or may come from an actual situation in the real world.

The problems we solve here are simple. You might use other methods to solve them, but it is better if you do not. Remember, you are learning how algebra works.

Example 1   What number plus 478.6 is 1019.2?

We translate the problem to an equation as follows:

$$\underbrace{\text{What number}}_{x} \underbrace{\text{plus}}_{+} \underbrace{478.6}_{478.6} \underbrace{\text{is}}_{=} \underbrace{1019.2}_{1019.2}?$$

The translation gives us the equation

$x + 478.6 = 1019.2.$

We solve it:

$x = 1019.2 - 478.6$ Subtracting 478.6

$x = 540.6.$

To check, we find out if 540.6 plus 478.6 is 1019.2:

$540.6 + 478.6 = 1019.2.$

We have the answer: 540.6.

DO EXERCISE 1.

Example 2   Solve this problem.

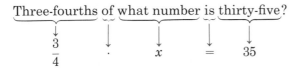

$$\underbrace{\text{Three-fourths}}_{\frac{3}{4}} \text{ of } \underbrace{\text{what number}}_{x} \underbrace{\text{is}}_{=} \underbrace{\text{thirty-five}}_{35}?$$

The translation gives us the equation

$\dfrac{3}{4}x = 35.$

We solve it:

$x = \dfrac{35}{\frac{3}{4}}$ Dividing by $\dfrac{3}{4}$

$x = 35 \cdot \dfrac{4}{3}$

$x = \dfrac{140}{3}.$

OBJECTIVE

After finishing Section 1.7, you should be able to:

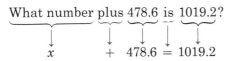 **Solve applied problems by translating to equations and solving.**

Translate to an equation. Then solve and check.

1. What number plus thirty-seven is seventy-three?

ANSWER ON PAGE A–2

Translate to an equation. Then solve and check.

2. Two-thirds of what number is forty-four?

To check, we find out if $\frac{3}{4}$ of this number is 35:

$$\frac{3}{4} \cdot \frac{140}{3} = \frac{3 \cdot 140}{4 \cdot 3} = \frac{3 \cdot 35 \cdot 4}{4 \cdot 3} = 35.$$

The number 140/3 is the answer.

> **Note that in translating, *is* translates to =. The word *of* translates to ·, and the unknown number translates to a variable.**

DO EXERCISE 2.

Sometimes it helps to reword a problem before translating.

Example 3 One year a person earned a salary of $23,400. This was $1700 more than last year. What was last year's salary?

Rewording: Last year's salary plus $1700 is this year's salary.

Translating: x +   1700 =   23,400

The translation gives us the equation

$x + 1700 = 23,400.$

We solve it:

$x = 23,400 - 1700$ Subtracting 1700

$x = 21,700.$

To check, we add $1700 to $21,700:

$\$1700 + \$21,700 = \$23,400.$

The answer is $21,700.

DO EXERCISE 3.

Translate to an equation. Then solve and check.

3. In Australia there are 145 million sheep. This is 132 million more than the number of people. How many people are in Australia?

Example 4 A solid-state color television set uses about 420 kilowatt hours (kWh) of electricity in a year. This is 3.5 times that used by a solid-state black and white set. How many kWh does the black and white set use each year?

Rewording: 3.5 times energy of the black and white set is 420.

Translating: 3.5 × y = 420

The translation gives us the equation

$3.5y = 420.$

We solve it:

$y = \dfrac{420}{3.5}$ Dividing by 3.5

$y = 120.$

To check, we multiply 120 by 3.5: 3.5(120) = 420. The answer is 120 kWh.

Translate to an equation. Then solve and check.

4. An investment was made that grew to $14,500 after one year. This was 1.16 times what was originally invested. How much was originally invested?

DO EXERCISE 4.

NAME   CLASS   ANSWERS

EXERCISE SET 1.7

▪●▪ Translate to equations. Then solve and check.

1. Two-thirds of what number is forty-eight?

              1. _____

2. One-eighth of what number is fifty-six?

              2. _____

3. What number plus five is twenty-two?

              3. _____

4. What number plus eight is sixty-three?

              4. _____

5. What number is 4 more than 5?

              5. _____

6. What number is 7 less than 10?

              6. _____

Translate to equations. Do not solve.

7. The area of Lake Superior is four times the area of Lake Ontario. The area of Lake Superior is 78,114 km². What is the area of Lake Ontario?

7. _____

8. The area of Alaska is about 483 times the area of Rhode Island. The area of Alaska is 1,519,202 km². What is the area of Rhode Island?

8. _____

9. Izzi Zlow's typing speed is 35 words per minute. This is two-fifths of Ty Preitter's speed. What is Preitter's speed?

9. _____

10. Walter Logged's body contains 57 kg of water. This is two-thirds of his weight. What is Logged's weight?

10. _____

11. The boiling point of ethyl alcohol is 78.3°C. This is 13.5°C more than the boiling point of methyl alcohol. What is the boiling point of methyl alcohol?

11. _____

12. The height of the Eiffel Tower is 295 m. This is about 203 m more than the height of the Statue of Liberty. What is the height of the Statue of Liberty?

12. _____

13. A color television set with tubes uses about 640 kilowatt hours of electricity in a year. This is 1.6 times that used by a solid-state color set. How many kilowatt hours does the solid-state color model use each year?

13. _____

14. The distance from the earth to the sun is about 150,000,000 km. This is about 391 times the distance from the earth to the moon. What is the distance from the earth to the moon?

14. _____

15. Recently, the average cost of having a baby in one Ohio hospital was $1175. This was about 1.8 times the average cost for a certain California hospital. What was the cost of having a baby in the California hospital?

15. _____

16. It takes a 60-watt bulb about 16.6 hours to use one kilowatt hour of electricity. This is about 2.5 times as long as it takes a 150-watt bulb to use one kilowatt hour. How long does it take a 150-watt bulb to use one kilowatt hour?

16. _____

17. Solve the equation in Exercise 7.

18. Solve the equation in Exercise 8.

19. Solve the equation in Exercise 9.

20. Solve the equation in Exercise 10.

21. Solve the equation in Exercise 11.

22. Solve the equation in Exercise 12.

ANSWERS

17. _____

18. _____

19. _____

20. _____

21. _____

22. _____

23. Solve the equation in Exercise 13.   **24.** Solve the equation in Exercise 14.

23. _____

24. _____

25. Solve the equation in Exercise 15.   **26.** Solve the equation in Exercise 16.

25. _____

26. _____

27. _____

These problems are impossible to solve because some piece of information is missing. Tell what you would need to know to solve the problem.

27. A person makes three times the salary of ten years ago. What was the salary ten years ago?

28. Records were on sale for 75¢ off the marked price. After buying four records, a person has $8.72 left. How much was there to begin with?

28. _____

1.8   PERCENT NOTATION

• CONVERTING TO DECIMAL NOTATION

There are other ways to name numbers of arithmetic besides using fractional and decimal notation. One other kind of notation uses the percent symbol %, which means "per hundred." We can regard the percent symbol as part of a numeral. For example,

$$37\% \quad \text{is defined to mean} \quad 37 \times 0.01 \quad \text{or} \quad 37 \times \frac{1}{100}.$$

In general,

$$n\% \text{ means} \quad n \times 0.01 \quad \text{or} \quad n \times \frac{1}{100}.$$

Example 1   Find decimal notation for 78.5%.

$$\begin{aligned}
78.5\% &= 78.5 \times 0.01 \qquad \text{Replacing \% by } \times 0.01 \\
&= 0.785
\end{aligned}$$

DO EXERCISES 1 AND 2.

•• CONVERTING TO FRACTIONAL NOTATION

Example 2   Find fractional notation for 88%.

$$88\% = 88 \times \frac{1}{100} \qquad \text{Replacing \% by } \times \frac{1}{100}$$

$$= \frac{88}{100} \qquad \text{You need not simplify.}$$

Example 3   Find fractional notation for 34.7%.

$$34.7\% = 34.7 \times \frac{1}{100} \qquad \text{Replacing \% by } \times \frac{1}{100}$$

$$= \frac{34.7}{100}$$

$$= \frac{34.7}{100} \cdot \frac{10}{10} \qquad \text{Multiplying by 1 to get a whole number in the numerator}$$

$$= \frac{347}{1000}$$

> There is a table of decimal and percent equivalents on the last page of this book. If you do not already know these facts, you should memorize them.

OBJECTIVES

After finishing Section 1.8, you should be able to:

• Convert from percent notation to decimal notation.

•• Convert from percent notation to fractional notation.

••• Convert from decimal to percent notation.

:: Convert from fractional to percent notation.

:·: Solve applied problems involving percents.

Find decimal notation.

1. 46.2%

2. 100%

Find fractional notation.

3. 67%

4. 45.6%

5. $\frac{1}{4}\%$

Find percent notation.

6. 6.77

7. 0.9944

Find percent notation.

8. $\frac{1}{4}$

9. $\frac{3}{8}$

10. $\frac{2}{3}$

●●● CONVERTING FROM DECIMAL TO PERCENT NOTATION

By applying the definition of % in reverse, we can convert from decimal notation to percent notation. We multiply by 1, naming it 100×0.01 .

Example 4   Find percent notation for 0.93.

$$0.93 = 0.93 \times 1$$
$$= 0.93 \times (\boxed{100 \times 0.01}) \quad\quad \text{Replacing 1 by } 100 \times 0.01$$
$$= (0.93 \times 100) \times 0.01 \quad\quad \text{Using associativity}$$
$$= 93 \times 0.01$$
$$= 93\% \quad\quad \text{Replacing } \times 0.01 \text{ by } \%$$

Example 5   Find percent notation for 0.002.

$$0.002 = 0.002 \times (\boxed{100 \times 0.01})$$
$$= (0.002 \times 100) \times 0.01$$
$$= 0.2 \times 0.01$$
$$= 0.2\% \quad\quad \text{Replacing } \times 0.01 \text{ by } \%$$

DO EXERCISES 6 AND 7.

■■ CONVERTING FROM FRACTIONAL TO PERCENT NOTATION

We can also convert from fractional to percent notation. Again, we multiply by 1, but this time we use $100 \times \frac{1}{100}$.

Example 6   Find percent notation for $\frac{5}{8}$.

$$\frac{5}{8} = \frac{5}{8} \times \left(\boxed{100 \times \frac{1}{100}} \right)$$
$$= \left(\frac{5}{8} \times 100 \right) \times \frac{1}{100}$$
$$= \frac{500}{8} \times \frac{1}{100}$$
$$= \frac{500}{8}\%, \quad \text{or} \quad 62.5\%$$

DO EXERCISES 8–10.

■■ SOLVING PROBLEMS INVOLVING PERCENTS

Let's solve some applied problems involving percents. Again, it is helpful to translate the problem situation to an equation and then solve the equation.

Example 7   What is 12% of 59?

$$x = 12\% \cdot 59 \qquad \text{Translating}$$

Solve:
$$x = (12 \times 0.01) \times 59 \qquad \begin{array}{l}\text{Replacing } 12\% \text{ by } 12 \times 0.01.\\ \text{This gets rid of the } \% \text{ sign.}\end{array}$$
$$x = 0.12 \times 59$$
$$x = 7.08$$

The solution is 7.08, so 7.08 is 12% of 59.

DO EXERCISES 11 AND 12.

Example 8   What percent of 45 is 15?

$$x \qquad \% \quad \cdot \ 45 = 15 \qquad \text{Translating}$$

Solve:
$$(x \times 0.01) \times 45 = 15$$
$$x(0.45) = 15$$
$$x = \frac{15}{0.45} \qquad \text{Dividing by } 0.45$$
$$x = \frac{15}{0.45} \times \frac{100}{100} = \frac{1500}{45} = 33\tfrac{1}{3}$$

The solution is $33\tfrac{1}{3}$, so $33\tfrac{1}{3}\%$ of 45 is 15.

DO EXERCISES 13 AND 14.

Example 9   3 is 16 percent of what?

$$3 = 16 \quad \% \quad \cdot \quad y \qquad \text{Translating}$$

Solve:
$$3 = 16 \times 0.01 \times y$$
$$3 = 0.16y$$
$$0.16y = 3$$
$$y = \frac{3}{0.16} \qquad \text{Dividing by } 0.16$$
$$y = \frac{3}{0.16} \cdot \frac{100}{100} = \frac{300}{16} = 18.75$$

16% of 18.75 is 3, so the solution is 18.75.

> Perhaps you have noticed in Examples 7–9 that to handle percents in such problems, you must first convert to decimal notation, and then go ahead.

DO EXERCISES 15 AND 16.

Translate and solve.

11. What is 23% of 48?

12. 25% of 40 is what?

Translate and solve.

13. What percent of 50 is 16?

14. 15 is what percent of 60?

Translate and solve.

15. 45 is 20 percent of what?

16. 120 percent of what is 60?

Translate to equations. Then solve.

17. The area of Arizona is 19% of the area of Alaska. The area of Alaska is 586,400 sq mi. What is the area of Arizona?

Sometimes it is helpful to reword the problem before translating.

Example 10   Blood is 90% water. The average adult has 5 quarts of blood. How much water is in the average adult's blood?

Rewording: 90% of 5 is what?

Translating: $90\% \cdot 5 = x$

Solve: $90 \times 0.01 \times 5 = x$

$0.90 \times 5 = x$ Converting 90% to decimal notation

$4.5 = x$

The number 4.5 checks in the problem. Thus there are 4.5 quarts of water in the average adult's blood.

Example 11   An investment is made at 8% simple interest for 1 year. It grows to $783. How much was originally invested (the principal)?

Rewording: (Principal) + (Interest) = Amount

Translating: x + $8\%x$ = 783 Interest is 8% of the principal

18. An investment is made at 7% simple interest for one year. It grows to $8988. How much was originally invested (the principal)?

Solve:

$$x + 8\%x = 783$$

$$x + 0.08x = 783 \quad \text{Converting}$$

$$1.08x = 783 \quad \text{Collecting like terms}$$

$$x = \frac{783}{1.08} \quad \text{Dividing by 1.08}$$

$$x = 725$$

The original investment (principal) was $725.

DO EXERCISES 17 AND 18.

Example 12   The price of an automobile rose 16% to a value of $10,393.60. What was the former price?

Rewording: (Former price) + (Increase) = New price

Translating: x + $16\%x$ = $\$10,393.60$

Translate to an equation. Then solve.

19. A person's salary increased 12% to an amount of $20,608. What was the former salary?

Solve:

$$x + 16\%x = 10,393.60$$

$$x + 0.16x = 10,393.60 \quad \text{Converting}$$

$$1.16x = 10,393.60 \quad \text{Collecting like terms}$$

$$x = \frac{10,393.60}{1.16} \quad \text{Dividing by 1.16}$$

$$x = 8960$$

The former price was $8960.

> A common error in a problem like this is to take 16% of the new price and subtract. This problem is easy with algebra. Without algebra it is not.

DO EXERCISE 19.

NAME   CLASS   ANSWERS

EXERCISE SET 1.8

Note: If you need extra practice on percent, do Exercise Set 1.8A on pp. 51–52.

• Find decimal notation.

1. 76%   **2.** 54%   **3.** 54.7%   **4.** 96.2%

• • Find fractional notation.

5. 20%   **6.** 80%   **7.** 78.6%   **8.** 12.5%

• • • Find percent notation.

9. 4.54   **10.** 1   **11.** 0.998   **12.** 0.751

• • Find percent notation.

13. $\dfrac{1}{8}$ **14.** $\dfrac{1}{3}$ **15.** $\dfrac{17}{25}$ **16.** $\dfrac{11}{20}$

ANSWERS

1. _____

2. _____

3. _____

4. _____

5. _____

6. _____

7. _____

8. _____

9. _____

10. _____

11. _____

12. _____

13. _____

14. _____

15. _____

16. _____

ANSWERS

Translate and solve.

17. What is 65% of 840?   **18.** 34% of 560 is what?

17. _____

18. _____

19. 24 percent of what is 20.4?   **20.** 45 is 30 percent of what?

19. _____

20. _____

21. What percent of 80 is 100?   **22.** 30 is what percent of 125?

21. _____

22. _____

Translate to equations. Then solve.

23. On a test of 88 items, a student got 76 correct. What percent were correct?

23. _____

24. A baseball player got 13 hits in 25 times at bat. What percent were hits?

24. _____

25. A family spent $208 one month for food. This was 26% of its income. What was their monthly income?

25. _____

26. The weight of the human brain is 2.7% of the body weight. A human's brain weighs 2.1 kilograms. What is its body weight?

26. _____

27. The sales tax rate in New York City is 5%. How much would be charged on a purchase of $428.86? How much would the total cost of the purchase be?

27. _____

Translate to equations. Then solve.

28. Water volume increases 9% when it freezes. If 400 cubic centimeters of water is frozen, how much would its volume increase? What would be the volume of the ice?

28. _____

29. An investment is made at 9% simple interest for 1 year. It grows to $8502. How much was originally invested?

29. _____

30. An investment is made at 8% simple interest for 1 year. It grows to $7776. How much was originally invested?

30. _____

31. A person earned $9600 one year. An 8% increase in salary was received, but the cost of living rose 7.4%. How much additional earning power was actually received?

31. _____

32. Due to inflation the price of an item rose 8%, which was 12¢. What was the old price? the new price?

32. _____

NAME   CLASS

EXERCISE SET 1.8A

This exercise set supplements Exercise Set 1.8 for those who need it.

■ □ Find decimal notation.

1. 38%   **2.** 72.1%   **3.** 65.4%   **4.** 3.25%

5. 8.24%   **6.** 0.61%   **7.** 0.012%   **8.** 0.023%

9. 0.73%   **10.** 0.0035%   **11.** 125%   **12.** 240%

■ ■ Find fractional notation.

13. 30%   **14.** 70%   **15.** 13.5%   **16.** 73.4%

17. 3.2%   **18.** 8.4%   **19.** 120%   **20.** 250%

21. 0.35%   **22.** 0.48%   **23.** 0.042%   **24.** 0.083%

■ ■ ■ Find percent notation.

25. 0.62   **26.** 0.73   **27.** 0.623   **28.** 0.812

ANSWERS

1. _____
2. _____
3. _____
4. _____
5. _____
6. _____
7. _____
8. _____
9. _____
10. _____
11. _____
12. _____
13. _____
14. _____
15. _____
16. _____
17. _____
18. _____
19. _____
20. _____
21. _____
22. _____
23. _____
24. _____
25. _____
26. _____
27. _____
28. _____

ANSWERS

29. _____

30. _____

31. _____

32. _____

33. _____

34. _____

35. _____

36. _____

37. _____

38. _____

39. _____

40. _____

41. _____

42. _____

43. _____

44. _____

45. _____

46. _____

47. _____

48. _____

49. _____

50. _____

51. _____

52. _____

53. _____

54. _____

29. 7.2   **30.** 3.5   **31.** 2   **32.** 5

33. 0.072   **34.** 0.013   **35.** 0.0057   **36.** 0.0068

Find percent notation.

37. $\dfrac{17}{100}$ **38.** $\dfrac{119}{100}$ **39.** $\dfrac{7}{10}$ **40.** $\dfrac{8}{10}$

41. $\dfrac{7}{20}$ **42.** $\dfrac{7}{25}$ **43.** $\dfrac{1}{2}$ **44.** $\dfrac{3}{4}$

45. $\dfrac{3}{5}$ **46.** $\dfrac{17}{50}$ **47.** $\dfrac{1}{3}$ **48.** $\dfrac{3}{8}$

Translate and solve.

49. What is 38 % of 250?   **50.** 37.2 % of 85 is what?

51. What percent of 80 is 20?   **52.** 35 is 20 percent of what?

53. 25 percent of what is 16?   **54.** 20 is what percent of 60?

1.9 GEOMETRIC FORMULAS

◖•◗ RECTANGLES

In this section we review some formulas from geometry. We first consider rectangles. The area of a rectangle is the number of *unit* squares it takes to fill it up. Unit squares look like these.

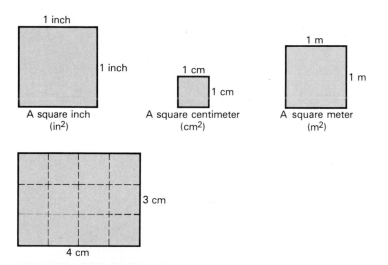

1 inch

1 inch

A square inch
(in²)

1 cm

1 cm

A square centimeter
(cm²)

1 m

1 m

A square meter
(m²)

3 cm

4 cm

The rectangle has length 4 cm and width 3 cm. It takes 12 unit squares to fill it up. Therefore, the area is 12 cm² (or 12 sq cm). In any rectangle we can find the area by multiplying the length by the width.

> **If a rectangle has length l and width w, the area is given by**
>
> $$A = lw. \quad \textbf{(Area is length times width)}$$

The *perimeter* of a rectangle is the distance around it. We can find the perimeter by adding the lengths of the four sides. Or, we can double the length and the width and then add.

> **The perimeter of a rectangle of length l and width w is given by**
>
> $$P = 2l + 2w, \quad \textbf{or} \quad 2(l + w).$$

Example 1 Find the area and perimeter of this rectangle.

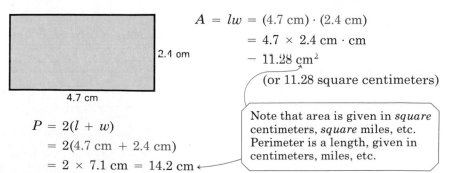

2.4 cm

4.7 cm

$$A = lw = (4.7 \text{ cm}) \cdot (2.4 \text{ cm})$$
$$= 4.7 \times 2.4 \text{ cm} \cdot \text{cm}$$
$$= 11.28 \text{ cm}^2$$

(or 11.28 square centimeters)

$$P = 2(l + w)$$
$$= 2(4.7 \text{ cm} + 2.4 \text{ cm})$$
$$= 2 \times 7.1 \text{ cm} = 14.2 \text{ cm}$$

Note that area is given in *square* centimeters, *square* miles, etc. Perimeter is a length, given in centimeters, miles, etc.

OBJECTIVES

After finishing Section 1.9, you should be able to:

◖•◗ Find the area and perimeter of a rectangle, given the length and width.

◖••◗ Find the area of a parallelogram, trapezoid, or triangle.

◖•••◗ Given two angle measures of a triangle, find the measure of the third angle.

◖••◖••◗ Find the volume of a rectangular solid.

Given the radius of a circle, find the diameter and the circumference.

Given the radius of a circle, find the area.

1. Find the area and perimeter of this rectangle.

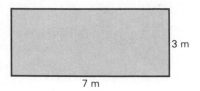

3 m

7 m

2. Find the area and perimeter of this square.

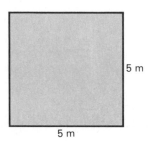

5 m

5 m

3. Find the area of this parallelogram.

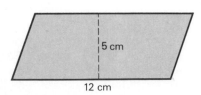

5 cm

12 cm

When we make computations with dimension symbols, such as inches (in.) or centimeters (cm), we can treat them as if they were numerals or variables.

Compare:

a) $3x \cdot 4x = 3 \cdot 4 \cdot x \cdot x = 12x^2$ with 3 in. $\cdot$ 4 in. $= 3 \cdot 4$ in. $\cdot$ in. $= 12$ in^2

b) $2x + 5x = (2 + 5)x = 7x$ with 2 cm $+$ 5 cm $= (2 + 5)$ cm $= 7$ cm

DO EXERCISE 1.

If the length and width of a rectangle are the same, then the rectangle is a square. In a square, all four sides are the same length. Suppose a square has sides of length s. Then its area is $s \cdot s$ or s^2 and its perimeter is $s + s + s + s$, or $4s$.

> **If a square has sides of length s, then the area is given by $A = s^2$ and the perimeter is given by $P = 4s$.**

DO EXERCISE 2.

●● PARALLELOGRAMS, TRAPEZOIDS, AND TRIANGLES

A *parallelogram* is a four-sided figure with two pairs of parallel sides. To find the area of a parallelogram, we can think of cutting off a part of it, as shown below. We can then place the part as shown to form a rectangle. The area of the rectangle is $b \cdot h$ (length of the base times height). This is also the area of the parallelogram.

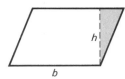   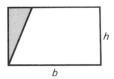

h

b

h

b

> **The area of a parallelogram is given by $A = b \cdot h$, where b is the length of the base and h is the height.**

DO EXERCISE 3.

A *trapezoid* is a four-sided figure with at least one pair of parallel sides. To find the area of a trapezoid, we can think of cutting out another one just like the given one and placing the two of them together, as shown below. This forms a parallelogram, whose area is $h \cdot (a + b)$.

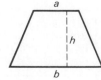   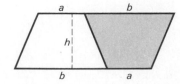

a

h

b

a   b

h

b   a

The trapezoid is half of this parallelogram, so its area is $\frac{1}{2}h \cdot (a + b)$. We call the parallel sides of the trapezoid the *bases*, so the area is half the product of the height and the sum of the bases.

> **If a trapezoid has bases of lengths a and b and has height h, its area is given by $A = \frac{1}{2} \cdot h \cdot (a + b)$.**

Example 2   Find the area of this trapezoid.

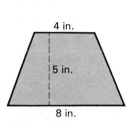

$$A = \frac{1}{2} \cdot h \cdot (a + b)$$

$$= \frac{1}{2} \cdot 5 \text{ in.} \cdot (4 \text{ in.} + 8 \text{ in.})$$

$$= \frac{1}{2} \cdot 5 \text{ in.} \cdot 12 \text{ in.}$$

$$= 30 \text{ in}^2$$

DO EXERCISE 4.

To find the area of a triangle, we can think of cutting out another one like the one given and placing the two of them together, as shown below.

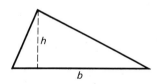 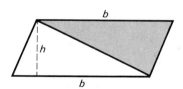

This forms a parallelogram, whose area is $b \cdot h$. The triangle is half of the parallelogram, so the area is half of $b \cdot h$.

> **If a triangle has a base of length b and has height h, then the area is given by $A = \frac{1}{2} \cdot b \cdot h$.**

Example 3   Find the area of this triangle.

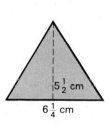

$$A = \frac{1}{2}bh$$

$$= \frac{1}{2} \cdot 6\frac{1}{4} \text{ cm} \cdot 5\frac{1}{2} \text{ cm}$$

$$= \frac{1}{2} \cdot \frac{25}{4} \cdot \frac{11}{2} \text{ cm}^2$$

$$= \frac{275}{16} \text{ cm}^2, \quad \text{or } 17\frac{3}{16} \text{ cm}^2$$

DO EXERCISE 5.

4. Find the area of this trapezoid.

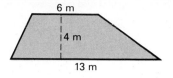

5. Find the area of this triangle.

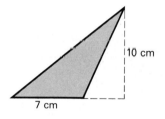

6. Find the missing angle measure.

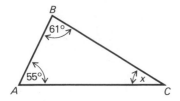

●●● ANGLES OF A TRIANGLE

The measures of the angles of a triangle add up to 180°. Thus, if we know the measures of two angles of a triangle we can calculate the third.

> **In any triangle, the sum of the measures of the angles is 180°:**
>
> $$m(\angle A) + m(\angle B) + m(\angle C) = 180°.$$

Example 4   Find the missing angle measure.

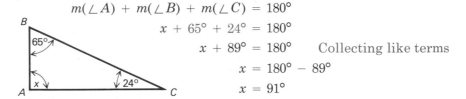

$$m(\angle A) + m(\angle B) + m(\angle C) = 180°$$
$$x + 65° + 24° = 180°$$
$$x + 89° = 180° \quad \text{Collecting like terms}$$
$$x = 180° - 89°$$
$$x = 91°$$

DO EXERCISE 6.

⬛⬛ VOLUMES OF RECTANGULAR SOLIDS

The volume of a rectangular solid is the number of *unit* cubes it takes to fill it up. Unit cubes look like these.

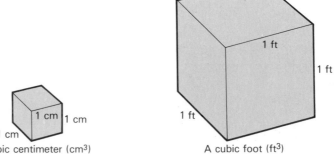

A cubic centimeter (cm³)   A cubic foot (ft³)

This rectangular solid has length 4 cm, width 3 cm, and height 2 cm. There are thus two layers of cubes, each having 3 × 4, or 12 cubes. The total volume is thus 24 cm³ (cubic centimeters).

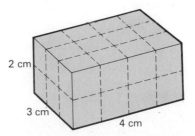

In any rectangular solid we can find the volume by multiplying the length by the width by the height.

> **If a rectangular solid has length l, width w, and height h, then the volume is given by $V = l \cdot w \cdot h$.**

Example 5   Find the volume of this solid.

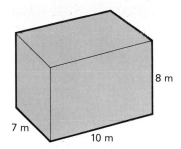

$V = lwh$

$\quad = 10 \text{ m} \cdot 7 \text{ m} \cdot 8 \text{ m}$

$\quad = 10 \cdot 56 \text{ m}^3 \text{ (cubic meters)}$

$\quad = 560 \text{ m}^3$

DO EXERCISE 7.

⁞⁞⁞ CIRCLES

Here is a circle, with center O. Also shown are a diameter d and a radius r. A diameter is twice as long as a radius.

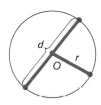

> **In any circle, if d is the diameter and r is the radius, then $d = 2 \cdot r$.**

The *circumference* of a circle is the distance around it. If we divide the circumference of a circle by the diameter, we always get the same number. This number is not a number of arithmetic, so we use the Greek letter π (pi) to name it. This number is approximately 3.14 or $\frac{22}{7}$. We use the letter C for circumference. Then $C/d = \pi$ in any circle, or $C = \pi d$.

> **If C is the circumference of a circle and r is the radius, then $C = \pi d$, or since $d = 2r$, $C = 2\pi r$.**

Example 6   Find the length of a diameter of a circle whose radius is 14 ft.

$d = 2r$

$\quad = 2 \cdot 14 \text{ ft} = 28 \text{ ft}$

Example 7   Find the circumference of a circle whose radius is 14 ft. Use $\frac{22}{7}$ for π.

$C = 2 \cdot \pi \cdot r$

$\quad \approx 2 \cdot \dfrac{22}{7} \cdot 14 \text{ ft} \qquad \approx \text{ means "approximately equal to"}$

$\quad \approx 88 \text{ ft}$

DO EXERCISES 8 AND 9.

---

7. Find the volume of this solid.

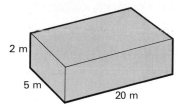

8. Find the length of a diameter of a circle whose radius is 9 meters.

9. Find the circumference of a circle whose radius is 9 cm. Use 3.14 for π.

10. Find the area of a circle with a 9-ft radius. Use 3.14 for π.

▦ AREAS OF CIRCLES

Suppose we take half a circular region, cut it into small wedges, and arrange the wedges as shown below. If the wedges are very small, the distance from A to B will be very nearly πr (half the circumference of the circle). Now if we cut the other half of the circle the same way and place the wedges together, we get what is almost a parallelogram, with area $\pi r \cdot r$.

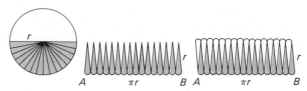

> **The area of a circle of radius r is given by $A = \pi r^2$.**

Example 8   Find the area of this circle. Use $\frac{22}{7}$ for π.

14 cm

$$A = \pi \cdot r \cdot r$$

$$A \approx \frac{22}{7} \cdot 14 \text{ cm} \cdot 14 \text{ cm}$$

$$A \approx 616 \text{ cm}^2$$

Caution! Be sure to use $\pi \cdot r \cdot r$ or πr^2 for area, not $2\pi r$ or πd.

The area is about 616 cm².

In summary, note that we use the symbols

$$\pi, \quad d, \quad r, \quad C, \quad \text{and} \quad A$$

in connection with circles; π is a number, about 3.14, d stands for the diameter, r stands for the radius, C stands for the circumference, and A stands for the area.

DO EXERCISE 10.

From time to time throughout this book, you will find some "extras," such as the one on this page. These are optional. You may find them of interest, but they can be skipped. Some require the use of a calculator.

SOMETHING EXTRA   ▦

CALCULATOR CORNER: ESTIMATING π

Each of the following is an approximation for π. Find decimal notation for each rounded to seven decimal places, using your calculator. Compare with the following approximation, correct to nine decimal places:

$$\pi \approx 3.141592653.$$

(Answers may vary due to capacity of calculator and possible rounding.)

1. $\pi \approx \dfrac{355}{113}$ **2.** $\pi \approx \dfrac{62,832}{20,000}$ **3.** $\pi \approx 2 \cdot \dfrac{2 \cdot 2 \cdot 4 \cdot 4 \cdot 6 \cdot 6 \cdot 8}{1 \cdot 3 \cdot 3 \cdot 5 \cdot 5 \cdot 7 \cdot 7}$

4. $\pi \approx 4\left(\dfrac{1}{2} + \dfrac{1}{3} - \dfrac{1}{15} + \dfrac{1}{35} - \dfrac{1}{63} + \dfrac{1}{99}\right)$

ANSWER ON PAGE A–3

NAME   CLASS   ANSWERS

EXERCISE SET 1.9

⚫ Find the area and perimeter of each rectangle.

1.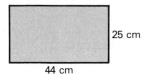
25 cm
44 cm

2.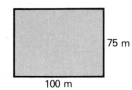
75 m
100 m

3.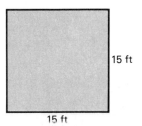
15 ft
15 ft

4.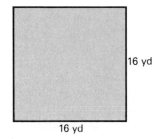
16 yd
16 yd

⚫⚫ Find the area of each figure.

5.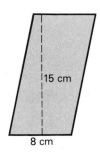
15 cm
8 cm

6.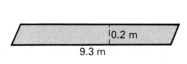
0.2 m
9.3 m

7.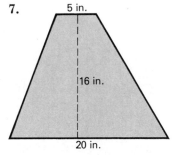
5 in.
16 in.
20 in.

1. _____

2. _____

3. _____

4. _____

5. _____

6. _____

7. _____

8.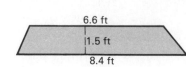

9.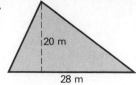

10.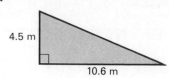

8. _____

9. _____

 Find the missing angle measures.

11.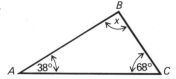

12.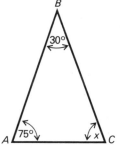

10. _____

11. _____

13.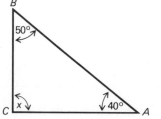

12. _____

14.

13. _____

14. _____

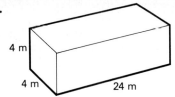 Find the volume of each solid.

15.

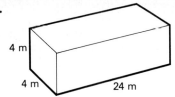

4 m

4 m   24 m

16.

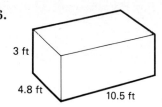

3 ft

4.8 ft   10.5 ft

15. _____

16. _____

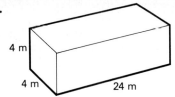 Find the diameter and circumference of each circle.

17. Use 3.14 for π.

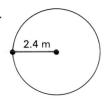

2.4 m

18. Use 3.14 for π.

10.2 cm

17. _____

19. Use $\dfrac{22}{7}$ for π.

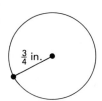

$\frac{3}{4}$ in.

20. Use $\dfrac{22}{7}$ for π.

$\frac{7}{2}$ ft

18. _____

19. _____

20. _____

ANSWERS

⠿ Find the area of the circle with given radius. Use 3.14 for π.

21. $r = 2.4$ m   **22.** $r = 10.2$ cm

21. _____

22. _____

23. _____

23. $r = 100$ ft   **24.** $r = 200$ yd

24. _____

25. _____

26. _____

25. ▦ The length of a diameter of a circle is 24.6 cm. What is the length of a radius?

26. ▦ The area of a rectangle is 49.2 ft² and the length is 10 ft. What is the width?

27. _____

27. ▦ The circumference of a circle is 22.608 yd. What is the length of a diameter? Use 3.14 for π.

28. ▦ The area of a parallelogram is 254.8 square meters and the base is 52 meters long. What is the height?

28. _____

EXTENSION EXERCISES

SECTION 1.2

Write "=" if the pair of expressions represents the same number; write "≠" otherwise.

1. $\dfrac{5n}{n}$, $5n^0$ $(n \neq 0)$

2. n^3, $\dfrac{n \cdot n \cdot n \cdot n}{n}$ $(n \neq 0)$

3. $\dfrac{4^5}{2}$, 2^5

4. 2^2, $2 \cdot 2 \cdot 2^0$

5. xy^2, $x \cdot x \cdot y \cdot y$

6. $4x$, $x \cdot x \cdot x \cdot x$

SECTION 1.3

7. Which of the following expressions represent the same number?

    a) $3 \cdot 4 + 2$ b) $3(4 + 2)$ c) $3 \cdot 4 + 3 \cdot 2$ d) $3 \cdot 2 + 4$ e) $3(2 + 4)$

8. Name the *two* laws illustrated by each sentence.

    a) $5 \cdot (4 \cdot 9) = (5 \cdot 9) \cdot 4$ b) $3 \cdot (4 + 7) = (7 + 4) \cdot 3$

    c) $2(a + b + c) = 2a + 2c + 2b$ d) $(a + b + c) + d = (c + d) + (a + b)$

9. Is there an associative law for division analogous to the associative law for multiplication? That is, is it *always* true that $(a \div b) \div c = a \div (b \div c)$ for all arithmetic numbers for which divisions are possible? Support your answer with an example.

SECTION 1.4

Factor.

10. $\dfrac{3}{2}x + \dfrac{5}{2}y$

11. $\dfrac{2}{3}x + \dfrac{2}{7}y + 2$

Collect like terms.

12. $\dfrac{1}{2}x + \dfrac{3}{4} + \dfrac{2}{3}x + \dfrac{1}{5}$

Collect like terms and then factor.

13. $3x + 10y + 2x + 5y + 25$

14. $\dfrac{1}{3}x + \dfrac{1}{4} + \dfrac{1}{6}y + \dfrac{1}{4}x + \dfrac{1}{6} + \dfrac{3}{4}y$

SECTION 1.5

15. Find two numbers between $\frac{3}{5}$ and $\frac{4}{5}$.

16. Find decimal notation for the reciprocal of each number.

    a) 0.5   b) 0.04   c) 0.32   d) 20   e) 0.1   f) 0.24

17. Arrange these numbers from smallest to largest.

    a) $\dfrac{7}{11}, \dfrac{7}{5}, \dfrac{7}{13}, \dfrac{7}{3}, \dfrac{7}{8}, \dfrac{7}{9}$

    b) $\dfrac{7}{15}, \dfrac{4}{15}, \dfrac{1}{15}, \dfrac{11}{15}, \dfrac{8}{15}, \dfrac{2}{15}$

    c) $\dfrac{1}{2}, \dfrac{2}{3}, \dfrac{6}{5}, \dfrac{7}{8}, \dfrac{8}{11}, \dfrac{9}{13}$

18. Use your answers to Exercise 17 to answer these questions.

    a) What happens to $\frac{a}{b}$ as a gets larger while b does not change?

    b) What happens to $\frac{a}{b}$ as a gets smaller while b does not change?

    c) What happens to $\frac{a}{b}$ as b gets larger while a does not change?

    d) What happens to $\frac{a}{b}$ as b gets smaller while a does not change?

    e) Does $\frac{a}{b}$ always get larger if both a and b get larger? Support your answer with an example.

SECTION 1.6

Write "yes" if the division is possible; write "no" otherwise.

19. $\dfrac{6}{0}$

20. $\dfrac{0}{5}$

21. $\dfrac{0}{0}$

22. $\dfrac{8}{7-7}$

23. $\dfrac{2-2}{4}$

24. $\dfrac{1}{x-x}$

Solve.

25. $2(x+1) = 12$

26. $3(y+2) + 3 = 15$

SECTION 1.7

27. If a number is doubled and then divided by 3, the result is 16. What is the number?

28. If the sum of 4 and a number is multiplied by 3, the result is 18. What is the number?

29. Machine A produces 90 bolts per hour. This is 4 more than twice the number Machine B produces per hour. How many bolts does Machine B produce per hour?

SECTION 1.8

30. a) What percent of 10 is 10?   b) 3 is 150% of what?   c) What is 200% of 6?

31. The price of a dress is marked down 20% on May 1. On May 25 the reduced price is marked down an additional 15% to $51. What percent of the original price is the final sale price? What is the original price?

32. A salesperson received a 10% commission on sales for the month. The salesperson also received a bonus in the amount of 2% of the commission. The commission and bonus totaled $6630. What was the amount of total sales?

33. In a marathon 30% of the runners did not complete the course. Of those remaining, 5% won ribbons. If 56 ribbons were awarded, how many runners entered the race?

SECTION 1.9

34. The measure of $\angle A$ of a triangle is 30°. The measure of $\angle B$ is $\frac{2}{3}$ the measure of $\angle C$. Find the measure of $\angle B$ and of $\angle C$.

35. The area of the bottom side of a box is 28 cm² and its volume is 84 cm³. Find its height.

36. The circumference of a circle is $\frac{66}{7}$ m. What is its area? (Use $\frac{22}{7}$ for π.)

37. Find the perimeter of each figure. (Use 3.14 for π.)   **38.** Find the area of each figure. (Use $\frac{22}{7}$ for π.)

a) 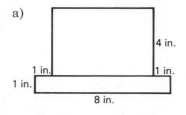   b) 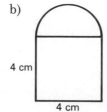   a) 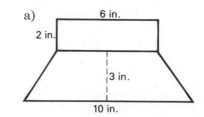   b)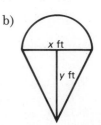

NAME   SCORE   ANSWERS

TEST OR REVIEW—CHAPTER 1

If you miss an item, review the indicated section and objective.

1. _____

[1.1, ●●●]   **1.** Multiply and simplify: $\dfrac{7}{8} \cdot \dfrac{16}{21}$.

2. _____

[1.1, ●●●]   **2.** Add and simplify: $\dfrac{7}{10} + \dfrac{5}{6}$.

3. _____

[1.1, ●●●]   **3.** Subtract and simplify: $\dfrac{3}{4} - \dfrac{15}{32}$.

4. _____

[1.2, ●●●]   **4.** What is the meaning of y^4?

5. _____

[1.2, ●]   **5.** Write fractional notation for 5.69.

6. _____

[1.2, ●●]   **6.** Write decimal notation for $\dfrac{3}{8}$.

7. _____

[1.3, ●]   **7.** Calculate: $(8 \times 7) + 6$.

[1.3, ●●]   **8.** What law is illustrated by this sentence? $6 \cdot (4 \cdot 3) = (6 \cdot 4) \cdot 3$

8. _____

[1.3, ●●●]   **9.** Factor $5x + 5y$. Then evaluate both the original and factored expressions when $x = 4$ and $y = 6$.

9. _____

[1.4, ●●]   Factor.

10. _____

    10. $18x + 6y$

    11. $36x + 16 + 4y$

11. _____

[1.4, ●]   **12.** Multiply: $8(5a + 3b + 2)$.

12. _____

[1.4, ●●●]   Collect like terms.

    13. $36y + 8a + 4y + 2a$

13. _____

    14. $9a + 0.23b + a + b$

14. _____

ANSWERS

[1.5, ● ●] **15.** Divide and simplify: $\dfrac{3}{7} \div \dfrac{9}{28}$.

15. _____

[1.5, ●● ●●] **16.** Use the proper symbol $>$, $<$, or $=$ between $\dfrac{5}{8}$ $\dfrac{6}{7}$.

16. _____

Solve and check.

[1.6, ● ●] **17.** $x + 3 = 11.4$

[1.6, ●●●] **18.** $5.4x = 32.4$

17. _____

[1.7, ●] **19.** Translate to an equation. Then solve: Three-fifths of what number is 18?

[1.8, ●] **20.** Find decimal notation for 45.8%.

18. _____

[1.8, ●● ●●] **21.** Find percent notation for $\dfrac{14}{25}$.

[1.8, ●●●] **22.** What percent of 75 is 5?

19. _____

Find the area of each figure.

20. _____

[1.9, ●] **23.**

8 ft

8 ft

21. _____

[1.9, ● ●] **24.**

5 m

4 m

8 m

22. _____

23. _____

[1.9, ●●●●●●] **25.** Find the area. Use 3.14 for π.

24. _____

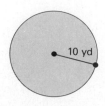

10 yd

25. _____

INTEGERS AND RATIONAL NUMBERS

2

1. Multiply and simplify:

$$\frac{21}{5} \cdot \frac{1}{7}$$

2. What is the meaning of w^4?

3. What law is illustrated by this sentence?

$$6 \cdot (4 + 8) = 6 \cdot 4 + 6 \cdot 8$$

4. Factor: $3x + 9 + 12y$.

5. Multiply: $7(3z + y + 2)$.

6. Divide and simplify: $\frac{7}{2} \div \frac{3}{8}$.

2.1 INTEGERS AND THE NUMBER LINE

● ADDITIVE INVERSES

The set of whole numbers is shown below, on a line.

To get numbers on the other side of 0 we can *reflect*. The reflection of 2 is named -2.

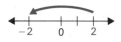

This is read "the additive inverse* of 2," or simply "the inverse of 2."

> **The inverse of any number x is named $-x$ (the inverse of x).**

Example 1   Find $-x$ when x is -3.

To find $-x$ we reflect x to the other side of 0.

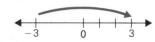

$-(-3) = 3$ (the inverse of the inverse of 3 is 3).

DO EXERCISES 1 AND 2.

Example 2   Find $-x$ when x is 0.

When we try to reflect 0 "to the other side of 0" we go nowhere:

$-x = 0$ when x is 0.

DO EXERCISE 3.

---

* Sometimes called the *opposite*.

OBJECTIVES

After finishing Section 2.1, you should be able to:

● Find the additive inverse of a number.

●● Tell which of two numbers is greater, using < or >.

●●● Find the absolute value of any integer.

In each case draw a number line, if necessary.

1. Find $-x$ when x is 1.

2. Find $-x$ when x is -2.

3. Find $-x$ when x is 0.

ANSWERS ON PAGE A–4

Example 3 Find $-(-x)$ when x is 2.

We replace x by 2. We wish to find $-(-2)$.

We see from the figure that $-(-2) = 2$.

Example 4 Find $-(-x)$ when x is -3.

We replace x by -3. We wish to find $-(-(-3))$.

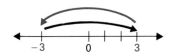

Reflecting -3 to the other side of 0 and then reflecting back gives us -3. Thus, $-(-x) = -3$ when x is -3.

DO EXERCISES 4–7.

Replacing a number by its additive inverse is sometimes called *changing the sign*.

Examples Change the sign. (Find the additive inverse.)

5. -3: $-(-3) = 3$

6. -10: $-(-10) = 10$

7. 0: $-0 = 0$

8. 14: $-(14) = -14$

DO EXERCISES 8–11.

4. Find $-(-x)$ when x is 4.

5. Find $-(-x)$ when x is 1.

6. Find $-(-x)$ when x is -2.

7. Find $-(-x)$ when x is -5.

Change the sign. (Find the additive inverse.)

8. -4

9. -13

10. 28

11. 0

Make true sentences using $<$ or $>$.

12. -5 ___ -7

13. 0 ___ -3

14. -5 ___ -2

15. 10 ___ 0

16. -3 ___ 4

17. -5 ___ 5

THE SET OF INTEGERS

The set of *integers* is made up of the whole numbers and all of their additive inverses. We call the natural numbers *positive*. The inverses of the positive integers are called *negative*. Zero is not considered either positive or negative.

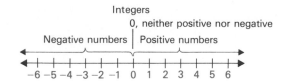

We often read -2 as "negative 2." We do not read $-x$ as "negative x," however, because we do not know whether or not it is negative.

Integers are associated with many situations in which numbers on both sides of 0 are considered.

Negative numbers can be associated with many situations.

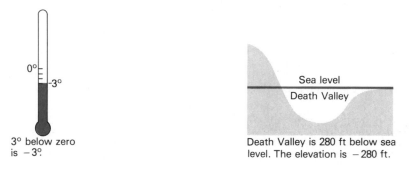

3° below zero
is $-3°$.

Death Valley is 280 ft below sea level. The elevation is -280 ft.

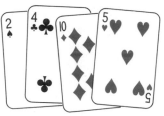

Getting set 21 points in a card game gives you -21 points.

●● ORDER AND INTEGERS

One integer is greater than another if it is to the right of the first on the number line.

Examples

9. $-2 < 1$ **10.** $-8 < -3$ **11.** $-5 < 0$ **12.** $5 > -5$

All negative integers are less than zero. All positive integers are greater than zero.

DO EXERCISES 12–17.

◆◆◆ ABSOLUTE VALUE

Let us consider the number line.

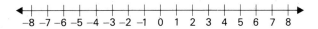

How far is 5 from 0? How far is -5 from 0? Since distance is always considered to be a nonnegative number (positive or zero), it follows that 5 is 5 units from 0 and -5 is 5 units from 0.

> **The *absolute value* of an integer can be thought of as its distance from 0 on the number line. The absolute value of an integer n can be named $|n|$.**

Examples   Simplify.

> *Simplify* means to rename with the least number of symbols.

13. $|-3| = 3$

14. $|25| = 25$

15. $|0| = 0$

The absolute value of an integer is never negative.

> *Caution!* Try not to make absolute value a difficult idea. It might help to keep in mind that the absolute value of a negative number is positive, whereas the absolute value of a positive number stays positive. The absolute value of 0 is 0.

DO EXERCISES 18–22.

Find the absolute value of each integer. Think of the number line and distance.

18. -8

19. 10

Simplify.

20. $|-29|$

21. $|6|$

22. $|0|$

SOMETHING EXTRA

CALCULATOR CORNER

1. *A number pattern.* Find each of the following using your calculator. Look for a pattern.

   1^3

   $1^3 + 2^3$

   $1^3 + 2^3 + 3^3$

   $1^3 + 2^3 + 3^3 + 4^3$

   Use the pattern to find each of the following without using your calculator.

   $1^3 + 2^3 + 3^3 + 4^3 + 5^3$

   $1^3 + 2^3 + 3^3 + 4^3 + 5^3 + 6^3$

2. Complete this table. Find decimal notation, rounding to four decimal places.

n	2^n	$\dfrac{1}{2^n}$
1		
2		
3		
4		
5		
6		
7		

NAME   CLASS

EXERCISE SET 2.1

▪● Find $-x$ when x is:

1. -6 **2.** -7 **3.** 6 **4.** 0

Find $-(-x)$ when x is:

5. -7 **6.** -8 **7.** 1 **8.** 2

Find $-x$ when x is:

9. -12 **10.** -19 **11.** 70 **12.** 80

Find $-(-x)$ when x is:

13. 0 **14.** -2 **15.** -34 **16.** -23

Change the sign. (Find the additive inverse.)

17. -1 **18.** -7 **19.** 7 **20.** 10

21. -14 **22.** -22 **23.** 0 **24.** 1

●● Make true sentences using $<$ or $>$.

25. 5 0 **26.** -9 0 **27.** -9 5 **28.** 8 -3

29. -6 6 **30.** -7 7 **31.** -8 -5 **32.** -3 -1

1.	
2.	
3.	
4.	
5.	
6.	
7.	
8.	
9.	
10.	
11.	
12.	
13.	
14.	
15.	
16.	
17.	
18.	
19.	
20.	
21.	
22.	
23.	
24.	
25.	
26.	
27.	
28.	
29.	
30.	
31.	
32.	

ANSWERS

33. _____

34. _____

35. _____

36. _____

37. _____

38. _____

39. _____

40. _____

41. _____

42. _____

43. _____

44. _____

45. _____

46. _____

47. _____

48. _____

49. _____

50. _____

51. _____

52. _____

53. _____

54. _____

55. _____

56. _____

57. _____

58. _____

33. -5 -11 **34.** -3 -4 **35.** -6 -5 **36.** -10 -14

●●● Simplify.

37. $|-3|$ **38.** $|-7|$ **39.** $|10|$ **40.** $|11|$

41. $|0|$ **42.** $|-4|$ **43.** $|-24|$ **44.** $|-36|$

45. $|53|$ **46.** $|54|$ **47.** $|-8|$ **48.** $|0|$

49. $|-6|$ **50.** $|-9|$ **51.** $|6|$ **52.** $|9|$

Simplify.

53. $-(-(-5))$ **54.** $-(-(-(-8)))$ **55.** $|-(-(-10))|$

56. $-(-x)$ **57.** $-(-(-x))$ **58.** $-(-(-(-x)))$

2.2 ADDITION OF INTEGERS

● ADDITION OF INTEGERS

To explain addition of integers we can use the number line.

> To do the addition $a + b$, we start at a, and move according to b.
> a) If b is positive, we move to the right.
> b) If b is negative, we move to the left.
> c) If b is 0, we stay at a.

Example 1   Add $3 + (-5)$.

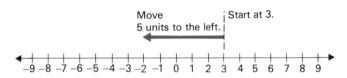

$3 + (-5) = -2$

Example 2   Add $-4 + (-3)$.

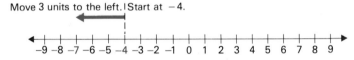

$-4 + (-3) = -7$

Example 3   Add $-4 + 9$.

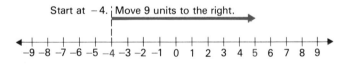

$-4 + 9 = 5$

DO EXERCISES 1–7.

You may have noticed some patterns in the previous examples and exercises. When we add positive integers and 0, the additions are, of course, additions of whole numbers. When we add two negative integers, the result is negative. We generalize as follows.

> To add two negative integers, we add their absolute values and change the sign (making the answer negative).

Examples   Add.

4. $-5 + (-7) = -12$ *Think:* Add the absolute values: $5 + 7 = 12$. Then change the sign.

5. $-8 + (-2) = -10$

DO EXERCISES 8–11.

OBJECTIVE

After finishing Section 2.2, you should be able to:

● Add integers without using the number line.

Add, using a number line.

1. $1 + (-4)$

2. $-3 + (-5)$

3. $-3 + 7$

4. $-5 + 5$

Write an addition sentence for each.

5.

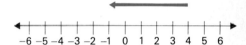

6.

7.

Add. Do not use a number line except as a check.

8. $-5 + (-6)$

9. $-9 + (-3)$

10. $-20 + (-14)$

11. $-11 + (-11)$

ANSWERS ON PAGE A–4

Add. Do not use a number line except as a check.

12. $-4 + 6$

13. $-7 + 3$

14. $5 + (-7)$

15. $10 + (-7)$

Add. Do not use a number line except as a check.

16. $5 + (-5)$

17. $-6 + 6$

18. $-10 + 10$

19. $89 + (-89)$

Add.

20. $\begin{array}{r} -35 \\ 17 \\ 14 \\ -27 \\ 31 \\ -12 \\ \hline \end{array}$

What happens when we add a positive and a negative integer with different absolute values? The answer is sometimes negative and sometimes positive, depending on which number has the greater absolute value (greater distance from 0). We can generalize as follows.

> **To add a positive and a negative integer, find the difference of their absolute values.**
> a) **If the negative integer has the greater absolute value, the answer is negative.**
> b) **If the positive integer has the greater absolute value, the answer is positive.**

Examples   Add.

6. $3 + (-5) = -2$ *Think:* Find the difference of the absolute values: $5 - 3 = 2$. Since -5 has the greater absolute value, make the answer negative.

7. $11 + (-8) = 3$ *Think:* Find the difference of the absolute values: $11 - 8 = 3$. Since 11 has the greater absolute value and it is positive, leave the answer positive.

8. $1 + (-5) = -4$ **9.** $-7 + 4 = -3$

10. $5 + (-10) = -5$ **11.** $7 + (-3) = 4$

12. $-6 + 10 = 4$ **13.** $5 + (-2) = 3$

DO EXERCISES 12–15.

We call $-a$ the *additive inverse of a* because adding any number to its additive inverse always gives 0—for example,
$$-8 + 8 = 0, \quad 14 + (-14) = 0, \quad \text{and} \quad 0 + 0 = 0.$$

> **For any integer a,**
> $$a + (-a) = -a + a = 0.$$

DO EXERCISES 16–19.

To add several integers, some positive and some negative, we add the positive ones. We add the negative ones. Then we add the results.

Example 14

$\begin{array}{r} 25 \\ -14 \\ -127 \\ 45 \\ 32 \\ -118 \\ 87 \\ \hline \end{array}$
First, add the positives.
$\begin{array}{r} 25 \\ 45 \\ 32 \\ 87 \\ \hline 189 \end{array}$
Next, add the negatives.
$\begin{array}{r} -14 \\ -127 \\ -118 \\ \hline -259 \end{array}$
Now add the results.
$\begin{array}{r} -259 \\ 189 \\ \hline -70 \end{array}$

DO EXERCISE 20.

NAME _____   CLASS _____

EXERCISE SET 2.2

⬤ Add. Do not use a number line except as a check.

1. $-9 + 2$ **2.** $2 + (-5)$ **3.** $-9 + 5$ **4.** $8 + (-3)$

5. $-6 + 6$ **6.** $-7 + 7$ **7.** $-8 + (-5)$ **8.** $-3 + (-1)$

9. $-5 + (-11)$ **10.** $-3 + (-4)$ **11.** $-6 + (-5)$ **12.** $-10 + (-14)$

13. $9 + (-9)$ **14.** $10 + (-10)$ **15.** $-2 + 2$ **16.** $-3 + 3$

17. $0 + 8$ **18.** $7 + 0$ **19.** $0 + (-8)$ **20.** $-7 + 0$

21. $-25 + 0$ **22.** $-43 + 0$ **23.** $0 + (-17)$ **24.** $0 + (-19)$

25. $17 + (-17)$ **26.** $-13 + 13$ **27.** $-18 + 18$ **28.** $11 + (-11)$

29. $8 + (-5)$ **30.** $-7 + 8$ **31.** $-4 + (-5)$ **32.** $0 + (-3)$

33. $0 + (-5)$ **34.** $10 + (-12)$ **35.** $13 + (-6)$ **36.** $-3 + 14$

ANSWERS

37. _____

38. _____

39. _____

40. _____

41. _____

42. _____

43. _____

44. _____

45. _____

46. _____

47. _____

48. _____

49. _____

50. _____

51. _____

52. _____

53. _____

54. _____

55. _____

56. _____

57. _____

58. _____

59. _____

60. _____

61. _____

62. _____

37. $-10 + 7$ **38.** $0 + (-9)$ **39.** $-6 + 6$ **40.** $-8 + 1$

41. $11 + (-16)$ **42.** $-7 + 15$ **43.** $-15 + (-6)$ **44.** $-8 + 8$

45. $11 + (-9)$ **46.** $-14 + (-19)$ **47.** $-20 + (-6)$ **48.** $17 + (-17)$

49. $-15 + (-7)$ **50.** $23 + (-5)$ **51.** $40 + (-8)$ **52.** $-23 + (-9)$

53. $-25 + 25$ **54.** $40 + (-40)$ **55.** $63 + (-18)$ **56.** $85 + (-65)$

57. $\begin{array}{r} -35 \\ -63 \\ -27 \\ -14 \\ -59 \\ \hline \end{array}$ **58.** $\begin{array}{r} -24 \\ -37 \\ -19 \\ -45 \\ -35 \\ \hline \end{array}$ **59.** $\begin{array}{r} 27 \\ -54 \\ -32 \\ 65 \\ 46 \\ \hline \end{array}$ **60.** $\begin{array}{r} -62 \\ 53 \\ -87 \\ 14 \\ -28 \\ \hline \end{array}$

Add.

61. ▦ $\begin{array}{r} -64374 \\ -27159 \\ 53690 \\ 39087 \\ 41646 \\ -11953 \\ \hline \end{array}$ **62.** ▦ $\begin{array}{r} -49752 \\ 28351 \\ -92265 \\ -40870 \\ 35649 \\ \hline \end{array}$

2.3 SUBTRACTION OF INTEGERS

● SUBTRACTION USING A NUMBER LINE (Optional)

Subtraction is defined as follows.

> **The difference $a - b$ is the number which when added to b gives a.**

To see that the definition is right, think of $5 - 2$. When you subtract, you get 3. To check, you add 3 to 2, to see if 3 is the number which when added to 2 gives 5.

Example 1 Subtract $10 - 4$.

We know that $10 - 4$ is the number which when added to 4 gives 10. So we start at 4 and go to 10. We go 6 units to the right (or $+6$) so the answer is 6.

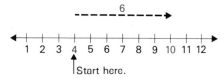

$10 - 4 = 6$

DO EXERCISE 1.

Example 2 Subtract $6 - 8$.

$6 - 8$ is the number which when added to 8 gives 6. So we start at 8 and go to 6. We go 2 units to the left (or -2), so the answer is -2.

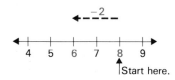

$6 - 8 = -2$

DO EXERCISE 2.

Example 3 Subtract $-1 - 5$. (We read this as "negative one minus five.")

$-1 - 5$ is the number which when added to 5 gives -1. So we start at 5 and go to -1. We go 6 units to the left (or -6), so the answer is -6.

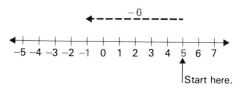

$-1 - 5 = -6$

DO EXERCISE 3.

OBJECTIVES

After finishing Section 2.3, you should be able to:
● Subtract using a number line.
●● Subtract integers by adding the inverse of the subtrahend.

Subtract. Use a number line.

1. $9 - 5$

Subtract. Use a number line.

2. $4 - 10$

Read the following. Then subtract. Use a number line.

3. $-2 - 6$

Read the following. Then subtract. Use number lines.

4. $-2 - (-3)$

5. $-7 - (-3)$

6. $-4 - 4$

7. $-6 - 2$

Add or subtract. Look for a pattern.

8. a) $4 - 6$

b) $4 + (-6)$

9. a) $-3 - 8$

b) $-3 + (-8)$

10. a) $-5 - (-9)$

b) $-5 + 9$

11. a) $-5 - (-3)$

b) $-5 + 3$

Read each of the following. Then subtract by adding the inverse of the subtrahend.

12. $2 - 8$

13. $10 - 3$

14. $-4 - 7$

15. $-5 - (-6)$

16. $-5 - (-1)$

Example 4   Subtract $-3 - (-7)$. (We read this as "negative three minus negative seven.")

$-3 - (-7)$ is the number which when added to -7 gives -3. So we start at -7 and go to -3. We go 4 units to the right (or $+4$), so the answer is 4.

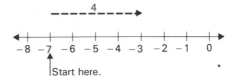

$-3 - (-7) = 4$

DO EXERCISES 4–7.

INVERSES AND SUBTRACTION

We want to find a faster way to subtract. Look for a pattern in the following examples.

Example 5

a) $3 - 8 = -5$ **b)** $3 + (-8) = -5$

Example 6

a) $-5 - 4 = -9$ **b)** $-5 + (-4) = -9$

Example 7

a) $-7 - (-10) = 3$ **b)** $-7 + 10 = 3$

Example 8

a) $-7 - (-2) = -5$ **b)** $-7 + 2 = -5$

DO EXERCISES 8–11.

Perhaps you have noticed in the preceding examples and exercises that we can subtract by adding an additive inverse. This can always be done.

> **For any integers a and b,**
>
> $$a - b = a + (-b).$$
>
> **(To subtract, we can add the inverse of the subtrahend.)**

Examples   Subtract.

9. $2 - 6 = 2 + (-6) = -4$ Changing the sign of 6 and changing the subtraction to addition

10. $-3 - 8 = -3 + (-8) = -11$

11. $10 - 7 = 10 + (-7) = 3$

12. $18 - 20 = 18 + (-20) = -2$

13. $-4 - (-9) = -4 + 9 = 5$ Changing the sign of -9 and changing the subtraction to addition

14. $-4 - (-3) = -4 + 3 = -1$

DO EXERCISES 12–16.

NAME   CLASS

EXERCISE SET 2.3

▮• Subtract using a number line.

1. $3 - 7$ **2.** $4 - 9$ **3.** $0 - 7$ **4.** $0 - 10$

5. $-8 - (-2)$ **6.** $-6 - (-8)$ **7.** $-10 - (-10)$ **8.** $-8 - (-8)$

▮• • Subtract by adding the inverse of the subtrahend.

9. $3 - 7$ **10.** $4 - 9$ **11.** $0 - 7$ **12.** $0 - 10$

13. $-8 - (-2)$ **14.** $-6 - (-8)$ **15.** $-10 - (-10)$ **16.** $-8 - (-8)$

17. $7 - 7$ **18.** $9 - 9$ **19.** $7 - (-7)$ **20.** $4 - (-4)$

21. $8 - (-3)$ **22.** $-7 - 4$ **23.** $-6 - 8$ **24.** $6 - (-10)$

25. $-4 - (-9)$ **26.** $-14 - 2$ **27.** $2 - 9$ **28.** $2 - 8$

29. $-6 - (-5)$ **30.** $-4 - (-3)$ **31.** $8 - (-10)$ **32.** $5 - (-6)$

33. $0 - 5$ **34.** $0 - 6$

ANSWERS

1. _____
2. _____
3. _____
4. _____
5. _____
6. _____
7. _____
8. _____
9. _____
10. _____
11. _____
12. _____
13. _____
14. _____
15. _____
16. _____
17. _____
18. _____
19. _____
20. _____
21. _____
22. _____
23. _____
24. _____
25. _____
26. _____
27. _____
28. _____
29. _____
30. _____
31. _____
32. _____
33. _____
34. _____

ANSWERS

35. _____

36. _____

37. _____

38. _____

39. _____

40. _____

41. _____

42. _____

43. _____

44. _____

45. _____

46. _____

47. _____

48. _____

49. _____

50. _____

51. _____

52. _____

53. _____

54. _____

55. _____

56. _____

35. $-5 - (-2)$ **36.** $-3 - (-1)$ **37.** $-7 - 14$ **38.** $-9 - 16$

39. $0 - (-5)$ **40.** $0 - (-1)$ **41.** $-8 - 0$ **42.** $-9 - 0$

43. $7 - (-5)$ **44.** $8 - (-3)$ **45.** $2 - 25$ **46.** $18 - 63$

47. $-42 - 26$ **48.** $-18 - 63$ **49.** $-71 - 2$ **50.** $-49 - 3$

51. $24 - (-92)$ **52.** $48 - (-73)$ **53.** $-50 - (-50)$ **54.** $-70 - (-70)$

Subtract.

55. ▦ $123,907 - 433,789$ **56.** ▦ $23,011 - (-60,432)$

2.4　MULTIPLICATION AND DIVISION OF INTEGERS

● ● MULTIPLICATION

Multiplication of integers is very much like multiplication of whole numbers. The only difference is that we must determine whether the answer is positive or negative. To see how this is done, consider the pattern in the following.

This number decreases by 1 each time.　　　　　　　This number decreases by 5 each time.

$$
\begin{aligned}
4 \cdot 5 &= 20 \\
3 \cdot 5 &= 15 \\
2 \cdot 5 &= 10 \\
1 \cdot 5 &= 5 \\
0 \cdot 5 &= 0 \\
-1 \cdot 5 &= -5 \\
-2 \cdot 5 &= -10 \\
-3 \cdot 5 &= -15
\end{aligned}
$$

DO EXERCISE 1.

According to this pattern, it looks as though the product of a negative and a positive integer is negative. This is the case.

> **To multiply a positive and a negative integer, multiply their absolute values. The answer is negative.**

Examples　Multiply.

1. $-4 \cdot 7 = -28$　　*Think:* Multiply the absolute values: $4 \cdot 7 = 28$. Make the answer negative.

2. $8 \cdot (-5) = -40$

3. $10 \cdot (-2) = -20$

4. $-7 \cdot 5 = -35$

DO EXERCISES 2–4.

How do we multiply two negative numbers? To see this we can again look for a pattern.

This number decreases by 1 each time.　　　　　　　This number increases by 5 each time.

$$
\begin{aligned}
4 \cdot (-5) &= -20 \\
3 \cdot (-5) &= -15 \\
2 \cdot (-5) &= 10 \\
1 \cdot (-5) &= -5 \\
0 \cdot (-5) &= 0 \\
-1 \cdot (-5) &= 5 \\
-2 \cdot (-5) &= 10 \\
-3 \cdot (-5) &= 15
\end{aligned}
$$

DO EXERCISE 5.

ANSWERS ON PAGE A–4

OBJECTIVES

After finishing Section 2.4, you should be able to:

● ● Multiply two integers.
● ● Divide two integers, where possible.

1. Complete, as in the example.

$$
\begin{aligned}
4 \cdot 10 &= 40 \\
3 \cdot 10 &= 30 \\
2 \cdot 10 &= \\
1 \cdot 10 &= \\
0 \cdot 10 &= \\
-1 \cdot 10 &= \\
-2 \cdot 10 &= \\
-3 \cdot 10 &=
\end{aligned}
$$

Multiply.

2. $-3 \cdot 6$

3. $20 \cdot (-5)$

4. $4 \cdot (-20)$

5. Complete, as in the example.

$$
\begin{aligned}
3 \cdot (-10) &= -30 \\
2 \cdot (-10) &= -20 \\
1 \cdot (-10) &= \\
0 \cdot (-10) &= \\
-1 \cdot (-10) &= \\
-2 \cdot (-10) &= \\
-3 \cdot (-10) &=
\end{aligned}
$$

Multiply.

6. $-3 \cdot (-4)$

7. $-16 \cdot (-2)$

8. $-7 \cdot (-5)$

Divide.

9. $6 \div (-3)$

10. $\dfrac{-15}{-3}$

Divide.

11. $-24 \div 8$

12. $\dfrac{-32}{-4}$

13. $\dfrac{30}{-5}$

To multiply two negative integers, multiply their absolute values. The answer is positive.

DO EXERCISES 6–8.

●● DIVISION

To divide integers we use the definition of division. $a \div b$ or a/b is the number which when multiplied by b gives a. Consider the division

$14 \div (-7)$.

We look for a number n such that $-7 \cdot n = 14$. We know from multiplication that -2 is the solution. That is,

$-7 \cdot (-2) = 14$ so $14 \div (-7) = -2$

Examples   Divide.

5. $-8 \div 2 = -4$ because $2 \cdot (-4) = -8$

6. $\dfrac{-10}{-2} = 5$ because $-2 \cdot 5 = -10$

DO EXERCISES 9 AND 10.

From these examples we see how to handle signs in division.

When we divide a positive integer by a negative, or a negative integer by a positive, the answer is negative. When we divide two negative integers, the answer is positive.

Examples   Divide.

7. $-24 \div 6 = -4$ *Think:* $\dfrac{24}{6} = 4$. Make the answer negative.

8. $\dfrac{-30}{-6} = 5$

9. $\dfrac{18}{-9} = -2$

There are some divisions of integers, such as $18/-5$, for which we do not get integers for answers. To get an answer, we need an expanded number system, which we will consider in the next section.

DO EXERCISES 11–13.

NAME   CLASS

EXERCISE SET 2.4

▪● Multiply.

1. $-8 \cdot 2$ **2.** $-2 \cdot 5$ **3.** $-7 \cdot 6$ **4.** $-9 \cdot 2$

5. $8 \cdot (-3)$ **6.** $9 \cdot (-5)$ **7.** $-9 \cdot 8$ **8.** $-10 \cdot 3$

9. $-8 \cdot (-2)$ **10.** $-2 \cdot (-5)$ **11.** $-7 \cdot (-6)$ **12.** $-9 \cdot (-2)$

13. $-8 \cdot (-3)$ **14.** $-9 \cdot (-5)$ **15.** $-9 \cdot (-8)$ **16.** $-10 \cdot (-3)$

17. $15 \cdot (-8)$ **18.** $12 \cdot (-10)$ **19.** $-25 \cdot (-40)$ **20.** $-22 \cdot (-50)$

1. _____

2. _____

3. _____

4. _____

5. _____

6. _____

7. _____

8. _____

9. _____

10. _____

11. _____

12. _____

13. _____

14. _____

15. _____

16. _____

17. _____

18. _____

19. _____

20. _____

ANSWERS

21. _____

22. _____

23. _____

24. _____

25. _____

26. _____

27. _____

28. _____

29. _____

30. _____

31. _____

32. _____

33. _____

34. _____

35. _____

36. _____

37. _____

38. _____

39. _____

40. _____

41. _____

21. $-6 \cdot (-15)$ **22.** $-8 \cdot (-22)$ **23.** $-25 \cdot (-8)$ **24.** $-35 \cdot (-4)$

● ● Divide.

25. $36 \div (-6)$ **26.** $\dfrac{28}{-7}$ **27.** $\dfrac{26}{-2}$ **28.** $26 \div (-13)$

29. $\dfrac{-16}{8}$ **30.** $-22 \div (-2)$ **31.** $\dfrac{-48}{-12}$ **32.** $\dfrac{-63}{-9}$

33. $\dfrac{-72}{9}$ **34.** $\dfrac{-50}{25}$ **35.** $-100 \div (-50)$ **36.** $\dfrac{-200}{8}$

37. $-108 \div 9$ **38.** $\dfrac{-64}{-8}$ **39.** $\dfrac{200}{-25}$ **40.** $-300 \div (-75)$

41. ▦ Calculate each of the following. Look for a pattern.

$(-1 \cdot 8) - 1$
$(-12 \cdot 8) - 2$
$(-123 \cdot 8) - 3$
$(-1234 \cdot 8) - 4$

Use the pattern to find the following without using your calculator.

$(-12345 \cdot 8) \ - 5$

2.5 RATIONAL NUMBERS

The set of *rational numbers* consists of all the numbers of arithmetic and their additive inverses. The rational numbers include all of the integers. There are positive and negative rational numbers, the same as for integers, but between any two integers there are many rational numbers.

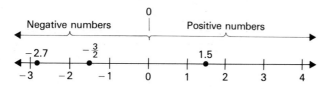

● INVERSES AND ABSOLUTE VALUE

Everything we have said about integers so far holds true for rational numbers. For example, all rational numbers have additive inverses, and absolute value means the same as for integers.

Examples   Find the additive inverse.

1. 4.2   The additive inverse is -4.2.

2. $-\dfrac{3}{2}$ The additive inverse is $\dfrac{3}{2}$.

3. -9.35 The additive inverse is 9.35.

DO EXERCISES 1–3.

Examples   Simplify.

4. $-\left(-\dfrac{4}{5}\right) = \dfrac{4}{5}$

5. $-(-9.9) = 9.9$

6. $-\left(\dfrac{3}{8}\right) = -\dfrac{3}{8}$

> "Simplify" means to rename with the least number of symbols.

DO EXERCISES 4–6.

Example 7   Find $-a$ when a is -6.

$-a$ is the inverse of a. When a is -6, we have

   $-(-6)$,   which is 6.

Example 8   Find $-a$ when a is $-\frac{7}{8}$.

$-a = -\left(-\dfrac{7}{8}\right) = \dfrac{7}{8}$

> *Remember:* $-a$ should not be read "negative a." As these examples again illustrate, we can't know that $-a$ is positive, negative, or 0 until we know what a represents.

DO EXERCISES 7–9.

OBJECTIVES

After finishing Section 2.5, you should be able to:

● Find the additive inverse and absolute value of any rational number and simplify additive inverse symbols.

●● Add rational numbers.

●●● Subtract rational numbers.

Find the additive inverse.

1. 8.7

2. $-\dfrac{8}{9}$

3. -7.74

Simplify.

4. $-\left(-\dfrac{10}{3}\right)$

5. $-(-8.32)$

6. $-\left(\dfrac{5}{4}\right)$

Find $-x$ when x is

7. -12

8. $-\dfrac{5}{6}$

9. 17.2

ANSWERS ON PAGE A–5

Simplify.

10. |4.1|

11. $\left|-\dfrac{8}{3}\right|$

12. |−3.5|

Add.

13. −8.6 + 2.4

14. $\dfrac{5}{9} + \left(-\dfrac{7}{9}\right)$

15. $-\dfrac{1}{5} + \left(-\dfrac{3}{4}\right)$

16. A lab technician had 500 milliliters of acid in a beaker. After pouring out 16.5 milliliters and then adding 27.3, how much was in the beaker?

Subtract.

17. −3 − 6

18. $\dfrac{7}{10} - \dfrac{9}{10}$

19. −8.8 − (−1.3)

20. $-\dfrac{2}{3} - \left(-\dfrac{5}{8}\right)$

Examples   Simplify.

9. |8.7| = 8.7   **10.** $\left|-\dfrac{3}{4}\right| = \dfrac{3}{4}$ **11.** |−17.3| = 17.3

DO EXERCISES 10–12.

●● ADDITION

Addition is done as with integers.

Examples   Add.

12. −9.2 + 3.1 = −6.1   **13.** $-\dfrac{3}{2} + \dfrac{9}{2} = \dfrac{6}{2} = 3$

14. $-\dfrac{2}{3} + \dfrac{5}{8} = -\dfrac{16}{24} + \dfrac{15}{24} = -\dfrac{1}{24}$

15. One day the value of a share of IBM stock was $253\frac{1}{4}$. That day it rose in value $\$\frac{5}{8}$. The next day it lost $\$\frac{3}{8}$. What was the value of the stock at the end of the two days?

$$253\frac{1}{4} + \frac{5}{8} + \left(-\frac{3}{8}\right) = 253\frac{2}{8} + \frac{5}{8} + \left(-\frac{3}{8}\right)$$

$$= 253\frac{4}{8}, \quad \text{or} \quad \$253\frac{1}{2}$$

DO EXERCISES 13–16.

●●● SUBTRACTION

Subtraction is done as with integers.

> **For any rational numbers a and b,**
>
> $$a - b = a + (-b).$$ **(To subtract we add the inverse of the subtrahend.)**

Examples   Subtract.

16. 3 − 5 = 3 + (−5) = −2

17. $\dfrac{1}{8} - \dfrac{7}{8} = \dfrac{1}{8} + \left(-\dfrac{7}{8}\right) = -\dfrac{6}{8}, \quad \text{or} \quad -\dfrac{3}{4}$

18. −4.6 − (−9.8) = −4.6 + 9.8 = 5.2

19. $-\dfrac{3}{4} - \dfrac{7}{5} = -\dfrac{15}{20} + \left(-\dfrac{28}{20}\right) = -\dfrac{43}{20}$

DO EXERCISES 17–20.

Addition of rational numbers is associative and commutative. Each rational number has an additive inverse.

NAME _____   CLASS _____

EXERCISE SET 2.5

⚫ Find the additive inverse.

1. -4.7

2. -5.2

3. $\dfrac{7}{2}$

4. $\dfrac{8}{3}$

5. -7

6. -9

7. -26.9

8. -17.83

Simplify.

9. $-\left(-\dfrac{1}{3}\right)$

10. $-\left(-\dfrac{1}{4}\right)$

11. $-\left(\dfrac{7}{6}\right)$

12. $-\left(\dfrac{9}{8}\right)$

13. $-(-9.3)$

14. $-(-8.6)$

15. $-(90.3)$

16. $-(78.8)$

Find $-x$ when x is

17. 12.4

18. 23.8

19. $-\dfrac{9}{10}$

20. $-\dfrac{11}{12}$

21. -34.8

22. 44.1

23. 567

24. 890

Simplify.

25. $|19.2|$

26. $|16.8|$

27. $\left|-\dfrac{2}{3}\right|$

28. $\left|-\dfrac{5}{6}\right|$

29. $|-89.3|$

30. $|-17.4|$

31. $\left|\dfrac{14}{3}\right|$

32. $\left|\dfrac{16}{5}\right|$

⚫⚫ Add.

33. $-6.5 + 4.7$

34. $-3.6 + 1.9$

35. $-2.8 + (-5.3)$

36. $-7.9 + (-6.5)$

ANSWERS
1. _____
2. _____
3. _____
4. _____
5. _____
6. _____
7. _____
8. _____
9. _____
10. _____
11. _____
12. _____
13. _____
14. _____
15. _____
16. _____
17. _____
18. _____
19. _____
20. _____
21. _____
22. _____
23. _____
24. _____
25. _____
26. _____
27. _____
28. _____
29. _____
30. _____
31. _____
32. _____
33. _____
34. _____
35. _____
36. _____

ANSWERS

37. _____

38. _____

39. _____

40. _____

41. _____

42. _____

43. _____

44. _____

45. _____

46. _____

47. _____

48. _____

49. _____

50. _____

51. _____

52. _____

53. _____

54. _____

55. _____

56. _____

57. _____

58. _____

59. _____

60. _____

61. _____

62. _____

63. _____

64. _____

65. _____

66. _____

67. _____

37. $-\dfrac{3}{5} + \dfrac{2}{5}$ **38.** $-\dfrac{4}{3} + \dfrac{2}{3}$ **39.** $-\dfrac{3}{7} + \left(-\dfrac{5}{7}\right)$ **40.** $-\dfrac{4}{9} + \left(-\dfrac{6}{9}\right)$

41. $-\dfrac{5}{8} + \dfrac{1}{4}$ **42.** $-\dfrac{5}{6} + \dfrac{2}{3}$ **43.** $-\dfrac{3}{7} + \left(-\dfrac{2}{5}\right)$ **44.** $-\dfrac{5}{8} + \left(-\dfrac{1}{3}\right)$

45. $-\dfrac{3}{5} + \left(-\dfrac{2}{15}\right)$ **46.** $-\dfrac{5}{9} + \left(-\dfrac{1}{18}\right)$

47. $-5.7 + (-7.2) + 6.6$ **48.** $-10.3 + (-7.5) + 3.1$

49. $-8.5 + 7.9 + (-3.7)$ **50.** $-9.6 + 8.4 + (-11.8)$

●●● Subtract.

51. $-9 - (-5)$ **52.** $-8 - (-6)$ **53.** $\dfrac{3}{8} - \dfrac{5}{8}$ **54.** $\dfrac{3}{9} - \dfrac{9}{9}$

55. $\dfrac{3}{4} - \dfrac{2}{3}$ **56.** $\dfrac{5}{8} - \dfrac{3}{4}$ **57.** $-\dfrac{3}{4} - \dfrac{2}{3}$ **58.** $-\dfrac{5}{8} - \dfrac{3}{4}$

59. $-\dfrac{5}{8} - \left(-\dfrac{3}{4}\right)$ **60.** $-\dfrac{3}{4} - \left(-\dfrac{2}{3}\right)$ **61.** $6.1 - (-13.8)$ **62.** $1.5 - (-3.5)$

63. $-3.2 - 5.8$ **64.** $-2.7 - 5.9$ **65.** $0.99 - 1$ **66.** $0.87 - 1$

Calculate.

67. ▦ $9.976 + (-8.732) + (-16.999) + 83.605$

2.6 MULTIPLICATION AND DIVISION OF RATIONAL NUMBERS

● MULTIPLICATION

Multiplication of rational numbers is like multiplication of integers.

> **To multiply a negative and a positive number, multiply their absolute values. The answer is negative.**

Examples   Multiply.

1. $-9 \cdot 4 = -36$

2. $-\dfrac{1}{3} \cdot \dfrac{5}{7} = -\dfrac{5}{21}$

3. $\dfrac{5}{6} \cdot \left(-\dfrac{1}{5}\right) = -\dfrac{5}{30} = -\dfrac{1}{6}$

4. $-8.66 \times 4.3 = -37.238$

DO EXERCISES 1–4.

> **To multiply two negative numbers, multiply their absolute values. The answer is positive.**

Examples   Multiply.

5. $-6(-8) = 48$

6. $-9.5(-7.4) = 70.3$

7. $-\dfrac{1}{2}\left(-\dfrac{1}{3}\right) = \dfrac{1}{6}$

8. $-\dfrac{3}{4}\left(-\dfrac{8}{9}\right) = \dfrac{24}{36} = \dfrac{2}{3}$

DO EXERCISES 5–8.

●● MULTIPLYING MORE THAN TWO NUMBERS

To multiply several numbers, we can multiply two at a time. We can do this because multiplication of rational numbers is also commutative and associative. This means we can group and order as we please.

Examples   Multiply.

9. $3 \cdot (-2) \cdot 4 = (3 \cdot 4) \cdot (-2)$ Changing grouping and ordering
$$= 12 \cdot (-2) = -24$$

10. $-5 \times (-3.8) \times 2.4 = 19 \times 2.4$ Multiplying the negatives
$$= 45.6$$

11. $-\dfrac{1}{3} \cdot \left(-\dfrac{3}{5}\right) \cdot \left(-\dfrac{15}{7}\right) = \dfrac{1}{5} \cdot \left(-\dfrac{15}{7}\right)$ Multiplying the first two numbers
$$= -\dfrac{3}{7}$$

12. $-5 \cdot (-2) \cdot (-3) \cdot (-6) = 10 \cdot 18$ Multiplying the first two numbers and the last two numbers
$$= 180$$

The product of three negative numbers is negative. The product of four negative numbers is positive.

OBJECTIVES

After finishing Section 2.6, you should be able to:
● Multiply two rational numbers.
●● Multiply several positive and negative numbers.
●●● Find the reciprocal of a rational number, and divide by multiplying by a reciprocal.

Multiply.

1. $6(-5)$

2. $\dfrac{2}{3}\left(-\dfrac{5}{9}\right)$

3. $-\dfrac{4}{5} \cdot \dfrac{7}{8}$

4. -4.23×7.1

Multiply.

5. $-16(-4)$

6. $-\dfrac{4}{7}\left(-\dfrac{5}{9}\right)$

7. $-\dfrac{3}{2}\left(-\dfrac{4}{9}\right)$

8. $-3.25(-4.14)$

Multiply.

9. $5 \cdot (-3) \cdot 2$

10. $-3 \times (-4.1) \times (-2.5)$

11. $-\frac{1}{2} \cdot \left(-\frac{4}{3}\right) \cdot \left(-\frac{5}{2}\right)$

12. $-2 \cdot (-5) \cdot (-4) \cdot (-3)$

Find the reciprocal.

13. $\frac{2}{3}$

14. $-\frac{5}{4}$

15. -3

16. $-\frac{1}{5}$

17. Complete the following table.

Number	Additive inverse	Reciprocal
$\frac{2}{3}$		
$-\frac{5}{4}$		
0		
1		
-4.5		

DO EXERCISES 9–12.

⚫⚫⚫ RECIPROCALS AND DIVISION

Two numbers are reciprocals* of each other if their product is 1. (See Section 1.5.)

Examples

13. The reciprocal of $\frac{7}{8}$ is $\frac{8}{7}$ because $\frac{7}{8} \cdot \frac{8}{7} = \frac{56}{56} = 1$.

14. The reciprocal of $-\frac{2}{3}$ is $-\frac{3}{2}$ because $-\frac{2}{3}\left(-\frac{3}{2}\right) = \frac{6}{6} = 1$.

15. The reciprocal of a nonzero number a is $\frac{1}{a}$ because $a \cdot \frac{1}{a} = \frac{a}{a} = 1$.

> **Any nonzero number a has a reciprocal $\frac{1}{a}$. The reciprocal of a negative number is negative. The reciprocal of a positive number is positive. The reciprocal of $\frac{a}{b}$ is $\frac{b}{a}$. The reciprocal of $-\left(\frac{b}{a}\right)$ is $-\left(\frac{a}{b}\right)$.**

DO EXERCISES 13–16.

It is important *not* to confuse *additive inverse* with *reciprocal*. Keep in mind that the additive inverse of a number is what we add to it to get 0, whereas a reciprocal is what we multiply the number by to get 1. Compare the following.

Number	Additive inverse	Reciprocal
$-\frac{3}{8}$	$\frac{3}{8}$	$-\frac{8}{3}$
19	-19	$\frac{1}{19}$
$\frac{18}{7}$	$-\frac{18}{7}$	$\frac{7}{18}$
-7.9	7.9	$-\frac{1}{7.9}$ or $-\frac{10}{79}$
0	0	Does not exist

DO EXERCISE 17.

* Also called *multiplicative inverses*.

We can divide by multiplying by the reciprocal of the divisor.

Examples   Divide.

16. $\dfrac{2}{3} \div \left(-\dfrac{5}{4}\right) = \dfrac{2}{3} \cdot \left(-\boxed{\dfrac{4}{5}}\right) = -\dfrac{8}{15}$

17. $-\dfrac{5}{6} \div \left(-\dfrac{3}{4}\right) = -\dfrac{5}{6} \cdot \left(-\boxed{\dfrac{4}{3}}\right) = \dfrac{20}{18} = \dfrac{10 \cdot 2}{9 \cdot 2} = \dfrac{10}{9} \cdot \dfrac{2}{2} = \dfrac{10}{9}$

> *Caution!* Do not change the sign of a number when taking its reciprocal.

18. $-\dfrac{3}{4} \div \dfrac{3}{10} = -\dfrac{3}{4} \cdot \left(\boxed{\dfrac{10}{3}}\right) = -\dfrac{30}{12} = -\dfrac{5}{2} \cdot \dfrac{6}{6} = -\dfrac{5}{2}$

19. $-6.3 \div 2.1 = -3$

DO EXERCISES 18–21.

We can also describe the rational numbers as quotients of integers.

> **The set of rational numbers consists of all numbers that can be named with fractional notation $\dfrac{a}{b}$, where a and b are integers and b is not 0.**

The following are examples of rational numbers:

$$\dfrac{5}{4}, \quad \dfrac{-3}{8}, \quad \dfrac{10}{1}, \quad 8, \quad -467, \quad 0, \quad \dfrac{13}{-5}, \quad -\dfrac{2}{3}.$$

A negative number divided by a negative number is positive. Thus,

$$\dfrac{-2}{-3} = \dfrac{2}{3}.$$

We can verify this by multiplying by 1 using $\dfrac{-1}{-1}$:

$$\dfrac{-2}{-3} = 1 \cdot \dfrac{-2}{-3} = \boxed{\dfrac{-1}{-1}} \cdot \dfrac{-2}{-3} = \dfrac{(-1)(-2)}{(-1)(-3)} = \dfrac{2}{3}.$$

Example 20   Divide: $\dfrac{-7}{-8}$.

$$\dfrac{-7}{-8} = \dfrac{7}{8}$$

DO EXERCISES 22 AND 23.

Divide.

18. $\dfrac{4}{7} \div \left(-\dfrac{3}{5}\right)$

19. $-\dfrac{8}{5} \div \dfrac{2}{3}$

20. $-\dfrac{12}{7} \div \left(-\dfrac{3}{4}\right)$

21. $21.7 \div (-3.1)$

Divide.

22. $\dfrac{-19}{-20}$

23. $\dfrac{-8}{-5}$

ANSWERS ON PAGE A–5

Divide.

24. $\dfrac{-10}{3}$

25. $\dfrac{5}{-6}$

Find two other names for each number by making appropriate sign changes.

26. $\dfrac{5}{-6}$

27. $-\dfrac{8}{7}$

28. $\dfrac{10}{-3}$

A negative number divided by a positive number is negative, and a positive number divided by a negative number is negative. Thus,

$$\frac{-2}{3} = \frac{2}{-3} = -\frac{2}{3}.$$

We can verify the preceding by working backwards from $-\dfrac{2}{3}$:

$$-\frac{2}{3} = -1 \cdot \frac{2}{3} = \frac{-1}{1} \cdot \frac{2}{3} = \frac{-1 \cdot 2}{1 \cdot 3} = \frac{-2}{3},$$

$$-\frac{2}{3} = -1 \cdot \frac{2}{3} = \frac{1}{-1} \cdot \frac{2}{3} = \frac{1 \cdot 2}{-1 \cdot 3} = \frac{2}{-3}.$$

Examples   Divide.

21. $\dfrac{-4}{5} = -\dfrac{4}{5}$

22. $\dfrac{23}{-8} = -\dfrac{23}{8}$

Example 23   Find two other names for $\dfrac{-4}{5}$ by making appropriate sign changes.

Two other names are $\dfrac{4}{-5}$ and $-\dfrac{4}{5}$.

DO EXERCISES 24–28.

NAME   CLASS

EXERCISE SET 2.6

● Multiply.

1. $9 \cdot (-8)$ **2.** $7 \cdot (-9)$ **3.** $4 \cdot (-3.1)$ **4.** $3 \cdot (-2.2)$

5. $-6 \cdot (-4)$ **6.** $-5 \cdot (-6)$ **7.** $-7 \cdot (-3.1)$ **8.** $-4 \cdot (-3.2)$

9. $\dfrac{2}{3} \cdot \left(-\dfrac{3}{5}\right)$ **10.** $\dfrac{5}{7} \cdot \left(-\dfrac{2}{3}\right)$ **11.** $-\dfrac{3}{8} \cdot \left(-\dfrac{2}{9}\right)$ **12.** $-\dfrac{5}{8} \cdot \left(-\dfrac{2}{5}\right)$

13. -6.3×2.7 **14.** -4.1×9.5 **15.** $-\dfrac{5}{9} \cdot \dfrac{3}{4}$ **16.** $-\dfrac{8}{3} \cdot \dfrac{9}{4}$

●● Multiply.

17. $6 \cdot (-5) \cdot 3$ **18.** $8 \cdot (-7) \cdot 6$

19. $7 \cdot (-4) \cdot (-3) \cdot 5$ **20.** $9 \cdot (-2) \cdot (-6) \cdot 7$

21. $-\dfrac{2}{3} \cdot \dfrac{1}{2} \cdot \left(-\dfrac{6}{7}\right)$ **22.** $-\dfrac{1}{8} \cdot \left(-\dfrac{1}{4}\right) \cdot \left(-\dfrac{3}{5}\right)$

23. $-3 \cdot (-4) \cdot (-5)$ **24.** $-2 \cdot (-5) \cdot (-7)$

25. $-2 \cdot (-5) \cdot (-3) \cdot (-5)$ **26.** $-3 \cdot (-5) \cdot (-2) \cdot (-1)$

1. _____

2. _____

3. _____

4. _____

5. _____

6. _____

7. _____

8. _____

9. _____

10. _____

11. _____

12. _____

13. _____

14. _____

15. _____

16. _____

17. _____

18. _____

19. _____

20. _____

21. _____

22. _____

23. _____

24. _____

25. _____

26. _____

●●●   Find the reciprocal.

27. -5 **28.** -7 **29.** $\dfrac{1}{4}$ **30.** $\dfrac{1}{8}$

31. $-\dfrac{7}{5}$ **32.** $-\dfrac{5}{3}$ **33.** $-\dfrac{4}{11}$ **34.** $-\dfrac{7}{12}$

Divide.

35. $\dfrac{3}{4} \div \left(-\dfrac{2}{3}\right)$ **36.** $\dfrac{7}{8} \div \left(-\dfrac{1}{2}\right)$ **37.** $-\dfrac{5}{4} \div \left(-\dfrac{3}{4}\right)$ **38.** $-\dfrac{5}{9} \div \left(-\dfrac{5}{6}\right)$

39. $-\dfrac{2}{7} \div \left(-\dfrac{4}{9}\right)$ **40.** $-\dfrac{3}{5} \div \left(-\dfrac{5}{8}\right)$ **41.** $-\dfrac{3}{8} \div \left(-\dfrac{8}{3}\right)$ **42.** $-\dfrac{5}{6} \div \left(-\dfrac{6}{5}\right)$

43. $\dfrac{5}{7} \div \left(-\dfrac{5}{7}\right)$ **44.** $\dfrac{7}{8} \div \left(-\dfrac{7}{8}\right)$ **45.** $-6.6 \div 3.3$ **46.** $-44.1 \div (-6.3)$

47. $\dfrac{-11}{-13}$ **48.** $\dfrac{-19}{20}$ **49.** $\dfrac{23}{-14}$ **50.** $\dfrac{-15}{-75}$

Evaluate.

51. $(-1)^4$ **52.** $(-1)^3$ **53.** $(-2)^3$ **54.** $(-2)^2$

2.7 DISTRIBUTIVE LAWS AND THEIR USE

● THE DISTRIBUTIVE LAWS

For rational numbers the distributive law of multiplication over addition holds. There is also another distributive law. It is the distributive law of multiplication over subtraction. When we multiply a number by a difference, we can either subtract and then multiply or multiply and then subtract.

> **For any rational numbers a, b, and c,**
>
> $a(b - c) = ab - ac.$ **(The distributive law of multiplication over subtraction)**

Examples Compute.

1. a) $3\,(4 - 2) = 3 \cdot 2 = 6$ **b)** $3 \cdot 4 - 3 \cdot 2 = 12 - 6 = 6$

2. a) $6\,(3 - 8) = 6 \cdot (-5)$ **b)** $6 \cdot 3 - 6 \cdot 8 = 18 - 48$
$= -30$ $= -30$

DO EXERCISES 1–4.

In an expression like $(ab) - (ac)$, we can omit the parentheses. In other words, in $ab - ac$ we *agree* that the multiplications are to be done first. The distributive laws are used in algebra in several ways. The basic ones are factoring, multiplying, and collecting like terms.

●● TERMS

What do we mean by the *terms* of an expression? When they are all separated by plus signs, it is easy to tell. If there are subtraction signs, we can rewrite using addition signs.

Example 3 What are the terms of $3x - 4y + 2z$?

$3x - 4y + 2z = 3x + (-4y) + 2z$ Separating parts with + signs

The terms are $3x$, $-4y$, and $2z$.

DO EXERCISES 5 AND 6.

●●● MULTIPLYING

The distributive laws are a basis for a procedure called *multiplying*. In an expression such as $8(a + 2b - 7)$, we multiply each term inside the parentheses by 8.

$8 \cdot (a + 2b - 7) = 8 \cdot a + 8 \cdot 2b - 8 \cdot 7 = 8a + 16b - 56$

Examples Multiply.

4. $4\,(x - 2) = 4 \cdot x - 4 \cdot 2 = 4x - 8$

5. $b\,(s - 3t + 2w) = b \cdot s - b \cdot 3t + b \cdot 2w = bs - 3bt + 2bw$

6. $-4\,(x - 2y + 3z) = -4 \cdot x - (-4) \cdot 2y + (-4) \cdot 3z$
$= -4x + 8y - 12z$

OBJECTIVES

After finishing Section 2.7, you should be able to:

● Apply the distributive law of multiplication over subtraction in computing.

●● Identify the terms of an expression.

●●● Multiply, where one expression has several terms.

:: Factor when terms have a common factor.

::: Collect like terms.

Compute.

1. a) $4\,(5 - 3)$

b) $4 \cdot 5 - 4 \cdot 3$

2. a) $-2 \cdot (5 - 3)$

b) $-2 \cdot 5 - (-2) \cdot 3$

3. a) $5(2 - 7)$

b) $5 \cdot 2 - 5 \cdot 7$

4. a) $-4[3 - (-2)]$

b) $-4 \cdot 3 - (-4)(-2)$

What are the terms of each?

5. $5x - 4y + 3$

6. $-4y - 2x + 3z$

ANSWERS ON PAGE A-5

Multiply.

7. $3(x - 5)$

8. $5(x - y + 4)$

9. $-2(x - 3)$

10. $b(x - 2y + 4z)$

Factor.

11. $4x - 8$

12. $3x - 6y + 9$

13. $bx + by - bz$

Collect like terms.

14. $6x - 3x$

15. $7x - x$

16. $x - 0.41x$

17. $5x + 4y - 2x - y$

18. $3x - 7x - 11 + 8y + 4 - 13y$

DO EXERCISES 7–10.

▌▌ FACTORING

Factoring is the reverse of multiplying. Look at Example 4. To *factor* $4x - 8$ we would write the product $4(x - 2)$.

When all the terms of an expression have a factor in common, we can "factor it out" using the distributive laws. Note that:

$4x$ has the following factors: $4, x, 2, -2, 1, -1, -4$;

-8 has the following factors: $8, -1, 2, -2, 4, -4, 1, -8$.

We usually remove the largest common factor. In the case of $4x - 8$, that factor is 4.

Examples Factor.

7. $5x - 10 = \boxed{5} \cdot x - \boxed{5} \cdot 2 = \boxed{5}(x - 2)$

8. $\boxed{a}\,x - \boxed{a}\,y + \boxed{a}\,z = \boxed{a}(x - y + z)$

9. $9x + 27y - 9 = \boxed{9} \cdot x + \boxed{9} \cdot 3y - \boxed{9} \cdot 1 = \boxed{9}(x + 3y - 1)$

> *Remember:* An expression is factored when it is written as a product.

> *Caution!* Note that $3(3x + 9y - 3)$ is also a factored form of $9x + 27y - 9$, but it is *not* the desired one. Factor out the largest common factor.

DO EXERCISES 11–13.

▌▌ COLLECTING LIKE TERMS

The process of collecting like terms is also based on the distributive laws.

Examples Collect like terms.

10. $4\,\boxed{x} - 2\,\boxed{x} = (4 - 2)\,\boxed{x} = 2x$ Factor out the x.

11. $2\,\boxed{x} + 3y - 5\,\boxed{x} - 2y = 2\,\boxed{x} - 5\,\boxed{x} + 3y - 2y$

$\qquad\qquad = (2 - 5)\,\boxed{x} + (3 - 2)y = -3\,\boxed{x} + y$

12. $3\,\boxed{x} - \boxed{x} = (3 - 1)\,\boxed{x} = 2x$

13. $x - 0.24x = 1 \cdot x - 0.24x = (1 - 0.24)x = 0.76x$

14. $x - 6x = 1 \cdot x - 6 \cdot x = (1 - 6)x = -5x$

15. $4x - 7y + 9x - 5 + 3y - 8 = 13x - 4y - 13$

DO EXERCISES 14–18.

NAME   CLASS   ANSWERS

EXERCISE SET 2.7

◼• Multiply.

1. $-3(3 - 7)$ **2.** $-5(9 - 14)$ **3.** $8(9 - 10)$ **4.** $6(13 - 14)$

◼◼ What are the terms of each?

5. $8x - 1.4y$ **6.** $12x - 1.2y$

7. $-5x + 3y - 14z$ **8.** $-8a - 10b + 18z$

◼◼◼ Factor.

9. $8x - 24$ **10.** $10x - 50$ **11.** $32 - 4x$ **12.** $24 - 6x$

13. $8x + 10y - 22$ **14.** $9x + 6y - 15$ **15.** $ax - 7a$ **16.** $bx - 9b$

17. $ax + ay - az$ **18.** $cx - cy - cz$

◼◼ ◼◼ Multiply.

19. $7(x - 2)$ **20.** $5(x - 8)$ **21.** $-7(y - 2)$ **22.** $-9(y - 7)$

ANSWERS

1. _____
2. _____
3. _____
4. _____
5. _____
6. _____
7. _____
8. _____
9. _____
10. _____
11. _____
12. _____
13. _____
14. _____
15. _____
16. _____
17. _____
18. _____
19. _____
20. _____
21. _____
22. _____

ANSWERS

23. _____

24. _____

25. _____

26. _____

27. _____

28. _____

29. _____

30. _____

31. _____

32. _____

33. _____

34. _____

35. _____

36. _____

37. _____

38. _____

39. _____

40. _____

41. _____

42. _____

43. _____

44. _____

45. _____

46. _____

23. $-3(7 - t)$ **24.** $-5(14 - n)$ **25.** $-4(x + 3y)$ **26.** $-3(a + 2b)$

27. $7(-2x - 4y + 3)$ **28.** $9(-5x - 6y + 8)$

⣷ Collect like terms.

29. $11x - 3x$ **30.** $17t - 9t$ **31.** $17y - y$ **32.** $6n - n$

33. $x - 12x$ **34.** $y - 14y$ **35.** $x - 0.83x$ **36.** $t - 0.02t$

37. $9x + 2y - 5x$ **38.** $8y + 3z - 4y$

39. $11x + 2y + 7 - 4x - 18 - y$ **40.** $13a + 9b - 14 + 10 - 2a - 4b$

41. $2.7x + 2.3y - 1.9x - 1.8y$ **42.** $6.7a + 4.3b - 4.1a - 2.9b$

43. $\dfrac{1}{5}x + \dfrac{4}{5}y + \dfrac{2}{5}x - \dfrac{1}{5}y$ **44.** $\dfrac{7}{8}x + \dfrac{5}{8}y + \dfrac{1}{8}x - \dfrac{3}{8}y$

---

45. A principal of P dollars was invested in a savings account at 8% simple interest. How much was in the account after 1 year?

46. The population of a town is P. After a 6% increase, what was the new population?

2.8 MULTIPLYING BY −1 AND SIMPLIFYING

● MULTIPLYING BY −1

What happens when we multiply a rational number by −1?

Examples

1. $-1 \cdot 7 = -7$ **2.** $-1 \cdot (-5) = 5$ **3.** $-1 \cdot 0 = 0$

When we multiply a number by −1, the result is the additive inverse of that number.

> **For any rational number** a,
>
> $$-1 \cdot a = -a.$$
>
> **That is, negative one times** a **is the additive inverse of** a.

In other words, multiplying by −1 changes the sign. This fact enables us to rename an additive inverse when an expression has one or more terms.

Example 4 Rename the additive inverse without parentheses.

$-(3x - 2y + 4) = -1 \cdot (3x - 2y + 4)$ Taking the additive inverse is the same as multiplying by −1. We replace − by −1.

$$= -1 \cdot 3x - (-1) \cdot 2y + (-1) \cdot 4$$
$$= -3x + 2y - 4 \quad \text{Replacing } -1 \text{ by } -$$

DO EXERCISES 1 AND 2.

Example 4 illustrates how we find the inverse when we have more than one term.

> **To find the additive inverse of an expression with more than one term, change the sign of** *every* **term.**

Examples Rename each additive inverse without parentheses.

5. $-(5 - y) = -5 + y$ **6.** $-(2a - 7b + 6) = -2a + 7b - 6$

DO EXERCISES 3–6.

●● REMOVING CERTAIN PARENTHESES

When parentheses follow a minus sign, they can be removed by renaming, as above. Then we collect like terms.

Examples Remove parentheses and simplify.

7. $3x - (4x + 2) = 3x - 4x - 2$ Subtracting is adding an inverse: $3x - (4x + 2) = 3x + [-(4x + 2)]$. Change the sign of *every* term in the parentheses.

$$= -x - 2 \quad \text{Collecting like terms}$$

OBJECTIVES

After finishing Section 2.8, you should be able to:

● Rename an additive inverse without parentheses, where an expression has several terms.

●● Simplify expressions by removing parentheses and collecting like terms.

●●● Simplify expressions with parentheses inside parentheses.

Rename each additive inverse without parentheses.

1. $-(x + 2)$

2. $-(5x - 2y - 8)$

Rename each additive inverse without parentheses. Try to do this in one step.

3. $-(6 - t)$

4. $-(x - y)$

5. $-(-4a + 3t - 10)$

6. $(18 \quad m \quad 2n + 4z)$

Remove parentheses and simplify.

7. $5x - (3x + 9)$

8. $5y - 2 - (2y - 4)$

Simplify.

9. $[24 \div (-2)] \div (-2)$

Simplify.

10. $3(4 + 2) - \{7 - [4 - (6 + 5)]\}$

Simplify.

11. $[3(x + 2) + 2x]$
     $- [4(y + 2) - 3(y - 2)]$

8. $3y - 2 - (2y - 4) = 3y - 2 - 2y + 4$
$$= y + 2$$

Caution! Be sure to change the signs of all terms in parentheses.

When we remove parentheses that follow a subtraction or additive inverse sign, we change the sign of *every* term inside. Then we collect like terms.

DO EXERCISES 7 AND 8.

●●● PARENTHESES WITHIN PARENTHESES

Sometimes parentheses occur within parentheses. When this happens we may use parentheses of different shapes, such as [ ], called "brackets," or { }, called "braces."

When parentheses occur within parentheses, the computations in the inner ones are to be done first.

Example 9 Simplify.

$$\left[(-4) \div \left(-\frac{1}{4}\right)\right] \div \frac{1}{4} = [(-4) \cdot (-4)] \div \frac{1}{4}$$

Working from the inside out

$$= 16 \div \frac{1}{4}$$
$$= 16 \cdot 4$$
$$= 64$$

DO EXERCISE 9.

Example 10 Simplify.

$4(2 + 3) - \{6 - [3 - (7 + 3)]\}$
$= 4 \cdot 5 - \{6 - [3 - 10]\}$ Working from the inside out
$= 20 - \{6 - [-7]\}$
$= 20 - \{13\}$
$= 7$

DO EXERCISE 10.

Example 11 Simplify.

$[5(x + 2) - 3x] - [3(y + 2) - 7(y + 2)]$
$= [5x + 10 - 3x] - [3y + 6 - 7y - 14]$ Working from the inside out

$= [2x + 10] - [-4y - 8]$ Collecting like terms inside
$= 2x + 10 + 4y + 8$ Removing parentheses by renaming
$= 2x + 4y + 18$ Collecting like terms. You need *not* factor the final expression.

DO EXERCISE 11.

NAME   CLASS   ANSWERS

EXERCISE SET 2.8

■ Rename each additive inverse without parentheses.

1. $-(2x + 7)$ **2.** $-(3x + 5)$

3. $-(5x - 8)$ **4.** $-(6x - 7)$

5. $-(4a - 3b + 7c)$ **6.** $-(5x - 2y + 3z)$

7. $-(6x + 8y + 5)$ **8.** $-(8x + 3y + 9)$

9. $-(3x + 5y - 6)$ **10.** $-(6a + 4b - 7)$

11. $-(-8x - 6y - 43)$ **12.** $-(-2a - 9b - 5c)$

■■ Remove parentheses and simplify. You need not factor the final expression.

13. $9x - (4x + 3)$ **14.** $7y - (2y + 9)$

15. $2a + (5a - 9)$ **16.** $11n + (3n - 7)$

17. $2x + 7x - (4x + 6)$ **18.** $3a + 2a - (4a + 7)$

19. $2x - 4y - (7x - 2y)$ **20.** $3a - 7b - (4a - 3b)$

21. $3x - y - (3x - 2y + 5z)$ **22.** $4a - b - (5a - 7b + 8c)$

ANSWERS
1.
2.
3.
4.
5.
6.
7.
8.
9.
10.
11.
12.
13.
14.
15.
16.
17.
18.
19.
20.
21.
22.

ANSWERS

●●● Remove parentheses and simplify.

23. _____

23. $[(-24) \div (-3)] \div \left(-\dfrac{1}{2}\right)$ **24.** $[32 \div (-2)] \div (-2)$

24. _____

25. _____

25. $8 \cdot [9 - 2(5 - 4)]$ **26.** $10 \cdot [6 - 5(8 - 4)]$

26. _____

27. _____

27. $[4(9 - 6) + 11] - [14 - (6 + 4)]$ **28.** $[7(8 - 4) + 16] - [15 - (7 + 3)]$

28. _____

29. _____

29. $[3(8 - 4) + 12] - [10 - (3 + 5)]$ **30.** $[9(7 - 3) + 13] - [11 - (6 + 9)]$

30. _____

31. _____

31. $[10(x + 3) - 4] + [2(x - 17) + 6]$ **32.** $[9(x + 5) - 7] + [4(x - 12) + 9]$

32. _____

33. _____

33. $[4(2x - 5) + 7] + [3(x + 3) + 5x]$ **34.** $[8(3x - 2) + 9] + [7(x + 4) + 6x]$

34. _____

35. _____

35. $[7(x + 5) - 19] - [4(x - 6) + 10]$ **36.** $[6(x + 4) - 12] - [5(x - 8) + 11]$

36. _____

37. _____

37. $3\{[7(x - 2) + 4] - [2(2x - 5) + 6]\}$ **38.** $4\{[8(x - 3) + 9] - [4(3x - 7) + 2]\}$

38. _____

39. _____

39. $4\{[5(x - 3) + 2] - 3[2(x + 5) - 9]\}$ **40.** $3\{[6(x - 4) + 5] - 2[5(x + 8) - 10]\}$

40. _____

2.9 INTEGERS AS EXPONENTS

⚬ NEGATIVE EXPONENTS

Negative numbers can be used as exponents. Look for a pattern in the following.

$$10^3 = 10 \cdot 10 \cdot 10 \qquad\qquad 8^3 = 8 \cdot 8 \cdot 8$$
$$10^2 = 10 \cdot 10 \qquad\qquad\quad 8^2 = 8 \cdot 8$$
$$10^1 = 10 \qquad\qquad\qquad\quad 8^1 = 8$$
$$10^0 = 1 \qquad\qquad\qquad\quad\; 8^0 = 1$$
$$10^{-1} = ? \qquad\qquad\qquad\; 8^{-1} = ?$$
$$10^{-2} = ? \qquad\qquad\qquad\; 8^{-2} = ?$$

In the first case we divided by 10 each time. In the second we divided by 8. Continuing the pattern, we have

$$10^{-1} = \frac{1}{10} = \frac{1}{10^1} \quad \text{and} \quad 8^{-1} = \frac{1}{8} = \frac{1}{8^1};$$
$$10^{-2} = \frac{1}{10 \cdot 10} = \frac{1}{10^2} \quad \text{and} \quad 8^{-2} = \frac{1}{8 \cdot 8} = \frac{1}{8^2}.$$

We make the following agreement. It is a definition.

> **If n is any positive integer,**
> $$b^{-n} \text{ is given the meaning } \frac{1}{b^n}.$$
> **In other words, b^n and b^{-n} are reciprocals.**

Example 1   Explain the meaning of 3^{-4} without using negative exponents.

$$3^{-4} \quad \text{means} \quad \frac{1}{3^4}, \quad \text{or} \quad \frac{1}{3 \cdot 3 \cdot 3 \cdot 3}, \quad \text{or} \quad \frac{1}{81}.$$ 3^{-4} is *not* a negative number.

Example 2   Rename $\frac{1}{5^2}$ using a negative exponent.

$$\frac{1}{5^2} = 5^{-2}$$

Example 3   Rename 4^{-3} using a positive exponent.

$$4^{-3} = \frac{1}{4^3}$$

DO EXERCISES 1–9.

⚬⚬ MULTIPLYING USING EXPONENTS

Consider an expression with exponents, such as $a^3 \cdot a^2$. To simplify it, recall the definition of exponents:

$$a^3 \cdot a^2 \quad \text{means} \quad (a \cdot a \cdot a)(a \cdot a)$$

and $(a \cdot a \cdot a)(a \cdot a) = a^5$. The exponent in a^5 is the sum of those in $a^3 \cdot a^2$. Suppose one exponent is positive and one is negative.

$$a^5 \cdot a^{-2} \quad \text{means} \quad (a \cdot a \cdot a \cdot a \cdot a) \cdot \left(\frac{1}{a \cdot a}\right),$$

OBJECTIVES

After finishing Section 2.9, you should be able to:

⚬ Rename a number with or without negative exponents.

⚬⚬ Use exponents in multiplying (adding the exponents).

⚬⚬⚬ Use exponents in dividing (subtracting the exponents).

⚬⚬ Use exponents in raising a power to a power (multiplying the exponents).

⚬⚬ Solve problems involving interest compounded annually.

Explain the meaning of each of the following without using negative exponents.

1. 4^{-3}

2. 5^{-2}

3. 2^{-4}

Rename, using negative exponents.

4. $\frac{1}{3^2}$

5. $\frac{1}{5^4}$

6. $\frac{1}{7^3}$

Rename, using positive exponents.

7. 5^{-3}

8. 7^{-5}

9. 10^{-4}

Multiply and simplify.

10. $3^5 \cdot 3^3$

11. $5^{-2} \cdot 5^4$

12. $6^{-3} \cdot 6^{-4}$

13. $x \cdot x^{-5}$

14. $y^2 \cdot y^{-4}$

15. $x^{-2} \cdot x^{-6}$

which simplifies as follows:

$$a \cdot a \cdot a \cdot \left(\frac{a \cdot a}{1}\right)\left(\frac{1}{a \cdot a}\right) = a \cdot a \cdot a \cdot \frac{a \cdot a}{a \cdot a} = a \cdot a \cdot a = a^3.$$

If we add the exponents, we again get the correct result. Next suppose that both exponents are negative:

$$a^{-3} \cdot a^{-2} \quad \text{means} \quad \frac{1}{a \cdot a \cdot a} \cdot \frac{1}{a \cdot a}.$$

This is equal to

$$\frac{1}{a \cdot a \cdot a \cdot a \cdot a}, \quad \text{or} \quad \frac{1}{a^5}, \quad \text{or} \quad a^{-5}.$$

Again, adding the exponents gives the correct result. The same is true if one or both exponents are zero.

> **In multiplication with exponential notation, we can add exponents if the bases are the same:**
> $$a^m \cdot a^n = a^{m+n}.$$

Examples   Multiply and simplify.

4. $8^4 \cdot 8^3 = 8^{4+3}$ Adding exponents
$= 8^7$

5. $7^{-3} \cdot 7^6 = 7^{-3+6}$
$= 7^3$

6. $x \cdot x^8 = x^{1+8}$
$= x^9$

7. $x^4 \cdot x^{-3} = x^{4+(-3)}$
$= x^1$
$= x$

DO EXERCISES 10–15.

⚫⚫⚫ DIVIDING USING EXPONENTS

Consider dividing with exponential notation.

$$\frac{5^4}{5^2} \text{ means } \frac{5 \cdot 5 \cdot 5 \cdot 5}{5 \cdot 5}, \text{ which is } 5 \cdot 5 \cdot \frac{5 \cdot 5}{5 \cdot 5} \text{ and we have } 5 \cdot 5, \text{ or } 5^2.$$

> **In division, we subtract exponents if the bases are the same:**
> $$\frac{a^m}{a^n} = a^{m-n}.$$

This is true whether the exponents are positive, negative, or zero.

Examples   Divide and simplify.

8. $\frac{5^4}{5^{-2}} = 5^{4-(-2)}$ Subtracting exponents
$= 5^6$

9. $\dfrac{x}{x^7} = x^{1-7}$

$\qquad = x^{-6}$

10. $\dfrac{b^{-4}}{b^{-5}} = b^{-4-(-5)}$

$\qquad - b^1$

$\qquad = b$

If you have trouble, follow this step-by-step procedure.

$\dfrac{x^{-3}}{x^{-5}} = x^{-3}$ ① Write the top exponent.

$\qquad = x^{-3-}$ ② Write a minus sign.

$\qquad = x^{-3-(\ \ )}$ ③ Write parentheses.

$\qquad = x^{-3-(-5)}$ ④ Put the bottom exponent in the parentheses.

$\qquad = x^2$ ⑤ Carry out the subtraction.

DO EXERCISES 16–21.

⠿ RAISING A POWER TO A POWER

Consider raising a power to a power.

Example 11

$(\ 3^2\ )^4$ means $\boxed{3^2} \cdot \boxed{3^2} \cdot \boxed{3^2} \cdot \boxed{3^2}$.

This is $\boxed{3\cdot 3} \cdot \boxed{3\cdot 3} \cdot \boxed{3\cdot 3} \cdot \boxed{3\cdot 3}$,   or 3^8.

We could have multiplied the exponents in $(3^2)^4$. Suppose the exponents are not positive.

Example 12

$(\ 5^{-2}\ )^3$ means $\boxed{\dfrac{1}{5^2}} \cdot \boxed{\dfrac{1}{5^2}} \cdot \boxed{\dfrac{1}{5^2}}$.

This is $\dfrac{1}{5\cdot 5} \cdot \dfrac{1}{5\cdot 5} \cdot \dfrac{1}{5\cdot 5}$,   or $\dfrac{1}{5^6}$,   or 5^{-6}.

Again, we could have multiplied the exponents. This works for any integer exponents.

> **To raise a power to a power we can multiply the exponents. For any exponents m and n,**
>
> $$(a^m)^n = a^{mn}.$$

Examples   Simplify.

13. $(3^5)^4 = 3^{5\cdot 4}$ Multiply exponents.
$\qquad = 3^{20}$

14. $(y^{-5})^7 = y^{-5\cdot 7}$
$\qquad = y^{-35}$

15. $(x^4)^{-2} = x^{4(-2)}$
$\qquad = x^{-8}$

16. $(a^{-4})^{-6} = a^{(-4)(-6)}$
$\qquad = a^{24}$

DO EXERCISES 22–25.

Divide and simplify.

16. $\dfrac{4^5}{4^2}$

17. $\dfrac{7^{-2}}{7^3}$

18. $\dfrac{a^2}{a^{-5}}$

19. $\dfrac{b^{-2}}{b^{-3}}$

20. $\dfrac{x}{x^{-3}}$

21. $\dfrac{x^8}{x}$

Simplify.

22. $(3^4)^5$

23. $(x^{-3})^4$

24. $(y^{-5})^{-3}$

25. $(x^{-4})^8$

ANSWERS ON PAGE A–6

Simplify.

26. $(2x^5y^{-3})^4$

27. $(5x^5y^{-6}z^{-3})^2$

28. $(3y^{-2}x^{-5}z^8)^3$

29. Suppose $2000 is invested at 16%, compounded annually. How much is in the account at the end of 3 years?

> **When several factors are in parentheses, raise each to the given power:**
> $$(a^mb^n)^t = (a^m)^t(b^n)^t = a^{mt}b^{nt}.$$

Caution! Be sure to raise *every* factor in parentheses to the power.

Examples Simplify.

17. $(5x^2y^{-2})^3 = 5^3(x^2)^3(y^{-2})^3 = 125x^6y^{-6}$

18. $(3x^3y^{-5}z^2)^4 = 3^4(x^3)^4(y^{-5})^4(z^2)^4 = 81x^{12}y^{-20}z^8$

DO EXERCISES 26–28.

⠭ APPLICATION: INTEREST COMPOUNDED ANNUALLY

Suppose we invest P dollars at an interest rate of 8%, compounded annually. The amount to which this grows at the end of one year is given by

$$P + 8\%P = P + 0.08P \qquad \text{By definition of percent}$$
$$= (1 + 0.08)P \qquad \text{Factoring}$$
$$= 1.08P. \qquad \text{Simplifying}$$

Going into the second year, the new principal is $(1.08)P$ dollars since interest has been added to the account. By the end of the second year, the following amount will be in the account:

$$(1.08)[\,(1.08)P\,], \quad \text{or} \quad (1.08)^2P. \qquad \text{New principal}$$

Going into the third year, the principal will be $(1.08)^2P$ dollars. At the end of the third year, the following amount will be in the account:

$$(1.08)[\,(1.08)^2P\,], \quad \text{or} \quad (1.08)^3P. \qquad \text{New principal}$$

Note the pattern: At the end of years 1, 2, and 3 the amount is

$$(1.08)P, \quad (1.08)^2P, \quad (1.08)^3P, \quad \text{and so on.}$$

> **If principal P is invested at interest rate r, compounded annually, in t years it will grow to the amount A given by**
> $$A = P(1 + r)^t.$$

Compare the formula $A = P(1 + r)^t$ with the formula for simple interest, $A = P(1 + rt)$.

Example 19 Suppose $1000 is invested at 8%, compounded annually. How much is in the account at the end of 3 years?

Substituting 1000 for P, 0.08 for r, and 3 for t, we get

$$A = P(1 + r)^t$$
$$= 1000(1 + 0.08)^3 = 1000(1.08)^3 = 1000(1.259712) \approx \$1259.71.$$

DO EXERCISE 29.

NAME _____   CLASS _____

EXERCISE SET 2.9

▪• Explain the meaning of each without using negative exponents.

1. 3^{-2} **2.** 2^{-3} **3.** 10^{-4} **4.** 5^{-6}

Rename, using negative exponents.

5. $\dfrac{1}{4^3}$ **6.** $\dfrac{1}{5^2}$ **7.** $\dfrac{1}{x^3}$ **8.** $\dfrac{1}{y^2}$

9. $\dfrac{1}{a^4}$ **10.** $\dfrac{1}{t^5}$ **11.** $\dfrac{1}{p^n}$ **12.** $\dfrac{1}{m^n}$

Rename, using positive exponents.

13. 7^{-3} **14.** 5^{-2} **15.** a^{-3} **16.** x^{-2}

17. y^{-4} **18.** t^{-7} **19.** z^{-n} **20.** h^{-m}

▪••• Multiply and simplify.

21. $2^4 \cdot 2^3$ **22.** $3^5 \cdot 3^2$ **23.** $3^{-5} \cdot 3^8$ **24.** $5^{-8} \cdot 5^9$

25. $x^{-2} \cdot x$ **26.** $x \cdot x$ **27.** $x^4 \cdot x^3$ **28.** $x^9 \cdot x^4$

29. $x^{-7} \cdot x^{-6}$ **30.** $y^{-5} \cdot y^{-8}$ **31.** $t^8 \cdot t^{-8}$ **32.** $m^{10} \cdot m^{-10}$

ANSWERS

1. _____
2. _____
3. _____
4. _____
5. _____
6. _____
7. _____
8. _____
9. _____
10. _____
11. _____
12. _____
13. _____
14. _____
15. _____
16. _____
17. _____
18. _____
19. _____
20. _____
21. _____
22. _____
23. _____
24. _____
25. _____
26. _____
27. _____
28. _____
29. _____
30. _____
31. _____
32. _____

ANSWERS

33. _____

34. _____

35. _____

36. _____

37. _____

38. _____

39. _____

40. _____

41. _____

42. _____

43. _____

44. _____

45. _____

46. _____

47. _____

48. _____

49. _____

50. _____

51. _____

52. _____

53. _____

54. _____

55. _____

56. _____

57. _____

58. _____

59. _____

60. _____

●●● Divide and simplify.

33. $\dfrac{7^5}{7^2}$ **34.** $\dfrac{4^7}{4^3}$ **35.** $\dfrac{x}{x^{-1}}$ **36.** $\dfrac{x^6}{x}$

37. $\dfrac{x^7}{x^{-2}}$ **38.** $\dfrac{t^8}{t^{-3}}$ **39.** $\dfrac{z^{-6}}{z^{-2}}$ **40.** $\dfrac{y^{-7}}{y^{-3}}$

41. $\dfrac{x^{-5}}{x^{-8}}$ **42.** $\dfrac{y^{-4}}{y^{-9}}$ **43.** $\dfrac{m^{-9}}{m^{-9}}$ **44.** $\dfrac{x^{-8}}{x^{-8}}$

●● Simplify.

45. $(2^3)^2$ **46.** $(3^4)^3$ **47.** $(5^2)^{-3}$ **48.** $(9^3)^{-4}$

49. $(x^{-3})^{-4}$ **50.** $(a^{-5})^{-6}$ **51.** $(x^4y^5)^{-3}$ **52.** $(t^5x^3)^{-4}$

53. $(x^{-6}y^{-2})^{-4}$ **54.** $(x^{-2}y^{-7})^{-5}$ **55.** $(3x^3y^{-8}z^{-3})^2$ **56.** $(2a^2y^{-4}z^{-5})^3$

●●● Solve.

57. Suppose $2000 is invested at 12%, compounded annually. How much is in the account at the end of 2 years?

58. Suppose $2000 is invested at 15%, compounded annually. How much is in the account at the end of 3 years?

_____

59. ▦ Suppose $10,400 is invested at 16.5%, compounded annually. How much is in the account at the end of 5 years?

60. ▦ Suppose $20,800 is invested at 20.5%, compounded annually. How much is in the account at the end of 6 years?

2.10 SCIENTIFIC NOTATION (Optional)

◼◻ SCIENTIFIC NOTATION

The following are examples of scientific notation, which is useful when calculations involve very large or very small numbers:

$$6.4 \times 10^{23}, \quad 4.6 \times 10^{-4}, \quad 10^{15}.$$

> *Scientific notation* **for a number consists of exponential notation for a power of 10 and, if needed, decimal notation for a number between 1 and 10, and a multiplication sign:**
>
> $$N \times 10^n \quad \textbf{or} \quad 10^n.$$

We can convert to scientific notation by multiplying by 1, choosing a name like $10^b \cdot 10^{-b}$ for the number 1.

Example 1 Light travels about 9,460,000,000,000 kilometers in one year. Write scientific notation for the number.

We want to move the decimal point 12 places, between the 9 and the 4, so we choose $10^{-12} \times 10^{12}$ as a name for 1. Then we multiply.

$$9,460,000,000,000 \times 10^{-12} \times 10^{12} \quad \text{Multiplying by 1}$$
$$= 9.46 \times 10^{12} \quad \text{The } 10^{-12} \text{ moved the decimal point 12 places to the left and we have scientific notation.}$$

DO EXERCISES 1 AND 2.

Example 2 Write scientific notation for 0.0000000000156.

We want to move the decimal point 11 places. We choose $10^{11} \times 10^{-11}$ as a name for 1, and then multiply.

$$0.0000000000156 \times 10^{11} \times 10^{-11} \quad \text{Multiplying by 1}$$
$$= 1.56 \times 10^{-11} \quad \text{The } 10^{11} \text{ moved the decimal point 11 places to the right and we have scientific notation.}$$

You should try to make conversions to scientific notation mentally as much as possible.

DO EXERCISES 3 AND 4.

Examples Convert mentally to scientific notation.

3. $78{,}000 = 7.8 \times 10^4 \qquad 7.8{,}000$

　　　　　　　　　　　　　　　　　　　4 places

Large number, so the exponent is positive.

4. $0.0000057 = 5.7 \times 10^{-6} \qquad 0.000005.7$

　　　　　　　　　　　　　　　　　　　6 places

Small number, so the exponent is negative.

DO EXERCISES 5 AND 6.

OBJECTIVES

After finishing Section 2.10, you should be able to:

◼◻ **Convert between scientific and ordinary decimal notation.**

◼◼ **Multiply and divide using scientific notation.**

1. Convert 460,000,000,000 to scientific notation.

2. The distance from the earth to the sun is about 93,000,000 miles. Write scientific notation for this number.

3. Convert 0.00000001235 to scientific notation.

4. The mass of a hydrogen atom is about 0.00000000000000000000000017 gram. Write scientific notation for this number.

Convert mentally to scientific notation.

5. 0.000314

6. 218,000,000

ANSWERS ON PAGE A–6

Convert to decimal notation.

7. 7.893×10^{11}

8. 5.67×10^{-5}

Multiply and write scientific notation for the answer.

9. $(1.12 \times 10^{-8})(5 \times 10^{-7})$

10. $(9.1 \times 10^{-17})(8.2 \times 10^{3})$

Examples   Convert mentally to decimal notation.

5. $7.893 \times 10^5 = 789,300$ $7.89300.$

   Positive exponent, so the answer is a large number.

6. $4.7 \times 10^{-8} = 0.000000047$ $0.00000004.7$

   Negative exponent, so the answer is a small number.

DO EXERCISES 7 AND 8.

**●●   MULTIPLYING AND DIVIDING USING
         SCIENTIFIC NOTATION**

Multiplying.   Consider the product

   $400 \cdot 2000 = 800,000.$

In scientific notation this would be

   $(4 \times 10^2) \cdot (2 \times 10^3) = 8 \times 10^5.$

Note that we could find this product by multiplying $4 \cdot 2$, to get 8, and $10^2 \cdot 10^3$ to get 10^5 (we do this by adding the exponents).

Example 7   Multiply: $(1.8 \times 10^6) \cdot (2.3 \times 10^{-4})$.

a) Multiply 1.8 and 2.3:

   $1.8 \times 2.3 = 4.14.$

b) Multiply 10^6 and 10^{-4}:

   $10^6 \cdot 10^{-4} = 10^{6+(-4)}$ Add the exponents.

   $\qquad\qquad = 10^2.$

c) The answer is

   $4.14 \times 10^2.$

Example 8   Multiply: $(3.1 \times 10^5) \cdot (4.5 \times 10^{-3})$.

a) Multiply 3.1 and 4.5:

   $3.1 \times 4.5 = 13.95.$

b) Multiply 10^5 and 10^{-3}:

   $10^5 \cdot 10^{-3} = 10^{5+(-3)} = 10^2.$

c) The answer at this stage is

   $13.95 \times 10^2,$

   but this is *not* scientific notation, since 13.95 is not a number between 1 and 10. To find scientific notation we convert 13.95 to scientific notation and simplify:

   $13.95 \times 10^2 = (1.395 \times 10^1) \times 10^2$

   $\qquad\qquad\qquad = 1.395 \times 10^3$ Add the exponents.

DO EXERCISES 9 AND 10.

Division. Consider the quotient

$$800{,}000 \div 400 = 2000.$$

In scientific notation this is

$$(8 \times 10^5) \div (4 \times 10^2) = 2 \times 10^3.$$

Note that we could find this product by dividing $8 \div 4$, to get 2, and $10^5 \div 10^2$, to get 10^3 (we do this by subtracting the exponents).

Example 9 Divide: $(3.41 \times 10^5) \div (1.1 \times 10^{-3})$.

a) Divide 3.41 by 1.1:

$$3.41 \div 1.1 = 3.1.$$

b) Divide 10^5 by 10^{-3}:

$$10^5 \div 10^{-3} = 10^{5-(-3)} \qquad \text{Subtract the exponents.}$$
$$= 10^8.$$

c) The answer is

$$3.1 \times 10^8.$$

Example 10 Divide: $(6.4 \times 10^{-7}) \div (8.0 \times 10^6)$.

a) Divide 6.4 by 8.0:

$$6.4 \div 8.0 = 0.8.$$

b) Divide 10^{-7} by 10^6:

$$10^{-7} \div 10^6 = 10^{-7-6} \qquad \text{Subtract the exponents.}$$
$$= 10^{-13}.$$

c) The answer at this stage is

$$0.8 \times 10^{-13},$$

but this is *not* scientific notation, since 0.8 is not a number between 1 and 10. To find scientific notation we convert 0.8 to scientific notation and simplify:

$$0.8 \times 10^{-13} = (8.0 \times 10^{-1}) \times 10^{-13}$$
$$= 8.0 \times 10^{-14}. \qquad \text{Add the exponents.}$$

DO EXERCISES 11 AND 12.

Divide and write scientific notation for the answer.

11. $\dfrac{4.2 \times 10^5}{2.1 \times 10^2}$

12. $\dfrac{1.1 \times 10^{-4}}{2.0 \times 10^{-7}}$

SOMETHING EXTRA

CALCULATOR CORNER: NUMBER PATTERNS

In each of the following, do the first four calculations using your calculator. Look for a pattern. Then do the last calculation without using your calculator.

1. 1^2
    11^2
    111^2
    1111^2
    $11{,}111^2$

2. 6×4
    66×44
    666×444
    6666×4444
    $66{,}666 \times 44{,}444$

3. 8×6
    68×6
    668×6
    6668×6
    $66{,}668 \times 6$

4. 9^2
    99^2
    999^2
    9999^2
    $99{,}999^2$

NAME	CLASS	ANSWERS

EXERCISE SET 2.10

> • Convert to scientific notation.

1. 78,000,000,000

2. 3,700,000,000,000

3. 907,000,000,000,000,000

4. 168,000,000,000,000

5. 0.00000374

6. 0.0000000000275

7. 0.000000018

8. 0.00000000002

9. 10,000,000

10. 100,000,000,000

11. 0.000000001

12. 0.0000001

Convert to decimal notation.

13. 7.84×10^8

14. 1.35×10^7

15. 8.764×10^{-10}

16. 9.043×10^{-3}

17. 10^8

18. 10^4

19. 10^{-4}

20. 10^{-7}

> • • Multiply or divide and write scientific notation for the result.

21. $(3 \times 10^4)(2 \times 10^5)$

22. $(1.9 \times 10^8)(3.4 \times 10^{-3})$

23. $(5.2 \times 10^5)(6.5 \times 10^{-2})$

24. $(7.1 \times 10^{-7})(8.6 \times 10^{-5})$

ANSWERS

1. _____
2. _____
3. _____
4. _____
5. _____
6. _____
7. _____
8. _____
9. _____
10. _____
11. _____
12. _____
13. _____
14. _____
15. _____
16. _____
17. _____
18. _____
19. _____
20. _____
21. _____
22. _____
23. _____
24. _____

25. $(9.9 \times 10^{-6})(8.23 \times 10^{-8})$

26. $(1.123 \times 10^4) \times 10^{-9}$

27. $\dfrac{8.5 \times 10^8}{3.4 \times 10^{-5}}$

28. $\dfrac{5.6 \times 10^{-2}}{2.5 \times 10^5}$

29. $(3.0 \times 10^6) \div (6.0 \times 10^9)$

30. $(1.5 \times 10^{-3}) \div (1.6 \times 10^{-6})$

31. $\dfrac{7.5 \times 10^{-9}}{2.5 \times 10^{12}}$

32. $\dfrac{4.0 \times 10^{-3}}{8.0 \times 10^{20}}$

Carry out the indicated operations. Write scientific notation for the result.

33. ▦ $\dfrac{(5.2 \times 10^6)(6.1 \times 10^{-11})}{1.28 \times 10^{-3}}$

34. ▦ $\dfrac{3.9 \times 10^{15}}{(8.0 \times 10^{-12})(3.2 \times 10^{19})}$

EXTENSION EXERCISES

SECTION 2.1

Tell which of the following statements are always true. If a statement is not always true, give an example for which it is false.

1. $|x| \geqslant 0$ **2.** $|-x| \geqslant -|x|$ **3.** $x < |x|$ **4.** $-|x| \leqslant |x^2|$

5. $|x| < |x|^2$ **6.** $-|x| < x$ **7.** $|x + y| = |x| + |y|$

SECTION 2.2

8. When is $-x$ negative?   **9.** When is $-x$ positive?

10. Tell whether each sum or difference is positive, negative, or zero.
  a) $n > 0, m > 0.\ n + m$ is _____.
  b) $n > 0, m < 0.\ n - m$ is _____.
  c) $n > 0, m < 0.\ -n + m$ is _____.
  d) $n < m, n > 0, m > 0.\ n - m$ is _____.
  e) $n = m, n < 0, m < 0.\ n - m$ is _____.
  f) $|n| > |m|, n < 0, m > 0.\ n + m$ is _____.
  g) $n < m, n > 0, m > 0.\ m - n$ is _____.
  h) $|n| > |m|, n > 0, m < 0.\ n + m$ is _____.

SECTION 2.3

Simplify.

11. $-(-3 + 2)$ **12.** $4 - (-6) + (-3) - (-5)$ **13.** $|3 - 7| - 4$

14. $-5|2| + 3|4|$ **15.** $-|39 - (-12) - 60|$

16. Is there a commutative law of subtraction? That is, is it *always* true that $a - b = b - a$? (Give an example to support your answer.) If not, under what circumstances is the equation $a - b = b - a$ true?

SECTION 2.4

Simplify.

17. $(-4)^2$ **18.** -4^2 **19.** $-4 \cdot 3 + 10 \cdot (-2)$ **20.** $9 - 6 \cdot (-5)$ **21.** $\dfrac{|-24|}{3 - 15}$

22. Tell whether each expression represents a positive or a negative number when n and m are positive.

  a) $\dfrac{-n}{m}$ b) $\dfrac{-n}{-m}$ c) $-\left(\dfrac{-n}{m}\right)$ d) $-\left(\dfrac{n}{-m}\right)$ e) $-\left(\dfrac{-n}{-m}\right)$

SECTION 2.5

23. Find two rational numbers between the given numbers.
  a) -6 and -7 b) 3 and π

24. Insert $>$, $<$, or $=$ between each pair of numbers.

  a) -8.6 and -9.0 b) $-2\dfrac{1}{8}$ and $\dfrac{-35}{16}$

25. Arrange from smallest to largest.

$$-\dfrac{1}{3},\ -\dfrac{1}{4},\ -\dfrac{5}{6},\ -\dfrac{1}{16},\ 0,\ -\dfrac{5}{12}$$

26. Simplify.
  a) $-(-3.8 + 7.2)$

  b) $-\dfrac{3}{4} - \left(-\dfrac{7}{8}\right) + 2\dfrac{5}{16}$

  c) $|-4.21 - 3.68|$

  d) $-\left|-\dfrac{1}{3} + \dfrac{5}{6}\right|$

SECTION 2.6

Evaluate.

27. $-2 \cdot (-3) \cdot (-1) \cdot (-4) \cdot (-7)$

28. $(-1) \cdot (-2) \cdot (-3) \cdot (-4) \cdot (-5) \cdot (-6)$

29. $(-2)^4$ **30.** -2^4 **31.** $(-1)^5$ **32.** -1^5

33. Fill in the blank with "positive" or "negative."
    a) The product of an even number of negative numbers is _____.
    b) The product of an odd number of negative numbers is _____.

Find the reciprocal.

34. $-x \; (x \neq 0)$

35. $\dfrac{-a}{a + b}$

SECTION 2.7

Factor.

36. $2\pi r + \pi rs$

37. $\dfrac{1}{2}ah + \dfrac{1}{2}bh$

Simplify.

38. $8x - 9 - 2(7 - 5x)$

39. $-9(y + 7) - 6(y + 3)$

40. $\dfrac{5x - 15}{5} + \dfrac{2x + 6}{2}$

41. $\dfrac{x^2 - 2x}{x} + \dfrac{3 - 9x}{3}$

SECTION 2.8

If $n > 0$, $m > 0$, and $n \neq m$, which of the following are true?

42. $-n + m = n - m$

43. $-n + m = -(n + m)$

44. $-n - m = -(n + m)$

45. $-n - m = -(n - m)$

46. $n(-n - m) = -n^2 + nm$

47. $-m(n - m) = -(mn + m^2)$

48. $-m(-n + m) = m(n - m)$

49. $-n(-n - m) = n(n + m)$

SECTION 2.9

50. Insert $<$, $>$, or $=$ between each pair of numbers.
    a) $3^5 \quad 3^4$ b) $4^{-2} \quad 4^{-3}$ c) $4^3 \quad 5^3$ d) $6^{-2} \quad 7^{-2}$ e) $5^2 \cdot 5^{-2} \quad 2^5 \cdot 2^{-5}$

51. If $x \neq 0$ and $y \neq 0$, tell whether each of the following is always true.
    a) $x^m \cdot y^n = (xy)^{mn}$ b) $x^m \cdot y^n = (xy)^{m+n}$ c) $x^m \cdot y^m = (xy)^{2m}$

    d) $x^m \cdot x^n = x^{mn}$ e) $\left(\dfrac{x}{y}\right)^{-m} = \left(\dfrac{y}{x}\right)^m$ f) $(x + y)^{-m} = \dfrac{1}{x^m} + \dfrac{1}{y^m}$

SECTION 2.10

52. Perform the operations indicated. Express the result in scientific notation.

$$\{2.1 \times 10^6 [(2.5 \times 10^{-3}) \div (5.0 \times 10^{-5})]\} \div (3.0 \times 10^{17})$$

53. Find the reciprocal and express in scientific notation.
    a) 6.25×10^{-3} b) 4.0×10^{10}

NAME   SCORE   ANSWERS

TEST OR REVIEW—CHAPTER 2

If you miss an item, review the indicated section and objective.

1. _____

Simplify.

[2.1, ●●●]   **1.** $|9|$

2. _____

[2.5, ●]   **2.** $\left| -\dfrac{3}{4} \right|$

3. _____

[2.5, ●]   **3.** $-\left(-\dfrac{5}{6} \right)$

4. _____

Add.

[2.2, ●]   **4.** $-12 + (-5)$

5. _____

[2.5, ●●]   **5.** $-2.7 + 8.5$

[2.5, ●●]   **6.** $-\dfrac{4}{7} + \dfrac{1}{3}$

6. _____

Subtract.

[2.3, ●●]   **7.** $-12 - (-5)$

7. _____

[2.5, ●●●]   **8.** $7.4 - (-12.5)$

8. _____

[2.5, ●●●]   **9.** $-\dfrac{7}{8} - \dfrac{3}{8}$

9. _____

Multiply.

[2.4, ●]   **10.** $-8 \cdot (-5)$

10. _____

[2.6, ●]   **11.** $\dfrac{3}{4} \cdot \left(-\dfrac{5}{6} \right)$

11. _____

[2.6, ●●]   **12.** $5 \cdot (-3) \cdot (-4) \cdot (-6)$

12. _____

Divide.

[2.4, ●●]   **13.** $\dfrac{42}{-21}$

13. _____

[2.6, ●●●]   **14.** $-\dfrac{3}{7} \div \left(-\dfrac{7}{9} \right)$

14. _____

ANSWERS

15. _____

16. _____

17. _____

18. _____

19. _____

20. _____

21. _____

22. _____

23. _____

24. _____

25. _____

26. _____

[2.7, ⚃] 15. Factor: $7x - 28$.

[2.7, ⚂] 16. Multiply: $-3(5x - 2)$.

[2.7, ⚅] 17. Collect like terms: $8a - 5b - 10a + 9b$.

Remove parentheses and simplify.

[2.8, ⚁] 18. $9y - (8 - 4y)$

[2.8, ⚂] 19. $[5(x + 3) - 2] - [6(y - 4) + 7]$

[2.9, ⚀] 20. Rename, using a negative exponent: $\dfrac{1}{5^5}$.

[2.9, ⚁] 21. Rename, using a positive exponent: 6^{-4}.

[2.9, ⚁] 22. Multiply and simplify: $y^{-7} \cdot y^3$.

[2.9, ⚂] 23. Divide and simplify: $\dfrac{a^5}{a^{-3}}$.

[2.9, ⚃] 24. Simplify: $(x^{-4})^3$

[2.9, ⚅] 25. Suppose \$1000 is invested at 13%, compounded annually. How much is in the account at the end of 3 years?

[2.10, ⚁] 26. Divide. Write scientific notation for the answer.

$$(1.242 \times 10^{11}) \div (5.4 \times 10^{15})$$

SOLVING
EQUATIONS

3

There are equations for predicting world running records.

1. Add: $-3 + (-8)$.

2. Subtract: $-7 - (-23)$.

3. Multiply: $-\dfrac{2}{3} \cdot \dfrac{5}{8}$.

4. Divide: $-\dfrac{3}{7} \div \left(-\dfrac{9}{7}\right)$.

5. Multiply: $3(x - 5)$.

6. Collect like terms: $x - 8x$.

7. Remove parentheses and simplify: $7w - 3 - (4w - 8)$.

OBJECTIVE

After finishing Section 3.1, you should be able to:

 ◐ Solve equations using the addition principle.

3.1   THE ADDITION PRINCIPLE

 ◖●◗ We saw in Section 1.6 that solving equations is an important part of algebra. In this chapter we study equation solving in more detail. We will also study it again later on. Recall the following.

> The replacements that make an equation true are called *solutions*. **To** *solve* **an equation means to find all its solutions.**

There are various ways to solve equations. We develop one now, using an idea called the *addition principle*.

An equation $a = b$ says that a and b stand for the same number. Suppose this is true and then add a number c to the number a. We get the same answer if we add c to b, because a and b are the same number.

> *The Addition Principle:*   **If an equation** $a = b$ **is true, then**
>
> $$a + c = b + c \text{ is true for any number } c.$$

The idea in using this principle is to obtain an equation for which you can *see* what the solution is.

Example 1   Solve: $x + 5 = -7$.

$$x + 5 = -7.$$

$$x + 5 + (\boxed{-5}) = -7 + (\boxed{-5}) \qquad \begin{array}{l}\text{Using the addition} \\ \text{principle; adding } -5 \\ \text{on both sides}\end{array}$$

$$x + 0 = -7 + (-5) \qquad \text{Simplifying}$$

$$x = -12$$

We can see that the solution of $x = -12$ is the number -12. To check the answer we try -12 in the original equation.

Check: $x + 5 = -7$
$$\frac{-12 + 5 \quad | \quad -7}{\qquad -7 \quad |}$$

The solution of the original equation is -12.

In Example 1, to get x alone, we added the inverse of 5. This "got rid of" the 5 on the left. The addition principle also allows us to subtract on both sides of an equation. This is because subtracting is the same as adding an inverse.

DO EXERCISE 1.

Example 2   Solve: $y - 8.4 = -6.5$.

$$y - 8.4 = -6.5$$

$$y - 8.4 + \boxed{8.4} = -6.5 + \boxed{8.4} \qquad \text{Adding 8.4 to get rid of } -8.4 \text{ on the left}$$

$$y = 1.9$$

Check: $y - 8.4 = -6.5$
$$\frac{1.9 - 8.4 \quad | \quad -6.5}{\quad -6.5 \quad |}$$

The solution is 1.9.

DO EXERCISES 2–4.

Solve, using the addition principle.

1. $x + 7 = 2$

Solve, using the addition principle.

2. $n - 4.5 = 8.7$

3. $x + \dfrac{1}{2} = -\dfrac{3}{2}$

4. $7.6 = -3.2 + y$

SOMETHING EXTRA

CALCULATOR CORNER: NUMBER PATTERNS

There are many interesting number patterns in mathematics. Look for a pattern in the following. We can use a calculator for the computations.

$$6^2 = 36$$
$$66^2 = 4356$$
$$666^2 = 443556$$
$$6666^2 = 44435556$$

Do you see a pattern? If so, find 66666^2 without the use of your calculator.

EXERCISES

In each of the following, do the first four calculations using your calculator. Look for a pattern. Use the pattern to do the last calculation without the use of your calculator.

1. 3^2
    33^2
    333^2
    3333^2
    33333^2

2. $9 \cdot 6$
    $99 \cdot 66$
    $999 \cdot 666$
    $9999 \cdot 6666$
    $99999 \cdot 66666$

3. $37 \cdot 3$
    $37 \cdot 33$
    $37 \cdot 333$
    $37 \cdot 3333$
    $37 \cdot 33333$

4. $37 \cdot 3$
    $37 \cdot 6$
    $37 \cdot 9$
    $37 \cdot 12$
    $37 \cdot 15$

NAME   CLASS   ANSWERS

EXERCISE SET 3.1

● Solve, using the addition principle. Don't forget to check.

1. $x + 2 = 6$ **2.** $x + 5 = 8$ **3.** $x + 15 = -5$

1. _____

2. _____

Check: _____   *Check:* _____   *Check:*

3. _____

4. $y + 25 = -6$ **5.** $r + \dfrac{2}{3} = 1$ **6.** $t + \dfrac{3}{4} = 1$

4. _____

5. _____

Check: _____   *Check:* _____   *Check:* _____

6. _____

7. $x - \dfrac{5}{6} = \dfrac{7}{8}$ **8.** $x - \dfrac{2}{3} = \dfrac{7}{3}$ **9.** $8 + y = 12$

7. _____

8. _____

Check: _____   *Check:* _____   *Check:* _____

9. _____

10. $5 + t = 7$ **11.** $\dfrac{1}{3} + a = \dfrac{5}{6}$ **12.** $-\dfrac{1}{5} + z = -\dfrac{1}{4}$

10. _____

11. _____

Check: _____   *Check:* _____   *Check:* _____

12. _____

13. $x - 2.3 = -7.4$ **14.** $x - 3.7 = 8.4$ **15.** $-2.6 + x = 8.3$

13. _____

14. _____

Check: _____   *Check:* _____   *Check:* _____

15. _____

ANSWERS

16. _____

17. _____

18. _____

19. _____

20. _____

21. _____

22. _____

23. _____

24. _____

25. _____

26. _____

27. _____

28. _____

29. _____

30. _____

31. _____

16. $-5.7 + t = 7.4$

17. $m + \dfrac{5}{6} = -\dfrac{11}{12}$

18. $x + \dfrac{2}{3} = -\dfrac{5}{6}$

Check: _____⊤_____

Check: _____⊤_____

Check: _____⊤_____

19. $-6 = -2 + y$

20. $-8 = y - 2$

21. $10 = y - 6$

Check: _____⊤_____

Check: _____⊤_____

Check: _____⊤_____

22. $20 = t - 7$

23. $-9.7 = -4.7 + y$

24. $-7.8 = 2.8 + x$

Check: _____⊤_____

Check: _____⊤_____

Check: _____⊤_____

25. $5\dfrac{1}{6} + x = 7$

26. $4\dfrac{2}{3} + x = 5\dfrac{1}{4}$

27. $p + \dfrac{2}{3} = 7\dfrac{1}{3}$

Check: _____⊤_____

Check: _____⊤_____

Check: _____⊤_____

28. $q + \dfrac{1}{3} = -\dfrac{1}{7}$

29. $22\dfrac{1}{7} = 30 + t$

30. $47\dfrac{1}{8} = -76 + z$

Check: _____⊤_____

Check: _____⊤_____

Check: _____⊤_____

Solve.

31. ▦ $-356.788 = -699.034 + t$

3.2   THE MULTIPLICATION PRINCIPLE

• • Suppose $a = b$ is true and we multiply a by some number c. We get the same answer if we multiply b by c because a and b are the same number.

> *The Multiplication Principle:*   **If an equation $a = b$ is true, then**
>
> $a \cdot c = b \cdot c$ **is true for any number c.**

Example 1   Solve: $3x = 9$.

$$3x = 9$$

$$\frac{1}{3} \cdot 3x = \frac{1}{3} \cdot 9 \qquad \text{Using the multiplication principle, multiplying by } \tfrac{1}{3}$$

$$1 \cdot x = 3 \qquad \text{Simplifying}$$

$$x = 3$$

It is easy to see that the solution of $x = 3$ is 3.

Check: $$\begin{array}{c|c} 3x = 9 \\ \hline 3 \cdot 3 & 9 \\ 9 & \end{array}$$

The solution of the original equation is 3.

In Example 1, to get x alone, we multiplied by the reciprocal of 3. When we multiplied we got $1 \cdot x$, which simplified to x. This enabled us to "get rid of" the 3 on the left. The multiplication principle also allows us to divide on both sides of an equation by a nonzero number. This is because division by a number c is the same as multiplying by a reciprocal.

DO EXERCISE 1.

Example 2   Solve: $\dfrac{3}{8} = -\dfrac{5}{4}x$.

$$\frac{3}{8} = -\frac{5}{4}x$$

$$-\frac{4}{5} \cdot \frac{3}{8} = -\frac{4}{5} \cdot \left(-\frac{5}{4}x\right) \qquad \begin{array}{l}\text{Multiplying by } -\tfrac{4}{5} \text{ to get rid of} \\ -\tfrac{5}{4} \text{ on the right (this is the same as} \\ \text{dividing by } -\tfrac{5}{4})\end{array}$$

$$-\frac{3}{10} = x \qquad \text{Simplifying}$$

Check: $$\begin{array}{c|c} \dfrac{3}{8} = -\dfrac{5}{4}x \\ \hline \dfrac{3}{8} & -\dfrac{5}{4}\left(-\dfrac{3}{10}\right) \\[2mm] & \dfrac{3}{8} \end{array}$$

The solution is $-\tfrac{3}{10}$.

DO EXERCISES 2 AND 3.

OBJECTIVE

After finishing Section 3.2, you should be able to:

• • Solve equations using the multiplication principle.

Solve.

1. $5x = 25$

Solve.

2. $4 = -\dfrac{1}{3}y$

3. $4x = -7$

Solve.

4. $1.12x = 8736$

Example 3 Solve: $1.16y = 9744$.

$$1.16y = 9744$$

$$\boxed{\frac{1}{1.16}} \cdot (1.16y) = \boxed{\frac{1}{1.16}} \cdot (9744)$$ Multiplying by $\frac{1}{1.16}$ (or dividing by 1.16)

$$y = \frac{9744}{1.16}$$

$$y = 8400$$

Check:

$$
\begin{array}{c|c}
1.16y = 9744 \\
\hline
1.16(8400) & 9744 \\
9744 &
\end{array}
$$

The solution is 8400.

DO EXERCISES 4 AND 5.

5. $-2.1y = 6.3$

SOMETHING EXTRA
CALCULATOR CORNER

Find the shortest path from A to G.

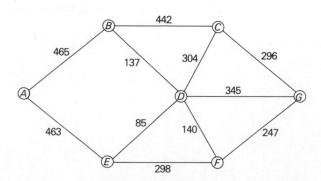

NAME

CLASS

ANSWERS

EXERCISE SET 3.2

⬤ Solve using the multiplication principle. Don't forget to check!

1. $6x = 36$

2. $3x = 39$

3. $9x = -36$

Check: ——————

Check: ——————

Check: ——————

4. $7x = -49$

5. $-12x = 72$

6. $-15x = -105$

Check: ——————

Check: ——————

Check: ——————

7. $\frac{1}{7}t = 9$

8. $\frac{1}{8}y = 11$

9. $\frac{1}{5} = \frac{1}{3}t$

Check: ——————

Check: ——————

Check: ——————

10. $\frac{1}{9} = \frac{1}{7}z$

11. $-2.7y = 54$

12. $-3.1y = 21.7$

Check: ——————

Check: ——————

Check: ——————

13. $\frac{3}{4}x = 27$

14. $\frac{4}{5}x = 16$

15. $-\frac{3}{5}r = -\frac{9}{10}$

Check: ——————

Check: ——————

Check: ——————

1. ——————

2. ——————

3. ——————

4. ——————

5. ——————

6. ——————

7. ——————

8. ——————

9. ——————

10. ——————

11. ——————

12. ——————

13. ——————

14. ——————

15. ——————

ANSWERS

16. _____

17. _____

18. _____

19. _____

20. _____

21. _____

22. _____

23. _____

24. _____

25. _____

26. _____

27. _____

28. _____

16. $-\dfrac{2}{5}y = -\dfrac{4}{15}$

17. $12 = \dfrac{6}{5}y$

18. $\dfrac{4}{5}y = 20$

Check: _____|_____

Check: _____|_____

Check: _____|_____

19. $-3.3y = 6.6$

20. $-6.3x = 44.1$

21. $38.7m = 309.6$

Check: _____|_____

Check: _____|_____

Check: _____|_____

22. $29.4x = 235.2$

23. $20.07 = \dfrac{3}{2}y$

24. $-\dfrac{9}{7}y = 12.06$

Check: _____|_____

Check: _____|_____

Check: _____|_____

25. $-\dfrac{3}{2}r = -\dfrac{27}{4}$

26. $\dfrac{5}{7}x = -\dfrac{10}{14}$

27. $-\dfrac{2}{3}y = -10.6$

Check: _____|_____

Check: _____|_____

Check: _____|_____

Solve.

28. ▦ $-0.2344m = 2028.732$

3.3 USING THE PRINCIPLES TOGETHER

● APPLYING BOTH PRINCIPLES

Let's solve an equation where we apply both principles. We usually apply the addition principle first.

Example 1   Solve $3x + 4 = 13$

$$3x + 4 = 13$$

$$3x + 4 + (\boxed{-4}) = 13 + (\boxed{-4}) \quad \text{Using the addition principle, adding } -4$$

$$3x = 9 \quad \text{Simplifying}$$

$$\boxed{\frac{1}{3}} \cdot 3x = \boxed{\frac{1}{3}} \cdot 9 \quad \text{Using the multiplication principle, multiplying by } \tfrac{1}{3}$$

$$x = 3 \quad \text{Simplifying}$$

Check:
$$\begin{array}{c|c} 3x + 4 = 13 \\ \hline 3 \cdot 3 + 4 & 13 \\ 9 + 4 & \\ 13 & \end{array}$$

The solution is 3.

DO EXERCISE 1.

Example 2   Solve: $-5x - 6 = 16$.

$$-5x - 6 = 16$$

$$-5x - 6 + \boxed{6} = 16 + \boxed{6} \quad \text{Adding 6}$$

$$-5x = 22$$

$$\boxed{-\frac{1}{5}} \cdot (-5x) = \boxed{-\frac{1}{5}} \cdot 22 \quad \text{Multiplying by } -\frac{1}{5}$$

$$x = -\frac{22}{5}, \quad \text{or} \quad -4\frac{2}{5}$$

Check:
$$\begin{array}{c|c} -5x - 6 = 16 \\ \hline -5\left(-\dfrac{22}{5}\right) - 6 & 16 \\ 22 - 6 & \\ 16 & \end{array}$$

The solution is $-\frac{22}{5}$.

DO EXERCISES 2 AND 3.

OBJECTIVES

After finishing Section 3.3, you should be able to:

● Solve simple equations using both the addition and multiplication principles.

●● Solve equations in which like terms are to be collected.

Solve.

1. $9x + 6 = 51$

Solve.

2. $8x - 4 = 28$

3. $-\dfrac{1}{2}x + 3 = 1$

Solve.

4. $-18 - x = -57$

Solve

5. $-4 - 8x = 8$

6. $41.68 = 4.7 - 8.6y$

Solve.

7. $4x + 3x = -21$

8. $x - 0.09x = 728$

Example 3 Solve: $45 - x = 13$.

$$45 - x = 13$$
$$-45 + 45 - x = -45 + 13 \qquad \text{Adding } -45$$
$$-x = -32$$
$$\left. \begin{array}{l} -1 \cdot x = -32 \\[2mm] x = \dfrac{-32}{-1} \\[3mm] x = 32 \end{array} \right\} \quad \begin{array}{l} -x = -1 \cdot x \\[2mm] \text{Dividing on both sides by } -1 \end{array}$$

You could have multiplied on both sides by -1 instead. That would change the sign on both sides.

Check:
$$\begin{array}{c|c} 45 - x = 13 \\ \hline 45 - 32 & 13 \\ 13 & \end{array}$$

The solution is 32.

DO EXERCISE 4.

Example 4 Solve: $16.3 - 7.2y = -8.18$.

$$16.3 - 7.2y = -8.18$$
$$-7.2y = -16.3 + (-8.18) \qquad \text{Adding } -16.3$$
$$-7.2y = -24.48$$
$$y = \frac{-24.48}{-7.2} \qquad \text{Dividing by } -7.2$$
$$y = 3.4$$

Check:
$$\begin{array}{c|c} 16.3 - 7.2y = -8.18 \\ \hline 16.3 - 7.2(3.4) & -8.18 \\ 16.3 - 24.48 & \\ -8.18 & \end{array}$$

The solution is 3.4.

DO EXERCISES 5 AND 6.

●● COLLECTING LIKE TERMS

If there are like terms on one side of the equation, we collect them before using the principles.

Example 5 Solve: $3x + 4x = -14$.

$$3x + 4x = -14$$
$$7x = -14 \qquad \text{Collecting like terms}$$
$$\frac{1}{7} \cdot 7x = \frac{1}{7} \cdot (-14)$$
$$x = -2$$

The number -2 checks, so the solution is -2.

DO EXERCISES 7 AND 8.

If there are like terms on opposite sides of the equation, we get them on the same side by using the addition principle. Then we collect them.

Example 6 Solve: $2x - 2 = -3x + 3$.

$$2x - 2 + \boxed{2} = -3x + 3 + \boxed{2} \quad \text{Adding 2}$$

$$2x = -3x + 5 \quad \text{Simplifying}$$

$$2x + \boxed{3x} = -3x + \boxed{3x} + 5 \quad \text{Adding } 3x$$

$$5x = 5 \quad \text{Collecting like terms and simplifying}$$

$$\boxed{\tfrac{1}{5}} \cdot 5x = \boxed{\tfrac{1}{5}} \cdot 5 \quad \text{Multiplying by } \tfrac{1}{5}$$

$$x = 1 \quad \text{Simplifying}$$

Check:

$$
\begin{array}{c|c}
2x - 2 = -3x + 3 & \\
\hline
2 \cdot 1 - 2 & -3 \cdot 1 + 3 \\
2 - 2 & -3 + 3 \\
0 & 0
\end{array}
$$

The solution is 1.

In Example 6, we used the addition principle to get all terms with a variable on one side and all numbers on the other side. Then we collected like terms and proceeded as before. If there are like terms on one side at the outset, they should be collected first.

DO EXERCISE 9.

Example 7 Solve: $6x + 5 - 7x = 10 - 4x + 3$.

$$6x + 5 - 7x = 10 - 4x + 3$$

$$-x + 5 = 13 - 4x \quad \text{Collecting like terms}$$

$$-x + 4x = 13 - 5 \quad \begin{array}{l}\text{Adding } 4x \text{ and subtracting 5 to get} \\ \text{all terms with variables on one side} \\ \text{and all other terms on the other}\end{array}$$

$$3x = 8 \quad \text{Collecting like terms}$$

$$\tfrac{1}{3} \cdot 3x = \tfrac{1}{3} \cdot 8 \quad \text{Multiplying by } \tfrac{1}{3}$$

$$x = \tfrac{8}{3} \quad \text{Simplifying}$$

The number $\frac{8}{3}$ checks, so it is the solution.

DO EXERCISES 10–12.

If we multiply on both sides of $\frac{1}{2}x = \frac{3}{4}$ by 4, we get $2x = 3$, which has no fractions. We have "cleared the fractions." If we multiply on both sides of $2.3x = 5$ by 10, we get $23x = 50$, which has no decimals. We have "cleared the decimals." In what follows we first multiply on both sides to "clear" or "get rid of" fractions or decimals. For fractions, the number we multiply by is either the product of all the denominators or any number that contains all the denominators as factors.

Solve.

9. $7y + 5 = 2y + 10$

Solve.

10. $5 - 2y = 3y - 5$

11. $7x - 17 + 2x = 2 - 8x + 15$

12. $3x - 15 = 5x + 2 - 4x$

Solve.

13. $\dfrac{7}{8}x - \dfrac{1}{4} + \dfrac{1}{2}x = \dfrac{3}{4} + x$

Example 8 Solve: $\dfrac{2}{3}x - \dfrac{1}{6} + \dfrac{1}{2}x = \dfrac{7}{6} + 2x$.

The number 6 contains all the denominators as factors. In fact, it is the least common denominator. We multiply on both sides by 6.

$$6\left(\dfrac{2}{3}x - \dfrac{1}{6} + \dfrac{1}{2}x\right) = 6\left(\dfrac{7}{6} + 2x\right) \qquad \text{Multiplying by 6 on both sides}$$

$$6 \cdot \dfrac{2}{3}x - 6 \cdot \dfrac{1}{6} + 6 \cdot \dfrac{1}{2}x = 6 \cdot \dfrac{7}{6} + 6 \cdot 2x \qquad \text{Using the distributive laws}$$

> *Caution!* Be sure to multiply *all* the terms by 6.

$$4x - 1 + 3x = 7 + 12x \qquad \text{Simplifying. Note that the fractions are cleared.}$$

$$7x - 1 = 7 + 12x \qquad \text{Collecting like terms}$$

$$7x - 12x = 7 + 1 \qquad \text{Subtracting } 12x \text{ and adding 1 to get all the terms with variables on one side and all the other terms on the other}$$

$$-5x = 8 \qquad \text{Collecting like terms}$$

$$-\dfrac{1}{5} \cdot (-5x) = -\dfrac{1}{5} \cdot 8 \qquad \text{Multiplying by } -\tfrac{1}{5} \text{ or dividing by } -5$$

$$x = -\dfrac{8}{5}$$

The number $-\tfrac{8}{5}$ checks and is the solution.

DO EXERCISE 13.

Here is a procedure for solving the equations given in this section.

> 1. **Multiply on both sides to clear of fractions or decimals.**
> 2. **Collect like terms on each side, if necessary.**
> 3. **Get all terms with variables on one side and all the other terms on the other side.**
> 4. **Collect like terms again, if necessary.**
> 5. **Divide to solve for the variable.**

Solve. Clear of decimals first.

14. $41.68 = 4.7 - 8.6y$

Let's repeat Example 4, but clear decimals first.

Example 9 Solve: $16.3 - 7.2y = -8.18$.

The greatest number of decimal places in any one number is two. Multiplying by 100, which has two 0's, will clear of decimals.

$$100(16.3 - 7.2y) = 100(-8.18) \qquad \text{Multiplying by 100}$$

$$100(16.3) - 100(7.2y) = 100(-8.18) \qquad \text{Using a distributive law}$$

$$1630 - 720y = -818 \qquad \text{Simplifying}$$

$$-720y = -818 - 1630 \qquad \text{Subtracting 1630}$$

$$-720y = -2448 \qquad \text{Collecting like terms}$$

$$y = \dfrac{-2448}{-720} = 3.4 \qquad \text{Dividing by } -720$$

The number 3.4 checks and is the solution.

DO EXERCISE 14.

NAME _____   CLASS _____   ANSWERS

EXERCISE SET 3.3

 Solve and check.

1. $5x + 6 = 31$ **2.** $3x + 6 = 30$ **3.** $4x - 6 = 34$

1. _____

2. _____

3. _____

Check: _____   *Check:* _____   *Check:* _____

4. $6x - 3 = 15$ **5.** $7x + 2 = -54$ **6.** $5x + 4 = -41$

4. _____

5. _____

6. _____

 Solve and check.

7. $5x + 7x = 72$ **8.** $4x + 5x = 45$ **9.** $4y - 2y = 10$

7. _____

8. _____

9. _____

10. $8y - 5y = 15$ **11.** $10.2y - 7.3y = -58$ **12.** $6.8y - 2.4y = -88$

10. _____

11. _____

12. _____

ANSWERS

13. $x + \dfrac{1}{3}x = 8$

14. $x + \dfrac{1}{4}x = 10$

15. $x + 0.08x = 9072$

13. _____

14. _____

15. _____

16. $x + 0.06x = 7738$

17. $8y - 35 = 3y$

18. $4x - 6 = 6x$

16. _____

17. _____

18. _____

19. $4x - 7 = 3x$

20. $y = 15 - 4y$

21. $x + 1 = 16 - 4x$

19. _____

20. _____

21. _____

22. $6y + 8 - 4y = 18$

23. $2x - 1 = 4 + x$

24. $5y - 2 = 28 - y$

22. _____

23. _____

24. _____

25. $5x + 2 = 3x + 6$

26. $6x + 3 = 2x + 11$

27. $5 - 2x = 3x - 7x + 25$

28. $10 - 3x = 2x - 8x + 40$

29. $4 + 3x - 6 = 3x + 2 - x$

30. $5 + 4x - 7 = 4x + 3 - x$

31. $4y - 4 + y = 6y + 20 - 4y$

32. $5y - 7 + y = 7y + 21 - 5y$

ANSWERS

33. $\dfrac{5}{2}x + \dfrac{1}{2}x = 3x + \dfrac{3}{2} + \dfrac{5}{2}x$

34. $\dfrac{7}{8}x - \dfrac{1}{4} + \dfrac{3}{4}x = \dfrac{1}{16} + x$

33. _____

34. _____

35. $2.1x + 45.2 = 3.2 - 8.4x$

36. $7\dfrac{1}{2}y - \dfrac{1}{2}y = \dfrac{15}{4}y + 39$

35. _____

37. $\dfrac{1}{5}t - 0.4 + \dfrac{2}{5}t = 0.6 - \dfrac{1}{10}t$

38. $1.7t + 8 - 1.62t = 0.4t - 0.32 + 8$

36. _____

37. _____

38. _____

Solve.

39. _____

39. 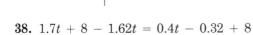 $0.008 + 9.62x - 42.8 = 0.944x + 0.0083 - x$

3.4   EQUATIONS CONTAINING PARENTHESES

•   Some equations containing parentheses can be solved by first multiplying to remove parentheses and then proceeding as before.

Example 1   Solve: $4x = 2(12 - 2x)$.

$$4x = 2(12 - 2x)$$

$4x = 24 - 4x$ \qquad Multiplying to remove parentheses

$4x + 4x = 24$ \qquad Adding $4x$ to get all x-terms on one side

$8x = 24$ \qquad Collecting like terms

$x = 3$ \qquad Multiplying by $\frac{1}{8}$

Check:

$$
\begin{array}{c|c}
4x & 2(12 - 2x) \\
\hline
4 \cdot 3 & 2(12 - 2 \cdot 3) \\
12 & 2(12 - 6) \\
 & 2 \cdot 6 \\
 & 12
\end{array}
$$

The solution is 3.

DO EXERCISES 1 AND 2.

Example 2   Solve: $3(x - 2) - 1 = 2 - 5(x + 5)$.

$3(x - 2) - 1 = 2 - 5(x + 5)$

$3x - 6 - 1 = 2 - 5x - 25$ \qquad Multiplying to remove parentheses

$3x - 7 = -5x - 23$ \qquad Simplifying

$3x + 5x = -23 + 7$ \qquad Adding $5x$ and also 7, to get all x-terms on one side and all other terms on the other side

$8x = -16$ \qquad Simplifying

$x = -2$ \qquad Multiplying by $\frac{1}{8}$

Check:

$$
\begin{array}{c|c}
3(x - 2) - 1 & 2 - 5(x + 5) \\
\hline
3(-2 - 2) - 1 & 2 - 5(-2 + 5) \\
3 \cdot (-4) - 1 & 2 - 5(3) \\
-12 - 1 & 2 - 15 \\
-13 & -13
\end{array}
$$

The solution is -2.

DO EXERCISES 3 AND 4.

OBJECTIVE

After finishing Section 3.4, you should be able to:

• Solve simple equations containing parentheses.

Solve.

1. $2(2y + 3) = 14$

2. $5(3x - 2) = 35$

Solve.

3. $3(7 + 2x) = 30 + 7(x - 1)$

4. $4(3 + 5x) - 4 = 3 + 2(x - 2)$

ANSWERS ON PAGE A–8

SOMETHING EXTRA
HANDLING DIMENSION SYMBOLS (PART I)

Speed is often measured by measuring a distance and a time, and then dividing the distance by the time (this is *average* speed). If a distance is measured in kilometers (km) and the time required to travel that distance is measured in hours, the speed will be computed in *kilometers per hour* (km/hr). For example, if a car travels 100 km in 2 hr, the average speed is

$$\frac{100 \text{ km}}{2 \text{ hr}}, \quad \text{or} \quad 50 \frac{\text{km}}{\text{hr}}.$$

The standard notation for km/hr is km/h.

The symbol

$$\frac{100 \text{ km}}{2 \text{ hr}}$$

makes it look as though we are dividing 100 km by 2 hr. It may be argued that we cannot divide 100 km by 2 hr (we can only divide 100 by 2). Nevertheless, it is convenient to treat dimension symbols such as *kilometers*, *meters*, *hours*, *feet*, *seconds*, and *pounds* much like numerals or variables, for the reason that correct results can thus be obtained mechanically. Compare, for example,

$$\frac{100x}{2y} = \frac{100}{2} \cdot \frac{x}{y} = 50\frac{x}{y} \quad \text{with} \quad \frac{100 \text{ km}}{2 \text{ hr}} = \frac{100}{2} \cdot \frac{\text{km}}{\text{hr}} = 50\frac{\text{km}}{\text{hr}}.$$

The analogy holds in other situations.

Example 1 Compare

$$3 \boxed{\text{ft}} + 2 \boxed{\text{ft}} = (3 + 2) \boxed{\text{ft}} = 5 \text{ ft}$$

with

$$3x + 2x = (3 + 2)x = 5x.$$

$$3 \text{ ft} + 2 \text{ yd} \quad \text{with} \quad 3x + 2y.$$

Note in this case that we cannot simplify further since the units are different. We could change units, but we will not do that here. (See pp. 374 and 478 for more on dimension symbols.)

EXERCISES

Distances and times are given. Use them to compute speed.

1. 45 mi, 9 hr	**2.** 680 km, 20 hr
3. 6.6 meters(m), 3 sec	**4.** 76 ft, 4 min

Add these measures.

5. 45 ft, 7 ft	**6.** 85 sec, 17 sec	**7.** 17 m, 14 m
8. 3 hr, 29 hr	**9.** $\frac{3}{4}$ lb, $\frac{2}{5}$ lb	**10.** 5 km, 7 km
11. 18 g, 4 g	**12.** 70 m/sec, 35 m/sec	

CLASS   ANSWERS

EXERCISE SET 3.4

■ Solve the following equations. Check.

1. $3(2y - 3) = 27$ **2.** $4(2y - 3) = 28$

1. _____

2. _____

3. $40 = 5(3x + 2)$ **4.** $9 = 3(5x - 2)$

3. _____

4. _____

5. $2(3 + 4m) - 9 = 45$ **6.** $3(5 + 3m) - 8 = 88$

5. _____

6. _____

ANSWERS

7. $5r - (2r + 8) = 16$

8. $6b - (3b + 8) = 16$

7. _____

8. _____

9. $3g - 3 = 3(7 - g)$

10. $3d - 10 = 5(d - 4)$

9. _____

10. _____

11. $6 - 2(3x - 1) = 2$

12. $10 - 3(2x - 1) = 1$

11. _____

12. _____

13. $5(d + 4) = 7(d - 2)$

14. $3(t - 2) = 9(t + 2)$

15. $3(x - 2) = 5(x + 2)$

16. $5(y + 4) = 3(y - 2)$

17. $8(2t + 1) = 4(7t + 7)$

18. $7(5x - 2) = 6(6x - 1)$

ANSWERS

19. $3(r - 6) + 2 = 4(r + 2) - 21$

20. $5(t + 3) + 9 = 3(t - 2) + 6$

19. _____

20. _____

21. $19 - (2x + 3) = 2(x + 3) + x$

22. $13 - (2c + 2) = 2(c + 2) + 3c$

21. _____

22. _____

23. $\frac{1}{4}(8y + 4) - 17 = -\frac{1}{2}(4y - 8)$

24. $\frac{1}{3}(6x + 24) - 20 = -\frac{1}{4}(12x - 72)$

23. _____

24. _____

Solve.

25. _____

25. ▦ $475(54x + 7856) + 9762 = 402(83x + 975)$

3.5 THE PRINCIPLE OF ZERO PRODUCTS

▪ The product of two numbers is 0 if one of the numbers is 0. Furthermore, *if any product is 0, then a factor must be 0.* For example, if $7x = 0$, then we know that $x = 0$. If $x(2x - 9) = 0$, then we know that $x = 0$ or $2x - 9 = 0$. If $(x + 3)(x - 2) = 0$, then we know that $x + 3 = 0$ or $x - 2 = 0$.

Example 1 Solve: $(x + 3)(x - 2) = 0$.

We have a product of 0. This equation will be true when either factor is 0. Hence it is true when

$$x + 3 = 0 \quad \text{or} \quad x - 2 = 0.$$

Here we have two simple equations, which we know how to solve:

$$x = -3 \quad \text{or} \quad x = 2.$$

There are two solutions, -3 and 2.

We have another principle to help in solving equations.

The Principle of Zero Products: **An equation with 0 on one side and with factors on the other can be solved by finding those numbers that make the factors 0.**

Example 2 Solve: $(5x + 1)(x - 7) = 0$. ⌐*Caution!* Don't multiply these.⌐

$$5x + 1 = 0 \quad \text{or} \quad x - 7 = 0 \quad \text{Using the principle of zero products}$$
$$5x = -1 \quad \text{or} \quad x = 7$$
$$x = -\frac{1}{5} \quad \text{or} \quad x = 7 \quad \text{Solving the two equations separately}$$

Check: For $-\frac{1}{5}$:

$$\frac{(5x + 1)(x - 7) = 0}{(5(-\frac{1}{5}) + 1)(-\frac{1}{5} - 7) \mid 0}$$
$$(-1 + 1)(-7\tfrac{1}{5})$$
$$0(-7\tfrac{1}{5})$$
$$0$$

For 7:

$$\frac{(5x + 1)(x - 7) = 0}{(5 \cdot 7 + 1)(7 - 7) \mid 0}$$
$$(35 + 1) \cdot 0$$
$$0$$

The solutions are $-\frac{1}{5}$ and 7.

The "possible solutions" we get by using the principle of zero products are actually always solutions, unless we have made an error in solving. Thus, when we use this principle, a check is not necessary, except to detect errors.

⌐*Caution!* Do not make the mistake of using this principle when there is not a zero on one side.⌐

DO EXERCISES 1–3.

OBJECTIVE

After finishing Section 3.5, you should be able to:

▪ Solve equations (already factored), using the principle of zero products.

Solve, using the principle of zero products.

1. $(x - 3)(x + 4) = 0$

2. $(x - 7)(x - 3) = 0$

3. $(4t + 1)(3t - 2) = 0$

Solve.

4. $y(3y - 17) = 0$

Example 3   Solve: $x(2x - 9) = 0$

$\dot{x} = 0$ or $2x - 9 = 0$ Using the principle of zero products

$x = 0$ or $2x = 9$

$x = 0$ or $x = \frac{9}{2}$

> When some factors have only one term, you can still use the principle of zero products in the same way.

The solutions are 0 and $\frac{9}{2}$.

DO EXERCISE 4.

SOMETHING EXTRA

CALCULATOR CORNER: NUMBER PATTERNS

1. Calculate each of the following using your calculator. Look for a pattern.

$$1$$
$$1 + 3$$
$$1 + 3 + 5$$
$$1 + 3 + 5 + 7$$
$$1 + 3 + 5 + 7 + 9$$
$$1 + 3 + 5 + 7 + 9 + 11$$

Use the pattern to find each of the following without using your calculator.

$$1 + 3 + 5 + 7 + 9 + 11 + 13$$
$$1 + 3 + 5 + 7 + 9 + 11 + 13 + 15$$

2. Verify each of the following using your calculator.

$$1 = \frac{1 \cdot 2}{2}$$

$$1 + 2 = \frac{2 \cdot 3}{2}$$

$$1 + 2 + 3 = \frac{3 \cdot 4}{2}$$

$$1 + 2 + 3 + 4 = \frac{4 \cdot 5}{2}$$

$$1 + 2 + 3 + 4 + 5 = \frac{5 \cdot 6}{2}$$

Use the pattern to find each of the following without using your calculator.

$$1 + 2 + 3 + 4 + 5 + 6 + 7 + 8$$
$$1 + 2 + 3 + 4 + 5 + 6 + 7 + 8 + 9$$

Can you give an expression for

$$1 + 2 + 3 + 4 + \cdots + n?$$

| NAME | CLASS | ANSWERS |

EXERCISE SET 3.5

■ Solve.

1. $(x + 8)(x + 6) = 0$ **2.** $(x + 3)(x + 2) = 0$

3. $(x - 3)(x + 5) = 0$ **4.** $(x + 9)(x - 3) = 0$

5. $(x - 12)(x - 11) = 0$ **6.** $(x - 13)(x - 53) = 0$

7. $y(y - 13) = 0$ **8.** $x(x - 4) = 0$

9. $0 = x(x + 21)$ **10.** $0 = y(y + 10)$

11. $(2x + 5)(x + 4) = 0$ **12.** $(2x + 9)(x + 8) = 0$

13. $(3x - 1)(x + 2) = 0$ **14.** $(5x + 1)(x - 3) = 0$

15. $2x(3x - 2) = 0$ **16.** $5x(8x - 9) = 0$

1. _____
2. _____
3. _____
4. _____
5. _____
6. _____
7. _____
8. _____
9. _____
10. _____
11. _____
12. _____
13. _____
14. _____
15. _____
16. _____

17. _____

18. _____

19. _____

20. _____

21. _____

22. _____

23. _____

24. _____

25. _____

26. _____

27. _____

28. _____

29. _____

30. _____

17. $\frac{1}{2}x\left(\frac{2}{3}x - 12\right) = 0$

18. $\frac{5}{7}x\left(\frac{3}{4}x - 6\right) = 0$

19. $0 = \left(\frac{1}{3} - 3x\right)\left(\frac{1}{5} - 2x\right)$

20. $0 = \left(\frac{1}{5} - 2x\right)\left(\frac{1}{9} - 3x\right)$

21. $\left(\frac{1}{3}y - \frac{2}{3}\right)\left(\frac{1}{4}y - \frac{3}{2}\right) = 0$

22. $\left(\frac{7}{4}x - \frac{1}{12}\right)\left(\frac{2}{3}x - \frac{12}{11}\right) = 0$

23. $2.5x(2.5x - 5) = 0$

24. $8.3x(4.5x - 9) = 0$

25. $(0.03x - 0.01)(0.05x - 1) = 0$

26. $(0.01x - 0.03)(0.04x - 2) = 0$

27. $\left(0.01x + \frac{1}{10}\right)\left(0.03x - \frac{1}{10}\right) = 0$

28. $\left(0.02x + \frac{1}{10}\right)\left(0.04x - \frac{1}{10}\right) = 0$

---

29. Check the numbers found below. What's wrong with the methods used? Try to find the solutions.

a) $(x - 3)(x + 4) = 8$
    $x - 3 = 0$ or $x + 4 = 8$
       $x = 3$ or $x = 4$

b) $(x - 3)(x + 4) = 8$
    $x - 3 = 2$ or $x + 4 = 4$
       $x = 5$ or $x = 0$

30. ▦ Solve $(0.00005x + 0.1)(0.0097x + 0.5) = 0$

3.6 SOLVING PROBLEMS

■ The first step in solving a problem is to translate it to mathematical language. Very often this means translating to an equation. Drawing a picture usually helps a great deal. Then we solve the equation and check to see if we have a solution to the problem.

Example 1 A 6-ft board is cut into two pieces, one twice as long as the other. How long are the pieces?

Drawing a picture

The picture can help in translating. Here is one way to do it.

$$\underbrace{\text{Length of one piece}}_{x} \underbrace{\text{plus}}_{+} \underbrace{\text{length of other}}_{2x} \underbrace{\text{is}}_{=} \underbrace{6}_{6}$$

(We use x for the length of one piece and $2x$ for the length of the other.) Now we solve:

$$x + 2x = 6$$
$$3x = 6 \qquad \text{Collecting like terms}$$
$$x = 2 \qquad \text{Multiplying by } \tfrac{1}{3}$$

Do we have an answer to the *problem*? If one piece is 2 ft long, the other, to be twice as long, must be 4 ft long, and the lengths of the pieces add up to 6 ft. This checks.

Note that we did not have to check the solution in the equation. Rather, we checked in the problem itself.

> *Steps to use in solving a problem:*
> 1. **Translate to an equation. (Always draw a picture if it makes sense to do so.)**
> 2. **Solve the equation.**
> 3. **Check the answer in the original problem.**

DO EXERCISE 1.

Example 2 Five plus three more than a number is nineteen. What is the number?

This time it does not make sense to draw a picture.

①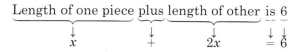

We have used x to represent the unknown number.

1. An 8-ft board is cut into two pieces. One piece is 2 ft longer than the other. How long are the pieces?

2. If five is subtracted from three times a certain number, the result is 10. What is the number?

3. Money is borrowed at 9% simple interest. After 1 year $6213 pays off the loan. How much was originally borrowed?

4. The sum of two consecutive even integers is 38. (Consecutive even integers are next to each other, such as 4 and 6. The larger is 2 plus the smaller.) What are the integers?

② Now solve: $5 + (x + 3) = 19$

$$x + 8 = 19 \qquad \text{Collecting like terms}$$
$$x = 11 \qquad \text{Adding } -8$$

③ *Check:* Three more than 11 is 14. Adding 5 to 14, we get 19. This checks and the answer is 11.

DO EXERCISE 2.

Always read the problem carefully. Make a list of the information in the problem.

Example 3 Money is invested in a savings account at 8% simple interest. After 1 year there is $8208 in the account. How much was originally invested?

Original investment plus interest is 8208
$$x \qquad + \qquad 8\%x \qquad = 8208$$

We have used x to represent the original investment.

$$x + 8\%x = 8208$$
$$1x + 0.08x = 8208$$
$$1.08x = 8208$$
$$x = \frac{8208}{1.08} = 7600$$

Check: 8% of 7600 is 608. Adding this to 7600, we get 8208. This checks so the original investment was $7600.

DO EXERCISE 3.

Example 4 The sum of two consecutive integers is 29. What are the integers?

(*Consecutive* integers are next to each other, such as 3 and 4. The larger is 1 plus the smaller.)

First integer + second integer = 29   Rewording
$$x \qquad + \qquad (x + 1) \qquad = 29 \qquad \text{Translating}$$

We have let x represent the first integer. Then $x + 1$ represents the second.

We solve: $x + (x + 1) = 29$

You should write down what your letters represent.

$$2x + 1 = 29$$
$$2x = 28$$
$$x = 14.$$

Check: Our answers are 14 and 15. These are consecutive integers. Their sum is 29, so the answers check in the *problem*.

DO EXERCISE 4.

Example 5   Acme Rent-a-Car rents an intermediate-size car (such as a Chevrolet, Ford, or Plymouth) at a daily rate of $14.95 plus 19¢ per mile. A businessperson is not to exceed a daily car rental budget of $50. What mileage will allow the businessperson to stay within budget?

We translate to an equation, using $0.19 for 19¢.

$$\underbrace{14.95}\ \underbrace{\text{plus}}\ \underbrace{19¢}\ \underbrace{\text{times}}\ \underbrace{\text{number of miles driven}}\ \underbrace{\text{is}}\ \underbrace{\text{budget}}$$

$$14.95\ +\ 0.19\ \cdot\ \qquad m \qquad\ =\ \ 50$$

We solve:

$$14.95 + 0.19m = 50$$
$$0.19m = 35.05 \qquad \text{Adding } -14.95$$
$$m = \frac{35.05}{0.19}$$
$$m \approx 184.5. \qquad \text{Rounded to the nearest tenth}$$

This checks in the original problem. In this case it is an approximation. At least the businessperson now knows to stay under this mileage.

DO EXERCISE 5.

Example 6   The perimeter of a rectangle is 150 cm. The length is 15 cm greater than the width. Find the dimensions.

We first draw a picture.

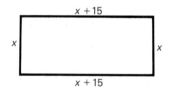

$$\underbrace{\text{Width}}\ +\ \underbrace{\text{Width}}\ +\ \underbrace{\text{Length}}\ +\ \underbrace{\text{Length}}\ =\ 150 \qquad \text{Rewording}$$

$$x\ \ +\ \ x\ \ +\ (x + 15)\ +\ (x + 15)\ =\ 150 \qquad \text{Translating}$$

We have let x represent the width. Then $x + 15$ represents the length.

$$4x + 30 = 150$$
$$4x = 120$$
$$x = 30$$

The length is $x + 15$, or 45.

Check: The perimeter is $30 + 30 + 45 + 45$, which is 150. This checks, so the width is 30 cm and the length is 45 cm.

DO EXERCISE 6.

Example 7   The second angle of a triangle is twice as large as the first. The measure of the third angle is 20° greater than that of the first angle. How large are the angles?

5. Acme also rents compact cars at $14.95 plus 17¢ per mile. What mileage will allow the businessperson to stay within a budget of $50?

6. The length of a rectangle is twice the width. The perimeter is 60 m. Find the dimensions.

7. The second angle of a triangle is 3 times as large as the first. The third angle measures 30° more than the first angle. Find the measures of the angles.

We draw a picture. We use x for the measure of the first angle. The second is twice as large, so its measure will be $2x$. The third angle is 20° greater than the first angle so its measure will be $x + 20$.

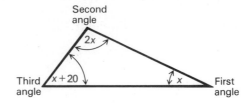

Now, to translate we need to recall a geometric fact. The measures of the angles of any triangle add up to 180°.

$$
\underbrace{\text{Measure of}\atop\text{first angle}}_{\displaystyle x} + \underbrace{\text{Measure of}\atop\text{second angle}}_{\displaystyle 2x} + \underbrace{\text{Measure of}\atop\text{third angle}}_{\displaystyle (x + 20)} = \underbrace{180°}_{\displaystyle 180}
$$

Now we solve:

$$x + 2x + (x + 20) = 180$$
$$4x + 20 = 180$$
$$4x = 160$$
$$x = 40.$$

The angles will have measures as follows:

First angle: $x = 40°$
Second angle: $2x = 80°$
Third angle: $x + 20 = 60°$.

These add up to 180° so they give the answer to the *problem*.

DO EXERCISE 7.

8. After a 30% reduction, an item is on sale for $8050. What was the marked price (the price before reduction)?

Example 8   After a 20% reduction, an item is on sale for $9600. What was the marked price (the price before reduction)?

$$
\underbrace{\text{(Marked price)}}_{\displaystyle x}\ \underbrace{\text{minus}}_{\displaystyle -}\ \underbrace{\text{(Reduction)}}_{\displaystyle 20\%x}\ \underbrace{\text{is}}_{\displaystyle =}\ \underbrace{\$9600}_{\displaystyle 9600}
$$

We have used x to represent the marked price:

$$x - 20\%x = 9600$$
$$1x - 0.2x = 9600$$
$$(1 - 0.2)x = 9600$$
$$0.8x = 9600$$
$$x = \frac{9600}{0.8}$$
$$x = 12,000$$

Check: 20% of $12,000 is $2400. Subtracting this from $12,000 we get $9600. This checks, so the marked price is $12,000.

DO EXERCISE 8.

NAME _____   CLASS _____

EXERCISE SET 3.6

■ Solve.

1. When 18 is subtracted from six times a certain number, the result is 96. What is the number?

1. _____

2. When 28 is subtracted from five times a certain number, the result is 232. What is the number?

2. _____

3. If you double a number and then add 16, you get $\frac{2}{5}$ of the original number. What is the original number?

3. _____

4. If you double a number and then add 85, you get $\frac{3}{4}$ of the original number. What is the original number?

4. _____

5. If you add two-fifths of a number to the number itself, you get 56. What is the number?

5. _____

6. If you add one-third of a number to the number itself, you get 48. What is the number?

6. _____

7. A 180-m rope is cut into three pieces. The second piece is twice as long as the first. The third piece is three times as long as the second. How long is each piece of rope?

7. _____

8. A 480-m wire is cut into three pieces. The second piece is three times as long as the first. The third piece is four times as long as the second. How long is each piece?

8. _____

9. Consecutive odd integers are next to each other, such as 5 and 7. The larger is 2 plus the smaller. The sum of two consecutive odd integers is 76. What are the integers?

9. _____

10. The sum of two consecutive odd integers is 84. What are the integers?

10. _____

11. Consecutive even integers are next to each other, like 6 and 8. The larger is 2 plus the smaller. The sum of two consecutive even integers is 114. What are the integers?

11. _____

12. The sum of two consecutive even integers is 106. What are the integers?

12. _____

13. The sum of three consecutive integers is 108. What are the integers?

14. The sum of three consecutive integers is 126. What are the integers?

13. _____

15. The sum of three consecutive odd integers is 189. What are the integers?

14. _____

16. The sum of three consecutive odd integers is 255. What are the integers?

15. _____

17. The perimeter of a rectangle is 310 m. The length is 25 m greater than the width. Find the width and length of the rectangle.

16. _____

18. The perimeter of a rectangle is 304 cm. The length is 40 cm greater than the width. Find the width and length of the rectangle.

17. _____

18. _____

19. The perimeter of a rectangle is 152 m. The width is 22 m less than the length. Find the width and the length.

19. _____

20. The perimeter of a rectangle is 280 m. The width is 26 m less than the length. Find the width and the length.

20. _____

21. The second angle of a triangle is four times as large as the first. The third angle is 45° less than the sum of the other two angles. Find the measure of the first angle.

21. _____

22. The second angle of a triangle is three times as large as the first. The third angle is 25° less than the sum of the other two angles. Find the measure of the first angle.

22. _____

23. Money is invested in a savings account at 7% simple interest. After 1 year there is $4708 in the account. How much was originally invested?

23. _____

24. Money is borrowed at 10% simple interest. After 1 year $7194 pays off the loan. How much was originally borrowed?

24. _____

25. After a 40% reduction, a shirt is on sale at $9.60. What was the marked price (the price before reduction)?

26. After a 34% reduction, a blouse is on sale at $9.24. What was the marked price?

25. _____

27. Badger Rent-a-Car rents an intermediate-size car at a daily rate of $14.95 plus 10¢ per mile. A businessperson is not to exceed a daily car rental budget of $60. What mileage will allow the businessperson to stay within budget?

26. _____

28. Badger also rents compact cars at $13.95 plus 10¢ per mile. What mileage will allow the businessperson to stay within the budget of $60?

27. _____

29. The second angle of a triangle is three times as large as the first. The measure of the third angle is 40° greater than that of the first angle. How large are the angles?

28. _____

30. The second angle of a triangle is five times as large as the first. The measure of the third angle is 10° greater than that of the first angle. How large are the angles?

29. _____

30. _____

31. The population of the world in 1976 was 4 billion. This was a 2% increase over the population 1 year earlier. What was the former population?

31. _____

32. The population of the United States in 1977 was 216 million. This was a 1% increase over the population 1 year earlier. What was the former population?

32. _____

33. The equation

$$R = -0.028t + 20.8$$

can be used to predict the world record in the 200-meter dash. R stands for the record in seconds, and t stands for the number of years since 1920. In what year will the record be 19.0 seconds?

33. _____

34. The equation

$$F = \frac{1}{4}N + 40$$

can be used to determine temperatures given how many times a cricket chirps per minute. F represents temperature in degrees and N is the number of chirps per minute. Determine the chirps per minute necessary for the temperature to be 80°.

34. _____

35. _____

35. Consumer experts advise us never to pay the sticker price for a car. A rule of thumb is to pay the sticker price minus 20% of the sticker price, plus $200. A car is purchased for $5320 using the rule. What was the sticker price?

36. At one time in Indiana, a sales tax of 4% was added on to the price of gasoline as registered on the pump. Suppose a driver pulled into a station and asked for $10 worth of regular. The attendant figured the pump should read $9.60, filled the tank with that amount, and charged the driver $10. Something was wrong! Use algebra to correct the error.

36. _____

3.7 FORMULAS

• A formula is a kind of recipe for doing a certain kind of calculation. Formulas are often given as equations. A familiar formula from electricity is

$$E = IR.$$

This formula tells us that to calculate the voltage E in an electric circuit we multiply the current I by the resistance R. Suppose we know the voltage E and the current I. We wish to calculate the resistance R. We can get R alone on one side, or "solve" the formula for R.

Example 1   Solve for R.

$$E = IR \qquad \text{We want this letter alone.}$$

$$\frac{1}{I} \cdot E = \frac{1}{I} \cdot IR \qquad \text{Multiplying on both sides by } \frac{1}{I}$$

$$\frac{E}{I} = R$$

Multiplying by $1/I$ is the same as dividing by I. The formula now says that we can find R by dividing E by I.

DO EXERCISE 1.

> *Caution!* Remember, formulas are equations. Use the same principles in solving that you use for any other equations.

To see how the principles apply to formulas, compare the following.

A. Solve.

$$5x + 2 = 12$$
$$5x = 12 - 2$$
$$5x = 10$$
$$x = \frac{10}{5}$$
$$x = 2$$

B. Solve.

$$5x + 2 = 12$$
$$5x = 12 - 2$$
$$x = \frac{12 - 2}{5}$$

C. Solve for x.

$$ax + b = c$$
$$ax = c - b$$
$$x = \frac{c - b}{a}$$

In (A) we solved as we did before. In (B) we did not carry out the calculations. In (C) we could not carry out the calculations because we had unknown numbers.

OBJECTIVE

After finishing Section 3.7, you should be able to:

• Solve formulas for a specified letter.

1. Solve for I: $E = IR$.

ANSWER ON PAGE A–9

2. Solve for D: $C = \pi D$.
(This is a formula for the circumference of a circle of diameter D.)

Example 2   Solve for r: $C = 2\pi r$.

This is a formula for the circumference C of a circle of radius r.

$$C = 2\pi r \quad \text{We want this letter alone.}$$

$$\boxed{\frac{1}{2\pi}} \cdot C = \boxed{\frac{1}{2\pi}} \cdot 2\pi r \quad \text{Multiplying by } \frac{1}{2\pi}$$

$$\frac{C}{2\pi} = \frac{2\pi}{2\pi} \cdot r$$

$$\frac{C}{2\pi} = r$$

DO EXERCISE 2.

With the formulas in this section we can use a procedure like that described in Section 3.3.

3. Solve for c: $A = \dfrac{a + b + c + d}{4}$.

> **To solve a formula for a given letter, identify the letter, and:**
> 1. **Multiply on both sides to clear of fractions or decimals.**
> 2. **Collect like terms on each side, if necessary.**
> 3. **Get all terms with the letter to be solved for on one side of the equation and all other terms on the other side.**
> 4. **Collect like terms again, if necessary.**
> 5. **Solve for the letter in question.**

Example 3   Solve for a: $A = \dfrac{a + b + c}{3}$.

This is a formula for the average A of three numbers a, b, and c.

$$A = \frac{a + b + c}{3} \quad \text{We want this letter alone.}$$

$$3A = a + b + c \quad \text{Multiplying by 3 to clear of the fraction}$$

$$3A - b - c = a$$

DO EXERCISE 3.

4. Solve for I: $A = \dfrac{9R}{I}$.
(This is a formula for computing the earned run average A of a pitcher who has given up R earned runs in I innings of pitching.)

Example 4   Solve for C: $Q = \dfrac{100M}{C}$.

This is a formula used in psychology for intelligence quotient Q, where M is mental age, and C is chronological, or actual, age.

$$Q = \frac{100M}{C} \quad \text{We want this letter alone.}$$

$$CQ = 100M \quad \text{Multiplying by } C \text{ to clear of the fraction}$$

$$C = \frac{100M}{Q} \quad \text{Multiplying by } \frac{1}{Q}$$

DO EXERCISE 4.

NAME   CLASS

EXERCISE SET 3.7

■ Solve:

1. $A = bh$, for b. (*an area formula*)   2. $A = bh$, for h.

3. $d = rt$, for r. (*a distance formula*)   4. $d = rt$, for t.

5. $I = Prt$, for P. (*an interest formula*)   6. $I = Prt$, for t.

7. $F = ma$, for a. (*a physics formula*)   8. $F = ma$, for m.

9. $P = 2l + 2w$, for w. (*a perimeter formula*)   10. $P = 2l + 2w$, for l.

11. $A = \pi r^2$, for r^2. (*an area formula*)   12. $A = \pi r^2$, for π.

13. $A = \frac{1}{2}bh$, for b. (*an area formula*)   14. $A = \frac{1}{2}bh$, for h.

1. _____

2. _____

3. _____

4. _____

5. _____

6. _____

7. _____

8. _____

9. _____

10. _____

11. _____

12. _____

13. _____

14. _____

ANSWERS

15. _____

16. _____

17. _____

18. _____

19. _____

20. _____

21. _____

22. _____

23. _____

24. _____

25. _____

26. _____

27. _____

28. _____

15. $E = mc^2$, for m. (*a relativity formula*)

16. $E = mc^2$, for c^2.

17. $A = \dfrac{a + b + c}{3}$, for b.

18. $A = \dfrac{a + b + c}{3}$, for c.

19. $v = \dfrac{3k}{t}$, for t.

20. $P = \dfrac{ab}{c}$, for c.

21. $A = \frac{1}{2}ah + \frac{1}{2}bh$, for b.

22. $A = \frac{1}{2}ah + \frac{1}{2}bh$, for a.

23. The formula

$$H = \dfrac{D^2 N}{2.5}$$

is used to find the horsepower H of an N-cylinder engine. Solve for D^2.

24. Solve for N:

$$H = \dfrac{D^2 N}{2.5}.$$

25. The area of a sector of a circle is given by

$$A = \dfrac{\pi r^2 S}{360},$$

where r is the radius and S is the angle measure of the sector. Solve for S.

26. Solve for r^2:

$$A = \dfrac{\pi r^2 S}{360}.$$

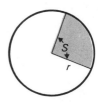

27. The formula

$$R = -0.0075t + 3.85$$

can be used to estimate the world record in the 1500-meter run t years after 1930. Solve for t.

28. The formula

$$F = \dfrac{9}{5}C + 32$$

can be used to convert from Celsius, or Centigrade, temperature C to Fahrenheit temperature F. Solve for C.

EXTENSION EXERCISES

SECTION 3.1

Solve for x.

1. $x - 4 = a$ **2.** $x + 7 = b + 10$ **3.** $1 - c = a + x$

SECTION 3.2

Solve for x.

4. $ax = 5a$ **5.** $3x = \dfrac{b}{c}$ **6.** $cx = a^2 + 1$ **7.** $\dfrac{a}{b}x = 4$

SECTION 3.3

Solve the first equation for x. Then substitute this number into the second equation and solve it for y.

8. $9x - 5 = 22$ **9.** $9x + 2 = -1$ **10.** $0.2x + 0.12 = 0.146$

     $4x + 2y = 2$ $4x - y = \dfrac{11}{3}$ $0.17x + 0.03y = 0.01238$

Solve for y.

11. $\dfrac{y - 2}{3} = \dfrac{2 - y}{5}$ **12.** $\dfrac{y}{a} - 3y = 1$

SECTION 3.4

Solve.

13. $-2[3(x - 2) + 4] = 4(1 - x) + 8$ **14.** $x(x - 4) = 3x(x + 1) - 2(x^2 + x - 5)$

SECTION 3.5

Solve.

15. $3(x - 1)(x + 2) = 0$ **16.** $\dfrac{(3x - 4)}{2} \cdot \dfrac{(5x + 1)}{6} = 0$ **17.** $(x - 1)(2x + 7)(3x - 5) = 0$

SECTION 3.6

18. The width of a rectangle is $\frac{3}{4}$ the length. The perimeter of the rectangle becomes 50 cm when the length and width are each increased by 2 cm. Find the length and width.

19. A student has an average score of 82 on three tests. The student's average score on the first two tests is 85. What was the score on the third test?

20. A merchant goes to the bank to get $20 worth of change and requests twice as many quarters as half dollars, twice as many dimes as quarters, three times as many nickels as dimes, and no pennies or dollars. How many of each coin did the merchant get?

21. The buyer of a piano priced at $2000 is given the choice of paying cash at the time of the purchase or $2150 at the end of one year. At what rate is the buyer being charged interest if payment is made at the end of one year?

SECTION 3.7

22. In $P = 2a + 2b$, P doubles. Do a and b both double?

23. In $A = lw$, l and w both double. What happens to A?

24. In $A = \frac{1}{2}bh$, b increases by 4 units, and h does not change. What happens to A?

25. In $T = 1.2a + 1.09b$, does an increase in a or an increase in b have more effect on T?

NAME   SCORE   ANSWERS

TEST OR REVIEW—CHAPTER 3

If you miss an item, review the indicated section and objective.

1. _____

Solve.

[3.1, ●]   **1.** $x + 5.1 = 11.2$

2. _____

[3.1, ●]   **2.** $n - 7 = -6$

[3.2, ●]   **3.** $4x = 48$

3. _____

[3.2, ●]   **4.** $5x = -35$

4. _____

[3.3, ●]   **5.** $\dfrac{1}{4}x - \dfrac{5}{8} = \dfrac{3}{8}$

5. _____

[3.3, ● ●]   **6.** $5t + 9 = 3t - 1$

6. _____

[3.4, ●]   **7.** $4(x + 3) = 36$

7. _____

[3.4, ●]   **8.** $-5x + 3(x + 8) = 36$

8. _____

ANSWERS

[3.5, ●]   **9.** $(x + 9)(x - 6) = 0$

9. _____

[3.5, ●]   **10.** $y(4y - 7) = 0$

10. _____

Solve.

[3.6, ●]   **11.** If 14 is added to three times a certain number, the result is 41. Find the number.

11. _____

[3.6, ●]   **12.** The sum of two consecutive odd integers is 116. Find the integers.

12. _____

[3.6, ●]   **13.** The perimeter of a rectangle is 56 cm. The width is 6 cm less than the length. Find the width and the length.

13. _____

[3.6, ●]   **14.** One angle of a triangle is three times as large as a second angle. The third angle is 10° larger than the second. How large is each angle?

14. _____

Solve:

[3.7, ●]   **15.** $P = 4s$, for s.

15. _____

[3.7, ●]   **16.** $V = \frac{1}{3}Bh$, for h.

16. _____

POLYNOMIALS

4

The cost of operating an automobile can be approximated by a polynomial.

4.1 POLYNOMIALS

OBJECTIVES

After finishing Section 4.1, you should be able to:

◖•◗ Evaluate a polynomial for a given value of a variable.

◖••◗ Identify the terms of a polynomial.

◖•••◗ Identify the like terms of a polynomial.

◖::◗ Collect the like terms of a polynomial.

1. Write three polynomials.

Evaluate each polynomial for $x = 3$.

2. $-4x - 7$

3. $-5x^3 + 7x + 10$

Evaluate each polynomial for $x = -4$.

4. $5x + 7$

5. $2x^2 + 5x - 4$

Expressions like these are called *polynomials:*

$$3x + 2, \qquad 2x^2 + 7x - 2, \qquad a^4 - 3a^2 + 0.5a + 7.$$

Each part to be added or subtracted is a number or a number times a variable to some power.

A polynomial can also be just one of the parts. These are also polynomials:

$$4x^2, \qquad -7a, \qquad \frac{3}{4}y^5, \qquad x, \qquad 5, \qquad -2, \qquad 0.$$

Polynomials are used often in algebra, especially in equation solving. We will learn how to add, subtract, and multiply polynomials.

DO EXERCISE 1.

◖•◗ EVALUATING POLYNOMIALS

When we replace the variable in a polynomial by a number, the polynomial then represents a number. Finding that number is called *evaluating the polynomial.*

Examples Evaluate each polynomial for $x = 2$.

1. $3x + 5$: $3 \cdot 2 + 5 = 6 + 5 = 11$

2. $2x^2 + 7x + 3$: $2 \cdot 2^2 + 7 \cdot 2 + 3 = 2 \cdot 4 + 14 + 3 = 8 + 14 + 3 = 25$

DO EXERCISES 2–5.

Example 3 The cost, in cents per mile, of operating an automobile at speed s, in mph, is approximated by the polynomial

$$0.005s^2 - 0.35s + 28.$$

Evaluate the polynomial for $s = 50$ in order to find the cost of operating an automobile at 50 mph.

$$0.005(50)^2 - 0.35(50) + 28 = 0.005(2500) - 17.5 + 28$$
$$= 12.5 - 17.5 + 28 = 23¢$$

The cost is 23¢ per mile.

DO EXERCISE 6.

●● IDENTIFYING TERMS

Subtractions can be rewritten as additions. We showed this in Section 2.3. Any polynomial can be rewritten using only additions. We can do this for polynomials because they represent numbers.

Examples   Write each polynomial using only additions.

4. $-5x^2 - x = -5x^2 + (-x)$
5. $4x^5 - 2x^6 - 4x = 4x^5 + (-2x^6) + (-4x)$

DO EXERCISES 7 AND 8.

When a polynomial has only additions, the parts being added are called *terms*.

Example 6   Identify the terms of the polynomial

$$4x^3 + 3x + 12 + 8x^3 + 5x.$$

Terms: $4x^3$, $3x$, 12, $8x^3$, $5x$

If there are subtractions you can think of the subtractions as additions.

Example 7   Identify the terms of the polynomial

$$3t^4 - 5t^6 - 4t + 2.$$

Terms: $3t^4$, $-5t^6$, $-4t$, 2

> *Caution!* Do not confuse terms (things added) with factors (things multiplied). In $3x + 5y$ the terms are $3x$ and $5y$. The first term has factors 3 and x. The second has factors 5 and y.

DO EXERCISES 9 AND 10.

●●● LIKE TERMS

Terms that have the same variable and the same exponent are called *like terms*, or *similar terms*.

Examples   Identify the like terms in each polynomial.

8. $4x^3 + 5x - 4x^2 + 2x^3 + x^2$

  Like terms: $4x^3$ and $2x^3$ Same exponent and variable
  Like terms: $-4x^2$ and x^2 Same exponent and variable

6. Evaluate the polynomial in Example 3 for $s = 70$ in order to find the cost of operating an automobile at 70 mph.

Write each polynomial using only additions.

7. $-9x^3 - 4x^5$

8. $-2x^3 + 3x^7 - 7x$

Identify the terms of each polynomial.

9. $3x^2 + 6x + \dfrac{1}{2}$

10. $-4y^5 + 7y^2 - 3y - 2$

Identify the like terms in each polynomial.

11. $4x^3 - x^3 + 2$

12. $4t^4 - 9t^3 - 7t^4 + 10t^3$

Collect like terms.

13. $3x^2 + 5x^2$

14. $4x^3 - 2x^3 + 2 + 5$

15. $\frac{1}{2}x^5 - \frac{3}{4}x^5 + 4x^2 - 2x^2$

Collect like terms.

16. $24 - 4x^3 - 24$

17. $5x^3 - 8x^5 + 8x^5$

18. $-2x^4 + 16 + 2x^4 + 9 - 3x^5$

Collect like terms.

19. $7x - x$

20. $5x^3 - x^3 + 4$

21. $\frac{3}{4}x^3 + 4x^2 - x^3 + 7$

22. $8x^2 - x^2 + x^3 - 1 - 4x^2 + 10$

9. $6 - 3a^2 + 8 - a - 5a$

Like terms: 6 and 8   No variable at all
Like terms: $-a$ and $-5a$

DO EXERCISES 11 AND 12.

∷ COLLECTING LIKE TERMS

We can often simplify polynomials by *collecting like terms*, or *combining similar terms*. To do this we use the distributive laws.

Examples Collect like terms.

10. $2\,\boxed{x^3}\, - 6\,\boxed{x^3}\, = (2 - 6)\,\boxed{x^3}$ Using a distributive law
$$= -4x^3$$

11. $5x^2 + 7 + 4x^4 + 2x^2 - 11 - 2x^4 = (5 + 2)x^2 + (4 - 2)x^4 + (7 - 11)$
$$= 7x^2 + 2x^4 - 4$$

DO EXERCISES 13–15.

In collecting like terms we may get zero.

Examples Collect like terms.

12. $5\,\boxed{x^3}\, - 5\,\boxed{x^3}\, = (5 - 5)\,\boxed{x^3}$
$$= 0x^3$$
$$= 0$$

13. $3x^4 - 3x^4 + 2x^2 = (3 - 3)x^4 + 2x^2$
$$= 0x^4 + 2x^2$$
$$= 2x^2$$

DO EXERCISES 16–18.

Multiplying a term of a polynomial by 1 does not change the polynomial, but it may make it easier to factor.

Examples Collect like terms.

14. $5x^2 + x^2 = 5x^2 + \boxed{1}\,x^2$ Replacing x^2 by $1x^2$
$$= (5 + 1)x^2$$ Using a distributive law
$$= 6x^2$$

15. $5x^4 - 6x^3 - x^4 = 5x^4 - 6x^3 - 1x^4$ $x^4 = 1x^4$
$$= (5 - 1)x^4 - 6x^3$$
$$= 4x^4 - 6x^3$$

DO EXERCISES 19–22.

NAME   CLASS

EXERCISE SET 4.1

⚫ Evaluate each polynomial for $x = 4$.

1. $-5x + 2$ **2.** $-3x + 1$ **3.** $2x^2 - 5x + 7$

4. $3x^2 + x + 7$ **5.** $x^3 - 5x^2 + x$ **6.** $7 - x + 3x^2$

The daily number of accidents involving a driver of age a is approximated by the polynomial

$$0.4a^2 - 40a + 1039.$$

7. Evaluate the polynomial for $a = 18$ to find the number of daily accidents involving an 18-year-old driver.

8. Evaluate the polynomial for $a = 20$ to find the number of daily accidents involving a 20-year-old driver.

Evaluate each polynomial for $x = -1$.

9. $3x + 5$ **10.** $6 - 2x$

11. $x^2 - 2x + 1$ **12.** $5x - 6 + x^2$

13. $-3x^3 + 7x^2 - 3x - 2$ **14.** $-2x^3 - 5x^2 + 4x + 3$

⚫⚫ Identify the terms of each polynomial.

15. $2 - 3x + x^2$ **16.** $2x^2 + 3x - 4$

⚫⚫⚫ Identify the like terms in each polynomial.

17. $5x^3 + 6x^2 - 3x^2$ **18.** $3x^2 + 4x^3 - 2x^2$

19. $2x^4 + 5x - 7x - 3x^4$ **20.** $-3t + t^3 - 2t - 5t^3$

⚫⚫ Collect like terms.

21. $2x - 5x$ **22.** $2x^2 + 8x^2$

23. $x - 9x$ **24.** $x - 5x$

ANSWERS

1. _____
2. _____
3. _____
4. _____
5. _____
6. _____
7. _____
8. _____
9. _____
10. _____
11. _____
12. _____
13. _____
14. _____
15. _____
16. _____
17. _____
18. _____
19. _____
20. _____
21. _____
22. _____
23. _____
24. _____

ANSWERS

25. _____

26. _____

27. _____

28. _____

29. _____

30. _____

31. _____

32. _____

33. _____

34. _____

35. _____

36. _____

25. $5x^3 + 6x^3 + 4$

26. $6x^4 - 2x^4 + 5$

27. $5x^3 + 6x - 4x^3 - 7x$

28. $3a^4 - 2a + 2a + a^4$

29. $6b^5 + 3b^2 - 2b^5 - 3b^2$

30. $2x^2 - 6x + 3x + 4x^2$

31. $\frac{1}{4}x^5 - 5 + \frac{1}{2}x^5 - 2x - 37$

32. $\frac{1}{3}x^3 + 2x - \frac{1}{6}x^3 + 4 - 16$

33. $6x^2 + 2x^4 - 2x^2 - x^4 - 4x^2$

34. $8x^2 + 2x^3 - 3x^3 - 4x^2 - 4x^2$

35. $\frac{1}{4}x^3 - x^2 - \frac{1}{6}x^2 + \frac{3}{8}x^3 + \frac{5}{16}x^3$

36. $\frac{1}{5}x^4 + \frac{1}{5} - 2x^2 + \frac{1}{10} - \frac{3}{15}x^4 + 2x^2 - \frac{3}{10}$

4.2   MORE ON POLYNOMIALS

◦ DESCENDING ORDER

This polynomial is arranged in *descending order:*

$$8x^4 - 2x^3 + 5x^2 - x + 3.$$

The term with the largest exponent is first. The term with the next largest exponent is second, and so on. The associative and commutative laws allow us to arrange the terms of a polynomial in descending order.

Examples   Arrange each polynomial in descending order.

1. $4x^5 + 4x^7 + x^2 + 2x^3 = 4x^7 + 4x^5 + 2x^3 + x^2$

2. $3 + 4x^5 - 4x^2 + 5x + 3x^3 = 4x^5 + 3x^3 - 4x^2 + 5x + 3$

We usually arrange polynomials in descending order. The opposite order is called *ascending.*

DO EXERCISES 1–3.

◦ ◦ COLLECTING LIKE TERMS AND DESCENDING ORDER

Example 3   Collect like terms and then arrange in descending order.

$$2x^2 - 4x^3 + 3 - x^2 - 2x^3 = x^2 - 6x^3 + 3 \qquad \text{Collecting like terms}$$
$$= -6x^3 + x^2 + 3 \qquad \text{Descending order}$$

DO EXERCISES 4 AND 5.

◦ ◦ ◦ DEGREES

The *degree* of a term is its exponent.

Example 4   Identify the degree of each term of $8x^4 + 3x + 7$.

The degree of $8x^4$ is 4.
The degree of $3x$ is 1.   Recall that $x = x^1$.
The degree of 7 is 0.   Think of 7 as $7x^0$. Recall that $x^0 = 1$.

The *degree of a polynomial* is its largest exponent.

Example 5   Identify the degree of $3x^4 - 6x^3 + 7$.

$3x^4 - 6x^3 + 7$ The largest exponent is 4.

The degree of the polynomial is 4 .

DO EXERCISE 6.

OBJECTIVES

After finishing Section 4.2, you should be able to:

◦ Arrange a polynomial in descending order.

◦ ◦ Collect the like terms of a polynomial and arrange in descending order.

◦ ◦ ◦ Identify the degrees of terms of polynomials and degrees of polynomials.

◦ ◦ Identify the coefficients of the terms of a polynomial.

◦ ◦ ◦ Identify the missing terms of a polynomial.

◦ ◦ ◦ Tell whether a polynomial is a monomial, binomial, trinomial, or none of these.

Arrange each polynomial in descending order.

1. $x + 3x^5 + 4x^3 + 5x^2 + 6x^7 - 2x^4$

2. $4x^2 - 3 + 7x^5 + 2x^3 - 5x^4$

3. $-14 + 7t^2 - 10t^5 + 14t^7$

Collect like terms and then arrange in descending order.

4. $3x^2 - 2x + 3 - 5x^2 - 1 - x$

5. $-x + \dfrac{1}{2} + 14x^4 - 7x - 1 - 4x^4$

Identify the degree of each term and the degree of the polynomial.

6. $-6x^4 + 8x^2 - 2x + 9$

Identify the coefficient of each term.

7. $5x^9 + 6x^3 + x^2 - x + 4$

Identify the missing terms in each polynomial.

8. $2x^3 + 4x^2 - 2$

9. $-3x^4$

10. $x^3 + 1$

11. $x^4 - x^2 + 3x + 0.25$

Tell whether each polynomial is a monomial, binomial, trinomial, or none of these.

12. $5x^4$

13. $4x^3 - 3x^2 + 4x + 2$

14. $3x^2 + x$

15. $3x^2 + 2x - 4$

⠒⠒ COEFFICIENTS

In this polynomial the colored numbers are the *coefficients*:

$$3x^5 - 2x^3 + 5x + 4.$$

Example 6 Identify the coefficient of each term in the polynomial

$$3x^4 - 4x^3 + 7x^2 + x - 8.$$

The coefficient of the first term is 3 .

The coefficient of the second term is -4 .

The coefficient of the third term is 7 .

The coefficient of the fourth term is 1 .

The coefficient of the fifth term is -8 .

DO EXERCISE 7.

⠖⠖ MISSING TERMS

If a coefficient is 0, we usually do not write the term. We say that we have a *missing term*.

Example 7 In

$$8x^5 - 2x^3 + 5x^2 + 7x + 8,$$

there is no term with x^4. We say that the x^4 term (or the *fourth-degree term*) is missing.

We could write missing terms with zero coefficients or leave space. For example,

$$3x^2 + 9 = 3x^2 + 0x + 9 = 3x^2 + \qquad 9,$$

but it is shorter not to write missing terms or to leave space.

DO EXERCISES 8–11.

⠿ MONOMIALS, BINOMIALS, AND TRINOMIALS

Polynomials with just one term are called *monomials*. Polynomials with just two terms are called *binomials*. Those with just three terms are called *trinomials*.

Example 8

Monomials	*Binomials*	*Trinomials*
$4x^2$	$2x + 4$	$3x^3 + 4x + 7$
9	$3x^5 + 6x$	$6x^7 - 7x^2 + 4$
$-23x^{19}$	$-9x^7 - 6$	$4x^2 - 6x - \frac{1}{2}$

DO EXERCISES 12–15.

NAME   CLASS

EXERCISE SET 4.2

■ Arrange each polynomial in descending order.

1. $x^5 + x + 6x^3 + 1 + 2x^2$ **2.** $3 + 2x^2 - 5x^6 - 2x^3 + 3x$

1. _____

2. _____

3. _____

3. $5x^3 + 15x^9 + x - x^2 + 7x^8$ **4.** $9x - 5 + 6x^3 - 5x^4 + x^5$

4. _____

5. _____

5. $8y^3 - 7y^2 + 9y^6 - 5y^8 + y^7$ **6.** $p^8 - 4 + p + p^2 - 7p^4$

6. _____

■ ■ Collect like terms and then arrange in descending order.

7. _____

7. $3x^4 - 5x^6 - 2x^4 + 6x^6$ **8.** $-1 + 5x^3 - 3 - 7x^3 + x^4 + 5$

8. _____

9. _____

9. $-2x + 4x^3 - 7x + 9x^3 + 8$ **10.** $-6x^2 + x - 5x + 7x^2 + 1$

10. _____

11. _____

11. $3x + 3x + 3x - x^2 - 4x^2$ **12.** $-2x - 2x - 2x + x^3 - 5x^3$

12. _____

13. $-x + \dfrac{3}{4} + 15x^4 - x - \dfrac{1}{2} - 3x^4$ **14.** $2x - \dfrac{5}{6} + 4x^3 + x + \dfrac{1}{3} - 2x$

13. _____

■ ■ ■ Identify the degree of each term of each polynomial and the degree of the polynomial.

14. _____

15. _____

15. $-7x^3 + 6x^2 + 3x + 7$ **16.** $5x^4 + x^2 - x + 2$

16. _____

17. _____

17. $x^2 - 3x + x^6 - 9x^4$ **18.** $8x - 3x^2 + 9 - 8x^3$

18. _____

ANSWERS

19. _____

20. _____

21. _____

22. _____

23. _____

24. _____

25. _____

26. _____

27. _____

28. _____

29. _____

30. _____

31. _____

32. _____

33. _____

34. _____

35. _____

36. _____

37. _____

38. _____

⠛ Identify the coefficient of each term of each polynomial.

19. $-3x + 6$

20. $3x^2 - 5x + 2$

21. $6x^3 + 7x^2 - 8x - 2$

22. $-2 + 8x - 3x^2 + 6x^3 - 5x^4$

⠶ Identify the missing terms in each polynomial.

23. $x^3 - 27$

24. $x^5 + x$

25. $x^4 - x$

26. $5x^4 - 7x + 2$

27. $2x^3 - 5x^2 + x - 3$

28. $-6x^3$

⠿ Tell whether each polynomial is a monomial, binomial, trinomial, or none of these.

29. $x^2 - 10x + 25$

30. $-6x^4$

31. $x^3 - 7x^2 + 2x - 4$

32. $x^2 - 9$

33. $4x^2 - 25$

34. $2x^4 - 7x^3 + x^2 + x - 6$

35. $40x$

36. $4x^2 + 12x + 9$

---

37. A 4-ft by 4-ft sandbox is placed on a square lawn x ft on a side. Express the area left over as a polynomial.

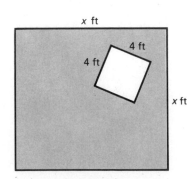

38. Express the colored area in the figure below as a polynomial.

4.3 ADDITION OF POLYNOMIALS

●

To *add* polynomials we can write a plus sign between them and collect like terms.

Example 1   Add: $-3x^3 + 2x - 4$ and $4x^3 + 3x^2 + 2$.

$(-3x^3 + 2x - 4) + (4x^3 + 3x^2 + 2)$

$= (-3 + 4)x^3 + 3x^2 + 2x + (-4 + 2)$ Collecting like terms. *No signs* are changed.

$= x^3 + 3x^2 + 2x - 2$

We usually arrange the polynomial in descending order.

Example 2   Add: $\frac{2}{3}x^4 + 3x^2 - 2x + \frac{1}{2}$ and $-\frac{1}{3}x^4 + 5x^3 - 3x^2 + 3x - \frac{1}{2}$.

$\left(\frac{2}{3}x^4 + 3x^2 - 2x + \frac{1}{2}\right) + \left(-\frac{1}{3}x^4 + 5x^3 - 3x^2 + 3x - \frac{1}{2}\right)$

$= \left(\frac{2}{3} - \frac{1}{3}\right)x^4 + 5x^3 + (3 - 3)x^2$

$\qquad + (-2 + 3)x + \left(\frac{1}{2} - \frac{1}{2}\right)$ Collecting like terms

$= \frac{1}{3}x^4 + 5x^3 + x$

We can add polynomials as we do because they represent numbers.

DO EXERCISES 1–4.

After some practice you will be able to add mentally.

Example 3   Add: $3x^2 - 2x + 2$ and $5x^3 - 2x^2 + 3x - 4$.

$(3x^2 - 2x + 2) + (5x^3 - 2x^2 + 3x - 4)$

$= 5x^3 + (3 - 2)x^2 + (-2 + 3)x + (2 - 4)$ You might do this step mentally.

$= 5x^3 + x^2 + x - 2$ Then you would write only this.

DO EXERCISES 5 AND 6.

We can also add polynomials by writing like terms in columns.

Example 4   Add: $9x^5 - 2x^3 + 6x^2 + 3$ and $5x^4 - 7x^2 + 6$ and $3x^6 - 5x^5 + x^2 + 5$.

Arrange the polynomials with like terms in columns.

$$\begin{array}{l}
9x^5 \qquad\quad -2x^3 + 6x^2 \quad\;\; + 3 \\
\qquad\quad 5x^4 \qquad\qquad -7x^2 + 6 \\
\underline{3x^6 - 5x^5 \qquad\qquad\qquad + \;\; x^2 + 5} \\
3x^6 + 4x^5 + 5x^4 - 2x^3 \qquad\quad + 14
\end{array}$$

We leave spaces for missing terms.

OBJECTIVES

After finishing Section 4.3, you should be able to:

● Add polynomials.

●● Solve problems using addition of polynomials.

Add.

1. $3x^2 + 2x - 2$ and $-2x^2 + 5x + 5$

2. $-4x^5 + 3x^3 + 4$ and $7x^4 + 2x^2$

3. $31x^4 + x^2 + 2x - 1$ and $-7x^4 + 5x^3 - 2x + 2$

4. $17x^3 - x^2 + 3x + 4$ and

$\qquad -15x^3 + x^2 - 3x - \frac{2}{3}$

Add mentally. Try to just write the answer.

5. $(4x^2 - 5x + 3) + (-2x^2 + 2x - 4)$

6. $(3x^3 - 4x^2 - 5x + 3) +$

$\qquad \left(5x^3 + 2x^2 - 3x - \frac{1}{2}\right)$

ANSWERS ON PAGE A-10

7. Add.

$$-2x^3 + 5x^2 - 2x + 4$$
$$x^4 \qquad + 6x^2 + 7x - 10$$
$$-9x^4 + 6x^3 + x^2 \qquad - 2$$

8. Add $-3x^3 + 5x + 2$ and
$x^3 + x^2 + 5$ and $x^3 - 2x - 4$.

9. Find the sum of the areas of these rectangles.

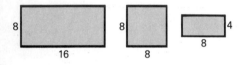

10. Find the sum of the areas of these rectangles.

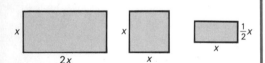

11. Find the sum of the areas in Exercise 9 by substituting 8 for x in the polynomial of Exercise 10.

DO EXERCISES 7 AND 8.

●● PROBLEMS

Suppose we want to find the sum of the areas of these rectangles.

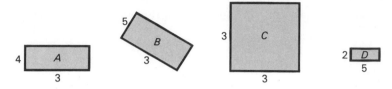

We can proceed as follows:

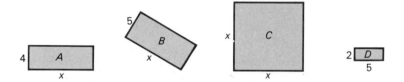

area of A	plus	area of B	plus	area of C	plus	area of D
$4 \cdot 3$	$+$	$5 \cdot 3$	$+$	$3 \cdot 3$	$+$	$2 \cdot 5$

$$= 12 + 15 + 9 + 10$$
$$= 46.$$

Now suppose certain sides were unknown, but represented by a variable x.

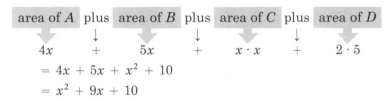

The sum of the areas is found as follows:

area of A	plus	area of B	plus	area of C	plus	area of D
$4x$	$+$	$5x$	$+$	$x \cdot x$	$+$	$2 \cdot 5$

$$= 4x + 5x + x^2 + 10$$
$$= x^2 + 9x + 10$$

Note that we could solve the first problem by replacing x by 3 in the preceding problem:

$$x^2 + 9x + 10 = 3^2 + 9 \cdot 3 + 10 = 9 + 27 + 10 = 46.$$

The polynomial is a formula for the sum of the areas of the rectangles with certain sides of length x. Thus we can substitute any length—say 8, 4, 78.6—for x and find the sum of the areas. This illustrates how we can manipulate unknown numbers in problems.

DO EXERCISES 9–11.

NAME   CLASS

ANSWERS

EXERCISE SET 4.3

● Add.

1. $3x + 2$ and $-4x + 3$

2. $5x^2 + 6x + 1$ and $-7x + 2$

3. $-6x + 2$ and $x^2 + x - 3$

4. $6x^4 + 3x^3 - 1$ and $4x^2 - 3x + 3$

5. $3x^5 + 6x^2 - 1$ and $7x^2 + 6x - 2$

6. $7x^3 + 3x^2 + 6x$ and $-3x^2 - 6$

7. $-4x^4 + 6x^2 - 3x - 5$ and $6x^3 + 5x + 9$

8. $5x^3 + 6x^2 - 3x + 1$ and $5x^4 - 6x^3 + 2x - 5$

9. $(7x^3 + 6x^2 + 4x + 1) + (-7x^3 + 6x^2 - 4x + 5)$

10. $(3x^4 - 5x^2 - 6x + 5) + (-4x^3 + 6x^2 + 7x - 1)$

ANSWERS

1. _____

2. _____

3. _____

4. _____

5. _____

6. _____

7. _____

8. _____

9. _____

10. _____

ANSWERS

11. _____

12. _____

13. _____

14. _____

15. _____

16. _____

17. _____

18. _____

19. _____

20. _____

21. _____

22. _____

23. _____

24. _____

25. _____

26. _____

11. $5x^4 - 6x^3 - 7x^2 + x - 1$ and $4x^3 - 6x + 1$

12. $8x^5 - 6x^3 + 6x + 5$ and $-4x^4 + 3x^3 - 7x$

13. $9x^8 - 7x^4 + 2x^2 + 5$ and $8x^7 + 4x^4 - 2x$

14. $4x^5 - 6x^3 - 9x + 1$ and $6x^3 + 9x^2 + 9x$

15. $\dfrac{1}{4}x^4 + \dfrac{2}{3}x^3 + \dfrac{5}{8}x^2 + 7$ and $-\dfrac{3}{4}x^4 + \dfrac{3}{8}x^2 - 7$

16. $\left(\dfrac{1}{3}x^9 + \dfrac{1}{5}x^5 - \dfrac{1}{2}x^2 + 7\right) + \left(-\dfrac{1}{5}x^9 + \dfrac{1}{4}x^4 - \dfrac{3}{5}x^5 + \dfrac{3}{4}x^2 + \dfrac{1}{2}\right)$

17. $0.02x^5 - 0.2x^3 + x + 0.08$ and $-0.01x^5 + x^4 - 0.8x - 0.02$

18. $(0.03x^6 + 0.05x^3 + 0.22x + 0.05) + \left(\dfrac{7}{100}x^6 - \dfrac{3}{100}x^3 + 0.5\right)$

19. $-3x^4 + 6x^2 + 2x - 1$
$\underline{ - 3x^2 + 2x + 1}$

20. $-4x^3 + 8x^2 + 3x - 2$
$\underline{ - 4x^2 + 3x + 2}$

21. $3x^5 - 6x^3 + 3x$
$\underline{ - 3x^4 + 3x^3 + x^2}$

22. $4x^5 - 5x^3 + 2x$
$\underline{ - 4x^4 + 2x^3 + 2x^2}$

23. $ - 3x^2 + x$
$5x^3 - 6x^2 + 1$
$\underline{ 3x - 8}$

24. $ - 4x^2 + 2x$
$3x^3 - 5x^2 + 3$
$\underline{ 5x - 5}$

25. $-\dfrac{1}{2}x^4 - \dfrac{3}{4}x^3 + 6x$
$ \dfrac{1}{2}x^3 + x^2 + \dfrac{1}{4}x$
$\underline{ \dfrac{3}{4}x^4 + \dfrac{1}{2}x^2 + \dfrac{1}{2}x + \dfrac{1}{4}}$

26. $-\dfrac{1}{4}x^4 - \dfrac{1}{2}x^3 + 2x$
$ \dfrac{3}{4}x^3 - x^2 + \dfrac{1}{2}x$
$\underline{ \dfrac{1}{2}x^4 + \dfrac{1}{2}x^2 + \dfrac{1}{2}x + \dfrac{1}{2}}$

27.

$$
\begin{array}{r}
-4x^2 \\
4x^4 - 3x^3 + 6x^2 + 5x \\
6x^3 - 8x^2 \qquad + 1 \\
-5x^4 \\
\hline
6x^2 - 3x \\
\end{array}
$$

28.

$$
\begin{array}{r}
3x^2 \\
5x^4 - 2x^3 + 4x^2 + 5x \\
5x^3 - 5x^2 \qquad + 2 \\
-7x^4 \\
\hline
3x^2 - 2x \\
\end{array}
$$

29.

$$
\begin{array}{r}
3x^4 - 6x^2 + 7x \\
3x^2 - 3x + 1 \\
-2x^4 + 7x^2 + 3x \\
5x - 2 \\
\hline
\end{array}
$$

30.

$$
\begin{array}{r}
5x^4 - 8x^2 + 4x \\
5x^2 - 2x + 3 \\
-3x^4 + 3x^2 + 5x \\
3x - 5 \\
\hline
\end{array}
$$

31.

$$
\begin{array}{r}
3x^5 - 6x^4 + 3x^3 \qquad - 1 \\
6x^4 - 4x^3 + 6x^2 \\
3x^5 \qquad + 2x^3 \\
- 6x^4 \qquad - 7x^2 \\
-5x^5 \qquad + 3x^3 \qquad + 2 \\
\hline
\end{array}
$$

32.

$$
\begin{array}{r}
4x^5 - 3x^4 + 2x^3 \qquad - 2 \\
6x^4 + 5x^3 + 3x^2 \\
5x^5 \qquad + 4x^3 \\
- 6x^4 \qquad - 5x^2 \\
-3x^5 \qquad + 2x^3 \qquad + 5 \\
\hline
\end{array}
$$

33.

$$
\begin{array}{r}
- x^3 + 6x^2 + 3x + 5 \\
x^4 \qquad - 3x^2 \qquad + 2 \\
- 5x + 3 \\
6x^4 \qquad + 4x^2 \qquad - 1 \\
- x^3 \qquad + 6x \\
\hline
\end{array}
$$

34.

$$
\begin{array}{r}
- 2x^3 + 3x^2 + 5x + 3 \\
x^4 \qquad - 5x^2 \qquad + 1 \\
- 7x + 4 \\
4x^4 \qquad + 6x^2 \qquad - 2 \\
- x^3 \qquad + 5x \\
\hline
\end{array}
$$

35.

$$
\begin{array}{r}
- 3x^4 + 6x^3 - 6x^2 + 5x + 1 \\
5x^5 \qquad - 3x^3 \qquad - 5x \\
4x^4 + 7x^3 \qquad + 3x + 1 \\
-2x^3 \qquad + 7x^2 \qquad - 8 \\
\hline
\end{array}
$$

ANSWERS

27. _____

28. _____

29. _____

30. _____

31. _____

32. _____

33. _____

34. _____

35. _____

ANSWERS

36.
$$
\begin{array}{r}
-5x^4 + 4x^3 - 7x^2 + 3x + 2 \\
3x^5 \qquad -7x^3 \qquad -6x \\
3x^4 + 5x^3 \qquad +5x - 3 \\
-5x^5 \qquad +10x^2 \qquad +4
\end{array}
$$

36. _____

37.
$$
\begin{array}{r}
0.15x^4 + 0.10x^3 - 0.9x^2 \\
-0.01x^3 + 0.01x^2 + x \\
1.25x^4 \qquad +0.11x^2 \qquad +0.01 \\
0.27x^3 \qquad +0.99 \\
-0.35x^4 \qquad +15x^2 \qquad -0.03
\end{array}
$$

37. _____

38.
$$
\begin{array}{r}
0.05x^4 + 0.12x^3 - 0.5x^2 \\
-0.02x^3 + 0.02x^2 + 2x \\
1.5x^4 \qquad +0.01x^2 \qquad +0.15 \\
0.25x^3 \qquad +0.85 \\
-0.25x^4 \qquad +10x^2 \qquad -0.04
\end{array}
$$

38. _____

●● Solve.

39. a) Express the sum of the areas of these rectangles as a polynomial.

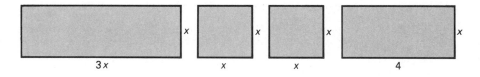

39. _____

b) Find the sum of the areas when $x = 3$ and $x = 8$.

40. a) Express the sum of the areas of these circles as a polynomial.

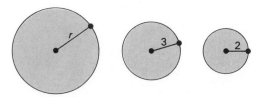

40. _____

b) Find the sum of the areas when $r = 5$ and $r = 11.3$.

41. _____

41. 🖩 Add: $(-20.344x^6 - 70.789x^5 + 890x) + (68.888x^6 + 69.994x^5)$.

4.4 SUBTRACTION OF POLYNOMIALS

● ADDITIVE INVERSES

Look for a pattern:

a) $2x + (-2x) = 0$;

b) $-6x^2 + 6x^2 = 0$;

c) $(5t^3 + 2) + (-5t^3 - 2) = 0$;

d) $(7x^3 - 6x^2 - x + 4) + (-7x^3 + 6x^2 + x - 4) = 0$.

If the sum of two polynomials is 0, they are additive inverses of each other.

The additive inverse of a polynomial is found by replacing each coefficient by its additive inverse.

Example 1 Find the additive inverse of

$$4x^5 - 7x^3 - 8x + \frac{5}{6}.$$

If we change the sign of *every* term, the inverse is $-4x^5 + 7x^3 + 8x - \frac{5}{6}$.

DO EXERCISES 1–4.

●● SYMBOLS FOR ADDITIVE INVERSES

The additive inverse of the polynomial $8x^2 - 4x + 3$ is the polynomial obtained by replacing each coefficient with its additive inverse. Thus, the inverse is

$$-8x^2 + 4x - 3.$$

We can also represent the additive inverse of $8x^2 - 4x + 3$ as follows:

$$-(8x^2 - 4x + 3).$$

Thus,

$$-(8x^2 - 4x + 3) = -8x^2 + 4x - 3.$$

Example 2 Simplify: $-\left(-7x^4 - \frac{5}{9}x^3 + 8x^2 - x + 67\right).$

$$-\left(-7x^4 - \frac{5}{9}x^3 + 8x^2 - x + 67\right) = 7x^4 + \frac{5}{9}x^3 - 8x^2 + x - 67$$

DO EXERCISES 5–7.

OBJECTIVES

After finishing Section 4.4, you should be able to:

● Find the additive inverse of a polynomial.

●● Rename an additive inverse such as $-(3x^2 + x - 2)$, by replacing each coefficient by its inverse.

●●● Subtract polynomials.

Find the additive inverse of each polynomial.

1. $12x^4 - 3x^2 + 4x$

2. $-4x^4 + 3x^2 - 4x$

3. $-13x^6 + 2x^4 - 3x^2 + x - \frac{5}{13}$

4. $-7y^3 + 2y^2 - y + 3$

Simplify.

5. $-(4x^3 - 6x + 3)$

6. $-(5x^4 + 3x^2 + 7x - 5)$

7. $-\left(14x^{10} - \frac{1}{2}x^5 + 5x^3 - x^2 + 3x\right)$

Subtract.

8. $(7x^3 + 2x + 4) - (5x^3 - 4)$

9. $(-3x^2 + 5x - 4) - (-4x^2 + 11x - 2)$

Subtract mentally. Try to write just the answer.

10. $(-6x^4 + 3x^2 + 6) - (2x^4 + 5x^3 - 5x^2 + 7)$

11. $\left(\dfrac{3}{2}x^3 - \dfrac{1}{2}x^2 + 0.3\right)$

  $\quad - \left(\dfrac{1}{2}x^3 + \dfrac{1}{2}x^2 + \dfrac{4}{3}x + 1.2\right)$

12. Subtract the second polynomial from the first. Use columns.

  $4x^3 + 2x^2 - 2x - 3,\ 2x^3 - 3x^2 + 2$

13. Subtract.

  $\quad\quad 2x^3 + x^2 - 6x + 2$
  $x^5 + 4x^3 - 2x^2 - 4x$

●●● SUBTRACTION OF POLYNOMIALS

Recall that we can subtract a rational number by adding its additive inverse: $a - b = a + (-b)$. This also works for polynomials because they represent numbers.

Example 3   Subtract: $(9x^5 + x^3 - 2x^2 + 4) - (2x^5 + x^4 - 4x^3 - 3x^2)$.

$(9x^5 + x^3 - 2x^2 + 4) - (2x^5 + x^4 - 4x^3 - 3x^2)$

$\quad = (9x^5 + x^3 - 2x^2 + 4)$

$\quad\quad + [-(2x^5 + x^4 - 4x^3 - 3x^2)]$ Adding an inverse

$\quad = (9x^5 + x^3 - 2x^2 + 4)$

$\quad\quad + (-2x^5 - x^4 + 4x^3 + 3x^2)$ Finding the inverse by changing the sign of *every* term

$\quad = 7x^5 - x^4 + 5x^3 + x^2 + 4$ Collecting like terms

DO EXERCISES 8 AND 9.

After some practice you will be able to subtract mentally.

Example 4   Subtract: $(9x^5 + x^3 - 2x) - (-2x^5 + 5x^3 + 6)$.

$(9x^5 + x^3 - 2x) - (-2x^5 + 5x^3 + 6)$

$\quad = (9x^5 + 2x^5) + (x^3 - 5x^3) - 2x - 6$ Subtract the like terms mentally.

$\quad = 11x^5 - 4x^3 - 2x - 6$ Write only this.

> *Caution!* If you make errors by trying to subtract mentally, then don't do it!

DO EXERCISES 10 AND 11.

We can use columns to subtract. We replace coefficients by their inverses, as shown in Example 2. You can also do it mentally.

Example 5   Subtract: $(5x^2 - 3x + 6) - (9x^2 - 5x - 3)$.

a) $5x^2 - 3x + 6$ Writing similar terms in columns
  $9x^2 - 5x - 3$

b) $\quad 5x^2 - 3x + 6$
  $\underline{-9x^2 \pm 5x \pm 3}$ Changing signs

c) $\quad 5x^2 - 3x + 6$
  $\underline{-9x^2 + 5x + 3}$ Adding
  $-4x^2 + 2x + 9$

If you can do so without error, you should skip step (b). Just write the answer.

Example 6   Subtract: $(x^3 + x^2 + 2x - 12) - (2x^3 + x^2 - 3x)$.

  $\quad x^3 + x^2 + 2x - 12$
  $\underline{2x^3 + x^2 - 3x}$
  $-x^3 \quad\quad\quad + 5x - 12$

DO EXERCISES 12 AND 13.

NAME   CLASS   ANSWERS

EXERCISE SET 4.4

⚫ Find the additive inverse of each polynomial.

1. $-5x$

2. $x^2 - 3x$

3. $-x^2 + 10x - 2$

4. $-4x^3 - x^2 - x$

5. $12x^4 - 3x^3 + 3$

6. $4x^3 - 6x^2 - 8x + 1$

⚫⚫ Simplify.

7. $-(3x - 7)$

8. $-(-2x + 4)$

9. $-(4x^2 - 3x + 2)$

10. $-(-6a^3 + 2a^2 - 9a + 1)$

11. $-\left(-4x^4 - 6x^2 + \dfrac{3}{4}x - 8\right)$

12. $-(-5x^4 + 4x^3 - x^2 + 0.9)$

⚫⚫⚫ Subtract.

13. $(5x^2 + 6) - (3x^2 - 8)$

14. $(7x^3 - 2x^2 + 6) - (7x^2 + 2x - 4)$

15. $(6x^5 - 3x^4 + x + 1) - (8x^5 + 3x^4 - 1)$

16. $\left(\dfrac{1}{2}x^2 - \dfrac{3}{2}x + 2\right) - \left(\dfrac{3}{2}x^2 + \dfrac{1}{2}x - 2\right)$

17. $(6x^2 + 2x) - (-3x^2 - 7x + 8)$

18. $7x^3 - (-3x^2 - 2x + 1)$

19. $\left(\dfrac{5}{8}x^3 - \dfrac{1}{4}x - \dfrac{1}{3}\right) - \left(-\dfrac{1}{8}x^3 + \dfrac{1}{4}x - \dfrac{1}{3}\right)$

20. $\left(\dfrac{1}{5}x^3 + 2x^2 - 0.1\right) - \left(-\dfrac{2}{5}x^3 + 2x^2 + 0.01\right)$

ANSWERS

1. _____

2. _____

3. _____

4. _____

5. _____

6. _____

7. _____

8. _____

9. _____

10. _____

11. _____

12. _____

13. _____

14. _____

15. _____

16. _____

17. _____

18. _____

19. _____

20. _____

ANSWERS

21. _____

22. _____

23. _____

24. _____

25. _____

26. _____

27. _____

28. _____

29. _____

30. _____

31. _____

32. _____

33. _____

21. $(0.08x^3 - 0.02x^2 + 0.01x) - (0.02x^3 + 0.03x^2 - 1)$

22. $(0.8x^4 + 0.2x - 1) - \left(\dfrac{7}{10}x^4 + \dfrac{1}{5}x - 0.1\right)$

Subtract.

23. $\begin{array}{l} x^2 + 5x + 6 \\ \underline{x^2 + 2x} \end{array}$

24. $\begin{array}{l} x^3 \qquad\; + 1 \\ \underline{x^3 + x^2} \end{array}$

25. $\begin{array}{l} x^4 \qquad\quad - 3x^2 + x + 1 \\ \underline{x^4 - 4x^3} \end{array}$

26. $\begin{array}{l} 3x^2 - 6x + 1 \\ \underline{6x^2 + 8x - 3} \end{array}$

27. $\begin{array}{l} \;\;5x^4 + \;\;\;6x^3 - 9x^2 \\ \underline{-6x^4 \quad\; -6x^3 \qquad\quad + 8x + 9} \end{array}$

28. $\begin{array}{l} 5x^4 \qquad\quad + 6x^2 - 3x + 6 \\ \underline{\quad\;\; 6x^3 + 7x^2 - 8x - 9} \end{array}$

29. $\begin{array}{l} \qquad\;\;\; 3x^4 + 6x^2 + 8x - 1 \\ \underline{4x^5 - 6x^4 \qquad\quad - 8x - 7} \end{array}$

30. $\begin{array}{l} \;\;6x^5 \qquad\quad + 3x^2 - 7x + 2 \\ \underline{10x^5 + 6x^3 - 5x^2 - 2x + 4} \end{array}$

31. $\begin{array}{l} x^5 \qquad\qquad\qquad - 1 \\ \underline{x^5 - x^4 + x^3 - x^2 + x - 1} \end{array}$

32. $\begin{array}{l} x^5 + x^4 - x^3 + x^2 - x + 2 \\ \underline{x^5 - x^4 + x^3 - x^2 - x + 2} \end{array}$

Subtract.

33. ▦ $(345.099x^3 - 6.178x) - (-224.508x^3 + 8.99x)$

4.5 MULTIPLICATION OF POLYNOMIALS

▪▫ MULTIPLYING MONOMIALS

To multiply two monomials, we multiply the coefficients and then use properties of exponents. We use parentheses to show multiplication. $(3x)(4x)$ means $(3x) \cdot (4x)$.

Examples   Multiply.

1. $(\boxed{3}\,x)(\boxed{4}\,x) = (\boxed{3 \cdot 4})(x \cdot x)$ Multiplying the coefficients

   $\qquad = 12x^2$ Simplifying

2. $(3x)(-x) = (3x)(-1x)$

   $\qquad = (3)(-1)(x \cdot x)$

   $\qquad = -3x^2$

3. $(-7x^5)(4x^3) = (-7 \cdot 4)(x^5 \cdot x^3)$

   $\qquad = -28x^{5+3}$

   $\qquad = -28x^8$ Adding exponents and simplifying

After some practice you can do this mentally. Multiply the coefficients and add the exponents. Write only the answer.

DO EXERCISES 1–8.

▪▪ MULTIPLYING A MONOMIAL AND A BINOMIAL

Multiplications are based on the distributive laws.

Example 4   Multiply: $2x$ and $5x + 3$.

$(\boxed{2x})(5x + 3) = (\boxed{2x})(5x) + (\boxed{2x})(3)$ Using a distributive law

$\qquad = 10x^2 + 6x$ Multiplying the monomials

DO EXERCISES 9 AND 10.

▪▪▪ MULTIPLYING TWO BINOMIALS

To multiply two binomials, we use the distributive laws more than once.

Example 5   Multiply: $x + 5$ and $x + 4$.

$(\boxed{x+5})(x+4) = (\boxed{x+5})x + (\boxed{x+5})4$ Using a distributive law

$\qquad\quad$ (a) $\qquad\qquad$ (b)

We now use a distributive law with parts (a) and (b).

OBJECTIVES

After finishing Section 4.5, you should be able to:

▪▫ Multiply two monomials.

▪▪ Multiply a monomial and a binomial.

▪▪▪ Multiply two binomials.

▪▪ ▪▪ Multiply a binomial and a trinomial.

▪▪▪ ▪▪▪ Multiply any polynomials.

Multiply.

1. $3x$ and -5

2. $-x$ and x

3. $-x$ and $-x$

4. $-x^2$ and x^3

5. $3x^5$ and $4x^2$

6. $4x^5$ and $-2x^6$

7. $-7y^4$ and $-y$

8. $7x^5$ and 0

Multiply.

9. $4x$ and $2x + 4$

10. $3x^2$ and $-5x + 2$

Multiply.

11. $x + 8$ and $x + 5$

12. $(x + 5)(x - 4)$

Multiply.

13. $5x + 3$ and $x - 4$

14. $(2x - 3)(3x - 5)$

a) $(x + 5) \boxed{x} = x \cdot \boxed{x} + 5 \cdot \boxed{x}$ Distributive law
   $\qquad\qquad = x^2 + 5x$ Multiplying the monomials
b) $(x + 5) \boxed{4} = x \cdot \boxed{4} + 5 \cdot \boxed{4}$ Distributive law
   $\qquad\qquad = 4x + 20$ Multiplying the monomials

Now we replace parts (a) and (b) in the original expression with their answers and collect like terms:

$$(x + 5)(x + 4) = (x^2 + 5x) + (4x + 20)$$
$$= x^2 + 9x + 20.$$

DO EXERCISES 11 AND 12.

Example 6   Multiply: $4x + 3$ and $x - 2$.

$$(\boxed{4x + 3})(x - 2) = (\boxed{4x + 3})x + (\boxed{4x + 3})(-2)$$
$$\qquad\qquad\qquad\quad \text{(a)} \qquad\qquad \text{(b)}$$

Now consider parts (a) and (b).

a) $(4x + 3)\boxed{x} = (4x)\boxed{x} + (3)\boxed{x}$
   $\qquad\qquad = 4x^2 + 3x$
b) $(4x + 3)(\boxed{-2}) = (4x)(\boxed{-2}) + (3)(\boxed{-2})$
   $\qquad\qquad\quad = -8x - 6$

Now we replace parts (a) and (b) in the original expression and collect like terms:

$$(4x + 3)(x - 2) = (4x^2 + 3x) + (-8x - 6)$$
$$= 4x^2 - 5x - 6.$$

DO EXERCISES 13 AND 14.

⬛⬛ MULTIPLYING A BINOMIAL AND A TRINOMIAL

Example 7   Multiply: $(x^2 + 2x - 3)(x^2 + 4)$.

$$(\boxed{x^2 + 2x - 3})(x^2 + 4) = (\boxed{x^2 + 2x - 3})(x^2) + (\boxed{x^2 + 2x - 3})(4)$$
$$\qquad\qquad\qquad\qquad\qquad \text{(a)} \qquad\qquad\qquad \text{(b)}$$

Consider parts (a) and (b).

a) $(x^2 + 2x - 3)\boxed{x^2} = (x^2)(\boxed{x^2}) + 2x(\boxed{x^2}) - 3(\boxed{x^2})$
   $\qquad\qquad\qquad = x^4 + 2x^3 - 3x^2$
b) $(x^2 + 2x - 3)\boxed{4} = (x^2)\boxed{4} + (2x)\boxed{4} + (-3)\boxed{4}$
   $\qquad\qquad\qquad = 4x^2 + 8x - 12$

Now we replace parts (a) and (b) in the original expression and collect like terms:

$$(x^2 + 2x - 3)(x^2 + 4) = (x^4 + 2x^3 - 3x^2) + (4x^2 + 8x - 12)$$
$$= x^4 + 2x^3 + x^2 + 8x - 12.$$

DO EXERCISES 15 AND 16.

▪▪ MULTIPLYING ANY POLYNOMIALS

Perhaps you have discovered the following.

To multiply two polynomials, multiply each term of one by every term of the other. Then add the results.

We may use columns for long multiplications. We multiply each term at the top by every term at the bottom. Then we add.

Example 8   Multiply: $(4x^2 - 2x + 3)(x + 2)$.

$$
\begin{array}{l}
4x^2 - 2x\ + 3 \\
\underline{x\ + 2} \\
4x^3 - 2x^2 + 3x \qquad \text{Multiplying the top row by } x \\
\qquad\quad 8x^2 - 4x + 6 \qquad \text{Multiplying the top row by } 2 \\
\overline{4x^3 + 6x^2 -\ x + 6} \qquad \text{Adding}
\end{array}
$$

Example 9   Multiply: $(5x^3 - 3x + 4)(-2x^2 - 3)$.

$$
\begin{array}{l}
5x^3 - 3x + 4 \\
\underline{-2x^2 - 3} \\
-10x^5 \quad\ + \ 6x^3 - 8x^2 \qquad \text{Multiplying by } -2x^2 \\
\qquad\quad - 15x^3 \qquad\ + 9x - 12 \qquad \text{Multiplying by } -3 \\
\overline{-10x^5 \qquad\ - \ 9x^3 - 8x^2 + 9x - 12} \qquad \text{Adding}
\end{array}
$$

> When we multiplied $-2x^2$ by $-3x$, the power dropped from x^5 to x^3 so we left a space for missing x^4 term.

In addition, we leave space for "missing terms." (Recall, from Section 4.2, that a *missing term* is a term with a 0 coefficient.)

DO EXERCISES 17 AND 18.

Example 10   Multiply: $(2x^2 + 3x - 4)(2x^2 - x + 3)$.

$$
\begin{array}{l}
2x^2 + 3x\ - 4 \\
\underline{2x^2 -\ x\ + 3} \\
4x^4 + 6x^3 - 8x^2 \qquad\qquad \text{Multiplying by } 2x^2 \\
\qquad\ - 2x^3 - 3x^2 +\ 4x \qquad \text{Multiplying by } -x \\
\qquad\qquad\quad 6x^2 +\ 9x - 12 \qquad \text{Multiplying by } 3 \\
\overline{4x^4 + 4x^3 - 5x^2 + 13x - 12} \qquad \text{Adding}
\end{array}
$$

DO EXERCISE 19.

Multiply.

15. $(x^2 + 3x - 4)(x^2 + 5)$

16. $(2x^3 - 2x + 5)(3x^2 - 7)$

Multiply.

17. $\begin{array}{l} 3x^2 - 2x + 4 \\ \underline{x + 5} \end{array}$

18. $\begin{array}{l} -5x^2 + 4x + 2 \\ \underline{-4x^2 - 8} \end{array}$

Multiply.

19. $\begin{array}{l} 3x^2 - 2x - 5 \\ \underline{2x^2 +\ x - 2} \end{array}$

SOMETHING EXTRA
EXPANDED NOTATION AND POLYNOMIALS

The number 6345 can be expressed with *expanded notation* as follows:

$$6345 = 6000 + 300 + 40 + 5$$
$$= 6 \cdot 1000 + 3 \cdot 100 + 4 \cdot 10 + 5$$
$$= 6 \cdot 10^3 + 3 \cdot 10^2 + 4 \cdot 10 + 5. \quad \text{Expanded notation}$$

Note that this is the result of replacing x in the related polynomial

$$6x^3 + 3x^2 + 4x + 5 \quad \text{Related polynomial}$$

by the number 10:

$$6 \cdot 10^3 + 3 \cdot 10^2 + 4 \cdot 10 + 5.$$

For each of the following, write expanded notation and the related polynomial.

1. 8762   **2.** 786   **3.** 16,432   **4.** 7063

NAME   CLASS   ANSWERS

EXERCISE SET 4.5

■ • Multiply.

1. $3x$ and -4 **2.** $4x$ and 5 **3.** $6x^2$ and 7

4. $-4x$ and -3 **5.** $-5x$ and -6 **6.** $3x^2$ and -7

7. x^2 and $-2x$ **8.** $-x^3$ and $-x$ **9.** x^4 and x^2

10. $-x^5$ and x^3 **11.** $3x^4$ and $2x^2$ **12.** $-\dfrac{1}{5}x^3$ and $-\dfrac{1}{3}x$

13. $-4x^4$ and 0 **14.** $7x^5$ and x^5 **15.** $-0.1x^6$ and $0.2x^4$

■ ■ Multiply.

16. $3x$ and $-x + 5$ **17.** $2x$ and $4x - 6$ **18.** $4x^2$ and $3x + 6$

19. $-6x^2$ and $x^2 + x$ **20.** $3x^2$ and $6x^4 + 8x^3$ **21.** $4x^4$ and $x^2 - 6x$

■■■ Multiply.

22. $(x + 6)(x + 3)$ **23.** $(x + 6)(-x + 2)$ **24.** $(x + 3)(x - 3)$

1. _____
2. _____
3. _____
4. _____
5. _____
6. _____
7. _____
8. _____
9. _____
10. _____
11. _____
12. _____
13. _____
14. _____
15. _____
16. _____
17. _____
18. _____
19. _____
20. _____
21. _____
22. _____
23. _____
24. _____

ANSWERS

25. _____

26. _____

27. _____

28. _____

29. _____

30. _____

31. _____

32. _____

33. _____

34. _____

35. _____

36. _____

37. _____

38. _____

39. _____

40. _____

25. $(x - 5)(2x - 5)$

26. $(2x + 5)(2x + 5)$

27. $(3x - 5)(3x + 5)$

28. $(3x + 1)(3x + 1)$

29. $\left(2x - \dfrac{1}{2}\right)\left(x + \dfrac{3}{2}\right)$

30. $(2x + 0.1)(3x - 0.1)$

■■ Multiply.

31. $(x^2 + 6x + 1)(x + 1)$

32. $(2x^3 + 6x + 1)(2x + 1)$

33. $(-5x^2 - 7x + 3)(2x + 1)$

34. $(2x^3 - 5x + 6)(2x + 2)$

■■■ Multiply.

35. $(3x^2 - 6x + 2)(x^2 - 3)$

36. $(x^2 + 6x - 1)(-3x^2 + 2)$

37. $(2t^2 - t - 4)(3t^2 + 2t - 1)$

38. $(3a^2 - 5a + 2)(2a^2 - 3a + 4)$

39. $(x^3 + x^2 + x + 1)(x - 1)$

40. $(x + 2)(x^2 - 2x + 4)$

4.6   SPECIAL PRODUCTS OF POLYNOMIALS

● PRODUCT OF A MONOMIAL AND ANY POLYNOMIAL

There is a quick way to multiply a monomial and any polynomial. Use the distributive law mentally, multiplying every term inside by the monomial. Just write the answer.

Example 1   Multiply: $5x(2x^2 - 3x + 4)$.

$$5x(2x^2 - 3x + 4) = 10x^3 - 15x^2 + 20x$$

DO EXERCISES 1 AND 2.

●● PRODUCTS OF TWO BINOMIALS

To multiply two binomials, we multiply each term of one by every term of the other. We can do it like this.

$$(A + B)(C + D) = AC + AD + BC + BD$$

1. Multiply First terms: AC

2. Multiply Outside terms: AD

3. Multiply Inside terms: BC

4. Multiply Last terms: BD

        ↓
     FOIL   This will help you remember the rule.

Example 2   Multiply: $(x + 8)(x^2 + 5)$.

$$\begin{array}{cccc} F & O & I & L \end{array}$$
$$(x + 8)(x^2 + 5) = x^3 + 5x + 8x^2 + 40$$

Often we can collect like terms after we multiply.

Examples   Multiply.

3. $(x + 6)(x - 6) = x^2 - 6x + 6x - 36$ Using FOIL
$$= x^2 - 36 \qquad \text{Collecting like terms}$$

4. $(x + 3)(x - 2) = x^2 - 2x + 3x - 6$
$$= x^2 + x - 6$$

5. $(x^3 + 5)(x^3 - 5) = x^6 - 5x^3 + 5x^3 - 25$
$$= x^6 - 25$$

6. $(4x^3 + 5)(3x^2 - 2) = 12x^5 - 8x^3 + 15x^2 - 10$

DO EXERCISES 3–9.

Multiply. Just write the answer.

1. $4x(2x^2 - 3x + 4)$

2. $2y^3(5y^3 + 4y^2 - 5y)$

Multiply mentally. Just write the answer.

3. $(x + 3)(x + 4)$

4. $(x + 3)(x - 5)$

5. $(2x + 1)(x + 4)$

6. $(2x^2 - 3)(x - 2)$

7. $(6x^2 + 5)(2x^3 + 1)$

8. $(y^3 + 7)(y^3 - 7)$

9. $(2x^4 + x^2)(-x^3 + x)$

Multiply.

10. $\left(x + \dfrac{4}{5}\right)\left(x - \dfrac{4}{5}\right)$

11. $(x^3 - 0.5)(x^2 + 0.5)$

12. $(2 + 3x^2)(4 - 5x^2)$

13. $(6x^3 - 3x^2)(5x^2 + 2x)$

Examples   Multiply.

7. $\left(x - \dfrac{2}{3}\right)\left(x + \dfrac{2}{3}\right) = x^2 + \dfrac{2}{3}x - \dfrac{2}{3}x - \dfrac{4}{9}$

$$= x^2 - \dfrac{4}{9}$$

8. $(x^2 - 0.3)(x^2 - 0.3) = x^4 - 0.3x^2 - 0.3x^2 + 0.09$
$$= x^4 - 0.6x^2 + 0.09$$

9. $(3 - 4x)(7 - 5x^3) = 21 - 15x^3 - 28x + 20x^4$
$$= 21 - 28x - 15x^3 + 20x^4$$

> If the original polynomials are in ascending order, it is natural to write the product in ascending order, but this is not a "must."

10. $(5x^4 + 2x^3)(3x^2 - 7x) = 15x^6 - 35x^5 + 6x^5 - 14x^4$
$$= 15x^6 - 29x^5 - 14x^4$$

DO EXERCISES 10–13.

 ANSWERS

EXERCISE SET 4.6

⬤ Multiply. Write only the answer.

1. $4x(x + 1)$

2. $3x(x + 2)$

3. $-3x(x - 1)$

4. $-5x(-x - 1)$

5. $x^2(x^3 + 1)$

6. $-2x^3(x^2 - 1)$

7. $3x(2x^2 - 6x + 1)$

8. $-4x(2x^3 - 6x^2 - 5x + 1)$

⬤⬤ Multiply. Write only the answer.

9. $(x + 1)(x^2 + 3)$

10. $(x^2 - 3)(x - 1)$

11. $(x^3 + 2)(x + 1)$

12. $(x^4 + 2)(x + 12)$

13. $(x + 2)(x - 3)$

14. $(x + 2)(x + 2)$

15. $(3x + 2)(3x + 3)$

16. $(4x + 1)(2x + 2)$

17. $(5x - 6)(x + 2)$

18. $(x - 8)(x + 8)$

19. $(3x - 1)(3x + 1)$

20. $(2x + 3)(2x + 3)$

21. $(4x - 2)(x - 1)$

22. $(2x - 1)(3x + 1)$

23. $\left(x - \dfrac{1}{4}\right)\left(x + \dfrac{1}{4}\right)$

24. $\left(x + \dfrac{3}{4}\right)\left(x + \dfrac{3}{4}\right)$

ANSWERS

1. _____

2. _____

3. _____

4. _____

5. _____

6. _____

7. _____

8. _____

9. _____

10. _____

11. _____

12. _____

13. _____

14. _____

15. _____

16. _____

17. _____

18. _____

19. _____

20. _____

21. _____

22. _____

23. _____

24. _____

25. _____

26. _____

27. _____

28. _____

29. _____

30. _____

31. _____

32. _____

33. _____

34. _____

35. _____

36. _____

37. _____

38. _____

39. _____

40. _____

41. _____

42. _____

43. _____

44. _____

45. _____

46. _____

47. _____

48. _____

25. $(x - 0.1)(x + 0.1)$

26. $(3x^2 + 1)(x + 1)$

27. $(2x^2 + 6)(x + 1)$

28. $(2x^2 + 3)(2x - 1)$

29. $(-2x + 1)(x + 6)$

30. $(3x + 4)(2x - 4)$

31. $(x + 7)(x + 7)$

32. $(2x + 5)(2x + 5)$

33. $(1 + 2x)(1 - 3x)$

34. $(-3x - 2)(x + 1)$

35. $(x^2 + 3)(x^3 - 1)$

36. $(x^4 - 3)(2x + 1)$

37. $(x^2 - 2)(x - 1)$

38. $(x^3 + 2)(x - 3)$

39. $(3x^2 - 2)(x^4 - 2)$

40. $(x^{10} + 3)(x^{10} - 3)$

41. $(3x^5 + 2)(2x^2 + 6)$

42. $(1 - 2x)(1 + 3x^2)$

43. $(8x^3 + 1)(x^3 + 8)$

44. $(4 - 2x)(5 - 2x^2)$

45. $(4x^2 + 3)(x - 3)$

46. $(7x - 2)(2x - 7)$

47. $(4x^4 + x^2)(x^2 + x)$

48. $(5x^6 + 3x^3)(2x^6 + 2x^3)$

4.7   MORE SPECIAL PRODUCTS

• MULTIPLYING SUMS AND DIFFERENCES OF TWO EXPRESSIONS

Look for a pattern.

a) $(x + 2)(x - 2) = x^2 - 2x + 2x - 4$
$$= x^2 - 4$$

b) $(3x - 5)(3x + 5) = 9x^2 + 15x - 15x - 25$
$$= 9x^2 - 25$$

DO EXERCISES 1 AND 2.

Perhaps you discovered the following.

> **The product of the sum and difference of two expressions is the square of the first expression minus the square of the second:**
> $$(A + B)(A - B) = A^2 - B^2.$$

Caution! You must memorize the rule. Otherwise, you will have trouble later on.

Examples   Multiply.

$$(A + B)(A - B) = A^2 - B^2$$

Carry out the rule. Say the words as you go.

1. $(x + 4)(x - 4) = x^2 - 4^2$ "The square of the first expression, x^2, minus the square of the second, 4^2."

$$= x^2 - 16$$ Simplifying

2. $(2w + 5)(2w - 5) = (2w)^2 - 5^2$
$$= 4w^2 - 25$$

3. $(3x^2 - 7)(3x^2 + 7) = (3x^2)^2 - 7^2$
$$= 9x^4 - 49$$

4. $(-4x - 10)(-4x + 10) = (-4x)^2 - 10^2$
$$= 16x^2 - 100$$

DO EXERCISES 3-6.

•• SQUARING BINOMIALS

In this special product we multiply a binomial by itself. This is also called "squaring a binomial." Look for a pattern.

a) $(x + 3)^2 = (x + 3)(x + 3) = x^2 + 3x + 3x + 9 = x^2 + 6x + 9$

b) $(3x + 5)^2 = (3x + 5)(3x + 5) = 9x^2 + 15x + 15x + 25$
$$= 9x^2 + 30x + 25$$

Multiply.

1. $(x + 5)(x - 5)$

2. $(2x - 3)(2x + 3)$

Multiply.

3. $(x + 2)(x - 2)$

4. $(x - 7)(x + 7)$

5. $(3t + 5)(3t - 5)$

6. $(2x^3 - 1)(2x^3 + 1)$

Multiply.

7. $(x + 8)(x + 8)$

8. $(x - 5)(x - 5)$

Multiply.

9. $(x + 2)^2$

10. $(a - 4)^2$

11. $(2x + 5)^2$

12. $(4x^2 - 3x)^2$

13. $(y + 9)(y + 9)$

14. $(3x^2 - 5)(3x^2 - 5)$

ANSWERS ON PAGE A–11

c) $(x - 3)^2 = (x - 3)(x - 3) = x^2 - 3x - 3x + 9 = x^2 - 6x + 9$

d) $(3x - 5)^2 = (3x - 5)(3x - 5) = 9x^2 - 15x - 15x + 25$
$$= 9x^2 - 30x + 25$$

DO EXERCISES 7 AND 8.

Perhaps you discovered a quick way to square a binomial.

> **The square of a binomial is the square of the first expression, plus or minus twice the product of the two expressions, plus the square of the last:**
> $$(A + B)^2 = A^2 + 2AB + B^2;$$
> $$(A - B)^2 = A^2 - 2AB + B^2.$$

Examples   Multiply.

$$(A + B)^2 = A^2 + 2 \quad A \quad B + B^2$$

5. $(x + 3)^2 = x^2 + 2 \cdot x \cdot 3 + 3^2$ $\qquad$ *Carry out the rule. Say the words as you go.*
$$= x^2 + 6x + 9$$

6. $(t - 5)^2 = t^2 - 2 \cdot t \cdot 5 + 5^2$
$$= t^2 - 10t + 25$$

7. $(2x + 7)^2 = (2x)^2 + 2 \cdot 2x \cdot 7 + 7^2$
$$= 4x^2 + 28x + 49$$

8. $(3x^2 - 5x)^2 = (3x^2)^2 - 2 \cdot 3x^2 \cdot 5x + (5x)^2$
$$= 9x^4 - 30x^3 + 25x^2$$

Caution! Note carefully in these examples that the square of a sum is *not* the sum of squares:

$$(A + B)^2 \neq A^2 + B^2.$$

To see this, note that

$$(20 + 5)^2 = 25^2 = 625,$$

but

$$20^2 + 5^2 = 400 + 25 = 425 \neq 625.$$

DO EXERCISES 9–14.

●●● MULTIPLICATIONS OF VARIOUS TYPES

Now that we have considered how to multiply quickly certain kinds of polynomials, let us try several kinds mixed together so we can learn to sort them out. When you multiply, first see what kind of multiplication you have. Then use the best method. The methods you have used so far are as follows.

1. $(A + B)(A + B) = (A + B)^2 = A^2 + 2AB + B^2$
2. $(A - B)(A - B) = (A - B)^2 = A^2 - 2AB + B^2$
3. $(A - B)(A + B) = A^2 - B^2$
4. FOIL
5. **The product of a monomial and any polynomial. Multiply each term of the polynomial by the monomial.**

Note that FOIL will work for any of the first three rules, but it is faster to learn to use them as they are given.

Example 9 Multiply: $(x + 3)(x - 3)$.

$(x + 3)(x - 3) = x^2 - 9$ Using method 3 (the product of the sum and difference of two expressions)

Example 10 Multiply: $(t + 7)(t - 5)$.

$(t + 7)(t - 5) = t^2 + 2t - 35$ Using method 4 (the product of two binomials, but neither the square of a binomial nor the product of the sum and difference of two expressions)

Example 11 Multiply: $(x + 7)(x + 7)$.

$(x + 7)(x + 7) = x^2 + 14x + 49$ Using method 1 (the square of a binomial sum)

Example 12 Multiply: $2x^3(9x^2 + x - 7)$.

$2x^3(9x^2 + x - 7) = 18x^5 + 2x^4 - 14x^3$ Using method 5 (the product of a monomial and a trinomial: multiply each term of the trinomial by the monomial)

Example 13 Multiply: $(3x^2 - 7x)^2$.

$(3x^2 - 7x)^2 = 9x^4 - 42x^3 + 49x^2$ Using method 2 (the square of a binomial difference)

Example 14 Multiply: $\left(3x + \dfrac{1}{4}\right)^2$.

$\left(3x + \dfrac{1}{4}\right)^2 = 9x^2 + 2(3x)\dfrac{1}{4} + \dfrac{1}{16}$

$= 9x^2 + \dfrac{3}{2}x + \dfrac{1}{16}$ Using method 1 (the square of a binomial sum. To get the middle term, we multiply $3x$ by $\frac{1}{4}$ and double.)

DO EXERCISES 15–20.

Multiply.

15. $(x + 5)(x + 6)$

16. $(t - 4)(t + 4)$

17. $4x^2(-2x^3 + 5x^2 + 10)$

18. $(9x^2 + 1)^2$

19. $(2a - 5)(2a + 8)$

20. $\left(5x + \dfrac{1}{2}\right)^2$

ANSWERS ON PAGE A–11

Multiply.

21. $\left(2x - \dfrac{1}{2}\right)^2$

Example 15 Multiply: $\left(4x - \dfrac{3}{4}\right)^2$.

$$\left(4x - \frac{3}{4}\right)^2 = 16x^2 - 2(4x)\frac{3}{4} + \frac{9}{16}$$

$$= 16x^2 - 6x + \frac{9}{16}$$

DO EXERCISE 21.

SOMETHING EXTRA

FACTORS AND SUMS

In the table, the top number has been factored in such a way that the sum of the factors is the bottom number. For example, in the first column 56 has been factored as $7 \cdot 8$, and $7 + 8 = 15$, the bottom number.

Product	56	63	36	72	140	96		168	110			
Factor	7									9	24	3
Factor	8						8	8		10	18	
Sum	15	16	20	38	24	20	14		21			24

EXERCISE

Find the missing numbers in the table.

NAME

CLASS

ANSWERS

1. _____

2. _____

3. _____

4. _____

5. _____

6. _____

7. _____

8. _____

9. _____

10. _____

11. _____

12. _____

13. _____

14. _____

15. _____

16. _____

17. _____

18. _____

19. _____

20. _____

21. _____

22. _____

23. _____

24. _____

25. _____

26. _____

27. _____

EXERCISE SET 4.7

● Multiply mentally.

1. $(x + 4)(x - 4)$

2. $(x + 1)(x - 1)$

3. $(2x + 1)(2x - 1)$

4. $(x^2 + 1)(x^2 - 1)$

5. $(5m - 2)(5m + 2)$

6. $(3x^4 + 2)(3x^4 - 2)$

7. $(2x^2 + 3)(2x^2 - 3)$

8. $(6x^5 - 5)(6x^5 + 5)$

9. $(3x^4 - 4)(3x^4 + 4)$

10. $(t^2 - 0.2)(t^2 + 0.2)$

11. $(x^6 - x^2)(x^6 + x^2)$

12. $(2x^3 - 0.3)(2x^3 + 0.3)$

13. $(x^4 + 3x)(x^4 - 3x)$

14. $\left(\dfrac{3}{4} + 2x^3\right)\left(\dfrac{3}{4} - 2x^3\right)$

15. $(x^{12} - 3)(x^{12} + 3)$

16. $(12 - 3x^2)(12 + 3x^2)$

17. $(2x^8 + 3)(2x^8 - 3)$

18. $\left(x - \dfrac{2}{3}\right)\left(x + \dfrac{2}{3}\right)$

● ● Multiply mentally.

19. $(x + 2)^2$

20. $(2x - 1)^2$

21. $(3x^2 + 1)^2$

22. $\left(3x + \dfrac{3}{4}\right)^2$

23. $\left(x - \dfrac{1}{2}\right)^2$

24. $\left(2x - \dfrac{1}{5}\right)^2$

25. $(3 + x)^2$

26. $(x^3 - 1)^2$

27. $(x^2 + 1)^2$

ANSWERS

28. _____

29. _____

30. _____

31. _____

32. _____

33. _____

34. _____

35. _____

36. _____

37. _____

38. _____

39. _____

40. _____

41. _____

42. _____

43. _____

44. _____

45. _____

46. _____

47. _____

48. _____

49. _____

28. $(8x - x^2)^2$ **29.** $(2 - 3x^4)^2$ **30.** $(6x^3 - 2)^2$

31. $(5 + 6x^2)^2$ **32.** $(3x^2 - x)^2$

●●● Multiply mentally.

33. $(3 - 2x^3)^2$ **34.** $(x - 4x^3)^2$

35. $4x(x^2 + 6x - 3)$ **36.** $8x(-x^5 + 6x^2 + 9)$

37. $\left(2x^2 - \frac{1}{2}\right)\left(2x^2 - \frac{1}{2}\right)$ **38.** $(-x^2 + 1)^2$

39. $(-1 + 3p)(1 + 3p)$ **40.** $(-3x + 2)(3x + 2)$

41. $3t^2(5t^3 - t^2 + t)$ **42.** $-6x^2(x^3 + 8x - 9)$

43. $(6x^4 + 4)^2$ **44.** $(8a + 5)^2$

45. $(3x + 2)(4x^2 + 5)$ **46.** $(2x^2 - 7)(3x^2 + 9)$

47. $(8 - 6x^4)^2$ **48.** $\left(\frac{1}{5}x^2 + 9\right)\left(\frac{3}{5}x^2 - 7\right)$

---

49. ▦ Multiply: $(67.58x + 3.225)^2$.

EXTENSION EXERCISES

SECTION 4.1

Collect like terms.

1. $3a^2b + 4b^2 - 9a^2b - 6b^2$

2. $3xy^2 + 4xy - 7xy^2 + 7xy + x^2y$

SECTION 4.2

3. A polynomial in the variable x has degree 3. The coefficient of x^2 is 3 less than the coefficient of x^3. The coefficient of x is 3 times the coefficient of x^2. The remaining coefficient is 2 more than the coefficient of x^3. The sum of the coefficients is -4. Find the polynomial.

4. The sum of a number and 2 is multiplied by the number, and then 3 is subtracted from the result. Express the final result as a polynomial.

SECTION 4.3

5. Addition of real numbers is commutative. That is, $a + b = b + a$, where a and b are any real numbers.
   a) Show that addition of binomials such as $(ax + b)$ and $(cx + d)$ is commutative.
   b) Show that addition of trinomials such as $(ax^2 + bx + c)$ and $(dx^2 + ex + f)$ is commutative.

SECTION 4.4

6. Show that subtraction of binomials is not commutative.

7. Is subtraction of binomials associative? Support your answer with an example.

SECTION 4.5

Compute.

8. $(x + 4)(x + 5) - (x + 2)(x + 1)$

9. $(x - 1)(x + 3)(x - 2)$

10. A box with a square bottom is to be made from a 12 in. square piece of cardboard. Squares with side x are cut out of the corners and the sides are folded up. (The box does not have a top.)

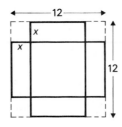

   a) Express the volume of the box as a polynomial.
   b) Express the surface area of the box as a polynomial.

11. An open wooden box is a cube with side x cm. The wood from which the box is made is 1 cm thick. Express the interior volume of the cube as a polynomial.

12. The height of a triangle is 4 ft longer than its base. The area of the triangle is 30 ft². Find its height and base.

13. A rectangular garden is twice as long as it is wide. It is surrounded by a sidewalk 4 ft wide. The area of the garden and sidewalk together is 256 ft² more than the area of the garden alone. Find the dimensions of the garden.

SECTION 4.6

Multiply.

14. $[(x + 1) - x^2] \cdot [(x - 2) + 2x^2]$

Solve.

15. $(x + 2)(x - 5) = (x + 1)(x - 3)$

16. $(x + 3)(x + 1) + (2x - 3)(x - 2) = (3x + 4)(x - 1)$

SECTION 4.7

Multiply. (Do not collect like terms before multiplying.)

17. $[(2x - 1)(2x + 1)](4x^2 + 1)$ **18.** $[(a + 5) + 1][(a + 5) - 1]$

19. $[3a - (2a - 3)][3a + (2a - 3)]$ **20.** $[(x + 3) + 2]^2$

Solve.

21. $(4x - 1)^2 - (3x + 2)^2 = (7x + 4)(x - 1)$

NAME   SCORE

TEST OR REVIEW—CHAPTER 4

If you miss an item, review the indicated section and objective.

[4.1, ▦] Collect like terms and then arrange in descending
[4.2, ▦] order.

    1. $3x^2 - 5x + 7x - 8x^2 + 1$

    2. $\frac{2}{5}a + 7 - 12a - \frac{3}{5} + \frac{1}{5}a^3 - \frac{32}{5}$

[4.3, ▪] Add.

    3. $(3x^4 - x^3 + x - 4) + (2x^5 - 5x^3 - 8x^2 + x - 7)$

    4. $(3x^5 - 4x^4 + x^3 - 3) + (-5x^4 + 2x^3 + 5)$

[4.4, ▪▪▪] Subtract.

    5. $(5x^2 - 4x + 1) - (-3x^2 + 7)$

    6. $(3x^5 - 4x^4 + 2x^2 + 3) - (2x^5 - 4x^4 + 3x^3 + 4x^2 - 5)$

Multiply.

[4.5, ▦]   **7.** $(2a + 3)(3a^2 - 2a - 1)$
[4.5, ▦]

[4.6, ▪]   **8.** $-6y^2(3y^2 - 5y - 2)$

1. _____

2. _____

3. _____

4. _____

5. _____

6. _____

7. _____

8. _____

ANSWERS

[4.7, ▨]   **9.** $(x + 9)(x - 9)$

9. _____

[4.7, ▨]   **10.** $(x - 9)(x - 9)$

10. _____

[4.6, ▨]   **11.** $(2x + 3)(x - 7)$

11. _____

[4.7, ▨]   **12.** $(7x - 1)^2$

12. _____

[4.6, ▨]   **13.** $(2p^2 + 3)(3p^3 - 5)$

13. _____

[4.7, ▨]   **14.** $(2x^3 - 7)(2x^3 + 7)$

14. _____

[4.7, ▨]   **15.** $(3x + 2)^2$

15. _____

[4.6, ▨]   **16.** $5x^4(3x^3 - 8x^2 + 10x + 2)$

16. _____

POLYNOMIALS AND FACTORING

5

There is a polynomial that gives the number of possible handshakes in a group of people.

1. Factor: $4y + 28 + 12z$.

2. Factor: $8x - 32$.

3. Solve: $4x + 9 = 17$.

4. Solve: $(4x + 3)(x - 7) = 0$.

Multiply.

5. $(-6x^8)(2x^5)$

6. $8x(2x^2 - 6x + 1)$

7. $(x + 6)(x - 4)$

8. $(7w + 6)(4w - 1)$

9. $(t - 9)^2$

10. $(5x + 3)^2$

11. $(p + 4)(p - 4)$

5.1 FACTORING POLYNOMIALS

To *factor* an expression means to write it as a product. To factor quickly, we study the quick methods of multiplication.

● FACTORING MONOMIALS

To factor a monomial we find two monomials whose product is that monomial. Compare.

	Multiplying	*Factoring*
a)	$(4x)(5x) = 20x^2$	$20x^2 = (4x)(5x)$
b)	$(2x)(10x) = 20x^2$	$20x^2 = (2x)(10x)$
c)	$(-4x)(-5x) = 20x^2$	$20x^2 = (-4x)(-5x)$
d)	$(x)(20x) = 20x^2$	$20x^2 = (x)(20x)$

The monomial $20x^2$ thus has many factorizations. There are still other ways to factor $20x^2$.

DO EXERCISES 1 AND 2.

To factor a monomial, factor the coefficient first. Then shift some of the letters to one factor and some to the other.

Example 1 Find three factorizations of $15x^3$.

a) $15x^3 = (3 \cdot 5)x^3$
$= (3x)(5x^2)$

b) $15x^3 = (3 \cdot 5)x^3$
$= (3x^2)(5x)$

c) $15x^3 = (-1) \cdot (-15)x^3$
$= (-x)(-15x^2)$

DO EXERCISES 3–5.

OBJECTIVES

After finishing Section 5.1, you should be able to:

● Factor monomials.

●● Factor polynomials when the terms have a common factor.

●●● Factor certain expressions with four terms by grouping.

1. a) Multiply: $(3x)(4x)$.

b) Factor: $12x^2$.

2. a) Multiply: $(2x)(8x^2)$.

b) Factor: $16x^3$.

Find three factorizations of each monomial.

3. $8x^4$

4. $21x^2$

5. $6x^5$

◖◗ FACTORING WHEN TERMS HAVE A COMMON FACTOR

To multiply a monomial and a polynomial with more than one term, we multiply each term by the monomial. To factor, we do the reverse.

Compare.

Multiply	*Factor*
$3x \ (x^2 + 2x - 4)$	$3x^3 + 6x^2 - 12x$
$= \boxed{3x} \cdot x^2 + \boxed{3x} \cdot 2x$	$= \boxed{3x} \cdot x^2 + \boxed{3x} \cdot 2x$
$+ \boxed{3x} \ (-4)$	$+ \boxed{3x} \cdot (-4)$
$= 3x^3 + 6x^2 - 12x$	$= \boxed{3x} \ (x^2 + 2x - 4)$

> *A common error:* $3x^3 + 6x^2 - 12x = 3 \cdot x \cdot x \cdot x + 6 \cdot x \cdot x - 4 \cdot 3x$. The parts, or terms, of the expression have been factored but the expression itself has not been factored.

DO EXERCISES 6 AND 7.

We are finding a factor in common with all the terms. There may not always be one other than 1. When there is, we generally use the factor with the largest possible coefficient and the largest exponent. In this way we "factor completely."

Example 2   Factor: $3x^2 + 6$.

$$3x^2 + 6 = \boxed{3} \cdot x^2 + \boxed{3} \cdot 2$$

$$= \boxed{3} \ (x^2 + 2) \qquad \text{Factoring out the common factor, 3}$$

Example 3   Factor: $16x^3 + 20x^2$.

$$16x^3 + 20x^2 = (\boxed{4x^2})(4x) + \boxed{4x^2} \cdot 5$$

$$= \boxed{4x^2} (4x + 5) \qquad \text{Factoring out } 4x^2$$

Example 4   Factor: $15x^5 - 12x^4 + 27x^3 - 3x^2$.

$$15x^5 - 12x^4 + 27x^3 - 3x^2$$

$$= (\boxed{3x^2})(5x^3) - (\boxed{3x^2})(4x^2) + (\boxed{3x^2})(9x) - (\boxed{3x^2})(1)$$

$$= 3x^2 (5x^3 - 4x^2 + 9x - 1) \qquad \text{Factoring out } 3x^2$$

If you can spot the common factor without factoring each term, you should write just the answer.

Example 5   Factor: $\dfrac{4}{5}x^2 + \dfrac{1}{5}x + \dfrac{2}{5}$.

$$\frac{4}{5}x^2 + \frac{1}{5}x + \frac{2}{5} = \frac{1}{5} (4x^2 + x + 2)$$

DO EXERCISES 8–12.

6. a) Multiply: $3(x + 2)$.

   b) Factor: $3x + 6$.

7. a) Multiply: $2x(x^2 + 5x + 4)$.

   b) Factor: $2x^3 + 10x^2 + 8x$.

Factor.

8. $x^2 + 3x$

9. $3x^6 - 5x^3 + 2x^2$

10. $9x^4 - 15x^3 + 3x^2$

11. $\dfrac{3}{4}x^3 + \dfrac{5}{4}x^2 + \dfrac{7}{4}x + \dfrac{1}{4}$

12. $35x^7 - 49x^6 + 14x^5 - 63x^3$

ANSWERS ON PAGE A–13

Factor.

13. $x^2 + 5x + 2x + 10$

14. $x^2 - 4x + 3x - 12$

15. $2x^2 + 8x - 3x - 12$

16. $16x^2 + 20x - 12x - 15$

•••   FACTORING BY GROUPING

Consider

$$x^2 + 3x + 4x + 12.$$

There is no factor common to all the terms other than 1. But we can factor $x^2 + 3x$ and $4x + 12$:

$x^2 + 3x = x(x + 3);$ Factoring $x^2 + 3x$

$4x + 12 = 4(x + 3).$ Factoring $4x + 12$

Then

$$x^2 + 3x + 4x + 12 = x(\boxed{x + 3}) + 4(\boxed{x + 3}).$$

Note the common *binomial* factor $\boxed{x + 3}$. We can use the distributive law again like this:

$$x(\boxed{x + 3}) + 4(\boxed{x + 3}) = (x + 4)(\boxed{x + 3}).$$ Factoring out the common factor, $\boxed{x + 3}$

Examples Factor. For purposes of learning this method, do not collect like terms.

6. $x^2 + 7x + 2x + 14$

   $= (x^2 + 7x) + (2x + 14)$ Separating into two binomials

   $= x(\boxed{x + 7}) + 2(\boxed{x + 7})$ Factoring each binomial

   $= (x + 2)(\boxed{x + 7})$ Factoring out the common factor, $\boxed{x + 7}$

7. $5x^2 - 10x + 2x - 4 = (5x^2 - 10x) + (2x - 4)$ Separating into two binomials

           $= 5x(x - 2) + 2(x - 2)$ Factoring each binomial

           $= (5x + 2)(x - 2)$ Factoring out the common factor, $x - 2$

8. $x^2 + 3x - x - 3 = (x^2 + 3x) + (-x - 3)$

           $= x(x + 3) - 1(x + 3)$ Factoring -1 out of the second binomial

           $= (x - 1)(x + 3)$

9. $2x^2 + 12x - 3x - 18 = (2x^2 + 12x) + (-3x - 18)$

              $= 2x(x + 6) - 3(x + 6)$

              $= (2x - 3)(x + 6)$

10. $12x^5 + 20x^2 - 21x^3 - 35 = (12x^5 + 20x^2) + (-21x^3 - 35)$

                 $= 4x^2(3x^3 + 5) - 7(3x^3 + 5)$

                 $= (4x^2 - 7)(3x^3 + 5)$

This method is called *factoring by grouping*. Not all expressions with four terms can be factored by this method.

DO EXERCISES 13–16.

NAME   CLASS

EXERCISE SET 5.1

⚫ Find three factorizations of each monomial.

1. $6x^3$ **2.** $9x^4$ **3.** $-9x^5$

4. $-12x^6$ **5.** $24x^4$ **6.** $15x^5$

⚫⚫ Factor.

7. $x^2 - 4x$ **8.** $2x^2 + 6x$

9. $x^3 + 6x^2$ **10.** $2x^2 + 2x - 8$

11. $8x^4 - 24x^2$ **12.** $x^5 + x^4 + x^3 - x^2$

13. $17x^5 + 34x^3 + 51x$ **14.** $6x^4 - 10x^3 + 3x^2$

15. $10x^3 + 25x^2 + 15x - 20$ **16.** $16x^4 - 24x^3 + 32x^2 + 64x$

1. _____

2. _____

3. _____

4. _____

5. _____

6. _____

7. _____

8. _____

9. _____

10. _____

11. _____

12. _____

13. _____

14. _____

15. _____

16. _____

ANSWERS

17. _____

18. _____

19. _____

20. _____

21. _____

22. _____

23. _____

24. _____

25. _____

26. _____

27. _____

28. _____

29. _____

30. _____

31. _____

32. _____

17. $\dfrac{5}{3}x^6 + \dfrac{4}{3}x^5 + \dfrac{1}{3}x^4 + \dfrac{1}{3}x^3$ **18.** $\dfrac{5}{7}x^7 + \dfrac{3}{7}x^5 - \dfrac{6}{7}x^3 - \dfrac{1}{7}x$

■■■ Factor.

19. $y^2 + 4y + y + 4$ **20.** $x^2 + 5x + 2x + 10$

21. $x^2 - 4x - x + 4$ **22.** $a^2 + 5a - 2a - 10$

23. $6x^2 + 4x + 9x + 6$ **24.** $3x^2 - 2x + 3x - 2$

25. $3x^2 - 4x - 12x + 16$ **26.** $24 - 18y - 20y + 15y^2$

27. $35x^2 - 40x + 21x - 24$ **28.** $8x^2 - 6x - 28x + 21$

29. $4x^2 + 6x - 6x - 9$ **30.** $2x^4 - 6x^2 - 5x^2 + 15$

31. $2x^4 + 6x^2 + 5x^2 + 15$ **32.** $4x^4 - 6x^2 - 6x^2 + 9$

5.2 FACTORING TRINOMIALS

■ • ■ FACTORING TRINOMIALS OF THE TYPE $x^2 + px + q$

Consider this multiplication:

$$F \quad O \quad I \quad L$$
$$(x + 2)(x + 5) = x^2 + \underbrace{5x + 2x}_{} + 10$$
$$= x^2 + \qquad 7x \qquad + 10.$$

The coefficient 7 is $2 + 5$ and the 10 is $2 \cdot 5$. In general,

$$(x + a)(x + b) = x^2 + (a + b)x + ab.$$

To factor we can use the preceding equation in reverse:

$$x^2 + (a + b)x + ab = (x + a)(x + b).$$

To factor

$$x^2 + 5x + 6,$$

we look for a pair of integers whose product is 6 and whose sum is 5

Pairs of factors	Sum of factors
1, 6	7
−1, −6	−7
−2, −3	−5
2, 3	5

Thus, the numbers are 2 and 3 . Then

$$x^2 + 5x + 6 = (x + 2)(x + 3).$$

You can check factoring by multiplying. Try to work mentally.

Example 1 Factor: $x^2 - 8x + 12$.

We look for two numbers whose product is 12 and whose sum is -8
They are -2 and -6 :

$$x^2 - 8x + 12 = (x - 2)(x - 6).$$

DO EXERCISES 1–3.

■ • • ■ FACTORING TRINOMIALS OF THE TYPE $ax^2 + bx + c$

Suppose the coefficient of the first term is not 1. Consider a multiplication.

OBJECTIVES

After finishing Section 5.2, you should be able to:

■ • ■ Factor trinomials of the type $x^2 + px + q$ by examining the last coefficient, q.

■ • • ■ Factor trinomials of the type $ax^2 + bx + c$.

Factor.

1. $x^2 + 7x + 12$
(*Hint:* Since all coefficients are positive, we need consider only positive factors.)

2. $x^2 - 12x + 35$
(*Hint:* Since the coefficient x is negative, and the others are positive, we will need to consider negative factors.)

3. $x^2 - x - 2$
(*Hint:* Since -2 is negative, and the x-coefficient is negative, we know one factor is positive and the other negative.)

Factor.

4. $6x^2 + 7x + 2$

$$
\begin{array}{ccccc}
 & \text{F} & \text{O} & \text{I} & \text{L} \\
(2x + 5)(3x + 4) = & 6x^2 + & 8x + & 15x + & 20
\end{array}
$$

$$
= 6x^2 + \quad 23x \quad + 20
$$

F	O + I	L
$2 \cdot 3$	$2 \cdot 4 + 5 \cdot 3$	$5 \cdot 4$

Now to factor $6x^2 + 23x + 20$, we do the reverse of what we just did:

$$
\begin{array}{cccc}
\text{F} & \text{O} + \text{I} & \text{L} \\
6x^2 + & 23x & + 20 \\
= (2x + 5)(3x + 4)
\end{array}
$$

We look for numbers (x-coefficients) whose product is 6—in this case, 2 and 3—and numbers whose product is 20—in this case, 4 and 5. The product of the outside terms plus the product of the inside terms must, of course, be 23.

In order to factor $ax^2 + bx + c$, we look for two binomials like this:

$$(\_x + \_)(\_x + \_),$$

where products of numbers in the blanks are as follows.

> 1. **The numbers in the *first* blanks of each binomial have product a.**
> 2. **The *outside* product and the *inside* product add up to b.**
> 3. **The numbers in the *last* blanks of each binomial have product c.**

Example 2   Factor: $3x^2 + 5x + 2$.

We look for two numbers whose product is 3. These are

$$1, 3 \quad \text{and} \quad -1, -3.$$

We have these possibilities:

$$(x + \_)(3x + \_) \quad \text{or} \quad (-x + \_)(-3x + \_).$$

Now we look for numbers whose product is 2. These are

$$1, 2 \quad \text{and} \quad -1, -2.$$

Here are some possibilities for factorizations. There are eight possibilities, but we have not listed all of them here.

$(x + 1)(3x + 2)$	$(-x + 1)(-3x + 2)$	$(x + 2)(3x + 1)$
$(x - 1)(3x - 2)$	$(-x - 1)(-3x - 2)$	$(x - 2)(3x - 1)$

When we multiply, we must get $3x^2 + 5x + 2$. When we multiply, we find that both of the shaded expressions are factorizations. We choose the one in which the first coefficients are positive. Thus the factorization is

$$(x + 1)(3x + 2).$$

DO EXERCISE 4.

We always look first for a common factor. If there is one, we remove that common factor before proceeding.

Example 3   Factor: $8x^2 + 8x - 6$.

a) First look for a factor common to all three terms. The number 2 is a common factor, so we factor it out:

$$2\,(4x^2 + 4x - 3).$$

b) Now we factor the trinomial $4x^2 + 4x - 3$. We look for pairs of numbers whose product is 4. These are

   $4, 1$ and $2, 2$.   Both positive

We then have these possibilities:

$$(4x + \text{\_\_})(x + \text{\_\_}) \text{ and } (2x + \text{\_\_})(2x + \text{\_\_}).$$

Next we look for pairs of numbers whose product is -3. They are

   $3, -1$ and $-3, 1$.

Then we have these possibilities for factorizations:

$$(4x + 3)(x - 1), \quad (2x + 3)(2x - 1),$$
$$(4x - 1)(x + 3), \quad (2x - 3)(2x + 1).$$
$$(4x + 1)(x - 3),$$
$$(4x - 3)(x + 1),$$

We usually do not write all of these. We multiply until we find the factors that give the product $4x^2 + 4x - 3$. We find that the factorization is

$$(2x + 3)(2x - 1).$$

c) But don't forget the common factor. We must include it to get a factorization of the original polynomial.

$$8x^2 + 8x - 6 = 2(2x + 3)(2x - 1)$$

> When we factor trinomials we must use trial and error. This is the way it's done. As you practice you will find that you can make better and better guesses. Don't forget: When factoring any polynomials, always look first for a common factor. Failing to do so is a common error.

DO EXERCISES 5–8.

Factor.

5. $6x^2 + 15x + 9$

6. $2x^2 + 4x - 6$

7. $4x^2 + 2x - 6$

8. $6x^2 - 5x + 1$

Factor.

9. $2x^2 - x - 15$

10. $12x^2 - 17x - 5$

FACTORING TRINOMIALS OF THE TYPE $ax^2 + bx + c$ (Optional)

Another way to factor $ax^2 + bx + c$ is as follows.

> **a) First look for a common factor.**
> **b) Multiply the first and last coefficients, a and c.**
> **c) Try to factor the product ac so that the sum of the factors is b. That is, find p and q such that $ac = pq$ and $p + q = b$.**
> **d) Write the middle term, bx, as a sum: $bx = px + qx$.**
> **e) Then factor by grouping. Regroup $ax^2 + px$ and $qx + c$.**

Example 4   Factor: $3x^2 - 10x - 8$.

a) First look for a common factor. There is none (other than 1).

b) Multiply the first and last coefficients, 3 and -8:

$$3(-8) = -24.$$

c) Try to factor -24 so that the sum of the factors is -10.

Some pairs of factors	Sums of factors
4, -6	-2
-4, 6	2
12, -2	10
-12, 2	-10

The desired factors are -12 and 2.

d) Write $-10x$ as a sum using the results of part (c):

$$-10x = -12x + 2x.$$

e) Factor by grouping:

$$3x^2 - 10x - 8 = 3x^2 - 12x + 2x - 8 \qquad \text{Substituting}$$
$$-12x + 2x \text{ for}$$
$$-10x, \text{ from (d)}$$
$$= (3x^2 - 12x) + (2x - 8)$$
$$= 3x(\,x - 4\,) + 2(\,x - 4\,)$$
$$= (3x + 2)(\,x - 4\,).$$

DO EXERCISES 9 AND 10.

Example 5   Factor: $15x^2 - 22x + 8$.

a) First look for a common factor. There is none (other than 1).

b) Multiply the first and last coefficients, 15 and 8:

$$15 \cdot 8 = 120.$$

c) Try to factor 120 so that the sum of the factors is -22.

Some pairs of factors	*Sum of factors*
30, 4	34
−30, −4	−34
3, 40	43
−3, −40	−43
6, 20	26
−6, −20	−26
10, 12	22
−10, −12	−22

The desired factors are -10 and -12.

d) Write $-22x$ as a sum using the results of part (c):

$$-22x = -10x - 12x.$$

e) Factor by grouping:

$$15x^2 - 22x + 8 = 15x^2 - 10x - 12x + 8 \quad \text{Substituting}$$
$$-10x - 12x \text{ for}$$
$$-22x, \text{ from (d)}$$

$$= (15x^2 - 10x) + (-12x + 8)$$

$$= 5x(\,3x - 2\,) - 4(\,3x - 2\,)$$

$$= (5x - 4)(\,3x - 2\,).$$

DO EXERCISES 11 AND 12.

This method of factoring is based on the following:

$$(ax + b)(cx + d) = acx^2 + adx + bcx + bd$$
$$= \underset{\downarrow}{\underline{ac}}x^2 + (ad + bc)x + \underline{bd}$$
$$(ac)(bd) = (ad)(bc)$$

We multiply the outside coefficients and factor them in such a way that we can write the middle term as a sum.

There are trinomials that are not factorable. An example is

$$x^2 - x + 5.$$

Factor.

11. $3x^2 - 19x + 20$

12. $20x^2 - 46x + 24$

SOMETHING EXTRA

CALCULATOR CORNER: NESTED MULTIPLICATION

To evaluate polynomials with a calculator, it helps to first use factoring to change the polynomial to *nested form*, as shown in the example below.

Example   Write nested form and evaluate for $x = 5.3$:

$$2x^4 + 4x^3 - x^2 + 7x - 9$$
$$= x\{2x^3 + 4x^2 - x + 7\} - 9 \qquad \text{Factoring out } x$$
$$= x\{x[2x^2 + 4x - 1] + 7\} - 9 \qquad \text{Factoring out } x \text{ again}$$
$$= x\{x[x(2x + 4) - 1] + 7\} - 9. \qquad \text{Factoring out } x \text{ again}$$

To evaluate $x\{x[x(2x + 4) - 1] + 7\} - 9$, start from the inside and work out. For $x = 5.3$, you enter 5.3, multiply by 2, add 4, multiply by 5.3, subtract 1, multiply by 5.3, add 7, multiply by 5.3, and subtract 9. The answer is 2173.6142.

EXERCISES

Write nested form for each polynomial. Then evaluate for $x = 5.3$, $x = 10.6$, and $x = -23$. Round to four decimal places.

1. $x^5 + 3x^4 - x^3 + x^2 - x + 9$

2. $5x^4 - 17x^3 + 2x^2 - x + 11$

3. $2x^4 - 3x^3 + 5x^2 - 2x + 18$

4. $-2x^5 + 4x^4 + 8x^3 - 4x^2 - 3x + 24$

NAME   CLASS   ANSWERS

EXERCISE SET 5.2

⬛ Factor.

1. $x^2 + 8x + 15$ **2.** $x^2 + 5x + 6$

3. $x^2 - 2x + 1$ **4.** $x^2 - 10x + 25$

5. $x^2 + 7x + 12$ **6.** $x^2 + x - 42$

7. $x^2 + 2x - 15$ **8.** $x^2 + 8x + 12$

9. $y^2 + 9y + 8$ **10.** $x^2 - 6x + 9$

11. $x^4 + 5x^2 + 6$ **12.** $y^2 + 16y + 64$

13. $x^2 + 3x - 28$ **14.** $x^2 - 11x + 10$

15. $16 + 8x + x^2$ **16.** $6 + 5x + x^2$

17. $a^2 - 12a + 11$ **18.** $c^2 - 10c + 21$

19. $x^2 - \frac{2}{5}x + \frac{1}{25}$ **20.** $x^2 + \frac{2}{3}x + \frac{1}{9}$

21. $y^2 - 0.2y - 0.08$ **22.** $t^2 - 0.3t - 0.10$

ANSWERS

1. _____

2. _____

3. _____

4. _____

5. _____

6. _____

7. _____

8. _____

9. _____

10. _____

11. _____

12. _____

13. _____

14. _____

15. _____

16. _____

17. _____

18. _____

19. _____

20. _____

21. _____

22. _____

ANSWERS

23. _____

24. _____

25. _____

26. _____

27. _____

28. _____

29. _____

30. _____

31. _____

32. _____

33. _____

34. _____

35. _____

36. _____

37. _____

38. _____

39. _____

40. _____

41. _____

42. _____

43. _____

44. _____

●● ●● Factor.

23. $3x^2 + 4x + 1$ **24.** $6x^2 + 13x + 6$

25. $12x^2 + 28x - 24$ **26.** $6x^2 + 33x + 15$

27. $2x^2 - x - 1$ **28.** $15x^2 - 19x + 6$

29. $9x^2 + 18x - 16$ **30.** $14x^2 + 35x + 14$

31. $15x^2 - 25x - 10$ **32.** $18x^2 - 3x - 10$

33. $12x^2 + 31x + 20$ **34.** $15x^2 + 19x - 10$

35. $14x^2 + 19x - 3$ **36.** $35x^2 + 34x + 8$

37. $9x^4 + 18x^2 + 8$ **38.** $6 - 13x + 6x^2$

39. $9x^2 - 42x + 49$ **40.** $15x^4 - 19x^2 + 6$

41. $6x^3 + 4x^2 - 10x$ **42.** $18x^3 - 21x^2 - 9x$

43. $x^2 + 3x - 7$ **44.** $x^2 + 11x + 12$

5.3   SQUARES OF BINOMIALS

◖●◗ RECOGNIZING SQUARES OF BINOMIALS

Some trinomials are squares of binomials; for example,

$$x^2 + 10x + 25 = (x + 5)^2.$$

A trinomial like this is called the *square of a binomial*, or a *trinomial square*.

Although the trial-and-error procedure we learned in Section 5.2 will work to factor trinomial squares, we want to develop a more efficient procedure. We use the equations for squaring a binomial in reverse to factor trinomial squares:

$$A^2 + 2AB + B^2 = (A + B)^2;$$
$$A^2 - 2AB + B^2 = (A - B)^2.$$

Use the following to help recognize squares of binomials.

a) Two of the terms, A^2 and B^2, must be squares, such as

  $4, \quad x^2, \quad 25x^4, \quad 16t^2.$

b) There must be no minus sign before A^2 or B^2.

c) If we multiply A and B (the square roots of these expressions) and double the result, we get the remaining term $2 \cdot A \cdot B$, or its additive inverse, $-2 \cdot A \cdot B$.

Example 1   Is $x^2 + 6x + 9$ the square of a binomial?

a) x^2 and 9 are squares.

b) There is no minus sign before x^2 or 9.

c) If we multiply the square roots, x and 3, and double the product, we get the remaining term: $\boxed{2} \cdot 3 \cdot x = 6x.$

Thus $x^2 + 6x + 9$ is the square of a binomial.

Example 2   Is $x^2 + 6x + 11$ the square of a binomial?

The answer is *no*, because only one term is a square.

Example 3   Is $16x^2 - 56x + 49$ a trinomial square?

a) $16x^2$ and 49 are squares.

b) There is no minus sign before $16x^2$ or 49.

c) If we multiply the square roots, $4x$ and 7, and double the product, we get the additive inverse of the remaining term: $\boxed{2} \cdot 4x \cdot 7 = 56x.$

Thus $16x^2 - 56x + 49$ is a trinomial square.

DO EXERCISE 1.

OBJECTIVES

After finishing Section 5.3, you should be able to:

◖●◗ Recognize squares of binomials (also called trinomial squares).

◖●●◗ Factor squares of binomials (or trinomial squares).

◖●●●◗ Recognize differences of two squares.

◖██◗ Factor differences of squares.

1. Which of the following are squares of binomials?

 a) $x^2 + 8x + 16$

 b) $x^2 - 10x + 25$

 c) $x^2 - 12x + 4$

 d) $4x^2 + 20x + 25$

 e) $5x^2 - 14x + 16$

 f) $16x^2 + 40x + 25$

 g) $x^2 + 6x - 9$

 h) $25x^2 - 30x - 9$

Factor.

2. $x^2 + 2x + 1$

3. $x^2 - 2x + 1$

4. $x^2 + 4x + 4$

5. $25x^2 - 70x + 49$

6. $16x^2 - 56x + 49$

7. Which of the following are differences of squares?

a) $x^2 - 25$

b) $x^2 - 24$

c) $x^2 + 36$

d) $4x^2 - 15$

e) $16x^4 - 49$

f) $9x^6 - 1$

●● FACTORING SQUARES OF BINOMIALS

To factor squares of binomials, we use the following equations:

$$A^2 + 2AB + B^2 = (A + B)^2;$$
$$A^2 - 2AB + B^2 = (A - B)^2.$$

We use the square roots of the squared terms and the sign of the remaining term.

Example 4   Factor: $x^2 + 6x + 9$.

$$x^2 + 6x + 9 = x^2 + 2 \cdot x \cdot 3 + 3^2 = (x + 3)^2$$

$$A^2 + 2AB + B^2 = (A + B)^2$$

Example 5   Factor: $x^2 - 14x + 49$.

$$x^2 - 14x + 49 = x^2 - 2 \cdot x \cdot 7 + 7^2$$
$$= (x - 7)^2$$

Example 6   Factor: $16x^2 - 40x + 25$.

$$16x^2 - 40x + 25 = (4x)^2 - 2 \cdot 4x \cdot 5 + 5^2 = (4x - 5)^2$$

$$A^2 - 2AB + B^2 = (A - B)^2$$

DO EXERCISES 2–6.

●●● RECOGNIZING DIFFERENCES OF TWO SQUARES

We can use the following equation to recognize a difference of squares:

$$A^2 - B^2 = (A - B)(A + B).$$

For a binomial to be a difference of squares:

a) There must be two expressions, both squares, such as

$$4x^2, \quad 9, \quad 25t^4, \quad 1, \quad x^6.$$

b) There must be a minus sign between them.

Example 7   Is $9x^2 - 64$ a difference of squares?

a) The first expression is a square: $9x^2 = (3x)^2$.
The second expression is a square: $64 = (8)^2$.

b) There is a minus sign between them.

So we have a difference of squares.

DO EXERCISE 7.

8·8 FACTORING DIFFERENCES OF SQUARES

To factor a difference of squares we can use the following equation:

$$A^2 - B^2 = (A - B)(A + B).$$

We find the square roots of the expressions A^2 and B^2. We write a plus sign one time and a minus sign one time.

Example 8 Factor: $x^2 - 4$.

$$x^2 - 4 = x^2 - 2^2 = (x - 2)(x + 2)$$
$$A^2 - B^2 = (A - B)(A + B)$$

Example 9 Factor: $18x^2 - 50x^6$.

Always look first for a factor common to all terms. This time there is one.

$$18x^2 - 50x^6 = \boxed{2x^2}\,(9 - 25x^4)$$
$$= 2x^2[3^2 - (5x^2)^2]$$
$$= 2x^2(3 - 5x^2)(3 + 5x^2)$$

Example 10 Factor: $49x^4 - 9x^6$.

$$49x^4 - 9x^6 = \boxed{x^4}\,(49 - 9x^2) = x^4(7 - 3x)(7 + 3x)$$

> *Caution!* Note carefully in these examples that a difference of squares is *not* the square of the difference; that is,
>
> $$A^2 - B^2 \neq (A - B)^2.$$
>
> For example,
>
> $$(45 - 5)^2 = 40^2 = 1600,$$
>
> but
>
> $$45^2 - 5^2 = 2025 - 25 = 2000.$$

DO EXERCISES 8–10.

Factor.

8. $x^2 - 9$

9. $x^2 - 64$

10. $5 - 20x^6$
　　[*Hint:* $1 = 1^2$, $x^6 = (x^3)^2$.]

SOMETHING EXTRA

CALCULATOR CORNER: A NUMBER PATTERN

1. Calculate each of the following. Look for a pattern.

$(20 \cdot 20) - (21 \cdot 19)$

$(34 \cdot 34) - (35 \cdot 33)$

$(1999 \cdot 1999) - (2000 \cdot 1998)$

$(5.8 \times 5.8) - (6.8 \times 4.8)$

$[0.7 \times 0.7] - [(1.7)(-0.3)]$

$(999 \cdot 999) - (1000)(998)$

2. Use the pattern to find the following without using a calculator.

$(3778 \cdot 3778) - (3779 \cdot 3777)$

$(14.78 \times 14.78) - (15.78 \times 13.78)$

$[0.375 \times 0.375] - [(1.375) \times (-0.625)]$

3. Find an equation that describes the pattern above. Verify it.

NAME   CLASS

EXERCISE SET 5.3

■ • Which are squares of binomials? Write "yes" or "no."

1. $x^2 - 14x + 49$ **2.** $x^2 - 16x + 64$

3. $x^2 + 16x - 64$ **4.** $8x^2 + 40x + 25$

■ • • Factor. Remember to look first for a common factor.

5. $x^2 + 6x + 9$ **6.** $y^2 - 6y + 9$ **7.** $2x^2 - 4x + 2$

8. $2y^2 - 40y + 200$ **9.** $x^3 - 18x^2 + 81x$ **10.** $144x + 24x^2 + x^3$

11. $y^2 - 12y + 36$ **12.** $81 + 18x + x^2$ **13.** $64 - 16y + y^2$

14. $a^4 + 14a^2 + 49$ **15.** $12x^2 + 36x + 27$ **16.** $2x^2 - 44x + 242$

17. $x^6 - 16x^3 + 64$ **18.** $y^6 + 26y^3 + 169$

19. $49 - 42x + 9x^2$ **20.** $20x^2 + 100x + 125$

21. $5y^4 + 10y^2 + 5$ **22.** $16x^8 - 8x^4 + 1$

■ • • • Which are differences of squares? Write "yes" or "no."

23. $x^2 - 35$ **24.** $x^2 + 36$

25. $16x^2 - 25$ **26.** $36t^2 - 1$

1. _____

2. _____

3. _____

4. _____

5. _____

6. _____

7. _____

8. _____

9. _____

10. _____

11. _____

12. _____

13. _____

14. _____

15. _____

16. _____

17. _____

18. _____

19. _____

20. _____

21. _____

22. _____

23. _____

24. _____

25. _____

26. _____

ANSWERS

27. _____

28. _____

29. _____

30. _____

31. _____

32. _____

33. _____

34. _____

35. _____

36. _____

37. _____

38. _____

39. _____

40. _____

41. _____

42. _____

43. _____

44. _____

⚃ Factor. Remember to look first for a common factor.

27. $x^2 - 4$ **28.** $x^2 - 36$ **29.** $8x^2 - 98$

30. $98x^4 - 2$ **31.** $4x^2 - 25$ **32.** $100a^2 - 9$

33. $16x - 81x^3$ **34.** $36b - 49b^3$ **35.** $16x^2 - 25x^4$

36. $x^{16} - 9x^2$ **37.** $49x^4 - 81$ **38.** $25a^2 - 9$

39. $a^{12} - 4a^2$ **40.** $81y^6 - 25$ **41.** $121a^8 - 100$

42. $x^6 - \dfrac{1}{9}$ **43.** $x^4 - 1$ **44.** $16 - t^4$

45. Determine whether the statement $(A - B)^2 = (A + B)(A - B)$ is true and show why.

5.4 FACTORING: A GENERAL STRATEGY

▪ ▪ HERE IS A GENERAL STRATEGY FOR FACTORING.

After finishing Section 5.4, you should be able to:

▪ ▪ Factor polynomials completely using any of the methods considered thus far in this chapter.

A. **Always look first for a common factor.**
B. **Then look at the number of terms.**

> **Four terms: Try factoring by grouping.**
>
> **Three terms: Determine whether the trinomial is a square. If so, you know how to factor. If not, try trial and error.**
>
> **Two terms: Determine whether you have a difference of squares. Do *not* try to factor a sum of squares: $A^2 + B^2$.**

C. **Always *factor completely*. If a factor with more than one term can still be factored, you should do so. When no factor can be factored further, you have finished.**

Example 1   Factor: $10x^3 - 40x$.

A. We look first for a common factor:

$$10x^3 - 40x = 10x(x^2 - 4).$$ Factoring out the largest common factor

B. The factor $x^2 - 4$ has only two terms. It is a difference of squares. We factor it:

$$10x(x - 2)(x + 2).$$

> We did not forget to include the common factor.

C. Have we factored completely? Yes, because no factor can be factored further.

Example 2   Factor: $t^4 - 16$.

A. We look for a common factor. There isn't one.

B. There are only two terms. It is a difference of squares: $(t^2)^2 - 4^2$. We factor it:

$$(t^2 + 4)(t^2 - 4).$$

We see that one of the factors is still a difference of squares. We factor it:

$$(t^2 + 4)(t - 2)(t + 2).$$

> This is a sum of squares. It cannot be factored!

C. We have factored completely because no factors can be factored further.

Factor completely.

1. $3m^4 - 3$

2. $x^6 + 8x^3 + 16$

3. $2x^4 + 8x^3 + 6x^2$

4. $3x^3 + 12x^2 - 2x - 8$

5. $8x^3 - 200x$

Example 3　Factor:　$2x^3 + 10x^2 + x + 5$.

A. We look for a common factor. There isn't one.

B. There are four terms. We try factoring by grouping.

$$2x^3 + 10x^2 + x + 5$$
$$= (2x^3 + 10x^2) + (x + 5) \qquad \text{Separating into two binomials}$$
$$= 2x^2(x + 5) + 1(x + 5) \qquad \text{Factoring each binomial}$$
$$= (2x^2 + 1)(x + 5) \qquad \text{Factoring out the common factor, } x + 5$$

C. No factor can be factored further, so we have factored completely.

Example 4　Factor:　$x^5 - 2x^4 - 35x^3$.

A. We look first for a common factor. This time there is one.

$$x^5 - 2x^4 - 35x^3 = x^3(x^2 - 2x - 35)$$

Don't forget to look for a common factor!

B. The factor $x^2 - 2x - 35$ has three terms, but it is not a trinomial square. We factor it using trial and error:

$$x^5 - 2x^4 - 35x^3 = x^3(x^2 - 2x - 35) = x^3(x - 7)(x + 5).$$

Don't forget to include the common factor in your final answer!

C. No factor can be factored further, so we have factored completely.

Example 5　Factor:　$x^4 - 10x^2 + 25$.

A. We look first for a common factor. There isn't one.

B. There are three terms. We see that this is a trinomial square. We factor it:

$$x^4 - 10x^2 + 25 = (x^2)^2 - 2 \cdot 5 \cdot x^2 + 5^2 = (x^2 - 5)^2.$$

C. No factor can be factored further, so we have factored completely.

DO EXERCISES 1–5.

NAME   CLASS   ANSWERS

EXERCISE SET 5.4

●   Factor completely.

1. $2x^2 - 128$ **2.** $3t^2 - 27$

1. _____

2. _____

3. $a^2 + 25 - 10a$ **4.** $y^2 + 49 + 14y$

3. _____

4. _____

5. $2x^2 - 11x + 12$ **6.** $8y^2 - 18y - 5$

5. _____

6. _____

7. $x^3 + 24x^2 + 144x$ **8.** $x^3 - 18x^2 + 81x$

7. _____

8. _____

9. $x^2 + 3x + 2x + 6$ **10.** $x^2 - 8x - 2x + 25$

9. _____

10. _____

11. $24x^2 - 54$ **12.** $8x^2 - 98$

11. _____

12. _____

13. $20x^3 - 4x^2 - 72x$ **14.** $9x^3 + 12x^2 - 45x$

13. _____

14. _____

15. $x^2 + 4$ **16.** $t^2 + 25$

15. _____

16. _____

ANSWERS

17. _____

18. _____

19. _____

20. _____

21. _____

22. _____

23. _____

24. _____

25. _____

26. _____

27. _____

28. _____

29. _____

30. _____

31. _____

32. _____

17. $x^4 + 7x^2 - 3x^2 - 21$

18. $m^4 + 8m^2 + 8m^2 + 64$

19. $x^5 - 14x^4 + 49x^3$

20. $2x^6 + 8x^5 + 8x^4$

21. $20 - 6x - 2x^2$

22. $45 - 3x - 6x^2$

23. $x^2 + 3x + 1$

24. $x^2 + 5x + 2$

25. $4x^4 - 64$

26. $5x^4 - 80$

27. $1 - y^8$

28. $t^8 - 1$

29. $x^5 - 4x^4 - 3x^3$

30. $x^6 - 2x^5 + 7x^4$

31. $36a^2 - 15a + \dfrac{25}{16}$

32. $\dfrac{1}{81}x^6 - \dfrac{8}{27}x^3 + \dfrac{16}{9}$

5.5 SOLVING EQUATIONS AND PROBLEMS BY FACTORING

● ● SOLVING EQUATIONS BY FACTORING

Let's recall the principle of zero products.

A product is 0 if and only if at least one of the factors is 0.

Remember from Section 3.5 how we solved an equation like

$(x + 3)(x - 2) = 0$.

It has a 0 on one side and a product on the other. To solve, we used the principle of zero products and set each factor equal to 0:

$x + 3 = 0$ or $x - 2 = 0$

$x = -3$ or $x = 2$.

The solutions are -3 and 2.

Now that we know how to factor, there are more equations that we can solve using this principle.

Example 1 Solve $x^2 + 5x + 6 = 0$.

We first factor the polynomial. Then we use the principle of zero products.

$x^2 + 5x + 6 = 0$

$(x + 2)(x + 3) = 0$ Factoring

$x + 2 = 0$ or $x + 3 = 0$ Using the principle of zero products

$x = -2$ or $x = -3$

Check:

$x^2 + 5x + 6 = 0$		$x^2 + 5x + 6 = 0$	
$(-2)^2 + 5(-2) + 6$	0	$(-3)^2 + 5(-3) + 6$	0
$4 - 10 + 6$		$9 - 15 + 6$	
$-6 + 6$		$-6 + 6$	
0		0	

The solutions are -2 and -3.

DO EXERCISE 1.

Example 2 Solve $x^2 - 8x = -16$.

We first add 16 to get 0 on one side.

$x^2 - 8x + 16 = 0$ Adding 16

$(x - 4)(x - 4) = 0$ Factoring

$x - 4 = 0$ or $x - 4 = 0$ Using the principle of zero products

$x = 4$ or $x = 4$

> You *must* have 0 on one side before you can use the principle of zero products. That is why it is called the principle of *zero* products.

There is only one solution, 4. The check is left to the student.

DO EXERCISES 2 AND 3.

Solve.

1. $x^2 - x - 6 = 0$

Solve.

2. $x^2 - 3x = 28$

3. $x^2 + 9 = 6x$

Solve.

4. $x^2 - 4x = 0$

5. $x^2 - 16 = 0$

Translate to equations. Then solve and check.

6. One more than a number times one less than the number is 24.

7. Seven less than a number times eight less than the number is 0.

Example 3   Solve $x^2 + 5x = 0$.

> When some factors have only one term, you can still use the principle of zero products in the same way.

$x(x + 5) = 0$ Factoring out a common factor

$x = 0$ or $x + 5 = 0$ Using the principle of zero products

$x = 0$ or $x = -5$

The solutions are 0 and -5. The check is left to the student.

DO EXERCISES 4 AND 5.

⬤⬤ APPLIED PROBLEMS

Recall that to solve a problem, we do the following.

> 1. **Translate the problem to an equation. (Very often a drawing helps.)**
> 2. **Solve the equation.**
> 3. **Check the answers in the problem.**

Example 4   Solve this problem.

One more than a number times one less than that number is 8.

$(x + 1)$ $(x - 1)$ $= 8$

Translating

Solve: $(x + 1)(x - 1) = 8$

$x^2 - 1 = 8$ Multiplying

$x^2 - 1 - 8 = 0$ Adding -8 to get 0 on one side

$x^2 - 9 = 0$

$(x - 3)(x + 3) = 0$ Factoring

$x - 3 = 0$ or $x + 3 = 0$ Using the principle of zero products

$x = 3$ or $x = -3$

Check for 3:   One more than 3 (this is 4) times one less than 3 (this is 2) is 8. Thus, 3 checks.

Check for -3:   Left to the student.

There are two such numbers, 3 and -3.

DO EXERCISES 6 AND 7.

Example 5 Solve this problem.

$$\underbrace{\text{The square of a number}}_{x^2} \; \underbrace{\text{minus}}_{-} \; \underbrace{\text{twice the number}}_{2x} \; \underbrace{\text{is}}_{=} \; \underbrace{48.}_{48} \quad \text{Translating}$$

$$x^2 - 2x = 48$$

$$x^2 - 2x - 48 = 0 \qquad \text{Adding } -48 \text{ to get 0 on one side}$$

$$(x - 8)(x + 6) = 0$$

$$x - 8 = 0 \quad \text{or} \quad x + 6 = 0 \qquad \text{Using the principle of zero products}$$

$$x = 8 \quad \text{or} \qquad x = -6$$

There are two such numbers, 8 and -6. They both check.

DO EXERCISE 8.

Sometimes it helps to reword a problem before translating.

Example 6 The height of a triangle is 7 cm more than the base. The area of the triangle is 30 cm². Find the height and the base. (Area is $\frac{1}{2} \cdot$ base $\cdot$ height.)

We first make a drawing.

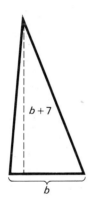

$$\underbrace{\tfrac{1}{2}}_{\tfrac{1}{2}} \; \underbrace{\text{times}}_{\cdot} \; \underbrace{\text{the base}}_{b} \; \underbrace{\text{times}}_{\cdot} \; \underbrace{\text{the base plus 7}}_{(b + 7)} \; \underbrace{\text{is}}_{=} \; \underbrace{30}_{30} \qquad \text{Rewording}$$

$$\frac{1}{2} \cdot b \cdot (b + 7) = 30 \qquad \text{Translating}$$

$$\frac{1}{2} \cdot b \cdot (b + 7) = 30$$

$$\frac{1}{2}(b^2 + 7b) = 30 \qquad \text{Multiplying}$$

$$b^2 + 7b = 60 \qquad \text{Multiplying by 2}$$

$$b^2 + 7b - 60 = 0 \qquad \text{Adding } -60 \text{ to get } 0 \text{ on one side}$$

$$(b + 12)(b - 5) = 0 \qquad \text{Factoring}$$

$$b + 12 = 0 \quad \text{or} \quad b - 5 = 0 \qquad \text{Using principle of zero products}$$

$$b = -12 \quad \text{or} \qquad b = 5$$

The solutions of the equation are -12 and 5. The base of a triangle cannot have a negative length. Thus, the base is 5 cm. The height is 7 cm more than the base, so the height is 12 cm. These numbers check.

DO EXERCISE 9.

Translate to an equation. Then solve and check.

8. The square of a number minus the number is 20.

9. The width of a rectangle is 2 cm less than the length. The area is 15 cm². Find the length and width.

10. Use $N = n^2 - n$ for the following.

a) A volleyball league has 19 teams. What is the total number of games to be played?

b) A slow-pitch softball league plays a total of 72 games. How many teams are in the league?

11. The product of two consecutive integers is 462. Find the integers.

Example 7   In a sports league of n teams in which each team plays every other team twice, the total number N of games to be played is given by

$$N = n^2 - n.$$

(a) A slow-pitch softball league has 17 teams. What is the total number of games to be played? **(b)** A basketball league plays a total of 90 games. How many teams are in the league?

a) We substitute 17 for n:

$$N = n^2 - n = 17^2 - 17 = 289 - 17 = 272.$$

b) We substitute 90 for N and solve for n:

$$n^2 - n = 90 \qquad \text{Substituting 90 for } N$$
$$n^2 - n - 90 = 0 \qquad \text{Adding } -90 \text{ to get 0 on one side}$$
$$(n - 10)(n + 9) = 0 \qquad \text{Factoring}$$
$$n - 10 = 0 \quad \text{or} \quad n + 9 = 0 \qquad \text{Principle of zero products}$$
$$n = 10 \quad \text{or} \qquad n = -9$$

Since the number of teams cannot be negative, -9 cannot be a solution. But 10 checks, so there are 10 teams in the league.

DO EXERCISE 10.

Example 8   The product of two consecutive integers is 156. Find the integers.

(*Consecutive* integers are next to each other, such as 49 and 50, or -6 and -5. The larger is 1 plus the smaller.)

$$\underbrace{\text{(First integer)}}_{x} \; \underbrace{\text{times}}_{\cdot} \; \underbrace{\text{(Second integer)}}_{(x + 1)} \; \underbrace{\text{is 156}}_{= 156} \qquad \text{Rewording}$$
$$\text{Translating}$$

We have let x represent the first integer. Then $x + 1$ represents the second.

We solve:

$$x(x + 1) = 156$$
$$x^2 + x = 156 \qquad \text{Multiplying}$$
$$x^2 + x - 156 = 0 \qquad \text{Adding } -156 \text{ to get 0 on one side}$$
$$(x - 12)(x + 13) = 0 \qquad \text{Factoring}$$
$$x - 12 = 0 \quad \text{or} \quad x + 13 = 0 \qquad \text{Using the principle of zero products}$$
$$x = 12 \quad \text{or} \qquad x = -13$$

The solutions of the equation are 12 and -13. When x is 12, then $x + 1$ is 13, and $12 \cdot 13 = 156$. So 12 and 13 are solutions to the problem. When x is -13, then $x + 1$ is -12, and $(-13)(-12) = 156$. Thus in this problem we have two pairs of solutions: 12 and 13, and -13 and -12. Both are pairs of consecutive integers whose product is 156.

DO EXERCISE 11.

NAME

CLASS

ANSWERS

EXERCISE SET 5.5

 Solve.

1. $x^2 + 6x + 5 = 0$

Check

1. _____

2. $x^2 + 7x + 6 = 0$

2. _____

3. $x^2 - 5x = 0$

3. _____

4. $4x^2 - 9 = 0$

4. _____

5. $x^2 + 6x + 9 = 0$

5. _____

6. $3t^2 + t = 2$

6. _____

7. $6y^2 - 4y - 10 = 0$

7. _____

ANSWERS

8. $5x^2 = 6x$

Check

8. _____

9. $3x^2 = 7x + 20$

9. _____

10. $0 = 2y^2 + 12y + 10$

10. _____

11. $-5x = -12x^2 + 2$

11. _____

12. $14 = x^2 - 5x$

12. _____

13. $0 = -3x + 5x^2$

13. _____

14. $x^2 - 5x = 18 + 2x$

14. _____

●● Solve.

15. If you subtract a number from four times its square, the result is three. Find the number.

15. _____

16. Eight more than the square of a number is six times the number. Find the number.

16. _____

17. The product of two consecutive integers is 182. Find the integers.

17. _____

18. The product of two consecutive even integers is 168. Find the integers.

18. _____

19. The product of two consecutive odd integers is 195. Find the integers.

19. _____

20. The length of a rectangle is 4 m greater than the width. The area of the rectangle is 96 m². Find the length and width.

20. _____

21. The area of a square is 5 more than the perimeter. Find the length of a side.

21. _____

22. The base of a triangle is 10 cm greater than the height. The area is 28 cm². Find the height and base.

22. _____

23. If the sides of a square are lengthened by 3 m, the area becomes 81 m². Find the length of a side of the original square.

23. _____

24. The sum of the squares of two consecutive odd positive integers is 74. Find the integers.

24. _____

Use $N = n^2 - n$ for Exercises 25–28.

25. A slow-pitch softball league has 23 teams. What is the total number of games to be played?

25. _____

26. A basketball league has 14 teams. What is the total number of games to be played?

26. _____

27. A slow-pitch softball league plays a total of 132 games. How many teams are in the league?

27. _____

28. A basketball league plays a total of 240 games. How many teams are in the league?

28. _____

The number N of possible handshakes within a group of n people is given by

$$N = \frac{1}{2}(n^2 - n).$$

29. At a meeting there are 40 people. How many handshakes are possible?

29. _____

30. At a party there are 100 people. How many handshakes are possible?

30. _____

31. Everyone shook hands at a party. There were 190 handshakes in all. How many were at the party? (*Hint:* Multiply on both sides by 2.)

31. _____

32. Everyone shook hands at a meeting. There were 300 handshakes in all. How many were at the meeting?

32. _____

5.6 COMPLETING THE SQUARE (Optional)

● FACTORING DIFFERENCES OF SQUARES

A difference of squares can have more than two terms. This can happen if one of the squares is a trinomial. We factor using a method important in many places in mathematics.

Example 1   Factor: $x^2 + 2x + 1 - 25$. For purposes of learning this method, do not collect like terms.

Since $x^2 + 2x + 1 = (x + 1)^2$, we have

$$A^2 - B^2 = (A - B)(A + B)$$

$$x^2 + 2x + 1 - 25 = (\boxed{x + 1})^2 - 5^2 = (\boxed{x + 1} - 5)(\boxed{x + 1} + 5)$$

$$= (x - 4)(x + 6).$$

> We used the factoring rule $A^2 - B^2 = (A - B)(A + B)$, where $A = \boxed{x + 1}$ and $B = 5$.

Example 2   Factor: $x^2 + 6x + 9 - 16$.

$$x^2 + 6x + 9 - 16 = (\boxed{x + 3})^2 - 4^2 \qquad \text{Rewriting as a difference of squares}$$

$$= (\boxed{x + 3} - 4)(\boxed{x + 3} + 4) \qquad \text{Factoring the difference of squares}$$

$$= (x - 1)(x + 7) \qquad \text{Simplifying}$$

> You could easily factor a different way, but do this in order to learn the new method.

DO EXERCISES 1–4.

● ● COMPLETING THE SQUARE

This is the square of a binomial:

$$x^2 + 10x + 25.$$

It is $(x + 5)^2$. We could find the 25 from $10x$, by taking half the coefficient of x and squaring it.

Example 3   Complete the square: $x^2 + 8x$.

What must be added to $x^2 + 8x$ to make it the square of a binomial? We find out as follows:

$$x^2 + 8x$$

$$\downarrow$$

$$\frac{8}{2} = 4 \qquad \text{Taking half the } x\text{-coefficient}$$

$$\downarrow$$

$$4^2 = 16 \qquad \text{Squaring}$$

$$\downarrow$$

$$x^2 + 8x + 16 \qquad \text{Adding}$$

Thus 16 must be added to $x^2 + 8x$ in order to make it the square of a binomial:

$$x^2 + 8x + 16 = (x + 4)^2.$$

We say that the number 16 *completes the square*.

OBJECTIVES

After finishing Section 5.6, you should be able to:

● Factor differences of squares when there are more than two terms.

● ● Complete the square of an expression $x^2 + bx$.

● ● ● Factor trinomials by completing the square.

Factor.

1. $x^2 + 2x + 1 - 16$

2. $x^2 - 2x + 1 - 64$

3. $x^2 + 8x + 16 - 9$

4. $x^2 + 16x + 64 - 100$

Complete the square.

5. $x^2 + 12x$

6. $x^2 - 6x$

7. $x^2 + 14x$

8. $x^2 - 22x$

Factor by completing the square. Show your work.

9. $x^2 + 2x - 3$

10. $x^2 + 6x + 8$

11. $x^2 - 22x + 120$

Example 4   Complete the square: $x^2 - 12x$.

We take half of -12, obtaining -6. We square: $(-6)^2 = 36$. Thus 36 must be added to $x^2 - 12x$ to make it the square of a binomial:

$$x^2 - 12x + \boxed{36} = (x - 6)^2.$$

DO EXERCISES 5–8.

⬤⬤⬤ FACTORING BY COMPLETING THE SQUARE

The method we learn now is important, even though the factoring can be done in other ways.

Example 5   Factor: $x^2 - 6x - 16$.

This trinomial is not a square. The third term would have to be 9. We complete the square by taking half the coefficient of x and squaring it. We add zero to the trinomial, naming it $\boxed{9 - 9}$.

$$
\begin{aligned}
x^2 - 6x - 16 &= x^2 - 6x + (\boxed{9 - 9}) - 16 && \text{Adding } 9 - 9 \\
&= (x^2 - 6x + 9) - 9 - 16 && \text{Grouping terms to} \\
& && \text{complete the square} \\
&= (x - 3)^2 - 25 && \text{Writing as a difference of} \\
& && \text{squares} \\
&= (\boxed{x - 3})^2 - 5^2 \\
&= (\boxed{x - 3} - 5)(\boxed{x - 3} + 5) && \text{Factoring the} \\
& && \text{difference of} \\
& && \text{squares} \\
&= (x - 8)(x + 2) && \text{Simplifying}
\end{aligned}
$$

This is factoring by *completing the square*.

Example 6   Factor $x^2 + 40x + 384$ by completing the square.

The third term would have to be 400 (half of 40 squared) in order for the trinomial to be a square. We add zero, naming it $\boxed{400 - 400}$.

$$
\begin{aligned}
& x^2 + 40x + 384 \\
&= x^2 + 40x + (\boxed{400 - 400}) + 384 && \text{Adding } 400 - 400 \\
&= (x^2 + 40x + 400) - 400 + 384 && \text{Grouping terms to} \\
& && \text{complete the square} \\
&= (x + 20)^2 - 16 && \text{Writing as a difference of squares} \\
&= (\boxed{x + 20})^2 - 4^2 \\
&= (\boxed{x + 20} - 4)(\boxed{x + 20} + 4) && \text{Factoring the difference} \\
& && \text{of squares} \\
&= (x + 16)(x + 24)
\end{aligned}
$$

You can factor many trinomials without completing the square. You could probably not factor $x^2 + 10.44x - 1201.9552$, however. Yet, by completing the square, it can be done (see Exercise 32 in Exercise Set 5.6). Completing the square is a method important in several ways.

DO EXERCISES 9–11.

NAME   CLASS   ANSWERS

EXERCISE SET 5.6

◼● Factor.

1. $x^2 + 12x + 36 - 49$ 2. $x^2 + 2x + 1 - 9$ 3. $x^2 + 14x + 49 - 16$

4. $x^2 - 16x + 64 - 100$ 5. $x^2 - 18x + 81 - 1$ 6. $x^2 + 14x + 49 - 121$

7. $x^4 + 20x^2 + 100 - 4$ 8. $x^2 - 16x + 64 - 36$ 9. $3x^6 + 6x^3 + 3 - 192$

10. $2x^2 + 4x + 2 - \dfrac{2}{9}$ 11. $a^4 - 6a^2 + 9 - 1$ 12. $4x^4 + 12x^2 + 9 - 144$

◼◼ Complete the square.

13. $x^2 + 4x$ 14. $x^2 - 8x$ 15. $x^2 - 14x$

16. $t^2 + 22t$ 17. $d^2 + 20d$ 18. $y^2 - 40y$

---

1. _____

2. _____

3. _____

4. _____

5. _____

6. _____

7. _____

8. _____

9. _____

10. _____

11. _____

12. _____

13. _____

14. _____

15. _____

16. _____

17. _____

18. _____

ANSWERS

19. _____

20. _____

21. _____

22. _____

23. _____

24. _____

25. _____

26. _____

27. _____

28. _____

29. _____

30. _____

31. _____

32. _____

●●● Factor by completing the square. Show your work.

19. $x^2 + 6x + 8$ **20.** $y^2 + 8y + 15$

21. $x^2 - 8x + 12$ **22.** $y^2 + 10y + 16$

23. $2x^2 - 16x - 256$ **24.** $2x^2 - 28x + 26$

25. $a^2 - 38a + 361$ **26.** $c^2 + 46c + 529$

27. $10b^2 - 50b - 840$ **28.** $10t^2 - 200t + 640$

29. $x^2 - 26x - 792$ **30.** $r^2 - 36r + 315$

---

31. ▦ What must be added to $x^2 -$ 75.836x in order to make it the square of a binomial?

32. ▦ Factor by completing the square: $x^2 + 10.44x - 1201.9552$.

EXTENSION EXERCISES

SECTION 5.1

Factor. (Do not collect like terms before factoring.)

1. $3x^3 + 3x^2 - 18x^2 - 18x$

2. $x^3 - x^2 + 2x^2 - 2x + 3x - 3$

3. Two polynomials are *relatively prime* if they have no common factors other than constants. Tell whether each pair is relatively prime.
  a) $5x, 6x^2$ b) $3x, 9x - 1$ c) $x + x^2, 3x^3$ d) $x - 6, x$ e) $7x, x$ f) $x^2 + 3, x$

SECTION 5.2

Factor.

4. $(x + 2)^2 - 3(x + 2) - 4$
6. $x^2(x - 1) + 2x(x - 1) + (x - 1)$

5. $6(x - 3)^2 + 5(x - 3) - 6$

Express the area of the shaded region as a polynomial in factored form. (Leave your answer in terms of π.)

7.

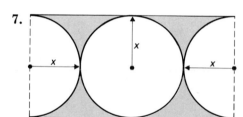

8.

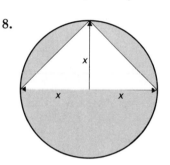

SECTION 5.3

Factor. (Do not collect like terms before factoring.)

9. $a^2 + 2a + 1 - 9$

10. $4x^4 + 4x^3 + x^2 - 25x^2$

11. $16 - 4x^2 - 4x - 1$

12. $y^2 + 6y + 9 - x^2 - 8x - 16$

Find c so that the polynomial will be the square of a binomial.

13. $a^2 - 6a + c$

14. $x^2 + cx + 25$

15. $cy^2 + 6y + 1$

16. Show that the difference of the squares of two consecutive integers is the sum of the integers.

SECTION 5.4

Factor completely.

17. $x^4 - 7x^2 - 18$

18. $3x^4 - 15x^2 + 12$

19. $x^3 - x^2 - 4x + 4$

20. $y^2(y + 1) - 4y(y + 1) - 21(y + 1)$

21. $y^2(y - 1) - 2y(y - 1) + (y - 1)$

SECTION 5.5

Solve.

22. $x^2 + 4x + 3 = 2x^2 - 3x + 13$

Find an equation that has the given numbers as its solutions.

23. 3 and -1 **24.** $\dfrac{1}{2}$ and $\dfrac{3}{4}$ **25.** 1, -2, and 3

26. The pages of a book are 15 cm by 20 cm. Margins of equal width surround the printing on each page and comprise one half the area of the page. Find the width of the margins.

27. A rectangular piece of cardboard is twice as long as it is wide. A 4 cm square is cut out of each corner, and the sides are turned up to make a box. The volume of the box is 616 cm³. Find the original dimensions of the cardboard.

28. An open rectangular gutter is made by turning up the sides of a piece of metal 20 in. wide. The area of the cross section of the gutter is 50 in². Find the depth of the gutter.

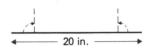

NAME   SCORE   ANSWERS

TEST OR REVIEW—CHAPTER 5

If you miss an item, review the indicated section and objective.

1. _____

Factor completely.

[5.1, ● ●]   **1.** $4x^4 + 20x^3$

2. _____

[5.1, ●●●]   **2.** $x^2 - 5x + 3x - 15$

[5.2, ●]   **3.** $x^2 - 8x + 15$

3. _____

[5.2, ●]   **4.** $x^3 - x^2 - 30x$

4. _____

[5.2, ●]   **5.** $3x^2 + 18x - 48$

5. _____

[5.2, ● ●]   **6.** $10x^2 + 101x + 10$

6. _____

[5.3, ● ●]   **7.** $x^2 - 6x + 9$

7. _____

[5.3, ● ●]   **8.** $12x^2 + 60x + 75$

8. _____

ANSWERS

[5.3, ⚅] **9.** $9a^2 - 4$

9. _____

[5.3, ⚅] **10.** $2x^2 - 50$

10. _____

[5.3, ⚅] **11.** $16t^4 - 1$
[5.5, ⚀]

11. _____

[5.5, ⚀] **12.** Solve: $x^2 + 2x - 35 = 0$.

12. _____

[5.5, ⚁] **13.** The product of two consecutive even integers is 288. Find the integers.

13. _____

[5.6, ⚂] **14.** Factor by completing the square. Show your work.

$$x^2 - 8x - 9$$

14. _____

SYSTEMS OF EQUATIONS AND GRAPHS

6

Problems involving distances traveled by trains can be solved using systems of equations.

READINESS CHECK: SKILLS FOR CHAPTER 6

Solve and check.

1. $\dfrac{5}{2} + y = \dfrac{1}{3}$ **2.** $-2x + 9 = -11$ **3.** $-8x + 3x = 25$

Evaluate each polynomial for $x = 4$.

4. $9 - 5x$ **5.** $3x^2 - x + 8$

OBJECTIVES

After finishing Section 6.1, you should be able to:

◧ Plot points associated with ordered pairs of numbers.

◧◧ Determine the quadrant in which a point lies.

◧◧◧ Find coordinates of a point on a graph.

Plot these points on the graph below.

1. $(4, 5)$ **2.** $(5, 4)$

3. $(-2, 5)$ **4.** $(-3, -4)$

5. $(5, -3)$ **6.** $(-2, -1)$

7. $(0, -3)$ **8.** $(2, 0)$

6.1 GRAPHS AND EQUATIONS

POINTS AND ORDERED PAIRS

On a number line each point is the graph of a number. On a plane each point is the graph of a number pair. We use two perpendicular number lines called *axes*. They cross at a point called the *origin*. It has coordinates (0, 0) but is usually labeled 0. The arrows show the positive directions.

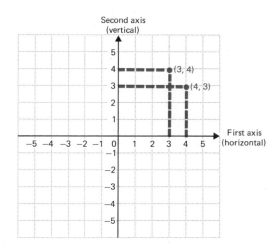

◧ PLOTTING POINTS

Note that (4, 3) and (3, 4) give different points. They are called *ordered pairs* of numbers because it makes a difference which number comes first.

Example 1   Plot the point $(-3, 4)$.

The first number, -3, is negative. We go -3 units in the first direction (3 units to the left). The second number, 4, is positive. We go 4 units in the second direction (up).

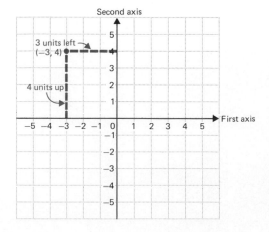

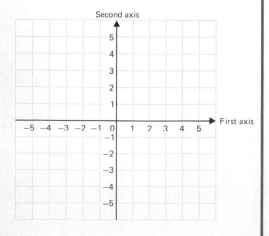

DO EXERCISES 1–8.

The numbers in an ordered pair are called *coordinates*. In $(-3, 4)$, the *first coordinate* is -3 and the *second coordinate** is 4.

●●○ QUADRANTS

This drawing shows some points and their coordinates. In region I (the first *quadrant*) both coordinates of any point are positive. In region II (the second quadrant) the first coordinate is negative and the second positive, and so on.

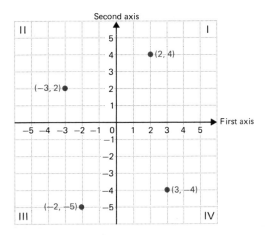

DO EXERCISES 9–14.

●●● FINDING COORDINATES

To find coordinates of a point, we see how far to the right or left of zero it is located and how far up or down.

Example 2   Find the coordinates of point A.

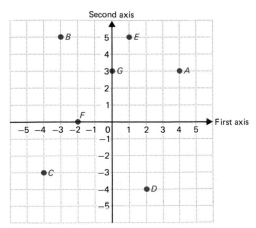

Point A is 4 units to the right (first direction) and 3 units up (second direction). Its coordinates are $(4, 3)$.

DO EXERCISE 15.

---

* The first coordinate of a point is sometimes called its *abscissa*, and the second coordinate its *ordinate*.

---

9. What can you say about the coordinates of a point in the third quadrant?

10. What can you say about the coordinates of a point in the fourth quadrant?

In which quadrant is each point located?

11. $(5, 3)$

12. $(-6, -4)$

13. $(10, -14)$

14. $(-13, 9)$

15. Find the coordinates of points B, C, D, E, F, and G in the drawing of Example 2.

SOMETHING EXTRA

AN APPLICATION: COORDINATES

Three-dimensional objects can also be coordinatized. 0° latitude is the equator. 0° longitude is a line from the North Pole to the South Pole through France and Spain. In the drawing below, the hurricane Clara is at a point about 260 miles northwest of Bermuda near latitude 36.0 North, longitude 69.0 West.

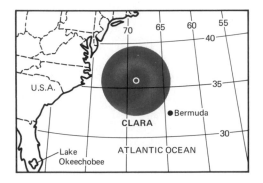

EXERCISES

1. Approximate the latitude and longitude of Bermuda.
2. Approximate the latitude and longitude of Lake Okeechobee.

NAME

CLASS

EXERCISE SET 6.1

 1. Plot these points.

2. Plot these points.

(2, 5)   (−1, 3)   (3, −2)   (−2, −4)

(4, 4)   (−2, 4)   (5, −3)   (−5, −5)

(0, 4)   (0, −5)   (5, 0)   (−5, 0)

(0, 4)   (0, −4)   (3, 0)   (−4, 0)

 1. See graph.

2. See graph.

3. _____

 In which quadrant is each point located?

3. (−5, 3)   **4.** (−12, 1)   **5.** (100, −1)   **6.** (35.6, −2.5)

7. (−6, −29)   **8.** (−3.6, −105.9)   **9.** (3.8, 9.2)   **10.** (1895, 1492)

4. _____

5. _____

6. _____

7. _____

8. _____

9. _____

10. _____

11. In quadrant III, first coordinates are always _____ and second coordinates are always _____ .

12. In quadrant II, _____ coordinates are always positive and _____ coordinates are always negative.

11. _____

12. _____

●●● **13.** Find the coordinates of points A, B, C, D, and E.

13. A: _____

B: _____

C: _____

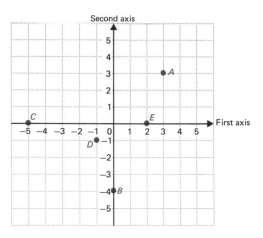

D: _____

14. Find the coordinates of points A, B, C, D, and E.

E: _____

14. A: _____

B: _____

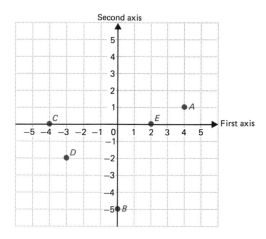

C: _____

D: _____

Use graph paper. Draw a first and second axis. Then plot these points. Attach the graphs to this page.

E: _____

15. $(0, -3)$, $(-1, -5)$, $(1, -1)$, $(2, 1)$ **16.** $(0, 1)$, $(1, 4)$, $(-1, -2)$, $(-2, -5)$

15. See graph. _____

16. See graph. _____

6.2 GRAPHING EQUATIONS

●｜ SOLUTIONS OF EQUATIONS

An equation with two variables has *pairs* of numbers for solutions. We usually take the variables in alphabetical order. Then we get *ordered* pairs for solutions.

Example 1   Determine whether $(3, 7)$ is a solution of $y = 2x + 1$.

$$y = 2x + 1$$

$$\begin{array}{c|l} 7 & 2 \cdot 3 + 1 \\ & 6 + 1 \\ & 7 \end{array}$$

We substitute 3 for x and 7 for y (alphabetical order of variables)

The equation becomes true: $(3, 7)$ is a solution.

Example 2   Determine whether $(-2, 3)$ is a solution of $2t = 4s - 8$.

$$2t = 4s - 8$$

$$\begin{array}{c|l} 2 \cdot 3 & 4(-2) - 8 \\ 6 & -8 - 8 \\ & -16 \end{array}$$

We substitute -2 for s and 3 for t.

The equation becomes false: $(-2, 3)$ is not a solution.

DO EXERCISES 1 AND 2.

●｜● GRAPHING EQUATIONS $y = mx$

> **To *graph* an equation means to make a drawing of its solutions.**

If an equation has a graph that is a line, we can graph it by plotting a few points and then drawing a line through them.

Example 3   Graph $y = x$.

We will use alphabetical order. Thus the first axis will be the x-axis and the second axis will be the y-axis. Next, we find some solutions of the equation. In this case it is easy. Here are a few:

$(0, 0)$, $(1, 1)$, $(5, 5)$, $(-2, -2)$, $(-4, -4)$.

Now we plot these points. We can see that if we were to plot a million solutions, the dots we draw would resemble a solid line. We see the pattern, so we can draw the line with a ruler. The line is the graph of the equation $y = x$. We label the line $y = x$ on the graph paper.

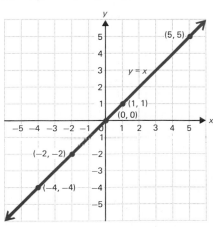

A graph of an equation is a picture of its solution set. Each point of the picture gives an ordered pair (a, b) that is a solution. No other points give solutions.

OBJECTIVES

After finishing Section 6.2, you should be able to:

●｜ Determine whether an ordered pair of numbers is a solution of an equation with two variables.

●｜● Graph equations of the type $y = mx$ and $y = mx + b$.

1. Determine whether $(2, 3)$ is a solution of $y = 2x + 3$.

2. Determine whether $(-2, 4)$ is a solution of $4q - 3p = 22$.

ANSWERS ON PAGE A–16

Graph.

3. $y = 3x$

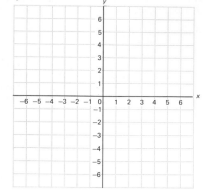

4. $y = \frac{1}{2}x$

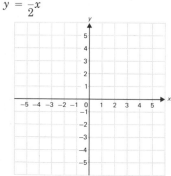

Graph.

5. $y = -x$ (or $-1 \cdot x$)

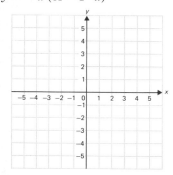

6. $y = -2x$

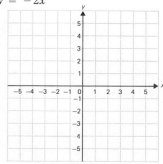

Example 4   Graph $y = 2x$.

We find some ordered pairs that are solutions, keeping the results in a table. We choose *any* number for x and then determine y by substitution. Suppose we choose 0 for x. Then $y = 2x = 2 \cdot 0 = 0$.

We get a solution: the ordered pair $(0, 0)$. Suppose we choose 3 for x. Then $y = 2x = 2 \cdot 3 = 6$.

We get a solution: the ordered pair $(3, 6)$. We make some negative choices for x, as well as some positive ones. If a number takes us off the graph paper, we usually do not use it. Continuing in this manner we get a table like the one shown below. In this case, since $y = 2x$, we get y by doubling x.

Now we plot these points. If we had enough of them, they would make a line. We draw it with a ruler and label it $y = 2x$.

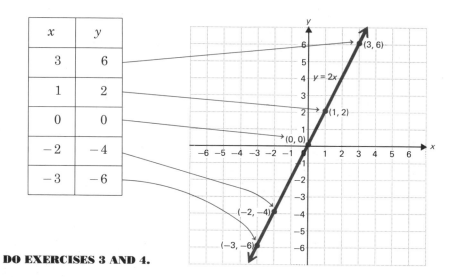

x	y
3	6
1	2
0	0
-2	-4
-3	-6

DO EXERCISES 3 AND 4.

Example 5   Graph $y = -3x$.

We make a table of solutions. Then we plot the points. If we had enough of them, they would make a line. We draw it with a ruler and label it $y = -3x$.

x	y
0	0
1	-3
-1	3
2	-6
-2	6

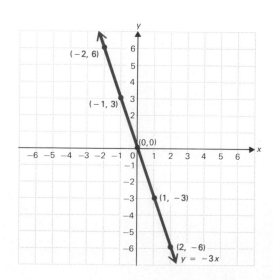

DO EXERCISES 5 AND 6.

Example 6   Graph $y = -\dfrac{5}{3}x$.

We make a table of solutions.

When $x = 0$, $y = -\dfrac{5}{3} \cdot 0 = 0$.

When $x = 3$, $y = -\dfrac{5}{3} \cdot 3 = -5$.

When $x = -3$, $y = -\dfrac{5}{3}(-3) = 5$.

When $x = 1$, $y = -\dfrac{5}{3} \cdot 1 = -\dfrac{5}{3}$.

Note that if we substitute multiples of 3 we can avoid fractions.

Next we plot the points. If we had enough of them, they would make a line.

x	y
0	0
3	-5
-3	5
1	$-\dfrac{5}{3}$

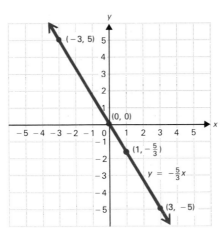

Every equation $y = mx$ has a graph that is a straight line. It contains the origin. The number m, called the *slope*, tells us how the line slants. For a positive slope a line slants up from left to right, as in Examples 3 and 4. For a negative slope a line slants down from left to right, as in Examples 5 and 6.

DO EXERCISES 7 AND 8.

We know that the graph of any equation $y = mx$ is a straight line through the origin, with slope m. What will happen if we add a number b on the right side to get an equation $y = mx + b$?

Graph.

7. $y = \frac{3}{4}x$

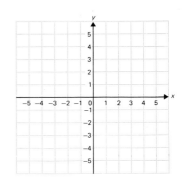

8. $y = -\frac{4}{5}x$

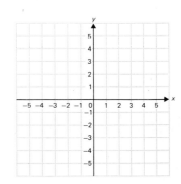

9. Graph $y = x + 3$ and compare it with $y = x$.

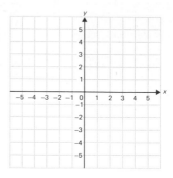

Example 7 Graph $y = x + 2$ and compare it with $y = x$.

We first make a table of values.

x	y (or $x + 2$)
0	2
1	3
-1	1
2	4
-2	0
3	5

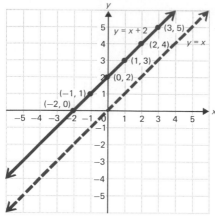

We plot these points. If we had enough of them they would make a line. We draw this line with a ruler and label it $y = x + 2$. The graph of $y = x$ is drawn for comparison. Note that the graph of $y = x + 2$ looks just like the graph of $y = x$, but it is moved up 2 units.

DO EXERCISES 9 AND 10.

10. Graph $y = x - 1$ and compare it with $y = x$.

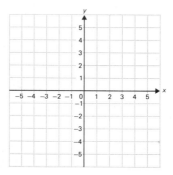

Example 8 Graph $y = 2x - 3$ and compare it with $y = 2x$.

We first make a table of values.

x	y (or $2x - 3$)
0	-3
1	-1
2	1
-1	-5

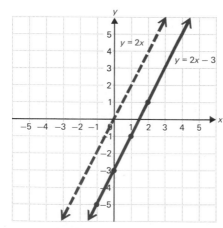

We draw the graph of $y = 2x - 3$. It looks just like the graph of $y = 2x$, but it is moved down 3 units.

11. Graph $y = 2x + 3$ and compare it with $y = 2x$.

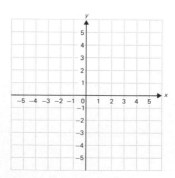

DO EXERCISE 11.

The graph of $y = mx$ goes through the origin $(0, 0)$. The graph of any equation $y = mx + b$ is also a line. It is parallel to $y = mx$, but moved up or down. It goes through the point $(0, b)$. That point is called the *y-intercept*. We may also refer to the number b as the *y*-intercept. The number m is still called the *slope*. It tells us how steeply the line slants.

Example 9   Graph $y = \dfrac{2}{5}x + 4$.

We first make a table of values. Using multiples of 5 avoids fractions.

When $x = 0$, $y = \dfrac{2}{5} \cdot 0 + 4 = 0 + 4 = 4$.

When $x = 5$, $y = \dfrac{2}{5} \cdot 5 + 4 = 2 + 4 = 6$.

When $x = -5$, $y = \dfrac{2}{5}(-5) + 4 = -2 + 4 = 2$.

Since two points determine a line, that is all you really need to graph a line, but you should always plot a third point as a check.

x	y
0	4
5	6
-5	2

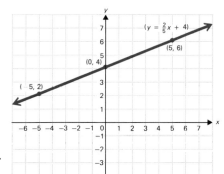

We draw the graph of $y = \dfrac{2}{5}x + 4$.

Example 10   Graph $y = -\dfrac{3}{4}x - 2$.

We first make a table of values.

When $x = 0$, $y = -\dfrac{3}{4} \cdot 0 - 2 = 0 - 2 = -2$.

When $x = 4$, $y = -\dfrac{3}{4} \cdot 4 - 2 = -3 - 2 = -5$.

When $x = -4$, $y = -\dfrac{3}{4}(-4) - 2 = 3 - 2 = 1$.

x	y
0	-2
4	-5
-4	1

We plot these points and draw a line through them. We plot this point for a check to see whether it is on the line.

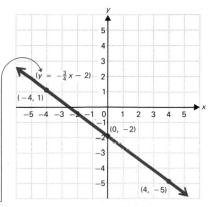

We draw the graph of $y = -\dfrac{3}{4}x - 2$.

Don't forget to label the graph.

DO EXERCISES 12–15.

Any equation $y = mx + b$ has a graph that is a straight line. It goes through the point $(0, b)$, the y-intercept, and has slope m.

Graph.

12. $y = \dfrac{3}{5}x + 2$

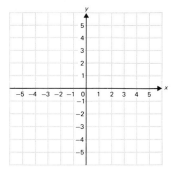

13. $y = \dfrac{3}{5}x - 2$

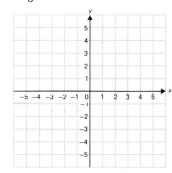

14. $y = -\dfrac{3}{5}x - 1$

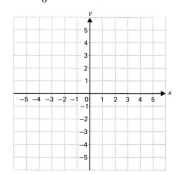

15. $y = -\dfrac{3}{5}x + 4$

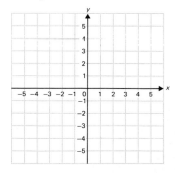

ANSWERS ON PAGE A–17

SOMETHING EXTRA

AN APPLICATION

The equation

$$P = 0.792t + 23$$

can be used to estimate the average percentage P of field goal completions by college basketball players t years from 1940. Since 1980 is 40 years from 1940, the average percentage of field goal completions in 1980 is given by

$$P = 0.792(40) + 23$$
$$P = 54.68\%.$$

EXERCISES

Find the average percentage of field goal completions in

1. 1984.   **2.** 1990.   **3.** 1995.   **4.** 2000.

NAME   CLASS   ANSWERS

EXERCISE SET 6.2

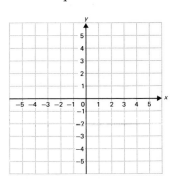 Determine whether the given point is a solution of the equation.

1. $(2, 5)$; $y = 3x - 1$

2. $(1, 7)$; $y = 2x + 5$

3. $(2, -3)$; $3x - y = 4$

4. $(-1, 4)$; $2x + y = 6$

5. $(-2, -1)$; $2a + 2b = -7$

6. $(0, -4)$; $4m + 2n = -9$

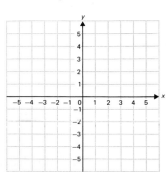 Graph.

7. $y = 4x$

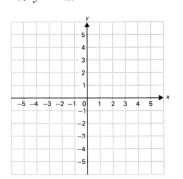

8. $y = 2x$

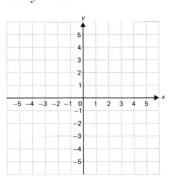

9. $y = -2x$

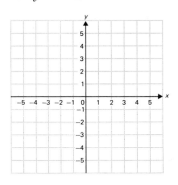

10. $y = -4x$

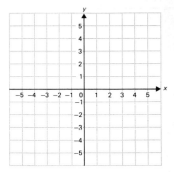

11. $y = \dfrac{1}{3}x$

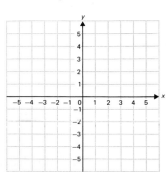

12. $y = \dfrac{1}{4}x$

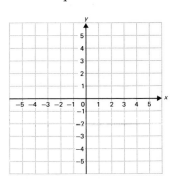

13. $y = -\dfrac{3}{2}x$

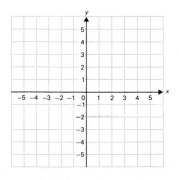

14. $y = -\dfrac{5}{4}x$

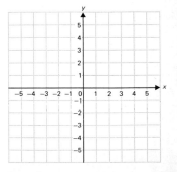

ANSWERS

1. _____

2. _____

3. _____

4. _____

5. _____

6. _____

15. $y = x + 1$

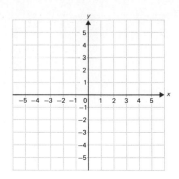

16. $y = -x + 1$

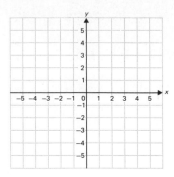

17. $y = 2x + 2$

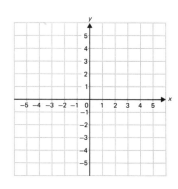

18. $y = 3x - 2$

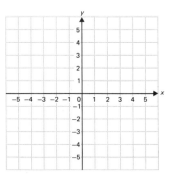

19. $y = \dfrac{1}{3}x - 1$

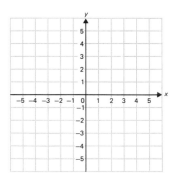

20. $y = \dfrac{1}{2}x + 1$

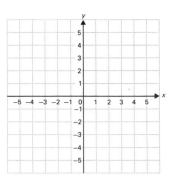

Use graph paper. Draw and label x- and y-axes. Graph these equations. Attach graphs to this page.

21. $y = -x - 3$

22. $y = -x - 2$

23. $y = \dfrac{5}{2}x + 3$

24. $y = \dfrac{5}{3}x - 2$

25. $y = -\dfrac{5}{2}x - 2$

26. $y = -\dfrac{5}{3}x + 3$

6.3 LINEAR EQUATIONS

● GRAPHING USING INTERCEPTS

The fastest method for graphing equations whose graphs are straight lines involves the use of intercepts. Look at the graph of $y - 2x = 4$ shown below. We could graph this equation by solving for y to get $y = 2x + 4$ and proceed as before, but we want to develop a faster method.

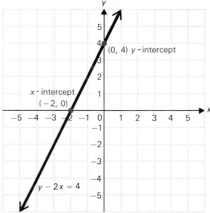

The y-intercept is $(0, 4)$. It occurs where the line crosses the y-axis and always has 0 as the first coordinate. The x-intercept is $(-2, 0)$. It occurs where the line crosses the x-axis and always has 0 as the second coordinate.

DO EXERCISE 1.

> The x-intercept is in the form $(a, 0)$. **To find a, let $y = 0$.**
> The y-intercept is in the form $(0, b)$. **To find b, let $x = 0$.**

Now let us draw a graph using intercepts.

Example 1   Graph $4x + 3y = 12$.

To find the x-intercept, let $y = 0$. Then

$$4x + 3 \cdot 0 = 12$$
$$4x = 12$$
$$x = 3.$$

Thus $(3, 0)$ is the x-intercept.

Note that this amounts to covering up the y-term and looking at the rest of the equation.

To find the y-intercept, let $x = 0$. Then

$$4 \cdot 0 + 3y = 12$$
$$3y = 12$$
$$y = 4.$$

Thus $(0, 4)$ is the y-intercept.

OBJECTIVES

After finishing Section 6.3, you should be able to:

● Graph using intercepts.
●● Graph equations of the type $x = a$ **or** $y = b$.

1. Look at the graph shown below.
   a) Find the coordinates of the x-intercept.
   b) Find the coordinates of the y-intercept.

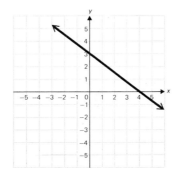

ANSWER ON PAGE A–19

Graph using intercepts.

2. $2x + 3y = 6$

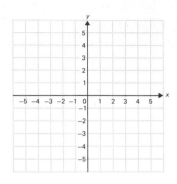

3. $3y - 4x = 12$

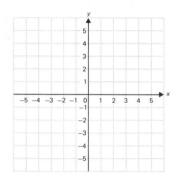

We plot these points and draw the line. A third point should be used as a check. We substitute any arbitrary value for x and solve for y.

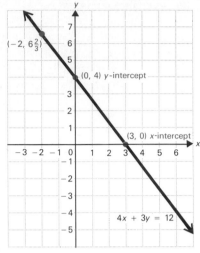

We let $x = -2$. Then

$$4(-2) + 3y = 12 \qquad \text{Substituting } -2 \text{ for } x$$
$$-8 + 3y = 12$$
$$3y = 12 + 8$$
$$3y = 20$$
$$y = \frac{20}{3}, \quad \text{or} \quad 6\frac{2}{3} \qquad \text{Solving for } y$$

We see that the point $(-2, 6\frac{2}{3})$ is on the graph, so the graph is probably correct.

DO EXERCISES 2 AND 3.

▄▄ EQUATIONS WITH A MISSING VARIABLE

Consider the equation $y = 3$. We can think of it as $y = 0 \cdot x + 3$. No matter what number we choose for x, we find that y is 3.

Example 2 Graph $y = 3$.

Any ordered pair $(x, 3)$ is a solution. So the line is parallel to the x-axis with y-intercept $(0, 3)$.

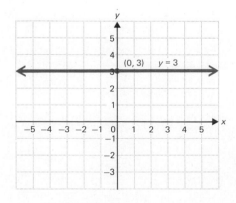

Example 3   Graph $x = -4$.

Any ordered pair $(-4, y)$ is a solution. So the line is parallel to the y-axis with x-intercept $(-4, 0)$.

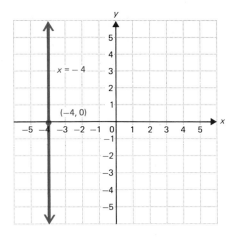

The graph of $y = b$ is a horizontal line. The graph of $x = a$ is a vertical line.

DO EXERCISES 4–7.

Equations whose graphs are straight lines are linear equations. We summarize the best procedure for graphing linear equations.

> *To Graph Linear Equations*
> 1. **Is the equation of the type $x = a$ or $y = b$? If so, the graph will be a line parallel to an axis.**
> 2. **If the line is not of the type $x = a$ or $y = b$, find the intercepts. Graph using the intercepts if this is feasible.**
> 3. **If the intercepts are too close together, choose another point farther from the origin.**
> 4. **In any case, use a third point as a check.**

Graph.

4. $x = 5$

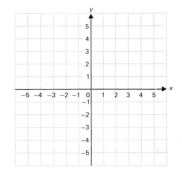

5. $y = -2$

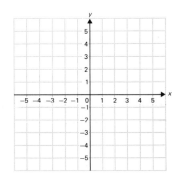

6. $x = 0$

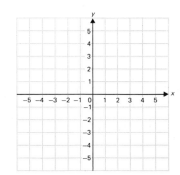

7. $x = -3$

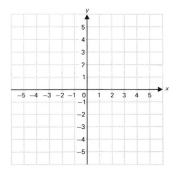

ANSWERS ON PAGE A–19

SOMETHING EXTRA

AN APPLICATION

DEPRECIATION: THE STRAIGHT-LINE METHOD

A company buys a machine for $5200. The machine is expected to last for 8 years at which time its trade-in, or scrap, value will be $1300. Over its lifetime it depreciates $5200 − $1300, or $3900. If the company figures the decline in value to be the same each year—that is, $\frac{1}{8}$, or 12.5% of $3900, which is $487.50—then they are using what is called *straight-line depreciation*. We see this in the graph below. We see that the book value after one year is $5200 − $487.50, or $4712.50. After two years it is $4712.50 − $487.50, or $4225. After 3 years it is $4225 − $487.50, or $3737.50, and so on.

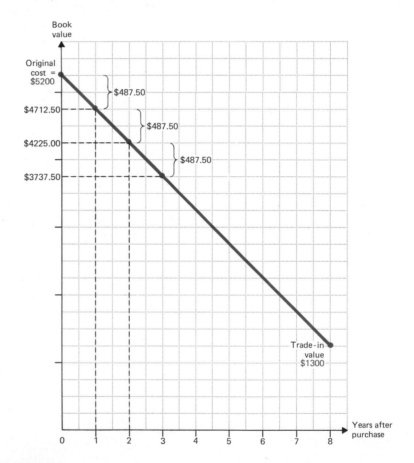

EXERCISE

Find the book values of the machine after each of the remaining years.

CLASS

ANSWERS

EXERCISE SET 6.3

■● Find the intercepts of each equation. Then graph.

1. $5x - 3y = 15$

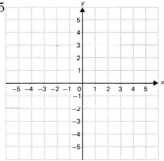

2. $2x - 4y = 8$

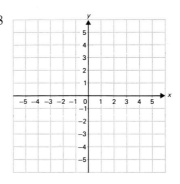

1. _____

2. _____

3. $4x + 2y = 8$

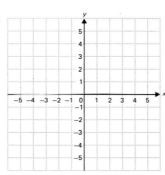

4. $3x + 5y = 15$

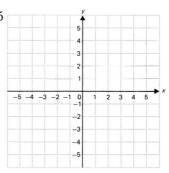

3. _____

4. _____

5. $x - 1 = y$

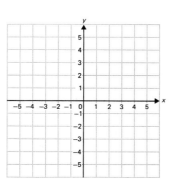

6. $x - 3 = y$

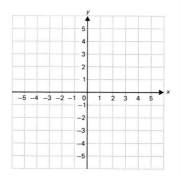

5. _____

6. _____

7. $2x - 1 = y$

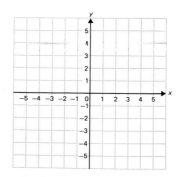

8. $3x - 2 = y$

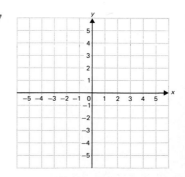

7. _____

8. _____

ANSWERS

9. _____

10. _____

11. _____

12. _____

13. _____

14. _____

15. See graph. _____

16. See graph. _____

17. See graph. _____

18. See graph. _____

19. See graph. _____

20. See graph. _____

21. See graph. _____

22. See graph. _____

23. See graph. _____

24. See graph. _____

Use graph paper. Find the intercepts of each equation. Then graph. Attach graphs to this page.

9. $4x - 3y = 12$ **10.** $2x - 5y = 10$ **11.** $7x + 2y = 6$

12. $3x + 3y = 5$ **13.** $y = -4 - 4x$ **14.** $y = -3 - 3x$

●● Graph.

15. $x = -2$

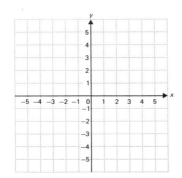

16. $x = -1$

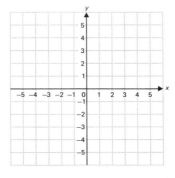

17. $y = 2$

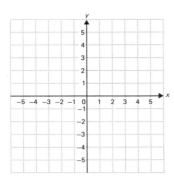

18. $y = 4$

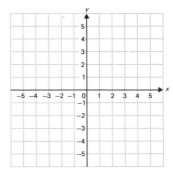

Use graph paper. Graph each of the following. Attach graphs to this page.

19. $x = 2$ **20.** $x = 3$ **21.** $y = 0$

22. $y = -1$ **23.** $x = \dfrac{3}{2}$ **24.** $x = -\dfrac{5}{2}$

6.4 TRANSLATING PROBLEMS TO EQUATIONS

● The first and often hardest part of solving a problem is translating it to mathematical language. Translating becomes easier in many cases if we translate to more than one equation. When we do this, we use more than one variable.

Example 1   The sum of two numbers is 15. One number is four times the other. Find the numbers.

There are two statements in this problem. We translate the first one.

Translate to a system of equations. Do not attempt to solve. Save for later use.

The sum of two numbers is 15.

$$x + y \qquad = 15$$

1. The sum of two numbers is 115. The difference is 21. Find the numbers.

We have used x and y for the numbers. Now we translate the second statement, remembering to use x and y.

One number is four times the other.

$$y \qquad = \quad 4 \quad \cdot \qquad x$$

For the second statement we could have also translated to $x = 4y$. This would also have been correct. The problem has been translated to a pair or *system of equations*. We list what the variables represent and then list the equations.

Let x represent one number and y represent the other number.

$$x + y = 15$$
$$y = 4x$$

DO EXERCISE 1.

Translate to a system of equations. Do not attempt to solve. Save for later use.

Example 2   Badger Rent-a-Car rents compact cars at a daily rate of $23.95 plus 20¢ per mile. Thirsty Rent-a-Car rents compact cars at a daily rate of $22.95 plus 22¢ per mile. For what mileage is the cost the same?

2. Acme Rent-a-Car rents a station wagon at a daily rate of $31.95 plus 33¢ per mile. Speedo Rentzit rents a wagon for $34.95 plus 29¢ per mile. For what mileage is the cost the same?

We translate the first statement, using $0.20 for 20¢.

23.95 plus 20¢ times the number of miles driven is cost.

$$23.95 \; + \; 0.20 \quad\cdot\qquad\qquad m \qquad\qquad = \; c$$

We have let m represent the mileage and c the cost. We translate the second statement, but again it helps to reword it first.

22.95 plus 22¢ times the number of miles driven is cost.

$$22.95 \; + \; 0.22 \quad\cdot\qquad\qquad m \qquad\qquad = \; c$$

We have now translated to a system of equations. We then let m represent the mileage and c represent the cost.

$$23.95 + 0.20m = c$$
$$22.95 + 0.22m = c$$

DO EXERCISE 2.

Translate to a system of equations. Do not attempt to solve. Draw a picture if helpful. Save for later use.

3. The perimeter of a rectangle is 76 cm. The length is 17 cm more than the width. Find the length and width.

The problem in Example 2 can also be translated to one equation with one variable, as follows:

$$23.95 + 0.20m = 22.95 + 0.22m.$$

When solving applied problems, use one variable if it is reasonable. Use two variables when it makes the translation easier.

Making a drawing is often helpful. Whenever it makes sense to do so you should make a drawing before trying to translate.

Example 3 The perimeter of a rectangle is 90 cm. The length is 20 cm greater than the width. Find the length and width.

From the drawing we see that the perimeter (the distance around) of the rectangle is $l + l + w + w$, or $2l + 2w$. We translate the first statement.

$$\underbrace{\text{The perimeter}}_{2l + 2w} \underbrace{\text{is}}_{=} \underbrace{\text{90 cm.}}_{90}$$

We translate the second statement.

$$\underbrace{\text{The length}}_{l} \underbrace{\text{is}}_{=} \underbrace{\text{20 cm greater than the width.}}_{20 + w}$$

We have translated to a system of equations. We have let w represent the width and l represent the length.

$$2w + 2l = 90$$
$$l = 20 + w$$

DO EXERCISE 3.

EXERCISE SET 6.4

▨ Translate to a system of equations. Do not attempt to solve. Save for later use.

1. The sum of two numbers is 58. The difference is 16. Find the numbers.

1. _____

2. The sum of two numbers is 26.4. One number is five times the other. Find the numbers.

2. _____

3. The perimeter of a rectangle is 400 m. The width is 40 m less than the length. Find the length and width.

3. _____

4. The perimeter of a rectangle is 76 cm. The width is 17 cm less than the length. Find the length and width.

4. _____

5. Badger Rent-a-Car rents a compact car at a daily rate of $23.95 plus 20¢ per mile. Hartz Rent-a-Car rents a compact car at a daily rate of $24.95 plus 27¢ per mile. For what mileage is the cost the same?

5. _____

6. Badger rents a basic car at a daily rate of $25.95 plus 20¢ per mile. Hartz rents a basic car at a daily rate of $25.95 plus 29¢ per mile. For what mileage is the cost the same?

6. _____

7. The difference between two numbers is 16. Three times the larger number is seven times the smaller. What are the numbers?

7. _____

8. The difference between two numbers is 18. Twice the smaller number plus three times the larger is 74. What are the numbers?

8. _____

9. Two angles are supplementary. One is 8° more than 3 times the other. Find the angles. (Supplementary angles are angles whose sum is 180°.)

9. _____

10. Two angles are supplementary. One is 30° more than 2 times the other. Find the angles.

10. _____

11. Two angles are complementary. Their difference is 34°. Find the angles. (Complementary angles are angles whose sum is 90°.)

11. _____

12. Two angles are complementary. One angle is 42° more than $\frac{1}{2}$ the other. Find the angles.

12. _____

13. In a vineyard a vintner uses 820 hectares to plant Chardonnay and Riesling grapes. The vintner knows that profits will be greatest by planting 140 hectares more of Chardonnay than Riesling. How many hectares of each grape should be planted?

13. _____

14. The Hayburner Horse Farm allots 650 hectares to plant hay and oats. The owners know that their needs are best met if they plant 180 hectares more of hay than oats. How many hectares of each should they plant?

14. _____

6.5   SYSTEMS OF EQUATIONS

● IDENTIFYING SOLUTIONS

OBJECTIVES

After finishing Section 6.5, you should be able to:

● Determine whether an ordered pair is a solution of a system of equations.

●● Solve systems of equations by graphing.

A set of equations such as

$$x + y = 8$$
$$2x - y = 1$$

is called a *system of equations*. A solution of a system of two equations is an ordered pair that makes both equations true. Consider the system listed above. Look at the graphs. Recall that a graph of an equation is a picture of its solution set. Each point on a graph corresponds to a solution. Which points (ordered pairs) are solutions of *both* equations? The graph shows that there is only one. It is the point P where the graphs cross. This point looks as if its coordinates are (3, 5). We check:

$$\frac{x + y = 8}{\begin{array}{c|c} 3 + 5 & 8 \\ \hline 8 & \end{array}}$$

$$\frac{2x - y = 1}{\begin{array}{c|c} 2 \cdot 3 - 5 & 1 \\ 6 - 5 & \\ 1 & \end{array}}$$

The point P is a picture of the set of common solutions.

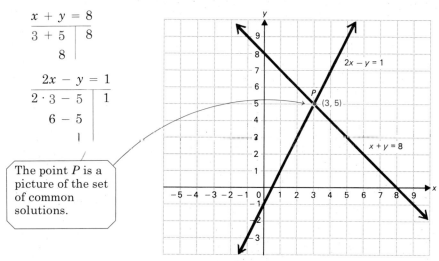

There is just one solution of the system of equations. It is (3, 5). In other words, $x = 3$ and $y = 5$.

Example 1   Determine whether (1, 2) is a solution of the system

$$y = x + 1$$
$$2x + y = 4.$$

We check:

$$\frac{y = x + 1}{\begin{array}{c|c} 2 & 1 + 1 \\ 2 & 2 \end{array}} \qquad \frac{2x + y = 4}{\begin{array}{c|c} 2 \cdot 1 + 2 & 4 \\ 2 + 2 & \\ 4 & \end{array}}$$

(1, 2) is a solution of the system.

Example 2   Determine whether $(-3, 2)$ is a solution of the system

$$a + b = -1$$
$$b + 3a = 4.$$

We check:

$$\frac{a + b = -1}{\begin{array}{c|c} -3 + 2 & -1 \\ -1 & \end{array}} \qquad \frac{b + 3a = 4}{\begin{array}{c|c} 2 + 3(-3) & 4 \\ 2 - 9 & \\ -7 & \end{array}}$$

Determine whether the given ordered pair is a solution of the system of equations.

1. $(2, -3);$ $x = 2y + 8$

$\qquad\qquad 2x + y = 1$

2. $(20, 40);$ $a = \dfrac{1}{2}b$

$\qquad\qquad b - a = 60$

Solve by graphing.

3. $2x + y = 1$

$\quad x = 2y + 8$

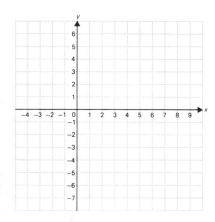

4. $y + 4 = x$

$\quad x - y = -2$

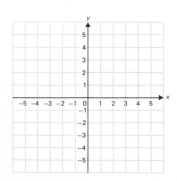

The point $(-3, 2)$ is not a solution of $b + 3a = 4$. Thus it is not a solution of the system.

DO EXERCISES 1 AND 2.

●● SOLVING SYSTEMS BY GRAPHING

> To solve a system of equations by graphing, we graph both equations and find coordinates of the point(s) of intersection. Then we check. If the lines are parallel, there is no solution.

Example 3   Solve by graphing: $x + 2y = 7$

$\qquad\qquad\qquad\qquad\qquad\quad x = y + 4.$

We graph the equations. Point P looks as if it has coordinates $(5, 1)$.

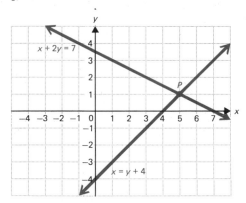

We check:

$$
\begin{array}{c|c}
x + 2y = 7 & \\
\hline
5 + 2 \cdot 1 & 7 \\
5 + 2 & \\
7 & \\
\end{array}
\qquad
\begin{array}{c|c}
x = y + 4 & \\
\hline
5 & 1 + 4 \\
5 & 5 \\
\end{array}
$$

The solution is $(5, 1)$.

Example 4   Solve by graphing: $y = 3x + 4$

$\qquad\qquad\qquad\qquad\qquad\quad y = 3x - 3.$

The graphs are parallel. There is no point where they cross, so the system has no solution.

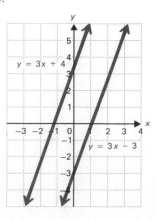

DO EXERCISES 3 AND 4.

NAME   CLASS   ANSWERS

EXERCISE SET 6.5

■ ● Determine whether the given ordered pair is a solution of the given system of equations. Remember to use alphabetical order of variables.

1. $(3, 2)$; $2x + 3y = 12$
　　　　　$x - 4y = -5$

2. $(1, 5)$; $5x - 2y = -5$
　　　　　$3x - 7y = -32$

1. _____

2. _____

3. $(-3, 4)$; $2x = -y - 2$
　　　　　　$y + 7x = 9$

4. $(3, -2)$; $3t - 2s = 0$
　　　　　　$t + 2s = 15$

3. _____

4. _____

5. $(2, -2)$; $b + 2a = 2$
　　　　　　$b - a = -4$

6. $(-1, -3)$; $3r + s = -6$
　　　　　　　$2r = 1 + s$

5. _____

6. _____

ANSWERS

● ● ● Use graph paper. Solve by graphing. Check on this paper.

7. $x + 2y = 10$
$3x + 4y = 8$

Check: $x + 2y = 10$ $3x + 4y = 8$

7. _____

8. $u = v$
$4u = 2v - 6$

Check: $u = v$ $4u = 2v - 6$

8. _____

9. $8x - y = 29$
$2x + y = 11$

Check: $8x - y = 29$ $2x + y = 11$

9. _____

10. $4x - y = 10$
$3x + 5y = 19$

Check: $4x - y = 10$ $3x + 5y = 19$

10. _____

11. $a - b = 6$
$a - 7 = b$

Check: $a - b = 6$ $a - 7 = b$

11. _____

12. $x - 2y = 6$
$2x - 3y = 5$

Check: $x - 2y = 6$ $2x - 3y = 5$

12. _____

6.6   SOLVING BY SUBSTITUTION

●   THE SUBSTITUTION METHOD

Graphing helps picture the solution of a system of equations, but solving by graphing is not fast or accurate. Let's learn better ways.

Example 1   Solve the system: $x + y = 6$
$$x = y + 2.$$

The second equation says that x and $y + 2$ name the same thing. Thus in the first equation, we can substitute $y + 2$ for x:

$$x + y = 6$$
$$(y + 2) + y = 6. \qquad \text{Substituting } y + 2 \text{ for } x$$

This last equation has only one variable. We solve it:

$$(y + 2) + y = 6$$
$$2y + 2 = 6 \qquad \text{Collecting like terms}$$
$$2y = 4$$
$$y = 2.$$

We return to the original pair of equations. We substitute 2 for y in either of them. We use the first equation.

$$x + y = 6$$
$$x + 2 = 6 \qquad \text{Substituting 2 for } y$$
$$x = 4.$$

The ordered pair $(4, 2)$ may be a solution. We check.

Check:

$x + y = 6$		$x = y + 2$	
4 ∣ 2	6	4	2 + 2
	6		4

Since $(4, 2)$ checks, we have the solution. We could also express the answer as $x = 4$, $y = 2$.

DO EXERCISE 1.

Example 2   Solve: $s = 13 - 3t$
$$s + t = 5.$$

We substitute $13 - 3t$ for s in the second equation:

$$s + t = 5$$
$$(13 - 3t) + t = 5. \qquad \text{Substituting } 13 - 3t \text{ for } s$$

Now we solve for t:

$$13 - 2t = 5 \qquad \text{Collecting like terms}$$
$$-2t = -8 \qquad \text{Adding } -13$$
$$t = \frac{-8}{-2}, \quad \text{or } 4 \qquad \text{Multiplying by } \frac{1}{-2}$$

OBJECTIVES

After finishing Section 6.6, you should be able to:

●   Solve a system of two equations by the substitution method when one of them has a variable alone on one side.

●●   Solve a system of two equations by the substitution method when neither equation has a variable alone on one side.

Solve by the substitution method. Do not graph.

1. $x + y = 5$
$$x = y + 1$$

Solve by the substitution method.

2. $a - b - 4$

$b = 2 - a$

Solve.

3. $x - 2y = 8$

$2x + y = 8$

Next we substitute 4 for t in the second equation of the original system:

$s + t = 5$

$s + 4 = 5$ Substituting 4 for t

$s = 1.$

Check:

	$s = 13 - 3t$		$s + t = 5$
1	$13 - 3 \cdot 4$	$1 + 4$	5
	$13 - 12$	5	
	1		

Since (1, 4) checks, it is the solution.

DO EXERCISE 2.

■ ■   **SOLVING FOR THE VARIABLE FIRST**

Sometimes neither equation of a pair has a variable alone on one side. Then we solve one equation for one of the variables and proceed as before.

Example 3   Solve: $x - 2y = 6$

$3x + 2y = 4.$

We solve one equation for one variable. Since the coefficient of x is 1 in the first equation, it is easier to solve it for x:

$x - 2y = 6$

$x = 6 + 2y.$

We substitute $6 + 2y$ for x in the second equation of the original pair and solve:

$3x + 2y = 4$

$3(6 + 2y) + 2y = 4$ Substituting $6 + 2y$ for x

$18 + 6y + 2y = 4$

$18 + 8y = 4$

$8y = -14$

$y = \dfrac{-14}{8}, \quad \text{or} \ -\dfrac{7}{4}.$

We go back to either of the original equations and substitute $-\frac{7}{4}$ for y. It will be easier to solve for x in the first equation:

$x - 2y = 6$

$x - 2\left(-\dfrac{7}{4}\right) = 6$

$x + \dfrac{7}{2} = 6$

$x = 6 - \dfrac{7}{2}$

$x = \dfrac{5}{2}$

Check:

	$x - 2y = 6$		$3x + 2y = 4$
	$\frac{5}{2} - 2(-\frac{7}{4})$ 6		$3 \cdot \frac{5}{2} + 2(-\frac{7}{4})$ 4
	$\frac{5}{2} + \frac{7}{2}$		$\frac{15}{2} - \frac{7}{2}$
	$\frac{12}{2}$		$\frac{8}{2}$
	6		4

Since $(\frac{5}{2}, -\frac{7}{4})$ checks, it is the solution.

DO EXERCISE 3.

EXERCISE SET 6.6

▗▄▖ Solve by the substitution method.

1. $x + y = 4$
$y = 2x + 1$

2. $x + y = 10$
$y = x + 8$

3. $y = x + 1$
$2x + y = 4$

4. $y = x - 6$
$x + y = -2$

5. $y = 2x - 5$
$3y - x = 5$

6. $y = 2x + 1$
$x + y = -2$

7. $x = -2y$
$x + 4y = 2$

8. $r = -3s$
$r + 4s = 10$

▗▄▄▖ Solve by the substitution method. Get one variable alone first.

9. $s + t = -4$
$s - t = 2$

10. $x - y = 6$
$x + y = -2$

1. _____

2. _____

3. _____

4. _____

5. _____

6. _____

7. _____

8. _____

9. _____

10. _____

ANSWERS

11. $y - 2x = -6$ **12.** $x - y = 5$ **13.** $2x + 3y = -2$ **14.** $x + 2y = 10$
$\quad\;\; 2y - x = 5$ $\quad\;\; x + 2y = 7$ $\quad\;\; 2x - y = 9$ $\quad\;\; 3x + 4y = 8$

11. _____

12. _____

13. _____

14. _____

15. $x - y = -3$ **16.** $3b + 2a = 2$ **17.** $r - 2s = 0$ **18.** $y - 2x = 0$
$\quad\;\; 2x + 3y = -6$ $\quad\;\; -2b + a = 8$ $\quad\;\; 4r - 3s = 15$ $\quad\;\; 3x + 7y = 17$

15. _____

16. _____

17. _____

18. _____

Solve by the substitution method.

19. $y - 2.35x = -5.97$
$\quad\;\; 2.14y - x = 4.88$

19. _____

6.7 THE ADDITION METHOD

⬛•⬛ SOLVING BY THE ADDITION METHOD

Another method of solving systems of equations is called the *addition method*.

Example 1   Solve: $x + y = 5$
$\qquad\qquad\qquad\qquad x - y = 1.$

We will use the addition principle for equations. According to the second equation, $x - y$ and 1 are the same thing. Thus we can add $x - y$ to the left side of the first equation and 1 to the right side:

$$\begin{array}{r} x + y = 5 \\ \underline{x - y = 1} \\ 2x + 0y = 6. \end{array}$$

We have made one variable "disappear." We have an equation with just one variable:

$$2x = 6.$$

We solve for x: $x = 3$. Next substitute 3 for x in either of the original equations:

$$\begin{array}{ll} x + y = 5 & \\ 3 + y = 5 & \text{Substituting 3 for } x \text{ in the first equation} \\ \quad\ y = 2 & \text{Solving for } y \end{array}$$

Check:

$$\begin{array}{c|c} x + y = 5 & x - y = 1 \\ \hline 3 + 2 \ \big|\ 5 & 3 - 2 \ \big|\ 1 \\ 5 \ \big| & 1 \ \big| \end{array}$$

Since (3, 2) checks, it is the solution.

DO EXERCISES 1 AND 2.

⬛•• USING THE MULTIPLICATION PRINCIPLE FIRST

The addition method allows us to eliminate a variable. We may need to multiply by -1 to make this happen.

Example 2   Solve: $2x + 3y = 8$
$\qquad\qquad\qquad\qquad x + 3y = 7.$

If we add, we do not eliminate a variable. However, if the $3y$ were $-3y$ in one equation we could. We multiply on both sides of the second equation by -1 and then add:

$$\begin{array}{rl} 2x + 3y = \ \ 8 & \\ \underline{-x - 3y = -7} & \text{Multiplying by } -1 \\ x \qquad\quad = \ \ 1. & \text{Adding} \end{array}$$

Solve using the addition method

1. $x + y = 5$
$\quad 2x - y = 4$

2. $3x - 3y = 6$
$\quad 3x + 3y = 0$

ANSWERS ON PAGE A–22

Solve. Multiply one equation by -1 first.

3. $5x + 3y = 17$
$5x - 2y = -3$

Solve.

4. $4a + 7b = 11$
$2a + 3b = 5$

Now we substitute 1 for x in one of the original equations:

$$x + 3y = 7$$
$$1 + 3y = 7 \qquad \text{Substituting 1 for } x \text{ in the second equation}$$
$$3y = 6$$
$$y = 2 \qquad \text{Solving for } y$$

Check:

$2x + 3y = 8$		$x + 3y = 7$	
$2 \cdot 1 + 3 \cdot 2$	8	$1 + 3 \cdot 2$	7
$2 + 6$		$1 + 6$	
8		7	

Since $(1, 2)$ checks, it is the solution.

DO EXERCISE 3.

In Example 2 we used the multiplication principle, multiplying by -1. We often need to multiply by something other than -1.

Example 3 Solve: $3x + 6y = -6$
$ 5x - 2y = 14.$

This time we multiply by 3 on both sides of the second equation. Then we add:

$$\begin{array}{ll} 3x + 6y = -6 & \\ \underline{15x - 6y = 42} & \text{Multiplying by 3} \\ 18x \phantom{{}+6y} = 36 & \text{Adding} \\ x \phantom{{}+18y} = 2 & \text{Solving for } x \end{array}$$

We go back to the first equation and substitute 2 for x:

$$3 \cdot 2 + 6y = -6 \qquad \text{Substituting}$$
$$6 + 6y = -6$$
$$6y = -12$$
$$y = -2 \qquad \text{Solving for } y$$

Check:

$3x + 6y = -6$		$5x - 2y = 14$	
$3 \cdot 2 + 6 \cdot (-2)$	-6	$5 \cdot 2 - 2 \cdot (-2)$	14
$6 + (-12)$		$10 - (-4)$	
-6		14	

Since $(2, -2)$ checks, it is the solution.

> *Caution!* Solving a *system* of equations in two variables requires finding an ordered *pair* of numbers. Once you have solved for one variable, don't forget the other.

DO EXERCISE 4.

Example 4   Solve: $3x + 5y = 6$

$\qquad\qquad\qquad\quad 5x + 3y = 4.$

We use the multiplication principle with both equations:

$3x + 5y = 6$

$5x + 3y = 4$

$15x + 25y = 30$ Multiply on both sides of the first equation by 5

$\underline{-15x - 9y = -12}$ Multiply on both sides of the second equation by -3

$\qquad\quad 16y = 18$ Adding

$\qquad\qquad y = \dfrac{18}{16},\quad\text{or } \dfrac{9}{8}.$

We substitute $\dfrac{9}{8}$ for y in one of the original equations:

$3x + 5y = 6$

$3x + 5 \cdot \dfrac{9}{8} = 6$ Substituting $\dfrac{9}{8}$ for y in the first equation

$3x + \dfrac{45}{8} = 6$

$3x = 6 - \dfrac{45}{8}$

$3x = \dfrac{48}{8} - \dfrac{45}{8}$

$3x = \dfrac{3}{8}$

$x = \dfrac{3}{8} \cdot \dfrac{1}{3},\quad\text{or } \dfrac{1}{8}$ Solving for x

Check:

$3x + 5y = 6$	$5x + 3y = 4$
$3 \cdot \frac{1}{8} + 5 \cdot \frac{9}{8}$ $\Big|$ 6	$5 \cdot \frac{1}{8} + 3 \cdot \frac{9}{8}$ $\Big|$ 4
$\frac{3}{8} + \frac{45}{8}$	$\frac{5}{8} + \frac{27}{8}$
$\frac{48}{8}$	$\frac{32}{8}$
6 $\Big|$	4 $\Big|$

The solution is $(\frac{1}{8}, \frac{9}{8})$.

DO EXERCISES 5 AND 6.

Solve.

5. $5x + 3y = 2$

$\phantom{\textbf{5.} }3x + 5y = -2$

6. $2x + 3y = 1$

$\phantom{\textbf{6.} }3x - 2y = 5$

Solve.

7. $2x + y = 15$
   $4x + 2y = 23$

Solve.

8. $\dfrac{1}{2}x + \dfrac{3}{10}y = \dfrac{1}{5}$

   $\dfrac{3}{10}x + \dfrac{1}{2}y = -\dfrac{1}{5}$

Some systems have no solution.

Example 5   Solve: $y - 3x = 2$
                                   $y - 3x = 1.$

We multiply by -1 on both sides of the second equation:

$$
\begin{array}{ll}
y - 3x = 2 & \\
\underline{-y + 3x = -1} & \text{Multiplying by } -1 \\
0 = 1. & \text{Adding}
\end{array}
$$

We obtain a false equation $0 = 1$, so there is no solution. The graphs of the equations are parallel lines. They do not intersect.

DO EXERCISE 7.

When decimals or fractions appear, multiply to clear them. Then proceed as before.

Example 6   Solve: $\dfrac{1}{4}x + \dfrac{5}{12}y = \dfrac{1}{2}$

                                   $\dfrac{5}{12}x + \dfrac{1}{4}y = \dfrac{1}{3}.$

The number 12 is a multiple of all of the denominators. We multiply on both sides of each equation by 12:

$$12\left(\dfrac{1}{4}x + \dfrac{5}{12}y\right) = 12 \cdot \dfrac{1}{2} \qquad 12\left(\dfrac{5}{12}x + \dfrac{1}{4}y\right) = 12 \cdot \dfrac{1}{3}$$

$$12 \cdot \dfrac{1}{4}x + 12 \cdot \dfrac{5}{12}y = 6 \qquad 12 \cdot \dfrac{5}{12}x + 12 \cdot \dfrac{1}{4}y = 4$$

$$3x + 5y = 6; \qquad\qquad 5x + 3y = 4.$$

The resulting system is

   $3x + 5y = 6$
   $5x + 3y = 4.$

The solution of this system is given in Example 4.

WHICH METHOD IS BETTER?

Although there are exceptions, the substitution method is probably better when a variable has a coefficient of 1, as in the following examples:

   $8x + 4y = 180$ $\qquad x + y = 561$
   $-x + y = 3;$ $\qquad\quad x = 2y.$

Otherwise, the addition method is better. Both methods work. When in doubt, use the addition method.

DO EXERCISE 8.

NAME

CLASS

EXERCISE SET 6.7

▮ • Solve using the addition method.

1. $x + y = 10$
$\quad x - y = 8$

2. $x - y = 7$
$\quad x + y = 3$

3. $x + y = 8$
$\quad 2x - y = 7$

4. $x + y = 6$
$\quad 3x - y = -2$

5. $3a + 4b = 7$
$\quad a - 4b = 5$

6. $7c + 5d = 18$
$\quad c - 5d = -2$

7. $8x - 5y = -9$
$\quad 3x + 5y = -2$

8. $7a - 3b = -12$
$\quad -4a + 3b = -3$

▮ • • Solve using the addition method. Use the multiplication principle first.

9. $-x - y = 8$
$\quad 2x - y = -1$

10. $x + y = -7$
$\quad 3x + y = -9$

11. $x + 3y = 19$
$\quad x + 3y = -1$

1. _____

2. _____

3. _____

4. _____

5. _____

6. _____

7. _____

8. _____

9. _____

10. _____

11. _____

ANSWERS

12. $4x - y = 1$
$4x - y = 7$

13. $3x - 2y = 10$
$5x + 3y = 4$

14. $2p + 5q = 9$
$3p - 2q = 4$

12. _____

13. _____

15. $2a + 3b = -1$
$3a + 5b = -2$

16. $5x - 2y = 0$
$2x - 3y = -11$

17. $0.3x + 0.2y = 0$
$x + 0.5y = -0.5$

14. _____

15. _____

16. _____

18. $0.4x + 0.1y = 0.1$
$0.6x + 0.2y = 0.3$

19. $\dfrac{3}{4}x + \dfrac{1}{3}y = 8$

$\dfrac{1}{2}x - \dfrac{5}{6}y = -1$

20. $\dfrac{2}{3}r - \dfrac{1}{5}s = 2$

$\dfrac{4}{3}r + 4s = -4$

17. _____

18. _____

19. _____

20. _____

Solve.

21. ▤ $4.05x + 2.53y = 1.23$
$1.12x + 1.43y = 2.81$

21. _____

6.8 SOLVING PROBLEMS

When solving problems involving systems, any method, substitution or addition, can be used.

● **Example 1** Badger Rent-a-Car rents compact cars at a daily rate of $23.95 plus 20¢ per mile. Thirsty Rent-a-Car rents compact cars at a daily rate of $22.95 plus 22¢ per mile. For what mileage is the cost the same?

(See Example 2 in Section 6.4.) If we let m represent the mileage and c represent cost, we see that the cost for Badger will be

$$23.95 + 0.20m = c.$$

The cost for Thirsty will be

$$22.95 + 0.22m = c.$$

We solve the system

$$23.95 + 0.20m = c$$
$$22.95 + 0.22m = c.$$

We use substitution since there is a variable alone on one side of an equation—in fact, both:

$$23.95 + 0.20m = 22.95 + 0.22m$$
$$1.00 + 0.20m = 0.22m \quad \text{Adding } -22.95$$
$$1.00 = 0.02m \quad \text{Adding } -0.20m$$
$$\frac{1.00}{0.02} = m$$
$$50 = m.$$

Thus if the cars are driven 50 miles, the cost will be the same.

DO EXERCISE 1.

Example 2 Howie Doon is 21 years older than Izzi Retyrd. In six years Howie will be twice as old as Izzi. How old are they now?

We translate the first statement.

Howie's age is 21 more than Izzi's age.   Rewording
x $= 21$ $+$ y Translating

Of course, x represents Howie's age and y represents Izzi's age. We translate the second statement.

Howie's age in six years will be twice Izzi's age in six years.   Rewording
$x + 6$ $=$ $2 \cdot$ $(y + 6)$ Translating

We now have translated to a system of equations:

$$x = 21 + y$$
$$x + 6 = 2(y + 6).$$

OBJECTIVE

After finishing Section 6.8, you should be able to:
● Solve problems by translating them to systems of equations.

1. Acme rents a station wagon at a daily rate of $21.95 plus 23¢ per mile. Speedo Rentzit rents a wagon for $24.95 plus 19¢ per mile. For what mileage is the cost the same?

2. Person A is 26 years older than Person B. In five years, A will be twice as old as B. How old are they now?

We use the addition method. We first add $-y$ on both sides of the first equation:

$$x - y = 21.$$

We also simplify the second equation:

$$x + 6 = 2y + 12$$
$$x - 2y = 6.$$

We solve the system

$$x - y = 21$$
$$x - 2y = 6.$$

We multiply on both sides of the second equation by -1 and add:

$$x - y = 21$$
$$\underline{-x + 2y = -6} \qquad \text{Multiplying by } -1$$
$$y = 15. \qquad \text{Adding}$$

We find x by substituting 15 for y in $x - y = 21$:

$$x - 15 = 21$$
$$x = 36.$$

We check in the original problem. Howie's age is 36, which is 21 more than 15, Izzi's age. In six years when Howie will be 42 and Izzi 21, Howie's age will be twice Izzi's age. The answer is that Howie is 36 and Izzi is 15.

DO EXERCISE 2.

Example 3 The perimeter of a rectangle is 90 cm. The length is 20 cm greater than the width. Find the length and width.

In Example 3, in Section 6.4, we translated this problem to the system,

$$2l + 2w = 90$$
$$l = 20 + w,$$

where we have let l represent the length and w represent the width. We solve, first adding $-w$ on both sides of $l = 20 + w$:

$$2l + 2w = 90 \qquad 2l + 2w = 90$$
$$l - w = 20 \qquad \underline{2l - 2w = 40} \qquad \text{Multiplying by 2}$$
$$4l = 130 \qquad \text{Adding}$$
$$l = 32.5$$

$$32.5 - w = 20 \qquad \text{Substituting in the equation } l - w = 20$$
$$-w = -12.5$$
$$w = 12.5$$

We leave the check to the student. The length is 32.5 cm and the width is 12.5 cm.

3. The perimeter of a rectangle is 76 cm. The length is 17 cm more than the width. Find the length and width.

DO EXERCISE 3.

NAME   CLASS   ANSWERS

EXERCISE SET 6.8

■ ● ■ Translate to systems of equations and solve. Many problems have already been translated in Exercise Set 6.4.

1. Badger rents an intermediate-size car at a daily rate of $14.95 plus 15¢ per mile. Another company rents an intermediate-size car for $17.95 plus 10¢ per mile. For what mileage is the cost the same?

1. _____

2. Badger rents a basic car at a daily rate of $15.95 plus 12¢ per mile. Another company rents a basic car for $16.95 plus 10¢ per mile. For what mileage is the cost the same?

2. _____

3. Sammy Tary is twice as old as his daughter. In four years Sammy's age will be three times what his daughter's age was six years ago. How old is each at present?

3. _____

4. Ann Shent is eighteen years older than her son. She was three times as old one year ago. How old is each?

4. _____

5. The difference between two numbers is 16. Three times the larger number is seven times the smaller. What are the numbers?

5. _____

6. The difference between two numbers is 18. Twice the smaller number plus three times the larger is 74. What are the numbers?

6. _____

7. Two angles are supplementary. One is 8° more than 3 times the other. Find the angles. (Supplementary angles are angles whose sum is 180°.)

7. _____

8. Two angles are supplementary. One is 30° more than 2 times the other. Find the angles.

8. _____

9. Two angles are complementary. Their difference is 34°. Find the angles. (Complementary angles are angles whose sum is 90°.)

10. Two angles are complementary. One angle is 42° more than $\frac{1}{2}$ the other. Find the angles.

9. _____

11. In a vineyard a vintner uses 820 hectares to plant Chardonnay and Riesling grapes. The vintner knows that profits will be greatest by planting 140 hectares more of Chardonnay than Riesling. How many hectares of each grape should be planted? (A *hectare* is a metric unit of area, about 2.47 acres.)

10. _____

12. The Hayburner Horse Farm allots 650 hectares to plant hay and oats. The owners know that their needs are best met if they plant 180 hectares more of hay than oats. How many hectares of each should they plant?

11. _____

12. _____

13. The perimeter of a rectangle is 400 m. The width is 40 m less than the length. Find the length and width.

13. _____

14. The perimeter of a rectangle is 76 cm. The width is 17 cm less than the length. Find the length and width.

14. _____

15. Several ancient Chinese books included problems that can be solved by translating to systems of equations. _Arithmetical Rules in Nine Sections_ is a book of 246 problems compiled by a Chinese mathematician, Chang Tsang, who died in 152 B.C. One of the problems is: Suppose there are a number of rabbits and pheasants confined in a cage. In all there are 35 heads and 94 feet. How many rabbits and how many pheasants are there? Solve the problem.

15. _____

6.9 MOTION PROBLEMS

▪️   Many problems deal with distance, time, and speed. A basic formula comes from the definition of speed.

$$\text{Speed} = \frac{\text{Distance}}{\text{Time}}, \qquad r = \frac{d}{t}, \quad \text{or} \quad d = rt.$$

From $r = d/t$ we get the formula $d = rt$ by multiplying by t.

In Section 3.6 we considered some general steps for problem solving. The following are also helpful when solving motion problems.

1. **Organize the information in a chart, keeping the formula $d = rt$ in mind.**
2. **Look for as many things as you can that are the same, so you can write equations.**

Example 1   A train leaves Podunk traveling east at 35 kilometers per hour (km/h). An hour later another train leaves Podunk on a parallel track at 40 km/h. How far from Podunk will the trains meet?

First make a drawing.

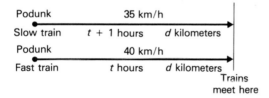

From the drawing we see that the distances are the same. Let's call the distance d. We don't know the times. Let t represent the time for the faster train. Then the time for the slower train will be $t + 1$. We can organize the information in a chart, keeping the formula $d = rt$ in mind:

	Distance	Speed	Time
Slow train	d	35	$t + 1$
Fast train	d	40	t

In these problems we look for things that are the same, so we can find equations. From each row of the chart we get an equation, $d = rt$. Thus we have two equations:

1. A car leaves Hereford traveling north at 56 km/h. Another car leaves Hereford one hour later traveling north at 84 km/h. How far from Hereford will the second car overtake the first? (*Hint:* The cars travel the same distance.)

We solve the system:

$$35(t + 1) = 40t \qquad \text{Using the substitution method}$$
$$35t + 35 = 40t \qquad \text{Multiplying}$$
$$35 = 5t \qquad \text{Adding } -35t$$
$$\frac{35}{5} = t \qquad \text{Multiplying by } \frac{1}{5}$$
$$7 = t.$$

The problem asks us to find how far from Podunk the trains meet. Thus we need to find d. We can do this by substituting 7 for t in the equation $d = 40t$:

$$d = 40(7) = 280.$$

Thus the trains meet 280 km from Podunk. Note also that $280 = 35(7 + 1)$, so 280 checks in the problem.

DO EXERCISE 1.

Example 2 A motorboat took 3 hr to make a downstream trip with a 6-km/h current. The return trip against the same current took 5 hr. Find the speed of the boat in still water.

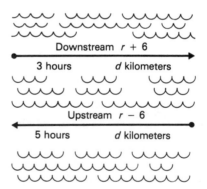

First make a drawing. From the drawing we see that the distances are the same. Let's call the distance d. Let r represent the speed of the boat in still water. Then, when the boat is traveling downstream, its speed is $r + 6$ (the current helps the boat along). When it is traveling upstream, its speed is $r - 6$ (the current holds the boat back some). We can organize the information in a chart, keeping the formula $d = rt$ in mind.

	Distance	Speed	Time
Downstream	d	$r + 6$	3
Upstream	d	$r - 6$	5

From each row of the chart, we get an equation: $d = rt$:

$$d = (r + 6)3,$$
$$d = (r - 6)5.$$

We solve the system:

$$(r + 6)3 = (r - 6)5 \qquad \text{Using substitution}$$

$$3r + 18 = 5r - 30 \qquad \text{Multiplying}$$

$$-2r + 18 = -30 \qquad \text{Adding } -5r$$

$$-2r = -48 \qquad \text{Adding } -18$$

$$r = \frac{-48}{-2}, \quad \text{or} \quad 24. \qquad \text{Multiplying by } -\frac{1}{2}$$

We check in the original problem. When $r = 24$, $r + 6 = 30$, and $30 \cdot 3 = 90$, the distance. When $r = 24$, $r - 6 = 18$, and $18 \cdot 5 = 90$. In both cases we get the same distance. Thus the speed in still water is 24 km/h.

DO EXERCISE 2.

Example 3   Two cars leave town at the same time going in opposite directions. One of them travels 60 mph and the other 30 mph. In how many hours will they be 150 miles apart?

We first make a drawing.

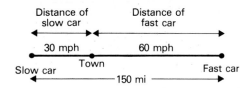

From the wording of the problem and the drawing, we see that the distances may *not* be the same. But the times the cars travel are the same, so we can just use t for time. We can organize the information in a chart.

	Distance	Speed	Time
Fast car	Distance of fast car	60	t
Slow car	Distance of slow car	30	t

2. An airplane flew for 5 hours with a 25-km/h tail wind. The return flight against the same wind took 6 hours. Find the speed of the airplane in still air. (*Hint:* The distance is the same both ways. The speeds are $r + 25$ and $r - 25$, where r is the speed in still air.)

ANSWER ON PAGE A–22

3. Two cars leave town at the same time in opposite directions. One travels 48 mph and the other 60 mph. How far apart will they be 3 hours later? (*Hint:* The times are the same. Be *sure* to make a drawing.)

From the drawing we see that

(Distance of fast car) + (Distance of slow car) = 150.

Then using $d = rt$ in each row of the table we get

$60t + 30t = 150.$

We solve the equation:

$90t = 150$ Collecting like terms

$t = \dfrac{150}{90}$, or $\dfrac{5}{3}$, or $1\dfrac{2}{3}$ hours.   Multiplying by $\dfrac{1}{90}$

This checks in the original problem, so in $1\frac{2}{3}$ hours the cars will be 150 miles apart.

DO EXERCISES 3 AND 4.

4. Two cars leave town at the same time in the same direction. One travels 35 mph and the other 40 mph. In how many hours will they be 15 miles apart? (*Hint:* The times are the same. Be *sure* to make a drawing.)

NAME   CLASS   ANSWERS

EXERCISE SET 6.9

▪ Solve.

1. Two cars leave town at the same time going in opposite directions. One travels 55 mph and the other travels 48 mph. In how many hours will they be 206 miles apart?

1. _____

2. Two cars leave town at the same time going in opposite directions. One travels 44 mph and the other travels 55 mph. In how many hours will they be 297 miles apart?

2. _____

3. Two cars leave town at the same time going in the same direction. One travels 30 mph and the other travels 46 mph. In how many hours will they be 72 miles apart?

3. _____

4. Two cars leave town at the same time going in the same direction. One travels 32 mph and the other travels 47 mph. In how many hours will they be 69 miles apart?

4. _____

5. A train leaves a station and travels east at 72 km/h. Three hours later a second train leaves on a parallel track and travels east at 120 km/h. When will it overtake the first train?

5. _____

6. A private airplane leaves an airport and flies due south at 192 km/h. Two hours later a jet leaves the same airport and flies due south at 960 km/h. When will the jet overtake the private plane?

6. _____

7. A canoeist paddled for 4 hours with a 6-km/h current to reach a campsite. The return trip against the same current took 10 hours. Find the speed of the canoe in still water.

7. _____

8. An airplane flew for 4 hours with a 20-km/h tail wind. The return flight against the same wind took 5 hours. Find the speed of the plane in still air.

8. _____

9. It takes a passenger train 2 hours less time than it takes a freight train to make the trip from Central City to Clear Creek. The passenger train averages 96 km/h while the freight train averages 64 km/h. How far is it from Central City to Clear Creek?

10. It takes a small jet plane 4 hours less time than it takes a propeller-driven plane to travel from Glen Rock to Oakville. The jet plane averages 637 km/h while the propeller plane averages 273 km/h. How far is it from Glen Rock to Oakville?

9. _____

11. An airplane took 2 hours to fly 600 km against a head wind. The return trip with the wind took $1\frac{2}{3}$ hours. Find the speed of the plane in still air.

10. _____

12. It took 3 hours to row a boat 18 km against the current. The return trip with the current took $1\frac{1}{2}$ hours. Find the speed of the rowboat in still water.

11. _____

12. _____

13. A motorcycle breaks down and the rider has to walk the rest of the way to work. The motorcycle was being driven at 45 mph and the rider walks at a speed of 6 mph. The distance from home to work is 25 miles and the total time for the trip was 2 hours. How far did the motorcycle go before it broke down?

13. _____

14. A student walks and jogs to college each day. The student averages 5 mph walking and 9 mph jogging. The distance from home to college is 8 miles and the student makes the trip in 1 hour. How far does the student jog?

14. _____

15. An airplane flew for 4.23 hours with a 25.5-km/h tail wind. The return flight against the same wind took 4.97 hours. Find the speed of the plane in still air.

15. _____

6.10 COIN AND MIXTURE PROBLEMS

■ **Example 1** A student has some nickels and dimes. The value of the coins is $1.65. There are 12 more nickels than dimes. How many of each kind of coin are there?

We will use d to represent the number of dimes and n to represent the number of nickels. We have one equation at once:

$$d + 12 = n.$$

The value of the nickels, in cents, is $5n$, since each is worth 5¢. The value of the dimes, in cents, is $10d$, since each is worth 10¢. The total value is given as $1.65. Since we have the values of the nickels and dimes *in cents*, we must use cents for the total value. This is 165. Now we have another equation:

$$10d + 5n = 165.$$

Thus we have a system of equations to solve:

$$d + 12 = n$$
$$10d + 5n = 165.$$

Since we have n alone on one side of one equation, we use the substitution method. We substitute $d + 12$ for n in the second equation:

$$10d + 5n = 165$$
$$10d + 5(d + 12) = 165 \quad \text{Substituting } d + 12 \text{ for } n$$
$$10d + 5d + 60 = 165 \quad \text{Multiplying to remove parentheses}$$
$$15d + 60 = 165 \quad \text{Collecting like terms}$$
$$15d = 105 \quad \text{Adding } -60$$
$$d = \frac{105}{15}, \quad \text{or} \quad 7. \quad \text{Multiplying by } \frac{1}{15}$$

We substitute 7 for d in either of the original equations to find n. We use the first equation:

$$d + 12 = n$$
$$7 + 12 = n$$
$$19 = n.$$

The solution of this system is $d = 7$ and $n = 19$. This checks, so the student has 7 dimes and 19 nickels.

DO EXERCISE 1.

OBJECTIVE

After finishing Section 6.10, you should be able to:

■ Solve coin and mixture problems.

1. On a table are 20 coins, quarters and dimes. Their value is $3.05. How many of each are there?

2. There were 166 paid admissions to a game. The price was \$2.10 for adults and \$0.75 for children. The amount taken in was \$293.25. How many adults and how many children attended?

Certain other types of problems are very much like coin problems. Although they are not about coins, they are solved in the same way.

Example 2 There were 411 people at a play. Admission for adults was \$1.00 and for children \$0.75. The receipts were \$395.75. How many adults and how many children attended?

We will use a for the number of adults and c for the number of children. Since the total number is 411, we have this equation:

$$a + c = 411.$$

The amount paid in by the adults is $1 \cdot a$, since each paid \$1.00. This amount is *in dollars*. In cents, the amount is $100a$.

The amount paid in by the children is $0.75c$, in dollars, or $75c$ in cents. If we use cents instead of dollars we can avoid decimal points. Then we have this equation:

$$100a + 75c = 39575,$$

since the total amount taken in is 39575, *in cents*.

Thus we have a system of equations to solve:

$$a + c = 411$$
$$100a + 75c = 39575.$$

This time we use the addition method. We multiply on both sides of the first equation by -100 and then add:

$$
\begin{array}{ll}
-100a - 100c = -41100 & \text{Multiplying by } -100 \\
\underline{100a + 75c = 39575} & \\
-25c = -1525 & \text{Adding} \\
c = \dfrac{-1525}{-25} & \text{Multiplying by } \dfrac{1}{-25} \\
c = 61. & \text{Solving for } c
\end{array}
$$

We go back to the first equation and substitute 61 for c:

$$
\begin{aligned}
a + c &= 411 \\
a + 61 &= 411 \\
a &= 350.
\end{aligned}
$$

The solution of the system is $a = 350$ and $c = 61$. This checks, so we know that 350 adults and 61 children attended.

DO EXERCISE 2.

Example 3　A chemist has one solution that is 80% acid and another that is 30% acid. What is needed is 200 liters of a solution that is 62% acid. The chemist will prepare it by mixing the two solutions on hand. How much of each should be used?

Suppose the chemist uses x liters of the first solution and y liters of the second. Since the total is to be 200 liters, we have

$x + y = 200.$

The *amount* of acid in the new mixture is to be 62% of 200 liters, or 124 liters. The amounts of acid from the two solutions are $80\%x$ and $30\%y$. Thus,

$80\%x + 30\%y = 124,$　or　$0.8x + 0.3y = 124.$

We clear the decimals by multiplying on both sides by 10:

$\boxed{10}\ (0.8x + 0.3y) = \boxed{10} \cdot 124$

$8x + 3y\ \ \ = 1240.$

We solve this system of equations:

$x + \ y = \ \ 200$
$8x + 3y = 1240.$

We use the addition method. We multiply on both sides of the first equation by -3 and then add:

$-3x - 3y = -600$　　　Multiplying by -3
$\underline{\ \ 8x + 3y = \ \ 1240}$
$\ \ \ 5x\ \ \ \ \ \ = \ \ \ 640$　Adding

$x = \dfrac{640}{5}$　　Multiplying by $\dfrac{1}{5}$

$x = 128.$　　Solving for x

We go back to the first equation and substitute 128 for x:

$x + y = 200$
$128 + y = 200$
$y = 72.$

The solution is $x = 128$ and $y = 72$.

Check:　The sum of 128 and 72 is 200. Now 80% of 128 is 102.4 and 30% of 72 is 21.6. These add up to 124. Thus the chemist should use 128 liters of the stronger acid and 72 liters of the other.

DO EXERCISE 3.

3. One solution is 50% alcohol and a second is 70% alcohol. How much of each should be used to make 30 liters of a solution that is 55% alcohol?

4. Grass seed A is worth $1.00 per pound and seed B is worth $1.35 per pound. How much of each would you use to make 50 lb of a mixture worth $1.14 per pound?

Example 4 A grocer wishes to mix some candy worth 45¢ per pound and some worth 80¢ per pound to make 350 pounds of a mixture worth 65¢ per pound. How much of each should be used?

We will use x and y for the amounts. Then

$$x + y = 350.$$

Our second equation will come from the values. The value of the first candy, in cents, is $45x$ (x pounds at 45¢ per pound). The value of the second is $80y$, and the value of the mixture is 350×65. Thus we have

$$45x + 80y = 350 \times 65.$$

We solve this system of equations:

$$x + y = 350$$
$$45x + 80y = 22750.$$

We use the addition method. We multiply on both sides of the first equation by -45 and then add:

$$-45x - 45y = -15750 \qquad \text{Multiplying by } -45$$
$$\underline{45x + 80y = \quad 22750}$$
$$35y = \quad 7000 \quad \text{Adding}$$
$$y = \frac{7000}{35}$$
$$y = 200.$$

We go back to the first equation and substitute 200 for y:

$$x + y = 350$$
$$x + 200 = 350$$
$$x = 150.$$

We get $x = 150$ lb and $y = 200$ lb. This checks, so the grocer mixes 150 lb of the 45¢ candy with 200 lb of the 80¢ candy.

DO EXERCISE 4.

NAME   CLASS   ANSWERS

EXERCISE SET 6.10

▮ ● ▮ Solve.

1. A collection of dimes and quarters is worth $15.25. There are 103 coins in all. How many of each are there?

1. _____

2. A collection of quarters and nickels is worth $1.25. There are 13 coins in all. How many of each are there?

2. _____

3. A collection of nickels and dimes is worth $25. There are three times as many nickels as dimes. How many of each are there?

3. _____

4. A collection of nickels and dimes is worth $2.90. There are 19 more nickels than dimes. How many of each are there?

4. _____

5. There were 429 people at a play. Admission was $1 for adults and 75¢ for children. The receipts were $372.50. How many adults and how many children attended?

6. The attendance at a school concert was 578. Admission was $2 for adults and $1.50 for children. The receipts were $985. How many adults and how many children attended?

5. _____

7. There were 200 tickets sold for a college basketball game. Tickets for students were $0.50 and for adults were $0.75. The total amount of money collected was $132.50. How many of each type of ticket were sold?

6. _____

8. There were 203 tickets sold for a wrestling match. For activity card holders the price was $1.25 and for non-card holders the price was $2. The total amount of money collected was $310. How many of each type of ticket were sold?

7. _____

8. _____

9. Solution A is 50% acid and solution B is 80% acid. How much of each should be used to make 100 grams of a solution that is 68% acid? (*Hint:* 68% of what is acid?)

9. _____

10. Solution A is 30% alcohol and solution B is 75% alcohol. How much of each should be used to make 100 liters of a solution that is 50% alcohol?

10. _____

11. Farmer Jones has 100 liters of milk that is 4.6% butterfat. How much skim milk (no butterfat) should be mixed with it to make milk that is 3.2% butterfat?

11. _____

12. A tank contains 8000 liters of a solution that is 40% acid. How much water should be added to make a solution that is 30% acid?

12. _____

13. A solution containing 30% insecticide is to be mixed with a solution containing 50% insecticide to make 200 liters of a solution containing 42% insecticide. How much of each solution should be used?

13. _____

14. A solution containing 28% fungicide is to be mixed with a solution containing 40% fungicide to make 300 liters of a solution containing 36% fungicide. How much of each solution should be used?

14. _____

15. The Nuthouse has 10 kg of mixed cashews and pecans worth $8.40 per kilogram. Cashews alone sell for $8.00 per kilogram and pecans sell for $9.00 per kilogram. How many kilograms of each are in the mixture?

15. _____

16. A coffee shop mixes Brazilian coffee worth $5 per kilogram with Turkish coffee worth $8 per kilogram. The mixture is to sell for $7 per kilogram. How much of each type of coffee should be used to make a mixture of 300 kg?

16. _____

EXTENSION EXERCISES

SECTION 6.1

In Exercises 1–4, tell in which quadrant(s) each of the following points could be located.

1. The first coordinate is positive.

2. The second coordinate is negative.

3. The first and second coordinates are the same.

4. The first coordinate is the additive inverse of the second coordinate.

5. The points $(-1, 1)$, $(4, 1)$ and $(4, -5)$ are three vertices of a rectangle. Find the coordinates of the fourth vertex.

6. Three parallelograms share the vertices $(-2, -3)$, $(-1, 2)$ and $(4, -3)$. Find the fourth vertex of each parallelogram.

SECTION 6.2

7. Connect the points $(10, 10)$, $(10, -10)$, $(-10, -10)$ and $(-10, 10)$ to form a square. Then graph the following equations, but only draw the part of the line that is inside the square. (This is called a *line segment*.)
   a) $y = x$ b) $y = -x$ c) $y = x - 7$ d) $y = -x + 7$ e) $y = x - 5$ f) $y = -x + 5$

8. Find an equation whose line segment has the same length inside the square (Ex. 7) as the segment of the graph of the equation $y = -3x + 5$.

SECTION 6.3

9. Find the value of m in $y = mx + 3$ so that the x-intercept of its graph will be $(2, 0)$.

10. Find the value of b in $2y = -5x + 3b$ so that the y-intercept of its graph will be $(0, -12)$.

SECTION 6.4

Translate to a system of equations. Do not attempt to solve. Save for later use.

11. Patrick's age is 20% of his father's age. Twenty years from now, Patrick's age will be 52% of his father's age. How old are Patrick and his father?

12. If 5 is added to a man's age and the total is divided by 5, the result will be his daughter's age. Five years ago the father's age was 8 times his daughter's age. Find their present ages.

13. When the base of a triangle is increased by 2 ft and the height is decreased by 1 ft, the height becomes $\frac{1}{3}$ of the base, and the area becomes 24 ft^2. Find the original dimensions of the triangle.

SECTION 6.5

14. If a system of two equations has more than one solution, how are the graphs of the equations related?

15. Determine whether the given ordered pair is a solution of each system of three equations.
   a) $(2, -3)$; $x + 3y = -7$
   $\qquad\qquad -x + y = -5$
   $\qquad\qquad 2x - y = 1$

   b) $(-1, -5)$; $4a - b = 1$
   $\qquad\qquad -a + b = -4$
   $\qquad\qquad 2a + 3b = -17$

16. Describe in words the graph of the system in Exercise 15(b).

SECTION 6.6

Solve by the substitution method.

17. $\frac{1}{4}(a - b) = 2$

$\frac{1}{6}(a + b) = 1$

18. $\frac{x}{2} + \frac{3y}{2} = 2$

$\frac{x}{5} - \frac{y}{2} = 3$

19. $0.4x + 0.7y = 0.1$

$0.5x - 0.1y = 1.1$

SECTION 6.7

20. The graphs of the equations $5y - 2x = 11, y - 4x = -5$ and $2y + x = -1$ intersect to form a triangle. Use the addition method to find the coordinates of the vertices of the triangle.

SECTION 6.8

21.–23. Solve the systems of equations you set up in Exercises 11–13.

SECTION 6.9

24. A car travels from one town to another at a speed of 32 mph. If it had gone 4 mph faster, it could have made the trip in $\frac{1}{2}$ hour less time. How far apart are the towns?

25. Two airplanes start at the same time and fly toward each other from points 1000 km apart at rates of 420 km/h and 330 km/h. When will they meet?

26. A truck and a car leave a service station at the same time and travel in the same direction. The truck travels at 55 mph and the car at 40 mph. They can maintain CB radio contact within a range of 10 miles. When will they lose contact?

SECTION 6.10

27. $27,000 is invested, part of it at 12% and part of it at 13%. The total yield is $3385. How much was invested at each rate?

28. A two-digit number is 6 times the sum of its digits. The ten's digit is 1 more than the unit's digit. Find the number.

29. The sum of the digits of a two-digit number is 12. When the digits are reversed, the number is decreased by 18. Find the original number.

30. An automobile radiator contains 16 liters of antifreeze and water. This mixture is 30% antifreeze. How much of this mixture should be drained and replaced with pure antifreeze so that the mixture will be 50% antifreeze?

31. An employer has a daily payroll of $325 when employing some workers at $20 per day and others at $25 per day. When the number of $20 workers is increased by 50% and the number of $25 workers is decreased by $\frac{1}{5}$, the new daily payroll is $400. Find how many were originally employed at each rate.

NAME	CLASS	ANSWERS

TEST OR REVIEW—CHAPTER 6

If you miss an item, review the indicated section and objective.

Graph.

[6.2, ●●]   **1.** $y = -\dfrac{2}{5}x + 3$

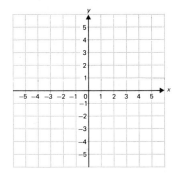

1. See graph.

Graph.

[6.3, ●]   **2.** $2x - 5y = 10$

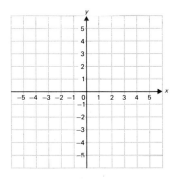

2. See graph.

[6.3, ●●]   **3.** $x = -3$

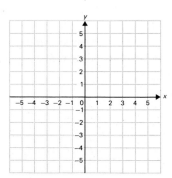

3. See graph.

[6.6, ●]   **4.** Solve by the substitution method:

$$y = 5 - x$$
$$3x - 4y = -20.$$

4.

ANSWERS

[6.7, ●]   **5.** $x + 2y = 9$
$3x - 2y = -5$

Solve by the addition method.

5. _____

[6.7, ● ●]   **6.** $2x + 3y = 8$
$5x + 2y = -2$

6. _____

Solve.

[6.8, ●]   **7.** The sum of two numbers is 2. The difference is 8. Find the numbers.

7. _____

[6.9, ●]   **8.** An airplane flew for 4 hours with a 15-km/h tail wind. The return flight against the same wind took 5 hours. Find the speed of the airplane in still air.

8. _____

[6.10, ●]   **9.** There were 508 people at an organ recital. Orchestra seats cost $5.00 per person with balcony seats costing $3.00. The total receipts were $2118. Find the number of orchestra and the number of balcony seats sold.

9. _____

[6.10, ●]   **10.** Solution A is 30% alcohol and solution B is 60% alcohol. How much of each is needed to make 80 liters of a solution that is 45% alcohol?

10. _____

POLYNOMIALS IN SEVERAL VARIABLES

7

The area of a beverage can is expressible by a polynomial in two variables.

READINESS CHECK: SKILLS FOR CHAPTER 7

1. Evaluate for $x = -2$:

$x^3 - x^2 + 3x + 3$.

2. Collect like terms:

$6x^3 - 5x^4 - 3x^3 + 8x^4$.

3. Add:

$(4x^2 - 2x + 2) + (-3x^2 + 5x - 7)$.

4. Subtract:

$(4x^2 - 2x + 2) - (-3x^2 + 5x - 7)$.

Multiply.

5. $(3x^2 + 4x - 2)(x - 5)$

6. $(x + 2)(x - 3)$

7. $(3t + 1)(3t - 1)$

8. $(p + 4)^2$

Factor.

9. $x^2 - 7x + 2x - 14$

10. $3x^2 - 5x - 2$

11. $9t^2 - 6t + 1$

12. $x^5 - 16x^3$

13. Solve:

$3(x + 2) - 4 = 3 - 4(x - 5)$

14. Solve $d = rt$, for r.

7.1 POLYNOMIALS IN SEVERAL VARIABLES*

Most of the polynomials you have studied so far have had only one variable. A *polynomial in several variables* is an expression like those you have already seen, but we allow that there can be more than one variable. Here are some examples:

$$3x + xy^2 + 5y + 4, \qquad 8xy^2z - 2x^3z - 13x^4y^2 + 5.$$

◼ EVALUATING POLYNOMIALS

Example 1   Evaluate the polynomial

$$4 + 3x + xy^2 + 8x^3y^3$$

for $x = -2$ and $y = 5$.

We replace x by -2 and y by 5:

$$4 + 3(-2) + (-2) \cdot 5^2 + 8(-2)^3 \cdot 5^3 = 4 - 6 - 50 - 8000 = -8052.$$

DO EXERCISES 1 AND 2.

Example 2   (*The magic number*) The Boston Red Sox are leading the New York Yankees for the Eastern Division championship of the American League. The magic number is 8. This means that any com-

OBJECTIVES

After finishing Section 7.1, you should be able to:

◼ Evaluate a polynomial in several variables for given values of the variables.

◼◼ Identify the coefficients and the degrees of the terms of a polynomial, and degrees of polynomials.

◼◼◼ Collect the like terms of a polynomial.

◼◼ Arrange a polynomial in ascending or descending order.

1. Evaluate the polynomial
$4 + 3x + xy^2 + 8x^3y^3$
for $x = 2$ and $y = -5$.

2. Evaluate the polynomial
$8xy^2 - 2x^3z - 13x^4y^2 + 5$
for $x = -1$, $y = 3$, and $z = 4$.

* *To the instructor:* This chapter is prerequisite only to Chapter 8. If time is short and that chapter is to be skipped, then this chapter might also be skipped.

bination of Red Sox wins and Yankee losses that totals 8 will ensure the championship for the Red Sox. The magic number is given by the polynomial

$$G - P - L + 1,$$

where G is the number of games in the season, P is the number of games the leading team has played, and L is the number of games ahead in the loss column.

Given the situation shown in the table and assuming a 162-game season, what is the magic number for the Philadelphia Phillies?

EASTERN DIVISION				
	W	L	Pct.	GB
Philadelphia	77	40	.658	——
Pittsburgh	65	53	.551	$12\frac{1}{2}$
New York	61	60	.504	18
Chicago	55	67	.451	$24\frac{1}{2}$
St. Louis	51	65	.440	$25\frac{1}{2}$
Montreal	41	73	.360	$34\frac{1}{2}$

We evaluate the polynomial for $G = 162$, $P = 77 + 40$, or 117, and $L = 53 - 40$, or 13:

$$162 - 117 - 13 + 1 = 33.$$

DO EXERCISE 3.

●●●○ COEFFICIENTS AND DEGREES

The *degree* of a term is the sum of the exponents of the variables. The *degree of a polynomial* is the degree of the term of highest degree.

Example 3   Identify the coefficient and degree of each term of

$$9x^2y^3 - 14xy^2z^3 + xy + 4y + 5x^2 + 7.$$

Term	*Coefficient*	*Degree*	
$9x^2y^3$	9	5	
$-14xy^2z^3$	-14	6	
xy	1	2	
$4y$	4	1	Think: $4y = 4y^1$
$5x^2$	5	2	
7	7	0	Think: $7 = 7x^0$

Example 4   What is the degree of $5x^3y + 9xy^4 - 8x^3y^3$?

The term of highest degree is $-8x^3y^3$. Its degree is 6. The degree of the polynomial is 6.

DO EXERCISES 4 AND 5.

---

3. Given the situation below, what is the magic number for the Cincinnati Reds? Assume $G = 162$.

WESTERN DIVISION				
	W	L	Pct.	GB
Cincinnati	77	44	.636	——
Los Angeles	65	54	.546	11
San Diego	60	64	.484	$18\frac{1}{2}$
Houston	59	64	.480	19
Atlanta	56	65	.463	21
San Francisco	52	70	.426	$25\frac{1}{2}$

4. Identify the coefficient of each term.

$$-3xy^2 + 3x^2y - 2y^3 + xy + 2$$

5. Identify the degree of each term and the degree of the polynomial.

$$4xy^2 + 7x^2y^3z^2 - 5x + 2y + 4$$

Collect like terms.

6. $4x^2y + 3xy - 2x^2y$

7. $-3pq - 5pqr^3 + 8pq + 5pqr^3 + 4$

8. Arrange in descending powers of y.

$3xy - 7xy^2 + 5xy^4 - 3xy^3$

9. Arrange in ascending powers of x.

$2x^2yz + 5xy^2z + 5x^3yz^2 - 2$

⠿ COLLECTING LIKE TERMS

Like terms (or *similar terms*) have exactly the same variables with exactly the same exponents. For example,

$3x^2y^3$ and $-7x^2y^3$ are like terms;

$9x^4z^7$ and $12x^4z^7$ are like terms.

But

$13xy^2$ and $-2x^2y$ are *not* like terms;

$3xyz^2$ and $4xy$ are *not* like terms.

Collecting like terms is based on the distributive law.

Examples   Collect like terms.

5. $5x^2y + 3xy^2 - 5x^2y - xy^2 = (5 - 5)x^2y + (3 - 1)xy^2 = 2xy^2$

6. $3xy - 5xy^2 + 3xy^2 + 9xy = -2xy^2 + 12xy$

7. $4ab^2 - 7a^2b^2 + 9a^2b^2 - 4a^2b = 4ab^2 + 2a^2b^2 - 4a^2b$

8. $3pq + 5pqr^3 - 8pq - 5pqr^3 - 4 = -5pq - 4$

DO EXERCISES 6 AND 7.

⠓ ASCENDING AND DESCENDING ORDER

We usually arrange polynomials in one variable in descending order and sometimes in ascending order. For polynomials in several variables we choose one of the variables and arrange the terms with respect to it.

Examples

9. This polynomial is arranged in descending powers of x:

$3x^5y - 5x^3 + 7xy + y^4 + 2.$ Note the descending powers of x.

10. This polynomial is arranged in ascending powers of y:

$3 + 13xy - x^3y^2 + 4xy^3.$

11. This polynomial is arranged in ascending powers of p:

$3pq - 2p^2qr + 7p^3q^5r^2 + 5p^4q^2r^4 - 3p^5r^2.$

12. This polynomial is arranged in descending powers of t:

$5s^5t^7 + 2s^6t^4 - 7s^6t^2 + 4s^3t + 6s^4.$

Example 13   Arrange the polynomial

$8u^3v - 9uv^3 + 7u^2v + 5u^2v^2$

in descending powers of u.

The answer is

$8u^3v + 5u^2v^2 + 7u^2v - 9uv^3.$

> Note that there are two different terms with u^2. In such a case, we arrange them so the term with v^2 is first and the term with v is second.

DO EXERCISES 8 AND 9.

NAME   CLASS   ANSWERS

EXERCISE SET 7.1

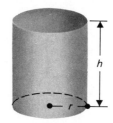 Evaluate each polynomial for $x = 3$ and $y = -2$.

1. $x^2 - y^2 + xy$ **2.** $x^2 + y^2 - xy$

Evaluate each polynomial for $x = 2$, $y = -3$, and $z = -1$.

3. $xyz^2 + z$ **4.** $xy - xz + yz$

An amount of money P is invested at interest rate r. In 3 years it will grow to an amount given by the polynomial

$$P + 3rP + 3r^2P + r^3P.$$

5. Evaluate the polynomial for $P = 10{,}000$ and $r = 0.08$ to find the amount to which \$10,000 will grow at 8% interest for 3 years.

6. Evaluate the polynomial for $P = 10{,}000$ and $r = 0.07$ to find the amount to which \$10,000 will grow at 7% interest for 3 years.

The area of a right circular cylinder is given by the polynomial

$$2\pi rh + 2\pi r^2,$$

where h is the height and r is the radius of the base.

7. A 12-oz beverage can has height 4.7 in. and radius 1.2 in. Evaluate the polynomial for $h = 4.7$ and $r = 1.2$ to find the area of the can. Use 3.14 for π.

8. A 16-oz beverage can has height 6.3 in. and radius 1.2 in. Evaluate the polynomial for $h = 6.3$ and $r = 1.2$ to find the area of the can. Use 3.14 for π.

●● Identify the coefficient and degree of each term of the following polynomials. Then find the degree of the polynomial.

9. $x^3y - 2xy + 3x^2 - 5$ **10.** $5y^3 - y^2 + 15y + 1$

11. $17x^2y^3 - 3x^3yz - 7$ **12.** $6 - xy + 8x^2y^2 - y^5$

●●● Collect like terms.

13. $a + b - 2a - 3b$ **14.** $y^2 - 1 + y - 6 - y^2$

ANSWERS

1. _____

2. _____

3. _____

4. _____

5. _____

6. _____

7. _____

8. _____

9. _____

10. _____

11. _____

12. _____

13. _____

14. _____

ANSWERS

15. _____

16. _____

17. _____

18. _____

19. _____

20. _____

21. _____

22. _____

23. _____

24. _____

25. _____

26. _____

27. _____

28. _____

29. _____

30. _____

31. _____

32. _____

15. $3x^2y - 2xy^2 + x^2$ **16.** $m^3 + 2m^2n - 3m^2 + 3mn^2$

17. $2u^2v - 3uv^2 + 6u^2v - 2uv^2$

18. $3x^2 + 6xy + 3y^2 - 5x^2 - 10xy - 5y^2$

19. $6au + 3av - 14au + 7av$

20. $3x^2y - 2z^2y + 3xy^2 + 5z^2y$

21. $a^3 - a^2b + ab^2 + a^2b - ab^2$

22. $x^3 + x^2y + xy^2 - x^2y - xy^2 - y^3$

⚅ Arrange in descending powers of y.

23. $-xy^2 - y^3 + 5x^2y$ **24.** $2xy^2 + 7xy + 4y^3$

25. $-xy^2 + x^3 - y^3$ **26.** $4y^3 - x^2y^2 - 2x^2y + xy^4 - y^5$

Arrange in ascending powers of m.

27. $5m^2n - 3mn^3 + 4n^2 - m^4n$

28. $5m^3n - 8mn^2 + 7m^2n - 3m^4n^3$

Arrange in descending powers of the second variable (using alphabetical order). Collect like terms if possible.

29. $x^2 + 3y^2 + 2xy$

30. $4n^2 - 3mn + m^2 - mn + n^2 + 5m^2$

31. $5uv - 8uv^2 + 7u^2v - 3uv^2$

32. $6r^2s + 11 + 3r^2s - 5r + 3$

7.2 CALCULATIONS WITH POLYNOMIALS

Calculations with polynomials in several variables are very much like those for polynomials in one variable. Additional practice is needed, however, and you should strive for as much speed as possible without losing accuracy.

◖•◗ ADDITION

To add polynomials in several variables, we collect like terms.

Example 1 Add: $-5x^3 + 3x - 5$ and $8x^3 + 4x^2 + 7$.

$$(-5x^3 + 3x - 5) + (8x^3 + 4x^2 + 7)$$
$$= (-5 + 8)x^3 + 4x^2 + 3x + (-5 + 7)$$
$$= 3x^3 + 4x^2 + 3x + 2$$

We still arrange the terms in descending order.

The use of columns is often helpful. To do this we write the polynomials one under the other, writing like terms under one another and leaving spaces for missing terms. Let us do the addition in Example 1 using columns.

$$
\begin{array}{rrrr}
-5x^3 &   & + 3x & - 5 \\
8x^3 & + 4x^2 &   & + 7 \\
\hline
3x^3 & + 4x^2 & + 3x & + 2
\end{array}
$$

Example 2 Add: $(3ax^2 + 4bx - 2) + (2ax^2 + 5bx + 4)$.

We can use columns. We write one polynomial under the other, keeping like terms in columns.

$$
\begin{array}{r}
3ax^2 + 4bx - 2 \\
2ax^2 + 5bx + 4 \\
\hline
5ax^2 + 9bx + 2
\end{array}
$$

Usually you should not use columns. Try to write only the answer.

Example 3 Add: $(5xy^2 - 4x^2y + 5x^3 + 2) + (3xy^2 - 2x^2y + 3x^3y - 5)$.

We look for like terms. The like terms are $5xy^2$ and $3xy^2$, $-4x^2y$ and $-2x^2y$, and 2 and -5. We collect these. There are no more like terms. The answer is

$$8xy^2 - 6x^2y + 5x^3 + 3x^3y - 3.$$

DO EXERCISES 1–3.

◖••◗ SUBTRACTION

We subtract a polynomial by adding its inverse. The additive inverse of a polynomial is found by replacing each coefficient by its additive inverse.

After finishing Section 7.2, you should be able to:

◖•◗ Add polynomials.

◖••◗ Subtract polynomials.

◖•••◗ Multiply polynomials.

Add.

1. $4x^3 + 4x^2 - 8x - 3$ and $-8x^3 - 2x^2 + 4x + 5$

2. $(13x^3y + 3x^2y - 5y)$ $+ (x^3y + 4x^2y - 3xy + 3y)$

3. $(-5p^2q^4 + 2p^2q^2 + 3q)$ $+ (6pq^2 + 3p^2q + 5)$

Subtract.

4. $(-4s^4t + s^3t^2 + 2s^2t^3)$
   $- (4s^4t - 5s^3t^2 + s^2t^2)$

5. $(-5p^4q + 5p^3q^2 - 3p^2q^3 - 7q^4)$
   $- (4p^4q - 5p^3q^2 + p^2q^3 + 2q^4)$

Multiply.

6. $(x^2y^3 + 2x)(x^3y^2 + 3x)$

7. $(p^4q - 2p^3q^2 + 3q^3)(p + 2q)$

The additive inverse of the polynomial

   $4x^2y - 6x^3y^2 + x^2y^2 - 5y$

is the polynomial obtained by replacing each coefficient by its additive inverse. Thus the inverse is

   $-4x^2y + 6x^3y^2 - x^2y^2 + 5y.$

We can also represent the additive inverse of $4x^2y - 6x^3y^2 + x^2y^2 - 5y$ as follows:

   $-(4x^2y - 6x^3y^2 + x^2y^2 - 5y).$

Thus,

   $-(4x^2y - 6x^3y^2 + x^2y^2 - 5y) = -4x^2y + 6x^3y^2 - x^2y^2 + 5y.$

Example 4   Subtract.

   $(4x^2y + x^3y^2 + 3x^2y^3 + 6y) - (4x^2y - 6x^3y^2 + x^2y^2 - 5y)$

$$= 4x^2y + x^3y^2 + 3x^2y^3 + 6y - 4x^2y$$
$$\quad + 6x^3y^2 - x^2y^2 + 5y \qquad \text{Adding the inverse}$$
$$= 7x^3y^2 + 3x^2y^3 - x^2y^2 + 11y \qquad \text{Collecting like terms (Try to write just the answer!)}$$

You can use columns as shown below, but you should avoid the use of columns wherever possible. Work for speed, with accuracy.

$$\begin{array}{l} 4x^2y + \ x^3y^2 + 3x^2y^3 \qquad\quad + \ 6y \\ \underline{4x^2y - 6x^3y^2 \qquad\qquad + x^2y^2 - \ 5y} \\ \qquad 7x^3y^2 + 3x^2y^3 - x^2y^2 + 11y \end{array}$$ Mentally change signs and add.

DO EXERCISES 4 AND 5.

●●● MULTIPLICATION

Multiplication of polynomials is based on the distributive laws. Recall that this means we multiply each term of one by every term of the other. For most polynomials in several variables having three or more terms, you will probably want to use columns.

Example 5   Multiply: $(p - 4q)(2p + 3q).$

$$\begin{array}{l} p \ - \ 4q \\ \underline{2p \ + \ 3q} \\ 2p^2 - 8pq \qquad\qquad \text{Multiplying by } 2p \\ \qquad\quad 3pq - 12q^2 \quad \text{Multiplying by } 3q \\ \underline{} \\ 2p^2 - 5pq - 12q^2 \quad \text{Adding} \end{array}$$

Example 6   Multiply: $(3x^2y - 2xy + 3y)(xy + 2y).$

$$\begin{array}{l} 3x^2y - 2xy \ + 3y \\ \underline{\qquad xy + 2y} \\ 3x^3y^2 - 2x^2y^2 + 3xy^2 \qquad\qquad \text{Multiplying by } xy \\ \qquad\quad 6x^2y^2 - 4xy^2 + 6y^2 \quad \text{Multiplying by } 2y \\ \underline{} \\ 3x^3y^2 + 4x^2y^2 - \ xy^2 + 6y^2 \quad \text{Adding} \end{array}$$

DO EXERCISES 6 AND 7.

NAME   CLASS

EXERCISE SET 7.2

◼• Add.

1. $(2x^2 - xy + y^2) + (-x^2 - 3xy + 2y^2)$

2. $(2z - z^2 + 5) + (z^2 - 3z + 1)$

3. $(r - 2s + 3) + (2r + s) + (s + 4)$

4. $(b^3a^2 - 2b^2a^3 + 3ba + 4) + (b^2a^3 - 4b^3a^2 + 2ba - 1)$

5. $(2x^2 - 3xy + y^2) + (-4x^2 - 6xy - y^2) + (x^2 + xy - y^2)$

◼•◼ Subtract.

6. $(x^3 - y^3) - (-2x^3 + x^2y - xy^2 + 2y^3)$

7. $(xy - ab) - (xy - 3ab)$

8. $(3y^4x^2 + 2y^3x - 3y) - (2y^4x^2 + 2y^3x - 4y - 2x)$

9. $(-2a + 7b - c) - (-3b + 4c - 8d)$

10. Find the sum of $2a + b$ and $3a - 4b$. Then subtract $5a + 2b$.

◼•◼• Multiply.

11. $(3z - u)(2z + 3u)$ **12.** $(a - b)(a^2 + b^2 + 2ab)$

13. $(a^2b - 2)(a^2b - 5)$ **14.** $(xy + 7)(xy - 4)$

15. $(a^2 + a - 1)(a^2 - y + 1)$ **16.** $(tx + r)(vx + s)$

ANSWERS
1. _____
2. _____
3. _____
4. _____
5. _____
6. _____
7. _____
8. _____
9. _____
10. _____
11. _____
12. _____
13. _____
14. _____
15. _____
16. _____

17. $(a^3 + bc)(a^3 - bc)$

18. $(m^2 + n^2 - mn)(m^2 + mn + n^2)$

19. $(y^4x + y^2 + 1)(y^2 + 1)$

20. $(a - b)(a^2 + ab + b^2)$

21. $(r + s)(r^2 + ry + s^2)$

22. $(a + 2b + 3c)(2a - b + 2c)$

Simplify.

23. $(xy + 1)(2xy - 3) + (xy + 1)(3xy - 1)$

24. $(2a + b)(3a - b) - (a - b)(2a - 3b)$

25. $(yz^2 + 2)(yz^2 - 2) + (2yz^2 + 1)(2yz^2 + 1)$

26. $(3a - d)(2a + 3d) - (4a + d)(2a - d)$

27. $(a - b)(a^3 + a^2b + ab^2 + b^3)$

28. $(c + d)(c^3 - c^2d + cd^2 - d^3)$

7.3 SPECIAL PRODUCTS OF POLYNOMIALS

● PRODUCTS OF TWO BINOMIALS

The methods of handling the special kinds of products you learned for polynomials in one variable are the same for polynomials in several variables.

Examples   Multiply.

$$\quad\quad\quad\quad\;\;\text{F}\quad\quad\text{O}\quad\quad\;\;\text{I}\quad\quad\;\;\text{L}$$

1. $(x^2y + 2x)(xy^2 + y^2) = x^3y^3 + x^2y^3 + 2x^2y^2 + 2xy^2$

2. $(p + 5q)(2p - 3q) = 2p^2 - 3pq + 10pq - 15q^2$
$$= 2p^2 + 7pq - 15q^2$$

DO EXERCISES 1 AND 2.

●● SQUARES OF BINOMIALS

> **The square of a binomial is the square of the first expression, plus or minus twice the product of the two expressions, plus the square of the last:**
> $$(A + B)^2 = A^2 + 2AB + B^2;$$
> $$(A - B)^2 = A^2 - 2AB + B^2.$$

When doing these multiplications, carry out the rule by saying the words as you go along.

Examples   Multiply.

$$(A + B)^2 = A^2 + 2\;A\;B + B^2$$

3. $(3x + 2y)^2 = (3x)^2 + 2(3x)(2y) + (2y)^2$
$$= 9x^2 + 12xy + 4y^2$$

> *Reminder!* $(A + B)^2 \neq A^2 + B^2$. The square of a sum is *not* the sum of the squares. There must be three terms.

4. $(2y^2 - 5x^2y)^2 = (2y^2)^2 - 2(2y^2)(5x^2y) + (5x^2y)^2$
$$= 4y^4 - 20x^2y^3 + 25x^4y^2$$

DO EXERCISES 3 AND 4.

●●● PRODUCTS OF SUMS AND DIFFERENCES

> **The product of the sum and difference of two expressions is the square of the first expression minus the square of the second:**
> $$(A + B)(A - B) = A^2 - B^2.$$

OBJECTIVES

After finishing Section 7.3, you should be able to:

● Multiply two binomials mentally.

●● Square a binomial mentally.

●●● Multiply the sum and difference of two expressions mentally.

Multiply.

1. $(3xy + 2x)(x^2 + 2xy^2)$

2. $(x - 3y)(2x - 5y)$

Multiply.

3. $(4x + 5y)^2$

4. $(3x^2 - 2xy^2)^2$

ANSWERS ON PAGE A–24

Multiply.

5. $(2xy^2 + 3x)(2xy^2 - 3x)$

Examples Multiply.

$$(A \ + \ B) \ (A \ - \ B) = \ A^2 \ - \ B^2$$

> Carry out the rule. Say the words as you go.

5. $(3x^2y + 2y)(3x^2y - 2y) = (3x^2y)^2 - (2y)^2$
$$= 9x^4y^2 - 4y^2$$

6. $(-2x^3y^2 + 5t)(2x^3y^2 + 5t) = (5t - 2x^3y^2)(5t + 2x^3y^2)$
$$= (5t)^2 - (2x^3y^2)^2$$
$$= 25t^2 - 4x^6y^4$$

6. $(3xy^2 + 4y)(-3xy^2 + 4y)$

DO EXERCISES 5 AND 6.

One of the expressions may have more than one term.

Examples Multiply.

$$(A \quad - \ B) \ (A \quad + \ B) = \quad A^2 \quad - \ B^2$$

7. $(\boxed{2x + 3} - 2y)(\boxed{2x + 3} + 2y) = (\boxed{2x + 3})^2 - (2y)^2$
$$= 4x^2 + 12x + 9 - 4y^2$$

Multiply.

7. $(3y + 4 - 3x)(3y + 4 + 3x)$

8. $(4x + 3y + 2)(4x - 3y - 2) = (4x + \boxed{3y + 2})(4x - [\boxed{3y + 2}])$
$$= 16x^2 - (\boxed{3y + 2})^2$$
$$= 16x^2 - (9y^2 + 12y + 4)$$
$$= 16x^2 - 9y^2 - 12y - 4$$

9. $(a + b + c)(a - b - c) = (a + \boxed{b + c})(a - [\boxed{b + c}])$
$$= a^2 - (\boxed{b + c})^2$$
$$= a^2 - (b^2 + 2bc + c^2)$$
$$= a^2 - b^2 - 2bc - c^2$$

DO EXERCISES 7 AND 8.

8. $(2a + 5b + c)(2a - 5b - c)$

NAME _____   CLASS _____

EXERCISE SET 7.3

◖•◗ Multiply.

1. $(3xy - 1)(4xy + 2)$

2. $(m^3n + 8)(m^3n - 6)$

3. $(3 - c^2d^2)(4 + c^2d^2)$

4. $(6x - 2y)(5x - 3y)$

5. $(m^2 - n^2)(m + n)$

6. $(pq + 0.2)(0.4pq - 0.1)$

7. $(x^5y^5 + xy)(x^4y^4 - xy)$

8. $(x - y^3)(x + 2y^3)$

◖••◗ Multiply.

9. $(x + h)^2$

10. $(3a + 2b)^2$

11. $(r^3t^2 - 4)^2$

12. $(3a^2b - b^2)^2$

13. $(p^4 + m^2n^2)^2$

14. $(ab + cd)^2$

15. $\left(2a^3 - \frac{1}{2}b^3\right)^2$

16. $-5x(x + 3y)^2$

17. $3a(a - 2b)^2$

18. $(a^2 + b + 2)^2$

◖•••◗ Multiply.

19. $(2a - b)(2a + b)$

20. $(x - y)(x + y)$

21. $(c^2 - d)(c^2 + d)$

22. $(p^3 - 5q)(p^3 + 5q)$

ANSWERS

1. _____

2. _____

3. _____

4. _____

5. _____

6. _____

7. _____

8. _____

9. _____

10. _____

11. _____

12. _____

13. _____

14. _____

15. _____

16. _____

17. _____

18. _____

19. _____

20. _____

21. _____

22. _____

23. _____

24. _____

25. _____

26. _____

27. _____

28. _____

29. _____

30. _____

31. _____

32. _____

33. _____

34. _____

35. _____

36. _____

37. _____

38. _____

39. _____

40. _____

23. $(ab + cd^2)(ab - cd^2)$ **24.** $(xy + pq)(xy - pq)$

25. $(x + y - 3)(x + y + 3)$ **26.** $(p + q + 4)(p + q - 4)$

27. $[x + y + z][x - (y + z)]$ **28.** $[a + b + c][a - (b + c)]$

29. $(a + b + c)(a - b - c)$ **30.** $(3x + 2 - 5y)(3x + 2 + 5y)$

Perform the indicated operations and simplify.

31. $(a - b)^2 - (a + b)(a - b)$ **32.** $(a - b)^2 + (a + b)^2$

33. $(a + b)^2 - (a - b)^2$ **34.** $9(x - 1)^2 - 25(x + 1)^2$

35. $y^2 - (y - 2)^2$ **36.** $(c^2 - cd + d^2)(c^2 + cd + d^2)$

37. $(a - b)^2 - (b - a)$ **38.** $(4x - y)(y - 4x)$

---

39. Find a formula for $(A + B)^3$.

40. Use the formula from Exercise 39 to multiply out the compound interest formula for $t = 3$:

$A = P(1 + r)^3$.

7.4 FACTORING

Factoring polynomials in several variables is quite similar to factoring polynomials in one variable.

■● TERMS WITH COMMON FACTORS

> **Whenever you factor polynomials, always look for the largest factor common to all the terms before trying any other kind of factoring.**

Example 1   Factor. $20x^3y + 12x^2y$.

$$20x^3y + 12x^2y = (\,4x^2y\,)(5x) + (\,4x^2y\,) \cdot 3$$

$$= 4x^2y\,(5x + 3) \qquad \text{Factoring out the largest common factor}$$

Try to write the answer directly, as follows.

Example 2   Factor: $6x^2y - 21x^3y^2 + 3x^2y^3$.

$$6x^2y - 21x^3y^2 + 3x^2y^3 = 3x^2y\,(2 - 7xy + y^2)$$

DO EXERCISES 1 AND 2.

●● FACTORING BY GROUPING

Sometimes a common factor is itself a binomial.

Example 3   Factor: $(p + q)(x + 2) + (p + q)(x + y)$.

$$(\,p + q\,)(x + 2) + (\,p + q\,)(x + y) = (\,p + q\,)[(x + 2) + (x + y)]$$

$$= (p + q)(2x + y + 2)$$

Sometimes pairs of terms have a common factor that can be removed.

Example 4   Factor: $px + py + qx + qy$.

$$px + py + qx + qy = p(\,x + y\,) + q(\,x + y\,)$$

$$= (p + q)(\,x + y\,)$$

DO EXERCISES 3 AND 4.

●●● TRINOMIAL SQUARES

If a trinomial is the square of a binomial, then it is easy to factor. So, whenever you have a trinomial to factor, you should check to see whether it is the square of a binomial.

Example 5   Factor: $25x^2 + 20xy + 4y^2$.

a) Find whether $25x^2 + 20xy + 4y^2$ is the square of a binomial.

The first term and the last term are squares:

$$25x^2 = (\,5x\,)^2 \quad \text{and} \quad 4y^2 = (\,2y\,)^2.$$

OBJECTIVES

After finishing Section 7.4, you should be able to:

■● Factor polynomials when the terms have a common factor.

●● Factor by grouping.

●●● Factor trinomials that are squares of binomials.

■■ Factor trinomials that are not squares.

Factor.

1. $x^4y^2 + 2x^3y + 3x^2y$

2. $10p^6q^2 - 4p^5q^3 + 2p^4q^4$

Factor.

3. $(a - b)(x + 5) + (a - b)(x + y^2)$

4. $ax^2 + ay + bx^2 + by$

Factor.

5. $x^4 + 2x^2y^2 + y^4$

6. $-4x^2 + 12xy - 9y^2$
(*Hint:* First factor out -1.)

Factor.

7. $x^2y^2 + 5xy + 4$

8. $2x^4y^6 + 6x^2y^3 - 20$

9. $t^5 - mt^4 - 2m^2t^3$

Twice the product of $5x$ and $2y$ should be the other term:

$$2 \cdot \boxed{5x} \cdot \boxed{2y} = 20xy.$$

Thus the trinomial is the square of a binomial.

b) We factor by writing the square roots of the square terms and the sign of the other term:

$$25x^2 + 20xy + 4y^2 = (5x + 2y)^2.$$

We can check by squaring $5x + 2y$.

DO EXERCISES 5 AND 6.

⠿ TRINOMIALS THAT ARE NOT SQUARES

If a trinomial is not a square, we use trial and error.

Examples Factor.

6. $p^2q^2 + 7pq + 12 = (pq)^2 + (3 + 4)pq + 3 \cdot 4$
$\qquad\qquad\qquad = (pq + 3)(pq + 4)$

7. $8x^4 - 20x^2y - 12y^2 = 4(2x^4 - 5x^2y - 3y^2)$ Removing a common factor
$\qquad\qquad\qquad = 4[(2x^2)(x^2) + (-6 + 1)x^2y + (-3y)y]$
$\qquad\qquad\qquad = 4(2x^2 + y)(x^2 - 3y)$

8. $a^4 + a^3b - 6a^2b^2 = a^2(a^2 + ab - 6b^2)$
$\qquad\qquad\qquad = a^2(a - 2b)(a + 3b)$

> **Remember always to look first for common factors. Remember, too, that some of the factors obtained may be factorable, and factor completely. Furthermore, not all trinomials can be factored.**

DO EXERCISES 7–9.

ANSWERS ON PAGE A–25

NAME

CLASS

EXERCISE SET 7.4

● ● Factor.

1. $12n^2 + 24n^3$

2. $ax^2 + ay^2$

3. $9x^2y^2 - 36xy$

4. $x^2y - xy^2$

5. $2\pi rh + 2\pi r^2$

6. $10p^4q^4 + 35p^3q^3 + 10p^2q^2$

● ● ● Factor.

7. $(a + b)(x - 3) + (a + b)(x + 4)$

8. $5c(a^3 + b) - (a^3 + b)$

9. $(x - 1)(x + 1) - y(x + 1)$

10. $x^2 + x + xy + y$

11. $n^2 + 2n + np + 2p$

12. $a^2 \quad 3a + ay - 3y$

13. $2x^2 - 4x + xz - 2z$

14. $6y^2 - 3y + 2py - p$

● ● ● Factor.

15. $x^2 - 2xy + y^2$

16. $a^2 - 4ab + 4b^2$

17. $9c^2 + 6cd + d^2$

18. $16x^2 + 24xy + 9y^2$

19. $49m^4 - 112m^2n + 64n^2$

20. $4x^2y^2 + 12xyz + 9z^2$

ANSWERS
1.
2.
3.
4.
5.
6.
7.
8.
9.
10.
11.
12.
13.
14.
15.
16.
17.
18.
19.
20.

21. $y^4 + 10y^2z^2 + 25z^4$

22. $0.01x^4 - 0.1x^2y^2 + 0.25y^4$

23. $\frac{1}{4}a^2 + \frac{1}{3}ab + \frac{1}{9}b^2$

24. $4p^2q + pq^2 + 4p^3$

⊞ Factor.

25. $a^2 - ab - 2b^2$

26. $3b^2 - 17ab - 6a^2$

27. $m^2 + 2mn - 360n^2$

28. $x^2y^2 + 8xy + 15$

29. $m^2n^2 - 4mn - 32$

30. $p^2q^2 + 7pq + 6$

31. $a^5b^2 + 3a^4b - 10a^3$

32. $m^2n^6 + 4mn^5 - 32n^4$

33. $a^5 + 4a^4b - 5a^3b^2$

34. $2s^6t^2 + 10s^3t^3 + 12t^4$

35. $x^6 + x^3y - 2y^2$

36. $a^4 + a^2bc - b^2c^2$

7.5   FACTORING COMPLETELY

● DIFFERENCES OF SQUARES

Differences of squares can be factored as before.

$$A^2 - B^2 = (A + B)(A - B)$$

Example 1   Factor: $36x^2 - 25y^6$.

$$36x^2 - 25y^6 = (6x)^2 - (5y^3)^2$$
$$= (6x + 5y^3)(6x - 5y^3)$$

DO EXERCISES 1 AND 2.

If a factor can be factored again, it should be factored.

Example 2   Factor: $32x^4y^4 - 50$.

$$32x^4y^4 - 50 = 2(16x^4y^4 - 25)$$
$$= 2(4x^2y^2 + 5)(4x^2y^2 - 5)$$

DO EXERCISE 3.

● ● FACTORING COMPLETELY

Let us review the general strategy for factoring, which we considered in Section 5.4.

> A. **Always look first for a common factor.**
>
> B. **Then look at the number of terms.**
>
> **Four terms: Try factoring by grouping.**
>
> **Three terms: Determine whether the trinomial is a square. If so, you know how to factor. If not, try trial and error.**
>
> **Two terms: Determine whether you have a difference of squares. Do *not* try to factor a sum of squares:** $A^2 + B^2$.
>
> C. **Always *factor completely*. If a factor with more than one term can still be factored, you should factor it. When no factor can be factored further, you have factored completely.**

Example 3   Factor: $5x^4 - 5y^4$.

A. We look first for a common factor:

$$5x^4 - 5y^4 = 5(x^4 - y^4).$$ Factoring out the largest common factor

OBJECTIVES

After finishing Section 7.5, you should be able to:

● Factor differences of squares.

● ● Factor polynomials completely.

Factor.

1. $9a^2 - 16x^4$

2. $25x^2y^4 - 4a^2$

Factor.

3. $5 - 5x^2y^6$

Factor completely.

4. $16y^2 - 81x^4y^2$

5. $m^4 + 10m^2t^2 + 25t^4$

6. $10p^2 - 25pq - 60q^2$

7. $3a^2 - 6ab + 5a^3 - 10a^2b$

B. The factor $x^4 - y^4$ has only two terms. It is a difference of squares. We factor it:

$$5(x^4 - y^4) = 5(x^2 + y^2)(x^2 - y^2)$$

> Do not forget to include the common factor.

We see that one of the factors is still a difference of squares. We factor it:

$$5(x^2 + y^2)(x + y)(x - y).$$

C. We have factored completely because no factors can be factored further.

Example 4 Factor: $2a^4 + 8a^3b + 6a^2b^2$.

A. We look first for a common factor:

$$2a^4 + 8a^3b + 6a^2b^2$$
$$= 2a^2(a^2 + 4ab + 3b^2). \quad \text{Factoring out the largest common factor}$$

B. The factor $a^2 + 4ab + 3b^2$ has three terms, but it is not a trinomial square. We factor it using trial and error:

$$2a^2(a^2 + 4ab + 3b^2) = 2a^2[a^2 + (1 + 3)ab + 3b^2]$$
$$= 2a^2(a + b)(a + 3b).$$

C. No factor can be factored further, so we have factored completely.

Example 5 Factor: $x^6 - 8x^3y + 16y^2$.

A. We look first for a common factor. There is none.

B. There are three terms. We see that this is a trinomial square and factor it:

$$x^6 - 8x^3y + 16y^2 = (x^3)^2 - 2x^3(4y) + (4y)^2$$
$$= (x^3 - 4y)^2$$

C. No factor can be factored further, so we have factored completely.

Example 6 Factor: $2x^2 + 4ax + 3a^2x + 6a^3$.

A. We look first for a common factor. There is none.

B. There are four terms. We try factoring by grouping.

$$2x^2 + 4ax + 3a^2x + 6a^3$$
$$= (2x^2 + 4ax) + (3a^2x + 6a^3) \quad \text{Separating into two binomials}$$
$$= 2x(x + 2a) + 3a^2(x + 2a) \quad \text{Factoring each binomial}$$
$$= (2x + 3a^2)(x + 2a) \quad \text{Factoring out the common factor, } x + 2a$$

C. No factor can be factored further, so we have factored completely.

DO EXERCISES 4–7.

CLASS   ANSWERS

EXERCISE SET 7.5

▪ Factor.

1. $x^2 - y^2$ **2.** $a^2 - h^2$

3. $a^2b^2 - 9$ **4.** $p^2q^2 - r^2$

5. $9x^4y^2 - b^2$ **6.** $36t^2 - 49p^2q^2$

7. $3x^2 - 48y^2$ **8.** $9s^4 - 9s^2$

9. $64z^2 - 25c^2d^2$ **10.** $5t^2 - 20m^2$

▪▪ Factor completely.

11. $7p^4 - 7q^4$ **12.** $a^4b^4 - 16$

13. $81a^4 - b^4$ **14.** $1 - 16x^{12}y^{12}$

15. $18m^4 + 12m^3 + 2m^2$ **16.** $75a^3 + 60a^2b + 12ab^2$

17. $xy^2 + 3y^2 - 4x - 12$ **18.** $ay^2 - a - y^2 + 1$

19. $p^3 - p^2t - 2pt^2$ **20.** $15x^2y - 20xy - 35y$

ANSWERS

1. _____

2. _____

3. _____

4. _____

5. _____

6. _____

7. _____

8. _____

9. _____

10. _____

11. _____

12. _____

13. _____

14. _____

15. _____

16. _____

17. _____

18. _____

19. _____

20. _____

21. $-a^2 - ab + 6b^2$
   (*Hint:* Factor out -1.)

22. $r^2 + 6rs + 9s^2$

23. $-k^2 + 36kl - 324l^2$
   (*Hint:* Factor out -1.)

24. $4a^2 - 4a^2b - 8a^2b^2$

25. $ab^3 - ab^2 - ab$

26. $x^4 + x^3 + x^2$

27. $3b^2 - 17ab - 6a^2$

28. $6a^2 - 47ab - 63b^2$

29. $a^2c^2 - 4a^2 - b^2c^2 + 4b^2$

30. $-16a^2b - 10a^2br - a^2br^2$

31. $-p^2 + 20pt - 100t^2$

32. $n^2 + 23nr - 420r^2$

33. $a^2 - b^2 + (a - b)^2$

34. $5ac - 5ad + 5bc - 5bd$

35. $k^2t^2 - 9s^2 - 9t^2 + k^2s^2$

36. $42m^2n - 24mn^2 - 18n^3$

37. Factor: $(x + h)^2 - x^2$

38. Use the answer to Exercise 39 to find
$(4.1)^2 - 4^2$ and $(4.01)^2 - 4^2$.

7.6 SOLVING EQUATIONS

Sometimes a letter represents a specific number, although we may not know what the number is. In this case it is called a *constant* rather than a variable. In this section we agree that letters near the end of the alphabet

x, y, z will be variables.

Letters at the beginning of the alphabet

a, b, c will be constants.

This is a fairly common agreement.

● EQUATIONS WITH UNKNOWN CONSTANTS

We treat unknown constants in the same way as known ones. The laws of numbers still apply.

Example 1   Solve for x: $cx + b^2 = 2$.

By our agreement x is a variable and b and c are constants. Thus we solve for x:

$$cx = 2 - b^2 \qquad \text{Adding } -b^2$$

$$x = \frac{2 - b^2}{c}. \qquad \text{Multiplying by } \frac{1}{c}$$

We treated c and b as if they were known. In this sense they are like the 2 in the equation. Note that when solving for a letter, we do the same things we would do to solve any equation. To see this, compare Example 1 with the following.

Example 1A   Solve the equation $3x + 7 = 2$ for x.

$$3x = 2 - 7 \qquad \text{Adding } -7$$

$$x = \frac{2 - 7}{3}. \qquad \text{Multiplying by } \frac{1}{3}$$

In this case, we can simplify further, whereas in Example 1 we could not.

DO EXERCISE 1.

OBJECTIVE

After finishing Section 7.6, you should be able to:

■ Solve equations where some letters are used as constants.

Solve for y.

1. $ay - b = 3$

Solve for x.

2. $2abx - 6a = abx$

Solve for y.

3. $3(t - 2y) + 7(2y + 1) = 36$

To solve an equation we try to get the variable alone on one side. If the variable appears in several terms, we usually collect like terms, or factor out the variable. We may need to multiply to remove parentheses.

Example 2   Solve for y: $cdy + 4d = 3cdy$.

$4d = 3cdy - cdy$ Getting the y-terms on one side

$4d = (3cd - cd)y$ Collecting like terms (or factoring out y)

$4d = 2cdy$

$\dfrac{4d}{2cd} = y$ Multiplying by $\dfrac{1}{2cd}$

$\dfrac{2}{c} = y$.   Simplifying

DO EXERCISE 2.

Example 3   Solve for x: $3(t + 2x) = 30 + 7(x - 1)$.

$3t + 6x = 30 + 7x - 7$ Multiplying to remove parentheses

$3t = 23 + 7x - 6x$ Adding $-6x$ and collecting like terms (the 30 and -7)

$3t - 23 = x$ Adding -23 and collecting x-terms

DO EXERCISE 3.

NAME   CLASS   ANSWERS

EXERCISE SET 7.6

■ ● Solve for x, y, or z.

1. $2a - 3x = 12$ **2.** $z - k = 3k$

3. $3b - 2y = 1$ **4.** $5a + 2x = 3b$

5. $3acy - 9a = acy$ **6.** $24a - 3x = 5x$

7. $4(x - a) + 4a = 7(b + 3) - 21$

8. $-3(c + y) + 3c = 6(c - 2) + 12$

9. $5(2a - z) + 6 = -2(z - 5a)$

10. $x(a + 2) - 3 = b(x + 7)$

1. _____

2. _____

3. _____

4. _____

5. _____

6. _____

7. _____

8. _____

9. _____

10. _____

11. $ax + b^2 = a^2 + 3x$

12. $m(m - x) = n(n - x)$

11. _____

12. _____

13. $m^2x = m + 1$

14. $3(x - a) = x + 3a$

13. _____

14. _____

15. $rxt = c$

16. $mx + 4 = x + 4m^2$

15. _____

16. _____

17. $ax + bx = 3ax - c$

18. $b^2x = 4b$

17. _____

18. _____

19. $3(m - x) = x - 7m$

20. $2(3 - x) + b = b(x - 5)$

19. _____

20. _____

7.7 FORMULAS

▪ • The use of formulas is important in many applications of mathematics. It is important to be able to solve a formula for a letter.

Example 1 (*Gravitational force*) The gravitational force f between planets of mass M and m, at a distance d from each other, is given by

$$f = \frac{kMm}{d^2},$$

where k is a constant. Solve for m.

$fd^2 = kMm$ Multiplying by d^2

$\dfrac{fd^2}{kM} = m$ Multiplying by $\dfrac{1}{kM}$

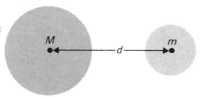

DO EXERCISE 1.

It is difficult, if not impossible, to look at a formula and tell which symbols are variables and which are constants. In fact, in a formula, a certain letter may be a variable at times and a constant at others, depending on how the formula is used.

Example 2 (*The area of a trapezoid*) The area A of a trapezoid is half the product of the height h and the sum of the lengths b_1 and b_2 of the parallel sides:

$$A = \frac{1}{2}(b_1 + b_2)h.$$

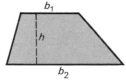

Solve for b_2.

We consider b_1 and b_2 to be different variables (or constants). The numbers 1 and 2 are called *subscripts*.

$2A = (b_1 + b_2)h$ Multiplying by 2

$2A = b_1h + b_2h$

$2A - b_1h = b_2h$ Adding $-b_1h$

$\dfrac{2A - b_1h}{h} = b_2$ Multiplying by $\dfrac{1}{h}$

Each of the following is also a correct answer:

$\dfrac{2A}{h} - b_1 = b_2$ and $\dfrac{1}{h}(2A - b_1h) = b_2.$

DO EXERCISE 2.

In both of the examples, the letter for which we solved was on the right side of the equation. Ordinarily we put the letter for which we solve on the left. This is a matter of choice, since all equations are reversible.

OBJECTIVE

After finishing Section 7.7, you should be able to:

▪ • **Solve a formula for a letter.**

1. Solve $f = \dfrac{kMm}{d^2}$ for M.

2. Solve $V = \dfrac{1}{6}\pi h(h^2 + 3a^2)$ for a^2.

ANSWERS ON PAGE A-25

SOMETHING EXTRA

APPLICATIONS

The following are some further applications of polynomials in several variables.

The area A of a rectangle of length l and width w:

$A = lw.$

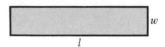

The volume V of a rectangular solid of length x, width y, and height h:

$V = xyh.$

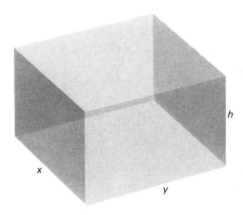

The total revenue R from the sale of x units of one product at \$4 each and y units of another product at \$7 each:

$R = 4x + 7y.$

The amount A that P dollars will grow to at interest rate r, compounded annually for 3 years:

$A = P + 3rP + 3r^2P + r^3P.$

The height S, in feet, of an object t seconds after being given an upward velocity of v feet per second from an altitude h:

$S = -16t^2 + vt + h.$

The approximate length L of a pulley belt around pulleys whose centers are D units apart and whose circumferences are C_1 and C_2:

$L = \dfrac{1}{2}C_1 + \dfrac{1}{2}C_2 + 2D.$

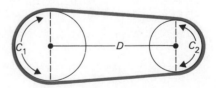

EXERCISE

Solve the pulley formula for D.

EXERCISE SET 7.7

• Solve for the letter indicated.

1. $S = 2\pi rh$; r

2. (*An interest formula*) $A = P(1 + rt)$; t

3. (*The area of a triangle*) $A - \frac{1}{2}bh$; b

4. $s = \frac{1}{2}gt^2$; g

5. $S = 180(n - 2)$; n

6. $S = \frac{n}{2}(a + l)$; a

7. $V = \frac{1}{3}k(B + b + 4M)$; b

8. $A = P + Prt$; P
   (*Hint:* Factor the right-hand side.)

9. $S(r - 1) = rl - a$; r

10. $T = mg - mf$; m
    (*Hint:* Factor the right-hand side.)

11. $A = \frac{1}{2}h(b_1 + b_2)$; h

12. (*The area of a right circular cylinder*)
    $S = 2\pi r(r + h)$; h

13. $r = \frac{v^2 pL}{a}$; a

14. $L = \frac{Mt - g}{t}$; M

ANSWERS

1. _____

2. _____

3. _____

4. _____

5. _____

6. _____

7. _____

8. _____

9. _____

10. _____

11. _____

12. _____

13. _____

14. _____

ANSWERS

15. $A = \frac{1}{2}h(b_1 + b_2)$; b_1 16. $l = a + (n - 1)d$; n

15. _____

16. _____

17. $A = \frac{\pi r^2 E}{180}$; E 18. $R = \frac{WL - x}{L}$; W

17. _____

18. _____   19. $V - -h(B + c + 4M)$; M 20. $W = I^2 R$; R

19. _____

21. $y = \frac{v^2 pL}{a}$; L 22. $V = \frac{1}{3}bh$; b

20. _____

21. _____

23. $r = \frac{v^2 pL}{a}$; p 24. $P = 2(l + w)$; l

22. _____

23. _____

25. $\frac{a}{c} = n + bn$; n 26. $C = \frac{Ka - b}{a}$; K

24. _____

(*Hint:* Factor the right-hand side.)

25. _____

26. _____   27. (*A temperature conversion formula*)   28. (*The volume of a sphere*)

$C = \frac{5}{9}(F - 32)$; F $V = \frac{4}{3}\pi r^3$; π

27. _____

28. _____

29. $f = \frac{gm - t}{m}$; g 30. $S = \frac{rl - a}{r - l}$; a

29. _____

30. _____

EXTENSION EXERCISES

SECTION 7.1

1. In the sixth-degree term of a polynomial in the variables x and y, the exponent of y is twice that of x. The coefficient of the term is 1 more than the exponent of y. Write the term.

SECTION 7.2

Express the area of the shaded region as a polynomial. (Leave results in terms of π where appropriate.)

2.

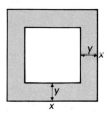

3.

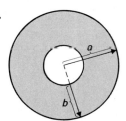

4.

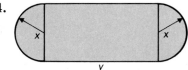

5.

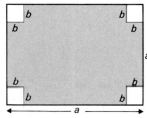

SECTION 7.3

6. Multiply. Look for patterns.
a) $(x + y)^2$ b) $(x - y)^2$ c) $(x + y)^3$ d) $(x - y)^3$ e) $(x + y)^4$
f) $(x - y)^4$ g) $(x + y)^5$ h) $(x - y)^5$ i) $(x + y)^6$ j) $(x - y)^6$

7. The expression you get when you multiply out $(x + y)^n$ is called the *expansion* of $(x + y)^n$. On the basis of the patterns observed in the expansions of the preceding exercise, answer the following questions:
a) How many terms are there in the expansion of $(x + y)^7$?
b) What is the coefficient of the xy^8 term in the expansion of $(x + y)^9$?
c) In the expansion of $(x - y)^{10}$, how many terms have negative coefficients, and how many have positive coefficients?
d) One term in the expansion of $(x + y)^{12}$ is $220x^3y^9$. What other term has the same coefficient?
e) What is the n in the term $6435x^7y^n$, if the term occurs in the expansion of $(x + y)^{15}$?

SECTION 7.4

Factor.

8. $6(x - 1)^2 + 7y(x - 1) - 3y^2$ **9.** $(y + 4)^2 + 2x(y + 4) + x^2$

10. $2(a + 3)^2 - (a + 3)(b - 2) - (b - 2)^2$

SECTION 7.5

Factor completely.

11. $2x^4 - 2x^2y^2 - 24y^4$ **12.** $a^2 + 2ab + b^2 - 16$ **13.** $25 - 9x^2 - 6xy - y^2$ **14.** $x^2y^2 - x^2 - 4y^2 + 4$

15. $y^2 + 2y - x^2 + 10x - 24$ (*Hint:* Rename -24 as the sum of two numbers. Then factor as the difference of two squares.)   **16.** $x^2y^3 + 4x^2y^2 + 4xy^2 + 16xy + 4y + 16$

SECTION 7.6

Solve for x, y, or z.

17. $(x - c)^2 = (x - d)^2$

18. $(y + a)^2 = (y + b)^2$

19. $(z + a)(z + 2b) = z^2 + a^2 + b^2$

20. $a^2x = a^6$

21. $\dfrac{y}{2c} = d + b$

22. $\dfrac{5x - a}{7} = \dfrac{4}{3}$

23. $\dfrac{a}{4} - \dfrac{3z}{bc} = 0$

24. $\dfrac{x}{d} - 2x = 5$

SECTION 7.7

25. In $V = \frac{4}{3}\pi r^3$, what is the effect on V when r is doubled?

26. The formula $C = \frac{5}{9}(F - 32)$ is used to convert Fahrenheit temperatures to Celsius temperatures. At what temperature are the Fahrenheit and Celsius readings the same?

27. For what value(s) of A will $\dfrac{B}{A} + \dfrac{4C - 1}{A + 3}$ be undefined?

28. In $N = \dfrac{a}{c}$, what is the effect on N when c increases? Assume a, c, and N are positive.

NAME   SCORE

TEST OR REVIEW—CHAPTER 7

If you miss an item, review the indicated section and objective.

[7.1, ⚏]   **1.** Collect like terms:
$$3x^2yz^3 - 2xy^2 + x^2yz^3 + x^2y^2 + 2xy^2.$$

1. _____

[7.1, ⚏]   **2.** Arrange in descending powers of a:
$$b^3 + 3a^2b + 3ab^2 + a^3.$$

2. _____

[7.2, ⚏]   **3.** Add: $(a^3b - 6a^2b^2 - ab^3 + 3) + (a^3b - 4a^2b^2 + 8)$

3. _____

[7.2, ⚏]   **4.** Subtract: $(7x - 4y + 6z) - (3x - 7y + z).$

4. _____

Multiply.

[7.2, ⚏]   **5.** $(y^3z^2 - 2yz - 3)(2yz - 1)$

5. _____

[7.3, ⚏]   **6.** $(a^2b - 3a)(ab^2 + 2a)$

6. _____

[7.3, ⚏]   **7.** $(a^2 + 2b)^2$

7. _____

ANSWERS

[7.3, ●●●]   **8.** $(c^2 + d)(c^2 - d)$

8. _____

Factor.

[7.4, ●]   **9.** $p^4q - p^3q - pq$

9. _____

[7.4, ●●]   **10.** $2uw - 6ux - 3vx + vw$

10. _____

[7.5, ●]   **11.** $16x^2y^2 - 1$

11. _____

[7.5, ●●]   **12.** $16c^4 - d^4$

12. _____

[7.5, ●]   **13.** $c^4 - c^2x^2$

13. _____

[7.4, ●●●]   **14.** $9p^2 + 42pq + 49q^2$

14. _____

[7.4, ●●]   **15.** $35x^2 - 22xy + 3y^2$

15. _____

[7.6, ●]   **16.** Solve for x: $abx - 3b = 5abx$.

16. _____

[7.7, ●]   **17.** Solve for P: $A - P = Prt$.

17. _____

FRACTIONAL EXPRESSIONS AND EQUATIONS

8

Certain problems involving the time required to complete a task can be solved using fractional equations.

READINESS CHECK: SKILLS FOR CHAPTER 8

Multiply and simplify.

1. $\dfrac{1}{6} \cdot \dfrac{3}{8}$

2. $\dfrac{7}{9} \cdot \dfrac{9}{7}$

3. Find the reciprocal of $\dfrac{2}{5}$.

4. Divide and simplify: $\dfrac{2}{5} \div \dfrac{4}{15}$.

5. Add and simplify: $\dfrac{4}{8} + \dfrac{2}{5}$.

6. Subtract and simplify: $\dfrac{13}{15} - \dfrac{2}{45}$.

7. Subtract:

$(x^2 + 6x + 8) - (x^2 - 3x - 4).$

8. Multiply:

$(x - 2)(x + 2).$

Factor.

9. $x^2 + 3x + 2$

10. $x^2 - 6x + 9$

11. $6x^2 + 4x$

12. $25t^2 - 4$

13. $2x^2 - 3x + 1$

14. $x^5 + x^4 - 2x^3$

Solve.

15. $3t + 2t = 12$

16. $x^2 - 5x + 6 = 0$

OBJECTIVES

After finishing Section 8.1, you should be able to:

⚬• Multiply fractional expressions.

•⚬• Multiply a fractional expression by 1, using an expression such as A/A.

••• Simplify fractional expressions by factoring numerator and denominator and removing factors of 1.

⦂⦂ Multiply fractional expressions and simplify.

8.1 MULTIPLYING AND SIMPLIFYING

These are *fractional expressions*:

$$\frac{3}{4}, \qquad \frac{5}{x + 2}, \qquad \frac{x^2 + 3x - 10}{7x^2 - 4}.$$

A fractional expression is a quotient of two polynomials and, therefore, indicates division. For example,

$$\frac{3}{4} \quad \text{means} \quad 3 \div 4$$

and

$$\frac{x^2 + 3x - 10}{7x^2 - 4} \quad \text{means} \quad (x^2 + 3x - 10) \div (7x^2 - 4).$$

◼◦ MULTIPLYING

For fractional expressions, multiplication is done as in arithmetic.

> **To multiply two fractional expressions, multiply numerators and multiply denominators.**

Examples   Multiply.

1. $\dfrac{x-2}{3} \cdot \dfrac{x+2}{x+3} = \dfrac{(x-2)(x+2)}{3(x+3)}$ Multiplying numerators and multiplying denominators

We could multiply out the numerator and the denominator to get $\dfrac{x^2-4}{3x+9}$, but it is best *not* to do it yet. We will see why later.

2. $\dfrac{-2}{2y+3} \cdot \dfrac{3}{y-5} = \dfrac{-2 \cdot 3}{(2y+3)(y-5)}$

DO EXERCISES 1 AND 2.

◼◼ MULTIPLYING BY 1

Any fractional expression with the same numerator and denominator is a symbol for 1:

$$\frac{x+2}{x+2} = 1, \qquad \frac{3x^2-4}{3x^2-4} = 1, \qquad \frac{-1}{-1} = 1.$$

It should be noted that certain replacements are not sensible. For example, in

$$\frac{x+2}{x+2}$$

we should not substitute -2 for x. We would get 0 for the denominator.

We can multiply by 1 to get an equivalent expression.

Examples   Multiply.

3. $\dfrac{3x+2}{x+1} \cdot \boxed{\dfrac{2x}{2x}} = \dfrac{(3x+2)2x}{(x+1)2x}$

4. $\dfrac{x+2}{x-1} \cdot \boxed{\dfrac{x+1}{x+1}} = \dfrac{(x+2)(x+1)}{(x-1)(x+1)}$

DO EXERCISES 3 AND 4.

Multiply.

1. $\dfrac{x+3}{5} \cdot \dfrac{x+2}{x+4}$

2. $\dfrac{-3}{2x+1} \cdot \dfrac{4}{2x-1}$

Multiply.

3. $\dfrac{2x+1}{3x-2} \cdot \boxed{\dfrac{x}{x}}$

4. $\dfrac{x+1}{x-2} \cdot \boxed{\dfrac{x+2}{x+2}}$

Multiply

5. $\dfrac{x-8}{x-y} \cdot \boxed{\dfrac{-1}{-1}}$

Simplify by removing a factor of 1.

6. $\dfrac{5y}{y}$

7. $\dfrac{8x^2}{24x}$

Example 5 Multiply.

$$\frac{2+x}{2-x} \cdot \boxed{\frac{-1}{-1}} = \frac{(2+x)(-1)}{(2-x)(-1)}$$

DO EXERCISE 5.

●●● SIMPLIFYING FRACTIONAL EXPRESSIONS

To simplify, we can do the reverse of multiplying. We factor numerator and denominator and "remove" a factor of 1.

Example 6 Simplify by removing a factor of 1: $\dfrac{3x}{x}$.

$$\frac{3x}{x} = \frac{3 \cdot x}{1 \cdot x} \qquad \text{Factoring numerator and denominator}$$

$$= \frac{3}{1} \cdot \boxed{\frac{x}{x}} \qquad \text{Factoring the fractional expression}$$

$$= \frac{3}{1} \cdot 1 \qquad \frac{x}{x} = 1$$

$$= 3 \qquad \text{We "removed" a factor of 1.}$$

In this example we supplied a 1 in the denominator. This can always be done whenever it is helpful.

Example 7 Simplify by removing a factor of 1: $\dfrac{7x^2}{14x}$.

$$\frac{7x^2}{14x} = \frac{7x \cdot x}{7x \cdot 2} \qquad \text{Factoring numerator and denominator}$$

$$= \boxed{\frac{7x}{7x}} \cdot \frac{x}{2} \qquad \text{Factoring the fractional expression}$$

$$= \frac{x}{2} \qquad \text{"Removing" a factor of 1}$$

DO EXERCISES 6 AND 7.

Examples   Simplify by removing a factor of 1.

8. $\dfrac{6a + 12}{7(a + 2)} = \dfrac{6(a + 2)}{7(a + 2)}$

$\qquad = \dfrac{6}{7} \cdot \boxed{\dfrac{a + 2}{a + 2}}$

$\qquad = \dfrac{6}{7}\qquad$ "Removing" the factor $\dfrac{a + 2}{a + 2}$

9. $\dfrac{6x^2 + 4x}{2x^2 + 2x} = \dfrac{2x(3x + 2)}{2x(x + 1)}\qquad$ Factoring numerator and denominator

$\qquad = \boxed{\dfrac{2x}{2x}} \cdot \dfrac{3x + 2}{x + 1}\qquad$ Factoring the fractional expression

$\qquad = \dfrac{3x + 2}{x + 1}\qquad$ "Removing" a factor of 1

10. $\dfrac{x^2 + 3x + 2}{x^2 - 1} = \dfrac{(x + 2)(x + 1)}{(x + 1)(x - 1)}$

$\qquad = \boxed{\dfrac{x + 1}{x + 1}} \cdot \dfrac{x + 2}{x - 1}$

$\qquad = \dfrac{x + 2}{x - 1}$

11. $\dfrac{5a + 15}{10} = \dfrac{5(a + 3)}{5 \cdot 2}$

$\qquad = \boxed{\dfrac{5}{5}} \cdot \dfrac{a + 3}{2}$

$\qquad = \dfrac{a + 3}{2}$

DO EXERCISES 8–11.

Simplify by removing a factor of 1.

8. $\dfrac{2x^2 + x}{3x^2 + 2x}$

9. $\dfrac{x^2 - 1}{2x^2 - x - 1}$

10. $\dfrac{7x + 14}{7}$

11. $\dfrac{12y + 24}{48}$

ANSWERS ON PAGE A–26

Multiply and simplify.

12. $\dfrac{a^2 - 4a + 4}{a^2 - 9} \cdot \dfrac{a + 3}{a - 2}$

⠒⠒ MULTIPLYING AND SIMPLIFYING

In the following examples we multiply and then we simplify.

Example 12 Multiply and simplify: $\dfrac{x^2 + 6x + 9}{x^2 - 4} \cdot \dfrac{x - 2}{x + 3}$.

$$\dfrac{x^2 + 6x + 9}{x^2 - 4} \cdot \dfrac{x - 2}{x + 3}$$

$$= \dfrac{(x^2 + 6x + 9)(x - 2)}{(x^2 - 4)(x + 3)} \quad \text{Multiplying numerators and also denominators}$$

$$= \dfrac{(x + 3)(x + 3)(x - 2)}{(x + 2)(x - 2)(x + 3)} \quad \text{Factoring numerator and denominator}$$

> Note that if we had multiplied out the numerator and denominator, we would have had to factor them again in order to simplify.

$$= \dfrac{(x + 3)(x - 2)}{(x + 3)(x - 2)} \cdot \dfrac{x + 3}{x + 2} \quad \text{Factoring the fractional expression}$$

$$= \dfrac{x + 3}{x + 2} \quad \text{Simplifying by removing a factor of 1}$$

Example 13 Multiply and simplify: $\dfrac{x^2 + x - 2}{15} \cdot \dfrac{5}{2x^2 - 3x + 1}$.

$$\dfrac{x^2 + x - 2}{15} \cdot \dfrac{5}{2x^2 - 3x + 1}$$

$$= \dfrac{(x^2 + x - 2)5}{15(2x^2 - 3x + 1)} \quad \text{Multiplying numerators and denominators}$$

$$= \dfrac{(x + 2)(x - 1)5}{(5)(3)(x - 1)(2x - 1)} \quad \text{Factoring numerator and denominator}$$

$$= \dfrac{(x - 1)5}{(x - 1)5} \cdot \dfrac{x + 2}{3(2x - 1)} \quad \text{Factoring the fractional expression}$$

$$= \dfrac{x + 2}{3(2x - 1)} \quad \text{Simplifying by removing a factor of 1}$$

> You need not carry out this multiplication.

13. $\dfrac{x^2 - 25}{6} \cdot \dfrac{3}{x + 5}$

DO EXERCISES 12 AND 13.

NAME ______________________ CLASS __________

EXERCISE SET 8.1

• Multiply. Do not carry out the multiplication in the numerator and denominator.

1. $\dfrac{x-2}{x-5} \cdot \dfrac{x-2}{x+5}$

2. $\dfrac{x-1}{x+2} \cdot \dfrac{x+1}{x+2}$

3. $\dfrac{c-3d}{c+d} \cdot \dfrac{c+3d}{c-d}$

4. $\dfrac{a+2b}{a+b} \cdot \dfrac{a-2b}{a-b}$

•• Multiply.

5. $\dfrac{2a-1}{2a-1} \cdot \dfrac{3a-1}{3a+2}$

6. $\dfrac{3x-2}{x+7} \cdot \dfrac{2x+5}{2x+5}$

••• Simplify by removing a factor of 1.

7. $\dfrac{x(3x+2)(x+1)}{x(x+1)(3x-2)}$

8. $\dfrac{x(3x+4)(5x+7)}{x(3x-4)(5x+7)}$

***9.** $\dfrac{8a+8b}{8a-8b}$

***10.** $\dfrac{6x-6y}{6x+6y}$

11. $\dfrac{t^2-25}{t^2+t-20}$

12. $\dfrac{a^2-9}{a^2+5a+6}$

13. $\dfrac{2x^2+6x+4}{4x^2-12x-16}$

14. $\dfrac{x^2-3x-4}{2x^2+10x+8}$

---

* Indicates exercises that can be omitted if Chapter 7 was not studied.

ANSWERS

1. ______________

2. ______________

3. ______________

4. ______________

5. ______________

6. ______________

7. ______________

8. ______________

9. ______________

10. ______________

11. ______________

12. ______________

13. ______________

14. ______________

15. $\dfrac{a^2 - 10a + 21}{a^2 - 11a + 28}$

16. $\dfrac{x^2 - 3x - 18}{x^2 - 2x - 15}$

15. _____

16. _____

17. $\dfrac{6x + 12}{x^2 - x - 6}$

18. $\dfrac{5y + 5}{y^2 + 7y + 6}$

17. _____

18. _____

19. $\dfrac{a^2 + 1}{a + 1}$

20. $\dfrac{t^2 - 1}{t + 1}$

19. _____

20. _____

:: Multiply and simplify.

21. $\dfrac{t^2}{t^2 - 4} \cdot \dfrac{t^2 - 5t + 6}{t^2 - 3t}$

(*Hint:* Factor t^2 as $t \cdot t$.)

22. $\dfrac{x^2 - 3x - 10}{(x - 2)^2} \cdot \dfrac{x - 2}{x - 5}$

[*Hint:* Factor $(x - 2)^2$ as $(x - 2)(x - 2)$.]

21. _____

22. _____

23. $\dfrac{24a^2}{3(a^2 - 4a + 4)} \cdot \dfrac{3a - 6}{2a}$

24. $\dfrac{5v + 5}{v - 2} \cdot \dfrac{v^2 - 4v + 4}{v^2 - 1}$

23. _____

24. _____

$\dfrac{ab - b^2}{2a} \cdot \dfrac{2a + 2b}{a^2b - b^3}$

26. $\dfrac{c^2 - 6c}{c - 6} \cdot \dfrac{c + 3}{c}$

25. _____

26. _____

This time carry out the multiplications in the numerator and denominator.

27. ▦ Multiply. $\dfrac{625}{x - 345.1} \cdot \dfrac{34.2}{x + 345.1}$.

27. _____

8.2 DIVISION AND RECIPROCALS

● FINDING RECIPROCALS

Two expressions are reciprocals of each other if their product is 1. The reciprocal of a fractional expression is found by interchanging numerator and denominator.

Examples

1. The reciprocal of $\dfrac{2}{5}$ is $\dfrac{5}{2}$. $\left(\text{This is because } \dfrac{2}{5} \cdot \dfrac{5}{2} = \dfrac{10}{10} = 1.\right)$

2. The reciprocal of $\dfrac{2x^2 - 3}{x + 4}$ is $\dfrac{x + 4}{2x^2 - 3}$.

3. The reciprocal of $x + 2$ is $\dfrac{1}{x + 2}$. $\left(\text{Think of } x + 2 \text{ as } \dfrac{x + 2}{1}.\right)$

DO EXERCISES 1–4.

●● DIVISION

> **To divide, we multiply by a reciprocal and simplify the result.**

(To review the reason for this, see Section 1.5.)

Examples　Divide.

4. $\dfrac{3}{4} \div \dfrac{2}{5} = \dfrac{3}{4} \cdot \boxed{\dfrac{5}{2}}$　　Multiplying by the reciprocal

$$= \dfrac{3 \cdot 5}{4 \cdot 2}$$

$$= \dfrac{15}{8}$$

5. $\dfrac{x + 1}{x + 2} \div \dfrac{x - 1}{x + 3} = \dfrac{x + 1}{x + 2} \cdot \boxed{\dfrac{x + 3}{x - 1}}$　　Multiplying by the reciprocal

$$= \dfrac{(x + 1)(x + 3)}{(x + 2)(x - 1)}$$　←　You need not carry out the multiplications in the numerator and denominator.

DO EXERCISES 5 AND 6.

OBJECTIVES

After finishing Section 8.2, you should be able to:

● Find the reciprocal of a fractional expression.

●● Divide fractional expressions and simplify.

Find the reciprocal.

1. $\dfrac{7}{2}$

2. $\dfrac{x^2 + 5}{2x^3 - 1}$

3. $x - 5$

4. $\dfrac{1}{x^2 - 3}$

Divide.

5. $\dfrac{3}{5} \div \dfrac{7}{2}$

6. $\dfrac{x - 3}{x + 5} \div \dfrac{x + 5}{x - 2}$

Divide and simplify.

7. $\dfrac{x - 3}{x + 5} \div \dfrac{x + 2}{x + 5}$

8. $\dfrac{x^2 - 5x + 6}{x + 5} \div \dfrac{x + 2}{x + 5}$

9. $\dfrac{y^2 - 1}{y + 1} \div \dfrac{y^2 - 2y + 1}{y + 1}$

Example 6   Divide and simplify: $\dfrac{x + 1}{x^2 - 1} \div \dfrac{x + 1}{x^2 - 2x + 1}$.

$$\dfrac{x + 1}{x^2 - 1} \div \dfrac{x + 1}{x^2 - 2x + 1}$$

$$= \dfrac{x + 1}{x^2 - 1} \cdot \dfrac{x^2 - 2x + 1}{x + 1} \qquad \text{Multiplying by the reciprocal}$$

$$= \dfrac{(x + 1)(x^2 - 2x + 1)}{(x^2 - 1)(x + 1)}$$

$$= \dfrac{(x + 1)(x - 1)(x - 1)}{(x - 1)(x + 1)(x + 1)} \qquad \begin{array}{l}\text{Factoring numerator and} \\ \text{denominator}\end{array}$$

$$= \dfrac{(x + 1)(x - 1)}{(x + 1)(x - 1)} \cdot \dfrac{x - 1}{x + 1} \qquad \begin{array}{l}\text{Factoring the fractional} \\ \text{expression}\end{array}$$

$$= \dfrac{x - 1}{x + 1} \qquad \text{Simplifying}$$

Example 7   Divide and simplify: $\dfrac{x^2 - 2x - 3}{x^2 - 4} \div \dfrac{x + 1}{x + 5}$.

$$\dfrac{x^2 - 2x - 3}{x^2 - 4} \div \dfrac{x + 1}{x + 5}$$

$$= \dfrac{x^2 - 2x - 3}{x^2 - 4} \cdot \dfrac{x + 5}{x + 1} \qquad \text{Multiplying by the reciprocal}$$

$$= \dfrac{(x^2 - 2x - 3)(x + 5)}{(x^2 - 4)(x + 1)}$$

$$= \dfrac{(x - 3)(x + 1)(x + 5)}{(x - 2)(x + 2)(x + 1)} \qquad \begin{array}{l}\text{Factoring numerator} \\ \text{and denominator}\end{array}$$

$$= \dfrac{x + 1}{x + 1} \cdot \dfrac{(x - 3)(x + 5)}{(x - 2)(x + 2)} \qquad \begin{array}{l}\text{Factoring the fractional} \\ \text{expression}\end{array}$$

$$= \dfrac{(x - 3)(x + 5)}{(x - 2)(x + 2)} \quad \text{Simplifying}$$

> You need not carry out the multiplications in the numerator and denominator.

DO EXERCISES 7–9.

EXERCISE SET 8.2

◖•◗ Find the reciprocal.

1. $\dfrac{4}{x}$

2. $\dfrac{a+3}{a-1}$

3. $x^2 - y^2$

4. $\dfrac{1}{a+b}$

5. $\dfrac{x^2 + 2x - 5}{x^2 - 4x + 7}$

6. $\dfrac{x^2 - 3xy + y^2}{x^2 + 7xy - y^2}$

◖••◗ Divide and simplify.

7. $\dfrac{2}{5} \div \dfrac{4}{3}$

8. $\dfrac{5}{6} \div \dfrac{2}{3}$

9. $\dfrac{2}{x} \div \dfrac{8}{x}$

10. $\dfrac{x}{2} \div \dfrac{3}{x}$

*11. $\dfrac{x^2}{y} \div \dfrac{x^3}{y^3}$

*12. $\dfrac{a}{b^2} \div \dfrac{a^2}{b^3}$

13. $\dfrac{a+2}{a-3} \div \dfrac{a-1}{a+3}$

14. $\dfrac{y+2}{4} \div \dfrac{y}{2}$

15. $\dfrac{x^2-1}{x} \div \dfrac{x+1}{x-1}$

16. $\dfrac{4y-8}{y+2} \div \dfrac{y-2}{y^2-4}$

17. $\dfrac{x+1}{6} \div \dfrac{x+1}{3}$

*18. $\dfrac{a}{a-b} \div \dfrac{b}{a-b}$

19. $\dfrac{x^2-9}{4x+12} \div \dfrac{x-3}{6}$

20. $\dfrac{c^2+3c}{c^2+2c-3} \div \dfrac{c}{c+1}$

ANSWERS

1. _____
2. _____
3. _____
4. _____
5. _____
6. _____
7. _____
8. _____
9. _____
10. _____
11. _____
12. _____
13. _____
14. _____
15. _____
16. _____
17. _____
18. _____
19. _____
20. _____

* Indicates exercises that can be omitted if Chapter 7 was not studied.

ANSWERS

21. _____

22. _____

23. _____

24. _____

25. _____

26. _____

27. _____

28. _____

*21. $\dfrac{x + y}{x - y} \div \dfrac{x^2 + y}{x^2 - y^2}$

*22. $\dfrac{x - b}{2x} \div \dfrac{x^2 - b^2}{5x^2}$

23. $\dfrac{x^2 - x - 20}{x^2 + 7x + 12} \div \dfrac{x^2 - 10x + 25}{x^2 + 6x + 9}$

24. $\dfrac{2y^2 - 7y + 3}{2y^2 + 3y - 2} \div \dfrac{6y^2 - 5y + 1}{3y^2 + 5y - 2}$

25. $\dfrac{c^2 + 10c + 21}{c^2 - 2c - 15} \div (c^2 + 2c - 35)$

26. $(1 - z) \div \dfrac{1 - z}{1 + 2z - z^2}$

27. $\dfrac{(t + 5)^3}{(t - 5)^3} \div \dfrac{(t + 5)^2}{(t - 5)^2}$

28. $\dfrac{(y - 3)^3}{(y + 3)^3} \div \dfrac{(y - 3)^2}{(y + 3)^2}$

8.3   ADDITION AND SUBTRACTION

● ADDITION WHEN DENOMINATORS ARE THE SAME

Addition is done as in arithmetic.

> **When denominators are the same, we add the numerators and keep the denominator.**

Examples   Add.

1. $\dfrac{x}{x+1} + \dfrac{2}{x+1} = \dfrac{x+2}{x+1}$

2. $\dfrac{2x^2 + 3x - 7}{2x + 1} + \dfrac{x^2 + x - 8}{2x + 1} = \dfrac{(2x^2 + 3x - 7) + (x^2 + x - 8)}{2x + 1}$

$$= \dfrac{3x^2 + 4x - 15}{2x + 1}$$

DO EXERCISES 1–3.

●● ADDITION WHEN DENOMINATORS ARE ADDITIVE INVERSES

When one denominator is the additive inverse of the other, we first multiply one expression by $-1/-1$.

Examples   Add.

3. $\dfrac{x}{2} + \dfrac{3}{-2} = \dfrac{x}{2} + \boxed{\dfrac{-1}{-1}} \cdot \dfrac{3}{-2}$ Multiplying by $\dfrac{-1}{-1}$

$$= \dfrac{x}{2} + \dfrac{(-1)3}{(-1)(-2)}$$

$$= \dfrac{x}{2} + \dfrac{-3}{2}$$ Denominators are now the same

$$= \dfrac{x + (-3)}{2} = \dfrac{x - 3}{2}$$

4. $\dfrac{3x + 4}{x - 2} + \dfrac{x - 7}{2 - x} = \dfrac{3x + 4}{x - 2} + \boxed{\dfrac{-1}{-1}} \cdot \dfrac{x - 7}{2 - x}$

> We could have chosen to multiply this expression by $-1/-1$. This would also work.

$$= \dfrac{3x + 4}{x - 2} + \dfrac{-1(x - 7)}{-1(2 - x)}$$ *Note.* $-1(2 - x) = -2 + x$
$$= x - 2$$

$$= \dfrac{3x + 4}{x - 2} + \dfrac{7 - x}{x - 2}$$

$$= \dfrac{(3x + 4) + (7 - x)}{x - 2} = \dfrac{2x + 11}{x - 2}$$

DO EXERCISES 4 AND 5.

OBJECTIVES

After finishing Section 8.3, you should be able to:

● Add fractional expressions having the same denominator.

●● Add fractional expressions whose denominators are additive inverses of each other.

●●● Subtract fractional expressions having the same denominator.

●●●● Subtract fractional expressions whose denominators are additive inverses of each other.

Add.

1. $\dfrac{5}{9} + \dfrac{2}{9}$

2. $\dfrac{3}{x - 2} + \dfrac{x}{x - 2}$

3. $\dfrac{4x + 5}{x - 1} + \dfrac{2x - 1}{x - 1}$

Add.

4. $\dfrac{x}{4} + \dfrac{5}{-4}$

5. $\dfrac{2x + 1}{x - 3} + \dfrac{x + 2}{3 - x}$

ANSWERS ON PAGE A–27

Subtract.

6. $\dfrac{7}{11} - \dfrac{3}{11}$

7. $\dfrac{2x^2 + 3x - 7}{2x + 1} - \dfrac{x^2 + x - 8}{2x + 1}$

Subtract.

8. $\dfrac{x}{3} - \dfrac{2x - 1}{-3}$

9. $\dfrac{3x}{x - 2} - \dfrac{x - 3}{2 - x}$

••• SUBTRACTION WHEN DENOMINATORS ARE THE SAME

Subtraction is done as in arithmetic.

> **When denominators are the same, we subtract the numerators and keep the denominator.**

Example 5   Subtract: $\dfrac{3x}{x + 2} - \dfrac{x - 2}{x + 2}$.

$$\frac{3x}{x + 2} - \frac{x - 2}{x + 2} = \frac{3x - (x - 2)}{x + 2}$$

> The parentheses are important to make sure that you subtract the entire numerator.

$$= \frac{3x - x + 2}{x + 2}$$

$$= \frac{2x + 2}{x + 2}$$

DO EXERCISES 6 AND 7.

⠒⠒ SUBTRACTION WHEN DENOMINATORS ARE ADDITIVE INVERSES

When one denominator is the additive inverse of the other, we first multiply one expression by $-1/-1$.

Example 6   Subtract: $\dfrac{x}{5} - \dfrac{3x - 4}{-5}$.

$$\frac{x}{5} - \frac{3x - 4}{-5} = \frac{x}{5} - \boxed{\frac{-1}{-1}} \cdot \frac{3x - 4}{-5}$$

$$= \frac{x}{5} - \frac{(-1)(3x - 4)}{(-1)(-5)} = \frac{x}{5} - \frac{4 - 3x}{5}$$

$$= \frac{x - (4 - 3x)}{5}$$

> Remember the parentheses!

$$= \frac{x - 4 + 3x}{5} = \frac{4x - 4}{5}$$

Example 7   Subtract: $\dfrac{5y}{y - 5} - \dfrac{2y - 3}{5 - y}$.

$$\frac{5y}{y - 5} - \frac{2y - 3}{5 - y} = \frac{5y}{y - 5} - \boxed{\frac{-1}{-1}} \cdot \frac{2y - 3}{5 - y}$$

$$= \frac{5y}{y - 5} - \frac{(-1)(2y - 3)}{(-1)(5 - y)} = \frac{5y}{y - 5} - \frac{3 - 2y}{y - 5}$$

$$= \frac{5y - (3 - 2y)}{y - 5}$$

> Remember the parentheses!

$$= \frac{5y - 3 + 2y}{y - 5} = \frac{7y - 3}{y - 5}$$

DO EXERCISES 8 AND 9.

EXERCISE SET 8.3

⬛⬤ Add. Simplify, if possible.

1. $\dfrac{5}{12} + \dfrac{7}{12}$

2. $\dfrac{3}{14} + \dfrac{5}{14}$

3. $\dfrac{1}{3 + x} + \dfrac{5}{3 + x}$

4. $\dfrac{4x + 1}{6x + 5} + \dfrac{3x - 7}{5 + 6x}$

5. $\dfrac{x^2 + 7x}{x^2 - 5x} + \dfrac{x^2 - 4x}{x^2 - 5x}$

*6. $\dfrac{a}{x + y} + \dfrac{b}{y + x}$

⬛⬛ Add. Simplify, if possible.

7. $\dfrac{7}{8} + \dfrac{5}{-8}$

8. $\dfrac{11}{6} + \dfrac{5}{-6}$

9. $\dfrac{3}{t} + \dfrac{4}{-t}$

10. $\dfrac{5}{-a} + \dfrac{8}{a}$

11. $\dfrac{2x + 7}{x - 6} + \dfrac{3x}{6 - x}$

12. $\dfrac{3x - 2}{4x - 3} + \dfrac{2x - 5}{3 - 4x}$

13. $\dfrac{y^2}{y - 3} + \dfrac{9}{3 - y}$

14. $\dfrac{t^2}{t - 2} + \dfrac{4}{2 - t}$

15. $\dfrac{b - 7}{b^2 - 16} + \dfrac{7 - b}{16 - b^2}$

16. $\dfrac{a - 3}{a^2 - 25} + \dfrac{a - 3}{25 - a^2}$

*17. $\dfrac{z}{(y + z)(y - z)} + \dfrac{y}{(z + y)(z - y)}$

[*Hint:* Multiply by $-1/-1$. Note that $(z + y)(z - y)(-1) = (z + y)(y - z)$.]

ANSWERS

1. _____
2. _____
3. _____
4. _____
5. _____
6. _____
7. _____
8. _____
9. _____
10. _____
11. _____
12. _____
13. _____
14. _____
15. _____
16. _____
17. _____

_____
* Indicates exercises that can be omitted if Chapter 7 was not studied.

*18. $\dfrac{a^2}{a - b} + \dfrac{b^2}{b - a}$

19. $\dfrac{x + 3}{x - 5} + \dfrac{2x - 1}{5 - x} + \dfrac{2(3x - 1)}{x - 5}$

18. _____

20. $\dfrac{3(x - 2)}{2x - 3} + \dfrac{5(2x + 1)}{2x - 3} + \dfrac{3(x - 1)}{3 - 2x}$

21. $\dfrac{2(4x + 1)}{5x - 7} + \dfrac{3(x - 2)}{7 - 5x} + \dfrac{-10x - 1}{5x - 7}$

19. _____

22. $\dfrac{5(x - 2)}{3x - 4} + \dfrac{2(x - 3)}{4 - 3x} + \dfrac{3(5x + 1)}{4 - 3x}$

20. _____

21. _____

23. $\dfrac{x + 1}{(x + 3)(x - 3)} + \dfrac{4(x - 3)}{(x - 3)(x + 3)} + \dfrac{(x - 1)(x - 3)}{(3 - x)(x + 3)}$

22. _____

24. $\dfrac{2(x + 5)}{(2x - 3)(x - 1)} + \dfrac{3x + 4}{(2x - 3)(1 - x)} + \dfrac{x - 5}{(3 - 2x)(x - 1)}$

23. _____

24. _____

●●● Subtract. Simplify, if possible.

25. $\dfrac{7}{8} - \dfrac{3}{8}$

26. $\dfrac{5}{y} - \dfrac{7}{y}$

27. $\dfrac{x}{x-1} - \dfrac{1}{x-1}$

28. $\dfrac{x^2}{x+4} - \dfrac{16}{x+4}$

29. $\dfrac{x+1}{x^2-2x+1} - \dfrac{5-3x}{x^2-2x+1}$

30. $\dfrac{2x-3}{x^2+3x-4} - \dfrac{x-7}{x^2+3x-4}$

●● Subtract. Simplify, if possible.

31. $\dfrac{11}{6} - \dfrac{5}{-6}$

32. $\dfrac{7}{8} - \dfrac{5}{-8}$

33. $\dfrac{5}{a} - \dfrac{8}{-a}$

34. $\dfrac{3}{t} - \dfrac{4}{-t}$

35. $\dfrac{x}{4} - \dfrac{3x-5}{-4}$

36. $\dfrac{2}{x-1} - \dfrac{2}{1-x}$

37. $\dfrac{3-x}{x-7} - \dfrac{2x-5}{7-x}$

38. $\dfrac{t^2}{t-2} - \dfrac{4}{2-t}$

39. $\dfrac{x-8}{x^2-16} - \dfrac{x-8}{16-x^2}$

ANSWERS

40. _____

41. _____

42. _____

43. _____

44. _____

45. _____

46. _____

47. _____

48. _____

49. _____

50. _____

40. $\dfrac{x-2}{x^2-25} - \dfrac{6-x}{25-x^2}$

41. $\dfrac{4-x}{x-9} - \dfrac{3x-8}{9-x}$

42. $\dfrac{3-x}{x-7} - \dfrac{2x-5}{7-x}$

43. $\dfrac{2(x-1)}{2x-3} - \dfrac{3(x+2)}{2x-3} - \dfrac{x-1}{3-2x}$

44. $\dfrac{3(x-2)}{2x-3} - \dfrac{5(2x+1)}{2x-3} - \dfrac{3(x-1)}{3-2x}$

Perform the indicated operations and simplify.

45. $\dfrac{3(2x+5)}{x-1} - \dfrac{3(2x-3)}{1-x} + \dfrac{6x-1}{x-1}$

46. $\dfrac{2x-y}{x-y} + \dfrac{x-2y}{y-x} - \dfrac{3x-3y}{x-y}$

47. $\dfrac{x-y}{x^2-y^2} + \dfrac{x+y}{x^2-y^2} - \dfrac{2x}{x^2-y^2}$

48. $\dfrac{x+y}{2(x-y)} - \dfrac{2x-2y}{2(x-y)} + \dfrac{x-3y}{2(y-x)}$

49. $\dfrac{10}{2y-1} - \dfrac{6}{1-2y} + \dfrac{y}{2y-1} + \dfrac{y-4}{1-2y}$

50. $\dfrac{(x+3)(2x-1)}{(2x-3)(x-3)} - \dfrac{(x-3)(x+1)}{(3-x)(3-2x)} + \dfrac{(2x+1)(x+3)}{(3-2x)(x-3)}$

8.4 LEAST COMMON MULTIPLES

▪ LEAST COMMON MULTIPLES

To add when denominators are different we first find a common denominator. For example, to add $\frac{5}{12}$ and $\frac{7}{30}$ we first look for a common multiple of both 12 and 30. Any multiple will do, but we prefer the smallest such number, the *Least Common Multiple* (LCM). To find the LCM, we factor.

$$12 = 2 \cdot 2 \cdot 3$$
$$30 = 2 \cdot 3 \cdot 5$$

The LCM is the number that has 2 as a factor twice, 3 as a factor once, and 5 as a factor once:

$$\text{LCM} = 2 \cdot 2 \cdot 3 \cdot 5, \quad \text{or} \quad 60.$$

> **To find the LCM, we use each factor the greatest number of times that it appears in any one factorization.**

Example 1   Find the LCM of 24 and 36.

$$\left. \begin{array}{l} 24 = \boxed{2 \cdot 2 \cdot 2} \cdot 3 \\ 36 = 2 \cdot 2 \cdot \boxed{3 \cdot 3} \end{array} \right\} \quad \text{LCM} = \boxed{2 \cdot 2 \cdot 2} \cdot \boxed{3 \cdot 3}, \quad \text{or} \quad 72.$$

DO EXERCISES 1–4.

▪▪ ADDING USING THE LCM

Let us finish adding $\frac{5}{12}$ and $\frac{7}{30}$.

$$\frac{5}{12} + \frac{7}{30} = \frac{5}{2 \cdot 2 \cdot 3} + \frac{7}{2 \cdot 3 \cdot 5}$$

The LCM is $2 \cdot 2 \cdot 3 \cdot 5$. To get the LCM in the first denominator we need a 5. To get the LCM in the second denominator we need another 2. We get these numbers by multiplying by 1.

$$\frac{5}{12} + \frac{7}{30} = \frac{5}{2 \cdot 2 \cdot 3} \cdot \boxed{\frac{5}{5}} + \frac{7}{2 \cdot 3 \cdot 5} \cdot \boxed{\frac{2}{2}} \qquad \text{Multiplying by 1}$$

$$= \frac{25}{2 \cdot 2 \cdot 3 \cdot 5} + \frac{14}{2 \cdot 3 \cdot 5 \cdot 2} \qquad \begin{array}{l}\text{Denominators are now}\\\text{the LCM}\end{array}$$

$$= \frac{39}{2 \cdot 2 \cdot 3 \cdot 5} \qquad \begin{array}{l}\text{Adding the numerators and keeping}\\\text{the LCM}\end{array}$$

$$= \frac{13}{20} \qquad \text{Simplifying}$$

Example 2   Add: $\dfrac{5}{12} + \dfrac{11}{18}$.

$$\left. \begin{array}{l} 12 = \boxed{2 \cdot 2} \cdot 3 \\ 18 = 2 \cdot \boxed{3 \cdot 3} \end{array} \right\} \quad \text{LCM} = \boxed{2 \cdot 2} \cdot \boxed{3 \cdot 3}, \quad \text{or} \quad 36.$$

OBJECTIVES

After finishing Section 8.4, you should be able to:

▪ Find the LCM of several numbers by factoring.

▪▪ Add fractions, first finding the LCM of the denominators.

▪▪▪ Find the LCM of algebraic expressions by factoring.

Find the LCM by factoring.

1. 16,   18

2. 6,   12

3. 2,   5

4. 24,   30,   20

Add, first finding the LCM of the denominators. Simplify, if possible.

5. $\dfrac{3}{16} + \dfrac{1}{18}$

6. $\dfrac{1}{6} + \dfrac{1}{12}$

7. $\dfrac{1}{2} + \dfrac{3}{5}$

8. $\dfrac{1}{24} + \dfrac{1}{30} + \dfrac{3}{20}$

Find the LCM.

9. $12xy^2, \quad 15x^3y$

10. $y^2 + 5y + 4, \quad y^2 + 2y + 1$

11. $t^2 + 16, \quad t - 2, \quad 7$

12. $x^2 + 2x + 1, \quad 3x - 3x^2, \quad x^2 - 1$

$$\frac{5}{12} + \frac{11}{18} = \frac{5}{2 \cdot 2 \cdot 3} \cdot \boxed{\frac{3}{3}} + \frac{11}{2 \cdot 3 \cdot 3} \cdot \boxed{\frac{2}{2}} = \frac{37}{2 \cdot 2 \cdot 3 \cdot 3} = \frac{37}{36}$$

DO EXERCISES 5–8.

●●● LCM'S OF ALGEBRAIC EXPRESSIONS

To find the LCM of two or more algebraic expressions, we factor them. Then we use each factor the greatest number of times it occurs in any one expression.

Example 3   Find the LCM of $12x$, $16y$, and $8xyz$.

$$\left. \begin{array}{l} 12x = 2 \cdot 2 \cdot \boxed{3 \cdot x} \\[4pt] 16y = \boxed{2 \cdot 2 \cdot 2 \cdot 2} \cdot \boxed{y} \\[4pt] 8xyz = 2 \cdot 2 \cdot 2 \cdot x \cdot y \cdot \boxed{z} \end{array} \right\} \quad \begin{array}{l} \text{LCM} = \boxed{2 \cdot 2 \cdot 2 \cdot 2} \cdot \boxed{3 \cdot x} \cdot \boxed{y} \cdot \boxed{z} \\[4pt] = 48xyz \end{array}$$

Example 4   Find the LCM of $x^2 + 5x - 6$ and $x^2 - 1$.

$$\left. \begin{array}{l} x^2 + 5x - 6 = \boxed{(x + 6)(x - 1)} \\[4pt] x^2 - 1 = \boxed{(x + 1)} \, (x - 1) \end{array} \right\} \quad \text{LCM} = \boxed{(x + 6)(x - 1)} \; \boxed{(x + 1)}$$

Example 5   Find the LCM of $x^2 + 4$, $x + 1$, and 5.

These expressions are not factorable, so the LCM is their product: $5(x^2 + 4)(x + 1)$.

The additive inverse of an LCM is also an LCM. For example, if $(x + 2)(x - 5)$ is an LCM, then $-(x + 2)(x - 5)$ is an LCM. We can name the latter $(x + 2)(-1)(x - 5)$, or $(x + 2)(5 - x)$. When finding LCMs, if factors that are additive inverses occur, we do not use them both. For example, if $(x - 5)$ occurs in one factorization and $(5 - x)$ occurs in another, since these are additive inverses, we do not use them both.

Example 6   Find the LCM of $x^2 - 25$ and $10 - 2x$.

$$\left. \begin{array}{l} x^2 - 25 = \boxed{(x + 5)(x - 5)} \\[4pt] 10 - 2x = \boxed{2} \, (5 - x) \end{array} \right\} \quad \begin{array}{l} \text{LCM} = \boxed{2} \; \boxed{(x + 5)} \, \boxed{(x - 5)} \leftarrow \\[4pt] \text{or} \quad 2(x + 5)(5 - x) \leftarrow \end{array}$$

$$\boxed{\text{We can use } x - 5 \text{ or } 5 - x \text{ but not both.}}$$

Example 7   Find the LCM of $x^2 - 4y^2$, $x^2 - 4xy + 4y^2$, and $x - 2y$.

$$\left. \begin{array}{l} x^2 - 4y^2 = \boxed{(x - 2y)} \, \boxed{(x + 2y)} \\[4pt] x^2 - 4xy + 4y^2 = \boxed{(x - 2y)(x - 2y)} \\[4pt] x - 2y = x - 2y \end{array} \right\} \quad \begin{array}{l} \text{LCM} = \\[4pt] \boxed{(x + 2y)} \; \boxed{(x - 2y)(x - 2y)} \\[4pt] = (x + 2y)(x - 2y)^2 \end{array}$$

DO EXERCISES 9–12.

NAME   CLASS

EXERCISE SET 8.4

◼● Find the LCM.

1._____

1. 12, 27   **2.** 10, 15   **3.** 8, 9

2._____

3._____

4. 12, 15   **5.** 6, 9, 21   **6.** 8, 36, 40

4._____

5._____

6._____

7. 24, 36, 40   **8.** 3, 4, 5   **9.** 28, 42, 60

7._____

8._____

◼◼ Add, first finding the LCM of the denominators. Simplify, if possible.

9._____

10. $\dfrac{7}{24} + \dfrac{11}{18}$ **11.** $\dfrac{7}{60} + \dfrac{6}{75}$ **12.** $\dfrac{1}{6} + \dfrac{3}{40} + \dfrac{2}{75}$

10._____

11._____

12._____

13. $\dfrac{5}{24} + \dfrac{3}{20} + \dfrac{7}{30}$ **14.** $\dfrac{2}{15} + \dfrac{5}{9} + \dfrac{3}{20}$ **15.** $\dfrac{1}{20} + \dfrac{1}{30} + \dfrac{2}{45}$

13._____

14._____

15._____

ANSWERS

●●● Find the LCM.

16. _____

16. $6x^2, 12x^3$ ***17.** $2a^2b, 8ab^2$ ***18.** $2x^2, 6xy, 18y^2$

17. _____

18. _____

*19. c^2d, cd^2, c^3d **20.** $2(y-3), 6(3-y)$ **21.** $4(x-1), 8(1-x)$

19. _____

20. _____

*22. $x^2 - y^2, 2x + 2y, x^2 + 2xy + y^2$ **23.** $a + 1; (a-1)^2; a^2 - 1$

21. _____

22. _____

24. $m^2 - 5m + 6; m^2 - 4m + 4$ **25.** $2 + 3k; 9k^2 - 4; 2 - 3k$

23. _____

24. _____

26. $10v^2 + 30v; -5v^2 - 35v - 60$ **27.** $9x^3 - 9x^2 - 18x; 6x^5 - 24x^4 + 24x^3$

25. _____

26. _____

27. _____

28. When is the LCM of two expressions the same as their product?

The planets Earth, Jupiter, Saturn, and Uranus revolve about the sun about once each 1, 12, 30, and 84 years, respectively.

28. _____

29. How often will Jupiter and Saturn appear in the same direction in the night sky as seen from Earth?

29. _____

30. How often will Jupiter, Saturn, and Uranus all appear in the same direction in the night sky as seen from Earth?

30. _____

* Indicates exercises that can be omitted if Chapter 7 was not studied.

8.5   ADDITION WITH DIFFERENT DENOMINATORS

● Now that we know how to find LCM's, we can add fractional expressions with different denominators. We first find the LCM of the denominators (the least common denominator) and then add.

Example 1   Add $\dfrac{3}{x+1} + \dfrac{5}{x-1}$.

The denominators do not factor, so the LCM is their product. We multiply by 1 to get the LCM in each expression.

$$\frac{3}{x+1} \cdot \frac{x-1}{x-1} + \frac{5}{x-1} \cdot \frac{x+1}{x+1} = \frac{3(x-1) + 5(x+1)}{(x-1)(x+1)}$$

$$= \frac{3x - 3 + 5x + 5}{(x-1)(x+1)}$$

$$= \frac{8x + 2}{(x-1)(x+1)}$$

The numerator and denominator have no common factor (other than 1), so we cannot simplify.

DO EXERCISE 1.

Example 2   Add $\dfrac{5}{x^2+x} + \dfrac{4}{2x+2}$.

First find the LCM of the denominators.

$$\left.\begin{array}{l} x^2 + x - x(x+1) \\[4pt] 2x + 2 = 2\,(x+1) \end{array}\right\} \qquad \text{LCM} = 2\ \ x(x+1)$$

Multiply by 1 to get the LCM in each expression. Then add and simplify.

$$\frac{5}{x(x+1)} \cdot \frac{2}{2} + \frac{4}{2(x+1)} \cdot \frac{x}{x} = \frac{10}{2x(x+1)} + \frac{4x}{2x(x+1)} \qquad \text{Multiplying by 1}$$

$$= \frac{10 + 4x}{2x(x+1)} \qquad \text{Adding}$$

$$= \frac{2(5 + 2x)}{2x(x+1)} \qquad \text{Factoring numerator}$$

$$= \frac{2}{2} \cdot \frac{5 + 2x}{x(x+1)} \qquad \text{Factoring the fractional expression}$$

$$= \frac{5 + 2x}{x(x+1)}$$

DO EXERCISES 2 AND 3.

After finishing Section 8.5, you should be able to:

● Add fractional expressions with different denominators and simplify the result.

Add.

1. $\dfrac{x}{x-2} + \dfrac{4}{x+2}$

Add.

2. $\dfrac{3}{x^3 - x} + \dfrac{4}{x^2 + 2x + 1}$

3. $\dfrac{5}{x^2 + 17x + 16} + \dfrac{3}{x^2 + 9x + 8}$

Add.

4. $\dfrac{a-1}{a+2} + \dfrac{a+3}{a-9}$

Add.

5. $\dfrac{x}{x^2+5x+6} + \dfrac{-2}{x^2+3x+2}$

Example 3 Add $\dfrac{x+4}{x-2} + \dfrac{x-7}{x+5}$.

First, find the LCM of the denominators. It is just the product.

$$\text{LCM} = (x-2)(x+5)$$

Multiply by 1 to get the LCM in each expression. Then add and simplify.

$$\frac{x+4}{x-2} \cdot \boxed{\frac{x+5}{x+5}} + \frac{x-7}{x+5} \cdot \boxed{\frac{x-2}{x-2}} = \frac{(x+4)(x+5)}{(x-2)(x+5)} + \frac{(x-7)(x-2)}{(x-2)(x+5)}$$

$$= \frac{x^2+9x+20}{(x-2)(x+5)} + \frac{x^2-9x+14}{(x-2)(x+5)}$$

$$= \frac{x^2+9x+20+x^2-9x+14}{(x-2)(x+5)}$$

$$= \frac{2x^2+34}{(x-2)(x+5)}$$

DO EXERCISE 4.

Example 4 Add $\dfrac{x}{x^2+11x+30} + \dfrac{-5}{x^2+9x+20}$.

$$\frac{x}{x^2+11x+30} + \frac{-5}{x^2+9x+20}$$

$$= \frac{x}{(x+5)(x+6)} + \frac{-5}{(x+5)(x+4)} \qquad \begin{array}{l}\text{Factoring denominators in}\\ \text{order to find the LCM. The}\\ \text{LCM is } (x+4)(x+5)(x+6).\end{array}$$

$$= \frac{x}{(x+5)(x+6)} \cdot \boxed{\frac{x+4}{x+4}} + \frac{-5}{(x+5)(x+4)} \cdot \boxed{\frac{x+6}{x+6}} \qquad \begin{array}{l}\text{Multiplying}\\ \text{by 1}\end{array}$$

$$= \frac{x(x+4)}{(x+5)(x+6)(x+4)} + \frac{-5(x+6)}{(x+5)(x+4)(x+6)}$$

$$= \frac{x(x+4)+(-5)(x+6)}{(x+4)(x+5)(x+6)} \qquad \text{Adding}$$

$$= \frac{x^2+4x-5x-30}{(x+4)(x+5)(x+6)}$$

$$= \frac{x^2-x-30}{(x+4)(x+5)(x+6)}$$

$$= \frac{(x-6)(x+5)}{(x+4)(x+5)(x+6)}$$

$$= \boxed{\frac{x+5}{x+5}} \cdot \frac{x-6}{(x+4)(x+6)}$$

$$= \frac{x-6}{(x+4)(x+6)}$$

Always simplify at the end if possible.

DO EXERCISE 5.

NAME   CLASS   ANSWERS

EXERCISE SET 8.5

▪   Add, and simplify if possible.

1. $\dfrac{2}{x} + \dfrac{5}{x^2}$ 2. $\dfrac{5}{6r} + \dfrac{7}{8r}$

1. _____

2. _____

*3. $\dfrac{x+y}{xy^2} + \dfrac{3x+y}{x^2y}$ *4. $\dfrac{2c-d}{c^2d} + \dfrac{c+d}{cd^2}$

3. _____

4. _____

5. $\dfrac{2}{x-1} + \dfrac{2}{x+1}$ 6. $\dfrac{3}{x+1} + \dfrac{2}{3x}$

5. _____

6. _____

7. $\dfrac{2x}{x^2-16} + \dfrac{x}{x-4}$ 8. $\dfrac{5}{z+4} + \dfrac{3}{3z+12}$

7. _____

8. _____

9. $\dfrac{3}{x-1} + \dfrac{2}{(x-1)^2}$ 10. $\dfrac{4a}{5a-10} + \dfrac{3a}{10a-20}$

9. _____

* Indicates exercises that can be omitted if Chapter 7 was not studied.

10. _____

ANSWERS

11. _____

12. _____

13. _____

14. _____

15. _____

16. _____

17. _____

18. _____

19. _____

20. _____

11. $\dfrac{x}{x^2 + 2x + 1} + \dfrac{1}{x^2 + 5x + 4}$

12. $\dfrac{7}{a^2 + a - 2} + \dfrac{5}{a^2 - 4a + 3}$

13. $\dfrac{x + 3}{x - 5} + \dfrac{x - 5}{x + 3}$

14. $\dfrac{3x}{2y - 3} + \dfrac{2x}{3y - 2}$

15. $\dfrac{a}{a^2 - 1} + \dfrac{2a}{a^2 - a}$

16. $\dfrac{3x + 2}{3x + 6} + \dfrac{x - 2}{x^2 - 4}$

*17. $\dfrac{6}{x - y} + \dfrac{4x}{y^2 - x^2}$

18. $\dfrac{a - 2}{3 - a} + \dfrac{4 - a^2}{a^2 - 9}$

19. $\dfrac{10}{x^2 + x - 6} + \dfrac{3x}{x^2 - 4x + 4}$

20. $\dfrac{2}{z^2 - z - 6} + \dfrac{3}{z^2 - 9}$

8.6 SUBTRACTION WITH DIFFERENT DENOMINATORS

● SUBTRACTION WITH DIFFERENT DENOMINATORS

Subtraction is like addition, except that we subtract numerators.

Example 1   Subtract: $\dfrac{x+2}{x-4} - \dfrac{x+1}{x+4}$.

LCM $= (x-4)(x+4)$

$$\frac{x+2}{x-4} \cdot \boxed{\frac{x+4}{x+4}} - \frac{x+1}{x+4} \cdot \boxed{\frac{x-4}{x-4}}$$

$$= \frac{(x+2)(x+4)}{(x-4)(x+4)} - \frac{(x+1)(x-4)}{(x-4)(x+4)}$$

$$= \frac{x^2+6x+8}{(x-4)(x+4)} - \frac{x^2-3x-4}{(x-4)(x+4)}$$

$$= \frac{x^2+6x+8 - (x^2-3x-4)}{(x-4)(x+4)} \qquad \text{Subtracting numerators}$$

$$= \frac{x^2+6x+8 - x^2+3x+4}{(x-4)(x+4)} \qquad \text{Don't forget parentheses.}$$

$$= \frac{9x+12}{(x-4)(x+4)}$$

$$= \frac{3(3x+4)}{(x-4)(x+4)}$$

DO EXERCISE 1.

●● SIMPLIFYING COMBINED ADDITIONS AND SUBTRACTIONS

Example 2   Simplify: $\dfrac{1}{x} - \dfrac{1}{x^2} + \dfrac{2}{x+1}$.

LCM $= x^2(x+1)$

$$\frac{1}{x} \cdot \boxed{\frac{x(x+1)}{x(x+1)}} - \frac{1}{x^2} \cdot \boxed{\frac{x+1}{x+1}} + \frac{2}{x+1} \cdot \boxed{\frac{x^2}{x^2}}$$

$$= \frac{x(x+1)}{x^2(x+1)} - \frac{x+1}{x^2(x+1)} + \frac{2x^2}{x^2(x+1)}$$

$$= \frac{x(x+1) - (x+1) + 2x^2}{x^2(x+1)}$$

$$= \frac{x^2+x-x-1+2x^2}{x^2(x+1)}$$

$$= \frac{3x^2-1}{x^2(x+1)}$$

DO EXERCISE 2.

OBJECTIVES

After finishing Section 8.6, you should be able to:

● Subtract fractional expressions with different denominators.

●● Simplify combined additions and subtractions of fractional expressions.

Subtract.

1. $\dfrac{x-2}{3x} - \dfrac{2x-1}{5x}$

Simplify.

2. $\dfrac{1}{x} - \dfrac{5}{3x} + \dfrac{2x}{x+1}$

SOMETHING EXTRA

AN APPLICATION: HANDLING DIMENSION SYMBOLS (PART II)

We can treat dimension symbols much like numerals and variables, since correct results can thus be obtained.

Example 1   Compare

$$4 \text{ m} \cdot 3 \text{ m} = 4 \cdot 3 \cdot \text{m} \cdot \text{m} = 12 \text{ m}^2 \text{ (square meters)}$$

with

$$4x \cdot 3x = 4 \cdot 3 \cdot x \cdot x = 12x^2.$$

Example 2   Compare

$$5 \text{ persons} \cdot 8 \text{ hr} = 5 \cdot 8 \text{ person-hr} = 40 \text{ person-hr}$$

with

$$5x \cdot 8y = 5 \cdot 8 \cdot x \cdot y = 40xy.$$

Example 3   Compare

$$480 \text{ cm} \cdot \frac{1 \text{ m}}{100 \text{ cm}} = \frac{480}{100} \text{ cm} \cdot \frac{\text{m}}{\text{cm}} = 4.8 \cdot \boxed{\frac{\text{cm}}{\text{cm}}} \cdot \text{m} = 4.8 \text{ m}$$

with

$$480x \cdot \frac{y}{100x} = \frac{480}{100}x \cdot \frac{y}{x} = 4.8 \cdot \boxed{\frac{x}{x}} \cdot y = 4.8y.$$

In each example, dimension symbols are treated as though they are variables or numerals, and as though a symbol like "3 ft" represents a product of "3" by "ft." A symbol like km/hr (standard notation is "km/h") is treated as if it represents a division of kilometers by hours.

Any two measures can be "multiplied" or "divided." For example,

$$6 \text{ ft} \cdot 4 \text{ lb} = 24 \text{ ft-lb}, \qquad 3 \text{ km} \cdot 4 \text{ sec} = 12 \text{ km-sec},$$

$$\frac{8 \text{ grams}}{4 \text{ min}} = 2\frac{\text{g}}{\text{min}}, \qquad \frac{3 \text{ in.} \cdot 8 \text{ days}}{6 \text{ lb}} = 4\frac{\text{in.-day}}{\text{lb}}.$$

EXERCISES

Perform these calculations and simplify if possible. *Do not* make unit changes.

1. $12 \text{ ft} \cdot \dfrac{1 \text{ yd}}{3 \text{ ft}}$

2. $6 \text{ lb} \cdot \dfrac{16 \text{ oz}}{1 \text{ lb}}$

3. $9\dfrac{\text{km}}{\text{hr}} \cdot 3 \text{ hr}$

4. $12\dfrac{\text{m}}{\text{sec}} \cdot 5 \text{ sec}$

5. $3 \text{ cm} \cdot \dfrac{2 \text{ g}}{2 \text{ cm}}$

6. $\dfrac{9 \text{ mi}}{3 \text{ days}} \cdot 6 \text{ days}$

7. $2347 \text{ m} \cdot \dfrac{1 \text{ km}}{1000 \text{ m}}$

8. $55 \text{ cm} \cdot \dfrac{10 \text{ mm}}{1 \text{ cm}}$

9. $\dfrac{3 \text{ kg}}{5 \text{ m}} \cdot \dfrac{7 \text{ kg}}{6 \text{ m}}$

10. $\dfrac{2000 \text{ lb} \cdot (6 \text{ mi/hr})^2}{100 \text{ ft}}$

11. $\dfrac{7 \text{ m} \cdot 8 \text{ kg/sec}}{4 \text{ sec}}$

NAME

CLASS

ANSWERS

EXERCISE SET 8.6

▪ Subtract, and simplify if possible.

1. $\dfrac{x-2}{6} - \dfrac{x+1}{3}$

2. $\dfrac{a+2}{2} - \dfrac{a-4}{4}$

3. $\dfrac{4z-9}{3z} - \dfrac{3z-8}{4z}$

4. $\dfrac{x-1}{4x} - \dfrac{2x+3}{x}$

*5. $\dfrac{4x+2t}{3xt^2} - \dfrac{5x-3t}{x^2t}$

*6. $\dfrac{5x+3y}{2x^2y} - \dfrac{3x+4y}{xy^2}$

7. $\dfrac{5}{x+5} - \dfrac{3}{x-5}$

8. $\dfrac{2z}{z-1} - \dfrac{3z}{z+1}$

9. $\dfrac{3}{2t^2-2t} - \dfrac{5}{2t-2}$

10. $\dfrac{8}{x^2-4} - \dfrac{3}{x+2}$

1. _____

2. _____

3. _____

4. _____

5. _____

6. _____

7. _____

8. _____

9. _____

10. _____

_____

* Indicates exercises that can be omitted if Chapter 7 was not studied.

ANSWERS

11. _____

12. _____

13. _____

14. _____

15. _____

16. _____

17. _____

18. _____

19. _____

20. _____

*11. $\dfrac{2s}{t^2 - s^2} - \dfrac{s}{t - s}$

12. $\dfrac{3}{12 + x - x^2} - \dfrac{2}{x^2 - 9}$

●● Simplify.

13. $\dfrac{4y}{y^2 - 1} - \dfrac{2}{y} - \dfrac{2}{y + 1}$

14. $\dfrac{x + 6}{4 - x^2} - \dfrac{x + 3}{x + 2} + \dfrac{x - 3}{2 - x}$

15. $\dfrac{2z}{1 - 2z} + \dfrac{3z}{2z + 1} - \dfrac{3}{4z^2 - 1}$

*16. $\dfrac{1}{x + y} + \dfrac{1}{x - y} - \dfrac{2x}{x^2 - y^2}$

17. $\dfrac{5}{3 - 2x} + \dfrac{3}{2x - 3} - \dfrac{x - 3}{2x^2 - x - 3}$

*18. $\dfrac{2r}{r^2 - s^2} + \dfrac{1}{r + s} - \dfrac{1}{r - s}$

19. $\dfrac{3}{2c - 1} - \dfrac{1}{c + 2} - \dfrac{5}{2c^2 + 3c - 2}$

20. $\dfrac{3y - 1}{2y^2 + y - 3} - \dfrac{2 - y}{y - 1}$

8.7   COMPLEX FRACTIONAL EXPRESSIONS

A *complex fractional expression* is one that has a fractional expression in its numerator or its denominator, or both.

Here are some examples of complex expressions.

$$\frac{1 + \dfrac{2}{x}}{3}, \quad \frac{\dfrac{x + y}{2}}{\dfrac{2x}{x + 1}}, \quad \frac{\dfrac{1}{3} + \dfrac{1}{5}}{\dfrac{2}{x} - \dfrac{x}{y}}.$$

To simplify a complex fractional expression we first add or subtract, if necessary, to get a single fractional expression in both numerator and denominator. Then we divide, by multiplying by the reciprocal of the denominator.

Example 1   Simplify.

$$\frac{\dfrac{1}{5} + \dfrac{2}{5}}{\dfrac{7}{3}} = \frac{\dfrac{3}{5}}{\dfrac{7}{3}} \qquad \text{Adding in the numerator}$$

$$= \frac{3}{5} \cdot \frac{3}{7} \qquad \text{Multiplying by the reciprocal of the denominator}$$

$$= \frac{9}{35}$$

Example 2   Simplify.

$$\frac{1 + \dfrac{2}{x}}{\dfrac{3}{4}} = \frac{1 \cdot \dfrac{x}{x} + \dfrac{2}{x}}{\dfrac{3}{4}} \qquad \text{Multiplying by } \dfrac{x}{x} \text{ to get a common denominator}$$

$$= \frac{\dfrac{x + 2}{x}}{\dfrac{3}{4}} \qquad \text{Adding in the numerator}$$

$$= \frac{x + 2}{x} \cdot \frac{4}{3} \qquad \text{Multiplying by the reciprocal of the denominator}$$

$$= \frac{4(x + 2)}{3x}$$

DO EXERCISES 1 AND 2.

Simplify.

1. $\dfrac{\dfrac{2}{7} + \dfrac{3}{7}}{\dfrac{3}{4}}$

2. $\dfrac{3 + \dfrac{x}{2}}{\dfrac{5}{4}}$

Simplify.

3. $\dfrac{\dfrac{x}{2} + \dfrac{2x}{3}}{\dfrac{1}{x} - \dfrac{x}{2}}$

Simplify.

4. $\dfrac{1 + \dfrac{1}{x}}{1 - \dfrac{1}{x^2}}$

Example 3   Simplify.

$$\frac{\dfrac{3}{x} + \dfrac{1}{2x}}{\dfrac{1}{3x} - \dfrac{3}{4x}} = \frac{\dfrac{3}{x} \cdot \dfrac{2}{2} + \dfrac{1}{2x}}{\dfrac{1}{3x} \cdot \dfrac{4}{4} - \dfrac{3}{4x} \cdot \dfrac{3}{3}} \left. \begin{array}{l} \\ \\ \end{array} \right\} \leftarrow \text{Finding the LCM, } 2x, \text{ and}$$

Finding the LCM, $2x$, and multiplying by 1 in the numerator

Finding the LCM, $12x$, and multiplying by 1 in the denominator

$$= \frac{\dfrac{6}{2x} + \dfrac{1}{2x}}{\dfrac{4}{12x} - \dfrac{9}{12x}}$$

$$= \frac{\dfrac{7}{2x}}{\dfrac{-5}{12x}} \quad \begin{array}{l} \leftarrow \text{Adding in the numerator,} \\ \swarrow \text{subtracting in the denominator} \end{array}$$

$$= \frac{7}{2x} \cdot \frac{12x}{-5} \quad \begin{array}{l} \text{Multiplying by the reciprocal of} \\ \text{the denominator} \end{array}$$

$$= \frac{2x}{2x} \cdot \frac{7 \cdot 6}{-5} \quad \text{Factoring}$$

$$= \frac{42}{-5} = -\frac{42}{5} \quad \text{Simplifying}$$

DO EXERCISE 3.

Example 4   Simplify.

$$\frac{1 - \dfrac{1}{x}}{1 - \dfrac{1}{x^2}} = \frac{\dfrac{x}{x} - \dfrac{1}{x}}{\dfrac{x^2}{x^2} - \dfrac{1}{x^2}} \left. \begin{array}{l} \\ \\ \end{array} \right\}$$

Finding the LCM, x, and multiplying by 1 in the numerator

Finding the LCM, x^2, and multiplying by 1 in the denominator

$$= \frac{\dfrac{x - 1}{x}}{\dfrac{x^2 - 1}{x^2}} \quad \begin{array}{l} \leftarrow \text{Subtracting in the numerator, and} \\ \swarrow \text{subtracting in the denominator} \end{array}$$

$$= \frac{x - 1}{x} \cdot \frac{x^2}{x^2 - 1} \quad \begin{array}{l} \text{Multiplying by the reciprocal of the} \\ \text{divisor} \end{array}$$

$$= \frac{(x - 1)x^2}{x(x^2 - 1)}$$

$$= \frac{x(x - 1)}{x(x - 1)} \cdot \frac{x}{x + 1} \quad \text{Factoring}$$

$$= \frac{x}{x + 1}.$$

DO EXERCISE 4.

EXERCISE SET 8.7

⚫ Simplify.

1. $\dfrac{1 + \dfrac{9}{16}}{1 - \dfrac{3}{4}}$

2. $\dfrac{9 - \dfrac{1}{4}}{3 + \dfrac{1}{2}}$

3. $\dfrac{1 - \dfrac{3}{5}}{1 + \dfrac{1}{5}}$

4. $\dfrac{\dfrac{5}{27} - 5}{\dfrac{1}{3} + 1}$

5. $\dfrac{\dfrac{1}{x} + 3}{\dfrac{1}{x} - 5}$

6. $\dfrac{\dfrac{3}{s} + s}{\dfrac{s}{3} + s}$

7. $\dfrac{\dfrac{1}{2} + \dfrac{3}{4}}{\dfrac{5}{8} - \dfrac{5}{6}}$

8. $\dfrac{\dfrac{2}{3} - \dfrac{5}{6}}{\dfrac{3}{4} + \dfrac{7}{8}}$

9. $\dfrac{\dfrac{2}{y} + \dfrac{1}{2y}}{y + \dfrac{y}{2}}$

1. _____

2. _____

3. _____

4. _____

5. _____

6. _____

7. _____

8. _____

9. _____

---

* Indicates exercises that can be omitted if Chapter 7 was not studied.

ANSWERS

10. $\dfrac{4 - \dfrac{1}{x^2}}{2 - \dfrac{1}{x}}$

11. $\dfrac{8 + \dfrac{8}{d}}{1 + \dfrac{1}{d}}$

12. $\dfrac{2 - \dfrac{3}{b}}{2 - \dfrac{b}{3}}$

10. _____

11. _____

12. _____

13. $\dfrac{\dfrac{1}{5} - \dfrac{1}{a}}{\dfrac{5 - a}{5}}$

14. $\dfrac{2 - \dfrac{1}{x}}{\dfrac{2}{x}}$

*15. $\dfrac{\dfrac{x}{x - y}}{\dfrac{x^2}{x^2 - y^2}}$

13. _____

14. _____

15. _____

*16. $\dfrac{\dfrac{x}{y} - \dfrac{y}{x}}{\dfrac{1}{y} + \dfrac{1}{x}}$

17. $\dfrac{x - 3 + \dfrac{2}{x}}{x - 4 + \dfrac{3}{x}}$

*18. $\dfrac{1 + \dfrac{a}{b - a}}{\dfrac{a}{a + b} - 1}$

16. _____

17. _____

18. _____

8.8 DIVISION OF POLYNOMIALS

◼● DIVISOR A MONOMIAL

Division can be shown by a fractional expression.

Example 1   Divide $x^3 + 10x^2 + 8x$ by $2x$.

We write a fractional expression to show division:

$$\frac{x^3 + 10x^2 + 8x}{2x}.$$

This is equivalent to

$\dfrac{x^3}{2x} + \dfrac{10x^2}{2x} + \dfrac{8x}{2x}.$ To see this, add and get the original expression.

Next, we do the separate divisions:

$$\frac{x^3}{2x} + \frac{10x^2}{2x} + \frac{8x}{2x} = \frac{1}{2}x^2 + 5x + 4.$$

DO EXERCISES 1 AND 2.

Example 2   Divide and check: $(5y^2 - 2y + 3) \div 2$.

$$\frac{5y^2 - 2y + 3}{2} = \frac{5y^2}{2} - \frac{2y}{2} + \frac{3}{2} = \frac{5}{2}y^2 - y + \frac{3}{2}$$

Check: $\dfrac{5}{2}y^2 - y + \dfrac{3}{2}$

$\dfrac{2}{5y^2 - 2y + 3}$ We multiply.
                    The answer checks.

Try to write only the answer.

> **To divide by a monomial, we can divide each term by the monomial.**

DO EXERCISES 3 AND 4.

◼●● DIVISOR NOT A MONOMIAL

When the divisor is not a monomial, we use long division very much as we do in arithmetic.

Example 3   Divide $x^2 + 5x + 6$ by $x + 2$.

$$
\begin{array}{r}
x \\
x + 2 \overline{\smash{)}x^2 + 5x + 6} \\
\underline{x^2 + 2x} \\
3x
\end{array}
$$

Divide first term by first term, to get x. Ignore the term 2.
Multiply x by divisor.
Subtract.

We now "bring down" the next term of the dividend, 6.

OBJECTIVES

After finishing Section 8.8, you should be able to:

◼● Divide a polynomial by a monomial and check the result.

◼●● Divide a polynomial by a divisor that is not a monomial and, if there is a remainder, express the result in two ways.

Divide.

1. $\dfrac{2x^3 + 6x^2 + 4x}{2x}$

2. $(6x^2 + 3x - 2) \div 3$

Divide and check.

3. $(8x^2 - 3x + 1) \div 2$

4. $\dfrac{2x^4 - 3x^3 + 5x^2}{x^2}$

Divide and check.

5. $(x^2 + x - 6) \div (x + 3)$

$$
\begin{array}{r}
x + 3 \leftarrow \\
x + 2 \overline{)x^2 + 5x + 6} \quad \text{Divide first term by first term to get 3.} \\
\underline{x^2 + 2x} \\
3x + 6 \\
\underline{3x + 6} \leftarrow \text{Multiply 3 by divisor.} \\
0 \quad \text{Subtract.}
\end{array}
$$

The quotient is $x + 3$.

To check, multiply the quotient by the divisor and add the remainder, if any, to see if you get the dividend:

$$(x + 2)(x + 3) = x^2 + 5x + 6. \quad \text{The division checks.}$$

Example 4 Divide and check: $(x^2 + 2x - 12) \div (x - 3)$.

$$
\begin{array}{r}
x + 5 \leftarrow \\
x - 3 \overline{)x^2 + 2x - 12} \\
\underline{x^2 - 3x} \\
5x - 12 \quad \text{Quotient} \\
\underline{5x - 15} \\
3 \leftarrow \text{Remainder}
\end{array}
$$

Check:
$$(x - 3)(x + 5) + 3 = x^2 + 2x - 15 + 3$$
$$= x^2 + 2x - 12$$

6. $x - 2 \overline{)x^2 + 2x - 8}$

The quotient is $x + 5$ and the remainder is 3. We can write this $x + 5$, R 3.

The answer can also be given in this way: $x + 5 + \dfrac{3}{x - 3}$.

DO EXERCISES 5 AND 6.

Example 5 Divide: $(x^3 + 1) \div (x + 1)$.

$$
\begin{array}{r}
x^2 - x \qquad + 1 \\
x + 1 \overline{)x^3 \qquad\qquad + 1} \quad \text{Leave space for missing terms.} \\
\underline{x^3 + x^2} \\
- x^2 \\
- x^2 - x \\
x + 1 \quad \text{The answer is } x^2 - x + 1. \\
\underline{x + 1} \quad \text{You need not write a 0 remainder.}
\end{array}
$$

Divide and check. Express your answer in two ways.

7. $x + 3 \overline{)x^2 + 7x + 10}$

Example 6 Divide: $(x^4 - 3x^2 + 1) \div (x - 4)$.

$$
\begin{array}{r}
x^3 + 4x^2 + 13x \; + 52 \\
x - 4 \overline{)x^4 \qquad\quad - 3x^2 \qquad\quad + 1} \\
\underline{x^4 - 4x^3} \\
4x^3 - 3x^2 \\
\underline{4x^3 - 16x^2} \\
13x^2 \\
\underline{13x^2 - 52x} \\
52x + 1 \\
\underline{52x - 208} \\
209
\end{array}
$$

8. $(x^3 - 1) \div (x - 1)$

The answer can be expressed as $x^3 + 4x^2 + 13x + 52$, R 209, or

$$x^3 + 4x^2 + 13x + 52 + \frac{209}{x - 4}.$$

DO EXERCISES 7 AND 8.

NAME   CLASS

EXERCISE SET 8.8

■● Divide and check.

1. $\dfrac{u - 2u^2 - u^5}{u}$

2. $\dfrac{50x^5 - 7x^4 + x^2}{x}$

3. $(15t^3 + 24t^2 - 6t) \div 3t$

4. $(25x^3 + 15x^2 - 30x) \div 5x$

5. $\dfrac{20x^6 - 20x^4 - 5x^2}{-5x^2}$

6. $\dfrac{24x^6 + 32x^5 - 8x^2}{-8x^2}$

7. $\dfrac{4x^4y - 8x^6y^2 + 12x^8y^6}{4x^4y}$

8. $\dfrac{9r^2s^2 + 3r^2s - 6rs^2}{-3rs}$

■●● Divide and check.

9. $(x^2 - 10x - 25) \div (x - 5)$

10. $(x^2 + 8x - 16) \div (x + 4)$

1. _____

2. _____

3. _____

4. _____

5. _____

6. _____

7. _____

8. _____

9. _____

10. _____

ANSWERS

11. $(x^2 + 4x + 4) \div (x + 2)$ 12. $(x^2 - 6x + 9) \div (x - 3)$

11. _____

12. _____

13. $(x^2 + 4x - 14) \div (x + 6)$ 14. $(x^2 + 5x - 9) \div (x - 2)$

13. _____

14. _____

15. $(x^5 - 1) \div (x - 1)$ 16. $(x^5 + 1) \div (x + 1)$

15. _____

16. _____

17. $(t^3 - t^2 + t - 1) \div (t - 1)$ 18. $(t^3 - t^2 + t - 1) \div (t + 1)$

17. _____

18. _____

19. $(x^6 - 13x^3 + 42) \div (x^3 - 7)$ 20. $(x^6 + 5x^3 - 24) \div (x^3 - 3)$

19. _____

20. _____

8.9 SOLVING FRACTIONAL EQUATIONS

●   *A fractional equation* is an equation containing one or more fractional expressions. Here are some examples:

$$\frac{2}{3} + \frac{5}{6} = \frac{x}{9}, \qquad x + \frac{6}{x} = -5, \qquad \frac{x^2}{x-1} = \frac{1}{x-1}.$$

To solve a fractional equation, multiply on both sides by the LCM of all the denominators. This is called *clearing of fractions*.

We have used clearing of fractions in Section 3.3 and Section 6.7.

Example 1   Solve $\frac{2}{3} + \frac{5}{6} = \frac{x}{9}$.

The LCM of all denominators is 18, or $2 \cdot 3 \cdot 3$. We multiply on both sides by $2 \cdot 3 \cdot 3$.

$$2 \cdot 3 \cdot 3 \left(\frac{2}{3} + \frac{5}{6} \right) = 2 \cdot 3 \cdot 3 \cdot \frac{x}{9} \qquad \text{Multiplying on both sides by the LCM}$$

$$2 \cdot 3 \cdot 3 \cdot \frac{2}{3} + 2 \cdot 3 \cdot 3 \cdot \frac{5}{6} = 2 \cdot 3 \cdot 3 \cdot \frac{x}{9} \qquad \text{Multiplying to remove parentheses}$$

$$2 \cdot 3 \cdot 2 + 3 \cdot 5 = 2 \cdot x \qquad \text{Simplifying}$$

$$12 + 15 = 2x$$

$$27 = 2x$$

$$\frac{27}{2} = x$$

Caution! Note that we did not use the LCM to add or subtract fractional expressions. We used it in such a way that all the denominators disappeared and the resulting equation was easier to solve.

When clearing an equation of fractions, be sure to multiply *all* terms in the equation by the LCM.

Example 2   Solve $\frac{1}{x} = \frac{1}{4-x}$.

The LCM is $x(4-x)$. We multiply on both sides by $x(4-x)$.

$$x(4-x) \cdot \frac{1}{x} = x(4-x) \cdot \frac{1}{4-x} \qquad \text{Multiplying on both sides by the LCM}$$

$$4 - x = x \qquad \text{Simplifying}$$

$$4 = 2x$$

$$x = 2$$

Check:

$$\frac{1}{x} = \frac{1}{4-x}$$

$$\frac{\dfrac{1}{2}}{} \;\Big|\; \frac{\dfrac{1}{4-2}}{}$$

$$\frac{1}{2}$$

This checks, so 2 is the solution.

DO EXERCISES 1 AND 2.

OBJECTIVE

After finishing Section 8.9, you should be able to:

● **Solve fractional equations.**

Solve.

1. $\dfrac{3}{4} + \dfrac{5}{8} = \dfrac{x}{12}$

2. $\dfrac{1}{x} = \dfrac{1}{6-x}$

Solve.

3. $\dfrac{x}{4} - \dfrac{x}{6} = \dfrac{1}{8}$

The next examples show how important it is to multiply *all* terms in an equation by the LCM.

Example 3   Solve $\dfrac{x}{6} - \dfrac{x}{8} = \dfrac{1}{12}$.

The LCM is 24. We multiply on both sides by 24.

$$\boxed{24}\left(\dfrac{x}{6} - \dfrac{x}{8}\right) = \boxed{24} \cdot \dfrac{1}{12} \qquad \begin{array}{l}\text{Multiplying on both sides by}\\\text{the LCM}\end{array}$$

$$\boxed{24} \cdot \dfrac{x}{6} - \boxed{24} \cdot \dfrac{x}{8} = \boxed{24} \cdot \dfrac{1}{12} \qquad \begin{array}{l}\text{Multiplying to remove}\\\text{parentheses}\end{array}$$

> Be sure to multiply *every* term by the LCM.

$$4x - 3x = 2 \qquad \text{Simplifying}$$
$$x = 2$$

Check:

$$\begin{array}{c|c}\dfrac{x}{6} - \dfrac{x}{8} = \dfrac{1}{12} \\ \hline \dfrac{2}{6} - \dfrac{2}{8} & \dfrac{1}{12} \\[2mm] \dfrac{1}{3} - \dfrac{1}{4} & \\[2mm] \dfrac{4}{12} - \dfrac{3}{12} & \\[2mm] \dfrac{1}{12} & \end{array}$$

This checks, so the solution is 2.

DO EXERCISE 3.

Example 4   Solve $\dfrac{2}{3x} + \dfrac{1}{x} = 10$.

The LCM is $3x$. We multiply on both sides by $3x$.

$$\boxed{3x}\left(\dfrac{2}{3x} + \dfrac{1}{x}\right) = \boxed{3x} \cdot 10 \qquad \begin{array}{l}\text{Multiplying on both sides by}\\\text{the LCM}\end{array}$$

$$\boxed{3x} \cdot \dfrac{2}{3x} + \boxed{3x} \cdot \dfrac{1}{x} = \boxed{3x} \cdot 10 \qquad \begin{array}{l}\text{Multiplying to remove}\\\text{parentheses}\end{array}$$

$$2 + 3 = 30x \qquad \text{Simplifying}$$
$$5 = 30x$$
$$\dfrac{5}{30} = x$$
$$\dfrac{1}{6} = x$$

This checks, so the solution is $\frac{1}{6}$.

DO EXERCISE 4.

Example 5   Solve $x + \frac{6}{x} = -5$.

The LCM is x. We multiply by x.

$$x\left(x + \frac{6}{x}\right) = -5\,x \qquad \text{Multiplying on both sides by } x$$

$$x^2 + x \cdot \frac{6}{x} = -5x \qquad \text{Note that \textbf{each term} on the left is now multiplied by } x.$$

$$x^2 + 6 = -5x \qquad \text{Simplifying}$$

$$x^2 + 5x + 6 = 0 \qquad \text{Adding } 5x \text{ to get a 0 on one side}$$

$$(x + 3)(x + 2) = 0 \qquad \text{Factoring}$$

$$x + 3 = 0 \quad \text{or} \quad x + 2 = 0 \qquad \text{Principle of zero products}$$

$$x = -3 \quad \text{or} \quad x = -2$$

Check:

$$x + \frac{6}{x} = -5 \qquad\qquad x + \frac{6}{x} = -5$$

$$\frac{-3 + \dfrac{6}{-3} \;\Big|\; -5}{\qquad\quad -5 \;\Big|\;} \qquad\qquad \frac{-2 + \dfrac{6}{-2} \;\Big|\; -5}{\qquad\quad -5 \;\Big|\;}$$

Both of these check, so the solutions are -3 and -2.

DO EXERCISE 5.

It is important *always* to check when solving fractional equations.

Example 6   Solve $\dfrac{x^2}{x - 1} = \dfrac{1}{x - 1}$.

The LCM is $x - 1$. We multiply by $x - 1$.

$$(x - 1) \cdot \frac{x^2}{x - 1} = (x - 1) \cdot \frac{1}{x - 1} \qquad \text{Multiplying on both sides by } x - 1$$

$$x^2 - 1 \qquad \text{Simplifying}$$

$$x^2 - 1 = 0 \qquad \text{Adding } -1 \text{ to get a 0 on one side}$$

$$(x - 1)(x + 1) = 0 \qquad \text{Factoring}$$

$$x - 1 = 0 \quad \text{or} \quad x + 1 = 0 \qquad \text{Principle of zero products}$$

$$x = 1 \quad \text{or} \quad x = -1$$

Possible solutions are 1 and -1.

Solve.

4. $\dfrac{1}{2x} + \dfrac{1}{x} = -12$

Solve.

5. $x + \dfrac{1}{x} = 2$

Solve.

6. $\dfrac{x^2}{x + 2} = \dfrac{4}{x + 2}$

Solve.

7. $\dfrac{4}{x - 2} + \dfrac{1}{x + 2} = \dfrac{26}{x^2 - 4}$

Check:

$$\frac{x^2}{x - 1} = \frac{1}{x - 1} \qquad\qquad \frac{x^2}{x - 1} = \frac{1}{x - 1}$$

$\dfrac{1^2}{1 - 1}$	$\dfrac{1}{1 - 1}$
$\dfrac{1}{0}$	$\dfrac{1}{0}$

$\dfrac{(-1)^2}{-1 - 1}$	$\dfrac{1}{-1 - 1}$
$-\dfrac{1}{2}$	$-\dfrac{1}{2}$

The number -1 is a solution, but 1 is not because it makes a denominator zero.

DO EXERCISE 6.

Example 7   Solve $\dfrac{3}{x - 5} + \dfrac{1}{x + 5} = \dfrac{2}{x^2 - 25}$.

The LCM is $(x - 5)(x + 5)$. We multiply by $(x - 5)(x + 5)$.

$$(x - 5)(x + 5)\left[\frac{3}{x - 5} + \frac{1}{x + 5}\right]$$

$$= (x - 5)(x + 5) \cdot \left[\frac{2}{x^2 - 25}\right] \qquad \text{Multiplying on both sides by the LCM}$$

$$(x - 5)(x + 5) \cdot \frac{3}{x - 5} + (x - 5)(x + 5) \cdot \frac{1}{x + 5}$$

$$= (x - 5)(x + 5) \cdot \frac{2}{x^2 - 25}$$

$$3(x + 5) + (x - 5) = 2 \qquad \text{Simplifying}$$
$$3x + 15 + x - 5 = 2 \qquad \text{Removing parentheses}$$
$$4x + 10 = 2$$
$$4x = -8$$
$$x = -2$$

Check: $\dfrac{3}{x - 5} + \dfrac{1}{x + 5} = \dfrac{2}{x^2 - 25}$

$\dfrac{3}{-2 - 5} + \dfrac{1}{-2 + 5}$	$\dfrac{2}{(-2)^2 - 25}$
$\dfrac{3}{-7} + \dfrac{1}{3}$	$\dfrac{2}{4 - 25}$
$\dfrac{9}{-21} + \dfrac{-7}{-21}$	$\dfrac{2}{-21}$
$\dfrac{2}{-21}$	

This checks, so the solution is -2.

DO EXERCISE 7.

ANSWERS ON PAGE A-28

NAME

EXERCISE SET 8.9

CLASS

● Solve.

1. $\dfrac{1}{4} + \dfrac{1}{6} = \dfrac{1}{t}$

2. $\dfrac{1}{3} - \dfrac{1}{4} = \dfrac{1}{x}$

3. $\dfrac{2}{3} - \dfrac{4}{5} = \dfrac{x}{15}$

4. $\dfrac{3}{4} + \dfrac{7}{8} = \dfrac{x}{2}$

5. $\dfrac{5}{x} = \dfrac{6}{x} - \dfrac{1}{3}$

6. $\dfrac{5}{3x} + \dfrac{3}{x} = 1$

7. $\dfrac{x-7}{x+2} = \dfrac{1}{4}$

8. $\dfrac{a-2}{a+3} = \dfrac{3}{8}$

1. _____

2. _____

3. _____

4. _____

5. _____

6. _____

7. _____

8. _____

ANSWERS

9. $\dfrac{2}{x + 1} = \dfrac{1}{x - 2}$

10. $\dfrac{5}{x - 1} = \dfrac{3}{x + 2}$

9. _____

10. _____

11. $\dfrac{x}{8} - \dfrac{x}{12} = \dfrac{1}{8}$

12. $\dfrac{t}{3} + \dfrac{t}{9} = \dfrac{5}{6}$

11. _____

12. _____

13. $\dfrac{a - 3}{3a + 2} = \dfrac{1}{5}$

14. $\dfrac{2}{x + 3} = \dfrac{5}{x}$

13. _____

14. _____

15. $\dfrac{x - 1}{x - 5} = \dfrac{4}{x - 5}$

16. $\dfrac{x - 7}{x - 9} = \dfrac{2}{x - 9}$

15. _____

16. _____

17. $x + \dfrac{4}{x} = -5$

18. $y + \dfrac{5}{y} = -6$

17. _____

19. $\dfrac{x+1}{3} - \dfrac{x-1}{2} = 1$

20. $\dfrac{y-1}{4} - \dfrac{y+2}{5} = 3$

18. _____

19. _____

21. $\dfrac{1}{x} + \dfrac{2}{x} + \dfrac{3}{x} = 2$

22. $4 - \dfrac{1}{y} - \dfrac{2}{y} = \dfrac{6}{y}$

20. _____

21. _____

23. $\dfrac{y+3}{y} = \dfrac{5}{4}$

24. $\dfrac{t-7}{t} = \dfrac{3}{4}$

22. _____

23. _____

24. _____

25. $\dfrac{x - 2}{x - 3} = \dfrac{x - 1}{x + 1}$

26. $\dfrac{2b - 3}{3b + 2} = \dfrac{2b + 1}{3b - 2}$

25. _____

26. _____

27. $\dfrac{6x - 2}{2x - 1} = \dfrac{9x}{3x + 1}$

28. $\dfrac{2a}{a + 1} = 2 - \dfrac{5}{2a}$

27. _____

29. $\dfrac{1}{x + 3} + \dfrac{1}{x - 3} = \dfrac{1}{x^2 - 9}$

30. $\dfrac{4}{x - 3} + \dfrac{2x}{x^2 - 9} = \dfrac{1}{x + 3}$

28. _____

29. _____

30. _____

31. $\dfrac{x}{x + 4} - \dfrac{4}{x - 4} = \dfrac{x^2 + 16}{x^2 - 16}$

32. $\dfrac{5}{y - 3} - \dfrac{30}{y^2 - 9} = 1$

31. _____

32. _____

8.10 APPLIED PROBLEMS AND FORMULAS

⬛◦ APPLIED PROBLEMS

OBJECTIVES

After finishing Section 8.10, you
should be able to:
⬛◦ Solve applied problems using
    fractional equations.
⬛◦◦ Solve a formula for a given
    letter when fractional expres-
    sions occur in the formula.

We now solve applied problems using fractional equations.

Example 1   The reciprocal of 2 less than a certain number is twice
the reciprocal of the number itself. What is the number?

First, translate to an equation. Let x = the number. Then 2 less than
the number is $x - 2$, and the reciprocal of the number is $1/x$.

$$\underbrace{\left(\begin{array}{c}\text{Reciprocal of 2}\\ \text{less than number}\end{array}\right)}_{\dfrac{1}{x-2}} \quad \text{is} \quad \underbrace{\left(\begin{array}{c}\text{Twice the reciprocal}\\ \text{of the number}\end{array}\right)}_{2 \cdot \dfrac{1}{x}} \qquad \text{Translating}$$

Now we solve.

$$\frac{1}{x-2} = \frac{2}{x} \qquad \text{The LCM is } x(x-2).$$

$$\frac{x(x-2)}{x-2} = \frac{2\ x(x-2)}{x} \qquad \text{Multiplying by LCM}$$

$$x = 2(x-2) \qquad \text{Simplifying}$$

$$x = 2x - 4$$

$$x = 4$$

Check: Go to the original problem. The number to be checked is 4.
Two less than 4 is 2. The reciprocal of 2 is $\frac{1}{2}$. The reciprocal of the
number itself is $\frac{1}{4}$. Now $\frac{1}{2}$ is twice $\frac{1}{4}$, so the conditions are satisfied.
Thus the solution is 4.

Solve.

1. The reciprocal of two more than a
   number is three times the reciprocal
   of the number. Find the number.

DO EXERCISE 1.

Example 2   One car travels 20 km/h faster than another. While one
of them goes 240 km, the other goes 160 km. Find their speeds.

First make a drawing. We really do not know the directions in which
the cars are traveling, but it does not matter.

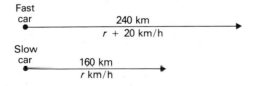

Let r represent the speed of the slow car. Then $r + 20$ is the speed of
the fast car. The cars travel the same length of time, so we can just
use t for time. We can organize the information in a chart, keeping in
mind the formula $d = rt$.

2. One car goes 10 km/h faster than another. While one car goes 120 km, the other goes 150 km. How fast is each car?

	Distance	Speed	Time
Slow car	160	r	t
Fast car	240	$r + 20$	t

If we solve the formula $d = rt$ for t, we get $t = d/r$. Then from the rows of the table, we get

$$t = \frac{160}{r} \quad \text{and} \quad t = \frac{240}{r + 20}.$$

Since these times are the same, we get the following equation:

$$\frac{160}{r} = \frac{240}{r + 20}.$$

We solve:

$$\text{LCM} = r(r + 20)$$

$$\frac{160 \cdot \boxed{r(r + 20)}}{r} = \frac{240 \cdot \boxed{r(r + 20)}}{r + 20} \qquad \text{Multiplying on both sides by the LCM, } r(r + 20)$$

$$160(r + 20) = 240r \qquad \text{Simplifying}$$

$$160r + 3200 = 240r \qquad \text{Removing parentheses}$$

$$3200 = 80r \qquad \text{Adding } -160r$$

$$\frac{3200}{80} = r \qquad \text{Multiplying by } \frac{1}{80}$$

$$40 = r$$

$$60 = r + 20$$

The speeds of 40 km/h for the slow car and 60 km/h for the fast car check in the problem.

DO EXERCISE 2.

Suppose it takes a person 4 hours to do a certain job. Then in 1 hour $\frac{1}{4}$ of the job is done.

> **If a job can be done in t hours (or days, or some other unit of time), then $1/t$ of it can be done in 1 hour (or day).**

Example 3   The head of a secretarial pool examines work records and finds that it takes Helen Huntinpeck 4 hr to type a certain report. It takes Willie Typitt 6 hr to type the same report. How long would it take them, working together, to type the same report?

It takes Helen 4 hr and Willie 6 hr. Then in 1 hr Helen types $\frac{1}{4}$ of the report and Willie types $\frac{1}{6}$ of the report. Working together, they can type

$$\frac{1}{4} + \frac{1}{6}$$

of the report in 1 hr. Let $t =$ the time it takes to type the report, if they work together. Then in 1 hr they can type

$$\frac{1}{t}$$

of the report. Thus,

$$\frac{1}{4} + \frac{1}{6} = \frac{1}{t}.$$

Solve:

$$\boxed{12t} \cdot \left(\frac{1}{4} + \frac{1}{6}\right) = \boxed{12t} \cdot \frac{1}{t} \qquad \text{The LCM is } 2 \cdot 2 \cdot 3t, \text{ or } 12t.$$

$$\frac{12t}{4} + \frac{12t}{6} = \frac{12t}{t}$$

$$3t + 2t = 12$$

$$5t = 12$$

$$t = \frac{12}{5}, \quad \text{or} \quad 2\frac{2}{5}.$$

Thus it takes $2\frac{2}{5}$ hr.

> Note that this answer, $2\frac{2}{5}$ hr, is less than the time it takes either person to do the job alone, 4 hr and 6 hr.

DO EXERCISE 3.

■ ■   FORMULAS

We solve some formulas containing fractional expressions.

Example 4   Solve for P: $A/P = 1 + r$. This is an interest formula.

The LCM is P. We multiply by this.

$$\boxed{P} \cdot \frac{A}{P} = \boxed{P}(1 + r) \qquad \text{Multiplying by } P$$

$$A = P(1 + r) \qquad \text{Simplifying the left side}$$

$$\frac{A}{1 + r} = \frac{P(1 + r)}{1 + r} \qquad \text{Multiplying by } \frac{1}{1 + r}$$

$$\frac{A}{1 + r} = P \qquad \text{Simplifying}$$

DO EXERCISE 4.

3. By checking work records, a contractor finds that it takes Red Bryck 6 hr to construct a wall of a certain size. It takes Lotta Mudd 8 hr to construct the same wall. How long would it take if they worked together?

4. Solve for p: $\dfrac{n}{p} = 2 - m$.

5. Solve for f: $\dfrac{1}{p} + \dfrac{1}{q} = \dfrac{1}{f}$.

(This is an optics formula.)

Example 5 If one person can do a job in a hr and another can do the same job in b hr, then working together they can do the same job in t hr, where a, b, and t are related by

$$\frac{1}{a} + \frac{1}{b} = \frac{1}{t}.$$

Solve for t.

The LCM is abt. We multiply by this.

$$\boxed{abt} \cdot \left(\frac{1}{a} + \frac{1}{b}\right) = \boxed{abt} \cdot \frac{1}{t} \qquad \text{Multiplying by } abt$$

$$\boxed{abt} \cdot \frac{1}{a} + \boxed{abt} \cdot \frac{1}{b} = \frac{abt}{t}$$

$$\frac{abt}{a} + \frac{abt}{b} = \frac{abt}{t}$$

$$bt + at = ab \qquad \text{Simplifying}$$

$$(b + a)t = ab \qquad \text{Factoring out } t$$

$$t = \frac{ab}{b + a} \qquad \text{Multiplying by } \frac{1}{b + a}$$

This answer can be used to find solutions to problems such as Example 3:

$$t = \frac{4 \cdot 6}{6 + 4} = \frac{24}{10} = 2\frac{2}{5}.$$

DO EXERCISE 5.

NAME   CLASS   ANSWERS

EXERCISE SET 8.10

▮●▮ Solve.

1. The reciprocal of 4 plus the reciprocal of 5 is the reciprocal of what number?

1. _____

2. The reciprocal of 3 plus the reciprocal of 8 is the reciprocal of what number?

2. _____

3. One number is 5 more than another. The quotient of the larger divided by the smaller is $\frac{4}{3}$. Find the numbers.

3. _____

4. One number is 4 more than another. The quotient of the larger divided by the smaller is $\frac{5}{2}$. Find the numbers.

4. _____

ANSWERS

5. One car travels 40 km/h faster than another. While one travels 150 km, the other goes 350 km. Find their speeds.

5. _____

6. One car travels 30 km/h faster than another. While one goes 250 km, the other goes 400 km. Find their speeds.

6. _____

7. A person traveled 120 mi in one direction. The return trip was accomplished at double the speed, and took 3 hr less time. Find the speed going.

7. _____

8. After making a trip of 126 mi, a person found that the trip would have taken 1 hr less time by increasing the speed by 8 mph. What was the actual speed?

8. _____

9. The speed of a freight train is 14 km/h slower than the speed of a passenger train. The freight train travels 330 km in the same time that it takes the passenger train to travel 400 km. Find the speed of each train.

10. The speed of a freight train is 15 km/h slower than the speed of a passenger train. The freight train travels 390 kilometers in the same time that it takes the passenger train to travel 480 kilometers. Find the speed of each train.

9. _____

11. It takes painter A 4 hours to paint a certain area of a house. It takes painter B 5 hours to do the same job. How long would it take them, working together, to do the painting job?

10. _____

12. By checking work records a carpenter finds that worker A can build a certain type of garage in 12 hours. Worker B can do the same job in 16 hours. How long would it take if they worked together?

11. _____

12. _____

13. By checking work records a plumber finds that worker A can do a certain job in 12 hours. Worker B can do the same job in 9 hours. How long would it take if they worked together?

13. _____

14. A tank can be filled in 18 hours by pipe A alone and 24 hours by pipe B alone. How long would it take to fill the tank if both pipes were working?

14. _____

15. _____

■ ● ● Solve each formula for the given letter.

16. _____

15. $\dfrac{1}{p} + \dfrac{1}{q} = \dfrac{1}{f};\quad p$

16. $\dfrac{1}{a} + \dfrac{1}{b} = \dfrac{1}{t};\quad b$

17. _____

17. $\dfrac{A}{P} = 1 + r;\quad A$

18. $\dfrac{2A}{h} = a + b;\quad h$

18. _____

19. $\dfrac{1}{R} = \dfrac{1}{r_1} + \dfrac{1}{r_2};\quad R$

19. _____

(An electricity formula)

20. $\dfrac{1}{R} = \dfrac{1}{r_1} + \dfrac{1}{r_2};\quad r_1$

20. _____

21. $\dfrac{A}{B} = \dfrac{C}{D};\quad D$

21. _____

22. $\dfrac{A}{B} = \dfrac{C}{D};\quad C$

22. _____

23. $h_1 = q\left(1 + \dfrac{h_2}{p}\right);\quad h_2$

24. $S = \dfrac{a - ar^n}{1 - r};\quad a$

23. _____

24. _____

8.11 RATIO, PROPORTION, AND VARIATION

⚫ RATIO

The *ratio* of two quantities is their quotient. For example, in the rectangle, the ratio of width to length is

$$\frac{2\text{ cm}}{3\text{ cm}},\quad\text{or}\quad\frac{2}{3}.$$

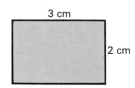

3 cm

2 cm

An older way to write this is 2:3, read "2 to 3." Percent notation is a ratio. For example, 37% is the ratio of 37 to 100, 37/100. The ratio of two different kinds of measure is called a *rate*.

Example 1 Betty Cuthbert of Australia set a world record in the 60-meter dash of 7.2 seconds. What was her rate, or *speed*, in meters per second?

$$\frac{60\text{ m}}{7.2\text{ sec}},\quad\text{or}\quad8.3\frac{\text{m}}{\text{sec}}\quad\text{(Rounded to the nearest tenth)}$$

DO EXERCISES 1–4.

⚫⚫ PROPORTIONS

In applied problems a single ratio is often expressed in two ways. For example, it takes 9 gallons of gas to drive 120 miles, and we wish to find how much will be required to go 550 miles. We can set up ratios:

$$\frac{9\text{ gal}}{120\text{ mi}}\quad\frac{x\text{ gal}}{550\text{ mi}}.$$

If we assume that the car uses gas at the same rate throughout the trip, the ratios are the same.

$$\begin{array}{c}\text{Gas}\longrightarrow\\\text{Miles}\longrightarrow\end{array}\frac{9}{120}=\frac{x}{550}\begin{array}{c}\longleftarrow\text{Gas}\\\longleftarrow\text{Miles}\end{array}$$

To solve, we multiply by 550 to get x alone on one side:

$$\boxed{550}\cdot\frac{9}{120}=\boxed{550}\cdot\frac{x}{550}$$

$$\frac{550\cdot9}{120}=x$$

$$41.25=x.$$

Thus 41.25 gallons will be required. (Note that we could have multiplied by the LCM of 120 and 550, which is 6600, but in this case, that would have been more complicated.)

> **An equality of ratios, $A/B = C/D$, is called a *proportion*. The numbers named in a true proportion are said to be *proportional*.**

1. Find the ratio of 145 km to 2.5 liters (L).

2. Recently, a baseball player got 7 hits in 25 times at bat. What was the rate, or batting average, in hits per times at bat?

3. Impulses in nerve fibers travel 310 km in 2.5 hours. What is the rate, or speed, in kilometers per hour?

4. A lake of area 550 square yards contains 1320 fish. What is the population density of the lake in fish per square yard?

402   Chapter 8   Fractional Expressions and Equations

5. A pitcher for the California Angels gave up 76 earned runs in 198 innings of pitching in a recent year. What was the earned run average?

Example 2   A pitcher gave up 71 earned runs in 285 innings in a recent year. At this rate, how many runs did the pitcher give up every 9 innings?

$$\text{Earned runs each 9 innings} \longrightarrow \frac{A}{9} = \frac{71}{285} \begin{array}{l} \longleftarrow \text{Earned runs} \\ \longleftarrow \text{Innings pitched} \end{array}$$

Solve:

$$\boxed{9} \cdot \frac{A}{9} = \boxed{9} \cdot \frac{71}{285}$$

$$A = \frac{9 \cdot 71}{285}$$

$$A = 2.24 \quad \text{(Rounded to the nearest hundredth)}$$

A stands for the *earned run average*. This means that on the average the pitcher gave up 2.24 earned runs for every 9 innings pitched.

DO EXERCISES 5 AND 6.

6. A sample of 184 light bulbs contained 6 defective bulbs. How many would you expect in 1288 bulbs?

Example 3   (*Estimating wildlife populations*) To determine the number of fish in a lake, a conservationist catches 225 fish, tags them, and throws them back into the lake. Later, 108 fish are caught. Fifteen of them are found to be tagged. Estimate how many fish are in the lake.

Let F = the number of fish in the lake. Then translate to a proportion.

$$\begin{array}{l} \text{Fish tagged originally} \longrightarrow \\ \text{Fish in lake} \longrightarrow \end{array} \frac{225}{F} = \frac{15}{108} \begin{array}{l} \longleftarrow \text{Tagged fish caught later} \\ \longleftarrow \text{Fish caught later} \end{array}$$

This time we multiply by the LCM, which is $108F$:

$$108F \cdot \frac{225}{F} = 108F \cdot \frac{15}{108}$$

$$108 \cdot 225 = F \cdot 15$$

$$\frac{108 \cdot 225}{15} = F$$

$$1620 = F.$$

Thus we estimate that there are about 1620 fish in the lake.

7. To determine the number of deer in a forest, a conservationist catches 612 deer, tags them, and lets them loose. Later 244 deer are caught. Seventy-two of them are tagged. Estimate how many deer are in the forest.

DO EXERCISE 7.

••• DIRECT VARIATION

A bicycle is traveling at 10 km/h. In 1 hour it goes 10 km. In 2 hours it goes 20 km. In 3 hours it goes 30 km, and so on. This gives rise to a set of pairs of numbers, all having the same ratio:

$$(1, 10), \quad (2, 20), \quad (3, 30), \quad (4, 40), \quad \text{and so on.}$$

The ratio of distance to time is always $\frac{10}{1}$, or 10.

ANSWERS ON PAGE A–28

Whenever a situation gives rise to pairs of numbers in which the ratio is constant, we say that there is *direct variation*. Here the *distance varies directly as the time*.

$$\frac{d}{t} = 10 \quad \text{(a constant)}, \quad \text{or} \quad d = 10t.$$

> **Whenever a situation gives rise to a relation among variables**
> $y = kx$, **where** k **is a constant, we say that there is** *direct variation*.
> **The number** k **is called the** *variation constant*.

Example 4   Consumer specialists have determined that the amount F that a family spends on food varies directly as its income I. A family making $9600 a year will spend $2496 on food. How much will a family making $10,500 spend on food?

a) First find an equation of variation.

$$F = kI$$

$$2496 = k \cdot 9600 \qquad \text{Substituting 2496 for } F \text{ and 9600 for } I$$

$$\frac{2496}{9600} = k$$

$$0.26 = k$$

The equation of variation is $F = 0.26I$.

b) Use the equation to find how much a family making $10,500 will spend on food.

$$F = 0.26I$$
$$F = 0.26(10{,}500)$$
$$F = 2730$$

The family will spend $2730 for food.

DO EXERCISES 8 AND 9.

We can solve the same kinds of problems with variation equations as with proportions. You should use the approach that makes the work easier.

8. The cost C of operating a TV varies directly as the number N of hours it is in operation. It costs $6.00 to operate a standard-size color TV continuously for 30 days. At this rate, how much would it cost to operate the TV for 1 day? 1 hour?

9. The number of servings S of meat that can be obtained from a turkey varies directly as its weight W. From a 15-lb turkey, one can get 20 servings of meat. How many servings can be obtained from an 8-lb turkey?

SOMETHING EXTRA

CALCULATOR CORNER: THE PRICE-EARNINGS RATIO

If a company has total earnings one year of \$5,000,000 and there are 100,000 shares of stock issued, the earnings per share are \$50. The price per share of IBM at one time was \$263$\frac{1}{8}$ and the earnings per share were \$17.60. The *price-earnings ratio*, P/E, is the price of the stock divided by the earnings per share. For the IBM stock the price-earnings ratio, P/E, is given by

$$\frac{P}{E} = \frac{263\frac{1}{8}}{17.60}$$

$$= \frac{263.125}{17.60} \quad \text{Converting to decimal notation}$$

$$\approx 15.0 \quad \text{Dividing and rounding to the nearest tenth}$$

EXERCISES

Calculate the price-earnings ratio for each stock.

	Stock	Price per share	Earnings
1.	General Motors	\$68$\frac{5}{8}$	\$11.50
2.	K-Mart	31	2.60
3.	United Airlines	18$\frac{5}{8}$	4.00
4.	AT&T	62	6.60

Exercise Set 8.11   405

EXERCISE SET 8.11

■ • Find the ratio of

1. 54 days, 6 days

2. 800 mi, 50 gallons

3. A black racer snake travels 4.6 kilometers in 2 hours. What is the speed in kilometers per hour?

4. Light travels 558,000 miles in 3 seconds. What is the speed in miles per second?

■ ■ Solve.

5. The coffee beans from 14 trees are required to produce 7.7 kilograms of coffee (this is the average that each person in the United States drinks each year). How many trees are required to produce 320 kilograms of coffee?

6. Last season a minor league baseball player got 240 hits in 600 times at bat. This season his ratio of hits to number of times at bat is the same. He batted 500 times. How many hits has he had?

7. A student traveled 234 kilometers in 14 days. At this same ratio, how far would the student travel in 42 days?

Copyright © 1983, by Addison-Wesley Publishing Company, Inc. All rights reserved.

8. In a bread recipe, the ratio of milk to flour is $\frac{4}{3}$. If 5 cups of milk are used, how many cups of flour are used?

8. _____

9. 10 cm³ of a normal specimen of human blood contains 1.2 grams of hemoglobin. How many grams would 16 cm³ of the same blood contain?

9. _____

10. The winner of an election for class president won by a vote of 3 to 2, with 324 votes. How many votes did the loser get?

10. _____

11. To determine the number of trout in a lake, a conservationist catches 112 trout, tags them, and throws them back into the lake. Later, 82 trout are caught; 32 of them are tagged. How many trout are in the lake?

11. _____

12. To determine the number of deer in a game preserve, a conservationist catches 318 deer, tags them, and lets them loose. Later, 168 deer are caught; 56 of them are tagged. How many deer are in the preserve?

12. _____

13. The ratio of the weight of an object on the moon to the weight of an object on earth is 0.16 to 1.

    a) How much would a 12-ton rocket weigh on the moon?

    b) IIow much would a 90-kilogram astronaut weigh on the moon?

13. _____

14. The ratio of the weight of an object on Mars to the weight of an object on earth is 0.4 to 1.

    a) How much would a 12-ton rocket weigh on Mars?

    b) How much would a 90-kilogram astronaut weigh on Mars?

14. _____

⦿⦿⦿ Solve.

15. The number B of bolts a machine can make varies directly as the time it operates. It can make 1000 bolts in 2 hours. How many can it make in 5 hours?

15. _____

16. A person's paycheck P varies directly as the number H of hours worked. For working 15 hours the pay is $48.75. Find the pay for 35 hours of work.

16. _____

17. The amount C that a family spends on car expenses varies directly as its income I. A family making $10,880 a year will spend $1632 a year for car expenses. How much will a family making $10,000 a year spend for car expenses?

17. _____

18. The number of kg of water W in a human body varies directly as the total body weight B. A person weighing 96 kg contains 64 kg of water. How many kilograms of water are in a person weighing 81 kg?

18. _____

19. The weight M of an object on Mars varies directly as its weight E on earth. A person who weighs 95 kg on earth weighs 38 kg on Mars. How much would an 80-kg person weigh on Mars?

19. _____

20. The number of servings S of meat that can be obtained from round steak varies directly as the weight W. From 9 kg of round steak one can get 70 servings of meat. How many servings can one get from 12 kg of round steak?

20. _____

EXTENSION EXERCISES

SECTION 8.1

Multiply and simplify.

1. $\dfrac{t^4 - 1}{t^4 - 81} \cdot \dfrac{t^2 + 9}{t^2 + 1}$

2. $\dfrac{x^2 + 5x + 6}{x^2 + x - 6} \cdot \dfrac{x^2 - 3x + 2}{x^2 + x - 2}$

3. $\dfrac{(t - 2)^3}{(t - 1)^3} \cdot \dfrac{t^2 - 2t + 1}{t^2 - 4t + 4}$

4. Determine which numbers of x are not sensible replacements in the following fractional expressions.

a) $\dfrac{x + 1}{x^2 + 4x + 4}$

b) $\dfrac{x^2 - 16}{x^2 + 2x - 3}$

5. What happens to the value of each expression as x becomes very large?

a) $\dfrac{1}{x - 3}$

b) $\dfrac{2x}{x - 1}$

c) $\dfrac{x + 1}{x^2}$

d) $\dfrac{2x}{x^2 - x}$

e) $\dfrac{x - 1}{x^2 - 1}$

SECTION 8.2

Divide and simplify.

6. $\dfrac{x^2 - x + xy - y}{x^2 + 8x + 7} \div \dfrac{x^2 + 2xy + y^2}{4x + 4y}$

7. $\dfrac{3x + 3y + 3}{9x} \div \dfrac{x^2 + 2xy + y^2 - 1}{x^4 + x^2}$

8. $\dfrac{z^2 - 8z + 16}{z^2 + 8z + 16} \div \dfrac{(z - 4)^5}{(z + 4)^5}$

9. $\left(\dfrac{y^2 + 5y + 6}{y^2} \cdot \dfrac{3y^3 + 6y^2}{y^2 - y - 12} \right) \div \dfrac{y^2 - 4}{y^2 + 7y + 10}$

SECTION 8.3

Perform the indicated operations. Simplify, if possible.

10. $\dfrac{x^2}{3x^2 - 5x - 2} - \dfrac{2x}{3x + 1} \cdot \dfrac{1}{x - 2}$

11. $\dfrac{3}{x + 4} \cdot \dfrac{2x + 11}{x - 3} - \dfrac{1}{4 + x} \cdot \dfrac{6x + 3}{3 - x}$

SECTION 8.4

12. Find the LCM.

a) $72, 90, 96$

b) $8x^2 - 8, 6x^2 - 12x + 6, 10 - 10x$

13. Two joggers leave the starting point of a circular course at the same time. One jogger completes one round in 6 minutes and the second jogger in 8 minutes. Assuming they continue to run at the same pace, after how many minutes will they meet again at the starting place?

14. If the LCM of two expressions is the same as one of the expressions, what is their relationship?

SECTION 8.5

Add, and simplify if possible.

15. $\dfrac{5}{z + 2} + \dfrac{4z}{z^2 - 4} + 2$

16. $\dfrac{-2}{y^2 - 9} + \dfrac{4y}{(y - 3)^2} + \dfrac{6}{3 - y}$

17. $\dfrac{3z^2}{z^4 - 4} + \dfrac{5z^2 - 3}{2z^4 + z^2 - 6}$

SECTION 8.6

Subtract, and simplify if possible.

18. $\dfrac{1}{2xy - 6x + ay - 3a} - \dfrac{ay + xy}{(a^2 - 4x^2)(y^2 - 6y + 9)}$

19. $\dfrac{x}{x^4 - y^4} - \dfrac{1}{x^2 + 2xy + y^2}$

SECTION 8.7

Simplify.

20. $\left[\dfrac{\dfrac{x+1}{x-1}+1}{\dfrac{x+1}{x-1}-1}\right]^5$

21. $1+\dfrac{1}{1+\dfrac{1}{1+\dfrac{1}{1+\dfrac{1}{x}}}}$

22. $\dfrac{\dfrac{z}{1-\dfrac{z}{2+2z}}-2z}{\dfrac{2z}{5z-2}-3}$

23. Find the reciprocal of $\dfrac{2}{x-1}-\dfrac{1}{3x-2}$ and simplify.

24. Show that the reciprocal of the reciprocal of a number is the number itself.

SECTION 8.8

Divide.

25. $(5x^7-3x^4+2x^2-3)\div(x^2-x+1)$

26. $(x^4+9x^2+20)\div(x^2+4)$

27. $(y^4+a^2)\div(y+a)$

28. If the remainder is 0 when one polynomial is divided by another, then the divisor is a factor of the dividend. Find the value(s) of c for which $x-1$ is a factor of each polynomial.
  a) x^2+4x+c b) $2x^2+3cx-8$ c) $c^2x^2-2cx+1$

SECTION 8.9

Solve.

29. $\dfrac{y}{y+0.2}-1.2=\dfrac{y-0.2}{y+0.2}$

30. $\dfrac{x^2}{x^2-4}=\dfrac{x}{x+2}-\dfrac{2x}{2-x}$

31. $\dfrac{x}{x^2+3x-4}+\dfrac{x+1}{x^2+6x+8}=\dfrac{2x}{x^2+x-2}$

SECTION 8.10

32. The denominator of a fraction is one more than the numerator. If 2 is subtracted from both the numerator and denominator, the resulting fraction is $\frac{1}{2}$. Find the original fraction.

33. Ann and Betty work together and complete a job in 4 hr. It would take Betty 6 hr longer, working alone, to do the job than it would Ann. How long would it take each of them to do the job working alone?

34. The speed of a boat in still water is 10 mph. It travels 24 mi upstream and 24 mi downstream in a total time of 5 hrs. What is the speed of the current?

35. Solve each formula for the given letter.

  a) $u=-F\left(E-\dfrac{P}{T}\right);T$ b) $1=a+(n-1)d;n$

SECTION 8.11

36. Divide 100 into two parts so that the ratio of one part, increased by 5, to the other part, decreased by 5, is 4.

37. Show that if p varies directly as q, then q varies directly as p.

38. The area of a circle varies directly as the square of the length of the radius. What is the variation constant?

NAME   SCORE   ANSWERS

TEST OR REVIEW—CHAPTER 8

If you miss an item, review the indicated section and objective.

[8.1, ●●●]   **1.** Simplify: $\dfrac{14x^2 - x - 3}{2x^2 - 7x + 3}$ 1. _____

[8.1, ▪▪]   **2.** Multiply and simplify: $\dfrac{a^2 - 36}{10a} \cdot \dfrac{2a}{a + 6}$

2. _____

[8.2, ●●○]   **3.** Divide and simplify: $\dfrac{4x^4}{x^2 - 1} \div \dfrac{2x^3}{x^2 - 2x + 1}$

3. _____

Add.

[8.3, ●●]   **4.** $\dfrac{x + 3}{x - 2} + \dfrac{x}{2 - x}$

4. _____

[8.5, ●]   **5.** $\dfrac{3}{3x - 9} + \dfrac{x - 2}{3 - x}$

5. _____

Subtract.

[8.3, ▪▪]   **6.** $\dfrac{x + 3}{x - 2} - \dfrac{x}{2 - x}$

6. _____

[8.6, ●]   **7.** $\dfrac{1}{x^2 - 25} - \dfrac{x - 5}{x^2 - 4x - 5}$

7. _____

[8.7, ▪]   **8.** Simplify: $\dfrac{\dfrac{1}{a} - 2}{\dfrac{1}{2a} - 1}$.

8. _____

[8.8, ▪▪]   **9.** Divide: $(2x^3 - 2x^2 + x - 4) \div (x + 3)$.

9. _____

[8.9, ▪]   **10.** Solve: $\dfrac{15}{x} = \dfrac{45}{x + 2}$.

10. _____

[8.10, ▪]   Solve.

11. One vehicle travels 90 km in the same time a car traveling 10 km/h slower travels 60 km. Find the speed of each.

11. _____

[8.10, ▪▪]   **12.** Solve for r: $\dfrac{E}{e} = \dfrac{R + r}{r}$.

12. _____

[8.11, ▪▪▪]   Solve.

13. A student's test score varies directly as the time T studied. A score of 80 is made after studying 4 hours. What score would be made after studying 5 hours?

13. _____

RADICAL EXPRESSIONS AND EQUATIONS

9

An expression involving square roots can be used to determine the speed of a car from the length of its skid marks.

1. What is the meaning of 5^2?

2. Multiply: $(-5)(-5)$.

3. Multiply: $\dfrac{4}{3} \cdot \dfrac{4}{3}$.

Simplify.

4. $\dfrac{18}{50}$

5. $\dfrac{81}{27}$

Write decimal notation.

6. $\dfrac{7}{8}$

7. $\dfrac{5}{11}$

8. Simplify: $|-8|$.

9. Multiply and simplify: $x^3 \cdot x^3$.

10. Factor: $x^3 - x^2$.

11. Factor: $x^2 + 2x + 1$.

12. Solve: $x + 1 = 2x - 5$.

9.1 RADICAL EXPRESSIONS

◼•◼ SQUARE ROOTS

The number c is a square root of a if $c^2 = a$.

Every positive number has two square roots. For example, the square roots of 25 are 5 and -5, because

$$5^2 = 25 \quad \text{and} \quad (-5)^2 = 25.$$

Zero has only one square root, 0 itself.

Example 1 Find the square roots of 81.

The square roots are 9 and -9, because $9^2 = 81$ and $(-9)^2 = 81$.

DO EXERCISES 1–3.

The symbol $\sqrt{}$ is called a *radical*. We use it in naming the nonnegative square root of a number. We call $\sqrt{a}$ the *principal square root* of a.

Examples

2. Find $\sqrt{25}$.

$\sqrt{25} = 5$ Remember, $\sqrt{}$ means to take the principal, or *nonnegative*, square root.

3. Find $\sqrt{225}$.

$\sqrt{225} = 15$

4. Find $\sqrt{0}$.

$\sqrt{0} = 0$

It would be most helpful for our work in this chapter to make a list of squares of numbers from 0 to 20, and to memorize the list.

DO EXERCISES 4–6.

OBJECTIVES

After finishing Section 9.1, you should be able to:

◼•◼ Find principal square roots and their additive inverses for squares of whole numbers from 0^2 to 20^2.

◼••◼ Identify radicands.

◼•••◼ Identify meaningless radical expressions, and determine whether a given number is a sensible replacement in a radical expression.

◼◼ Simplify radical expressions with a perfect square radicand.

Find the square roots of each number.

1. 36

2. 64

3. 225

Find.

4. $\sqrt{16}$

5. $\sqrt{49}$

6. $\sqrt{100}$

To name the negative square root of a number, we use $-\sqrt{}$.

Examples

5. Find $-\sqrt{25}$.

$$-\sqrt{25} = -5 \qquad \sqrt{25} = 5, \text{ so } -\sqrt{25} = -5$$

6. Find $-\sqrt{64}$.

$$-\sqrt{64} = -8$$

DO EXERCISES 7–9.

●● RADICAL EXPRESSIONS

When an expression is written under a radical, we have a *radical expression*. These are radical expressions:

$$\sqrt{14}, \quad \sqrt{x}, \quad \sqrt{x^2 + 4}, \quad \sqrt{\frac{x^2 - 5}{2}}.$$

The expression written under the radical is called the *radicand*.

Examples   Identify the radicand in each expression.

7. $\sqrt{x}$;   The radicand is x.

8. $\sqrt{y^2 - 5}$;   The radicand is $y^2 - 5$.

DO EXERCISES 10 AND 11.

●●● MEANINGLESS EXPRESSIONS

Negative numbers do not have square roots. This is because the square of any negative number is positive. For example,

$$(-5)^2 = 25, \quad (-11)^2 = 121.$$

Thus the following expressions are meaningless:

$$\sqrt{-100}, \quad \sqrt{-49}, \quad -\sqrt{-3}.$$

Any number that makes a radicand negative is a nonsensible replacement.

Example 9   Determine whether 6 is a sensible replacement in the expression $\sqrt{1 - y}$.

If we replace y by 6 we get

$$\sqrt{1 - 6} - \sqrt{-5},$$

which is meaningless because the radicand is negative. Thus 6 is not a sensible replacement.

DO EXERCISES 12–17.

Find.

7. $-\sqrt{16}$

8. $-\sqrt{49}$

9. $-\sqrt{169}$

Identify the radicand in each expression.

10. $\sqrt{45 + x}$

11. $\sqrt{\dfrac{x}{x + 2}}$

Is the expression meaningless? Write "yes" or "no."

12. $\sqrt{-25}$ **13.** $-\sqrt{25}$

14. $-\sqrt{-36}$ **15.** $-\sqrt{36}$

16. Determine whether 8 is a sensible replacement in $\sqrt{x}$.

17. Determine whether 10 is a sensible replacement in $\sqrt{4 - x}$.

ANSWERS ON PAGE A–29

Find the following. Assume that letters can represent any real numbers.

18. $\sqrt{(xy)^2}$

19. $\sqrt{x^2y^2}$

20. $\sqrt{(x-1)^2}$

21. $\sqrt{x^2 + 8x + 16}$

22. $\sqrt{25x^2}$

■■ PERFECT SQUARE RADICANDS

In the expression $\sqrt{x^2}$, the radicand is a perfect square. Since squares are never negative, all replacements are sensible.

Suppose $x = 3$. Then we have $\sqrt{3^2}$, which is $\sqrt{9}$, or 3.

Suppose $x = -3$. Then we have $\sqrt{(-3)^2}$, which is $\sqrt{9}$, or 3.

In any case, $\sqrt{x^2} = |x|$.

> **Any radical expression $\sqrt{A^2}$ can be simplified to $|A|$.**

Examples   Find the following. Assume that letters can represent any real numbers.

10. $\sqrt{(3x)^2} = |3x|$ (Don't forget the absolute values!)

11. $\sqrt{a^2b^2} = \sqrt{(ab)^2} = |ab|$

12. $\sqrt{x^2 + 2x + 1} = \sqrt{(x+1)^2} = |x+1|$

We can sometimes simplify absolute value notation. In Example 10, $|3x|$ simplifies to $|3| \cdot |x|$ or $3|x|$. In Example 11, we can change $|ab|$ to $|a| \cdot |b|$ if we wish. The absolute value of a product is always the product of the absolute values.

> **For any numbers a and b, $|a \cdot b| = |a| \cdot |b|$.**

DO EXERCISES 18–22.

EXERCISE SET 9.1

◼• Find the square roots of each number.

1. 1   **2.** 4   **3.** 16   **4.** 9

5. 100   **6.** 121   **7.** 169   **8.** 144

Find.

9. $\sqrt{4}$ **10.** $\sqrt{1}$ **11.** $-\sqrt{9}$ **12.** $-\sqrt{25}$

13. $-\sqrt{64}$ **14.** $-\sqrt{81}$ **15.** $\sqrt{225}$ **16.** $\sqrt{400}$

17. $\sqrt{361}$ **18.** $\sqrt{441}$ **19.** $\sqrt{324}$ **20.** $\sqrt{196}$

◼•◼• Identify the radicand in each expression.

21. $\sqrt{a-4}$ **22.** $\sqrt{t+3}$ **23.** $5\sqrt{t^2+1}$

24. $8\sqrt{x^2+5}$ **25.** $x^2y\sqrt{\dfrac{3}{x+2}}$ **26.** $ab^2\sqrt{\dfrac{a}{a-b}}$

ANSWERS

1. _____
2. _____
3. _____
4. _____
5. _____
6. _____
7. _____
8. _____
9. _____
10. _____
11. _____
12. _____
13. _____
14. _____
15. _____
16. _____
17. _____
18. _____
19. _____
20. _____
21. _____
22. _____
23. _____
24. _____
25. _____
26. _____

ANSWERS

27. _____

28. _____

29. _____

30. _____

31. _____

32. _____

33. _____

34. _____

35. _____

36. _____

37. _____

38. _____

39. _____

40. _____

41. _____

42. _____

43. _____

44. _____

45. _____

46. _____

47. _____

48. _____

●●● Which of these expressions are meaningless? Write "yes" or "no."

27. $\sqrt{-16}$ **28.** $\sqrt{-81}$ **29.** $-\sqrt{81}$ **30.** $-\sqrt{64}$

Determine whether the given number is a sensible replacement in the given radical expression.

31. $4; \sqrt{y}$ **32.** $-8; \sqrt{m}$ **33.** $-11; \sqrt{1+x}$ **34.** $-11; \sqrt{2-x}$

⋮⋮ Find the following. Assume that letters can represent any real number.

35. $\sqrt{x^2}$ **36.** $\sqrt{t^2}$ **37.** $\sqrt{4a^2}$

38. $\sqrt{9x^2}$ **39.** $\sqrt{(-5)^2}$ **40.** $\sqrt{(-7)^2}$

41. $\sqrt{(-3b)^2}$ **42.** $\sqrt{(-4d)^2}$ **43.** $\sqrt{(x-7)^2}$

44. $\sqrt{(x+3)^2}$ **45.** $\sqrt{x^2+2x+1}$ **46.** $\sqrt{a^2-10a+25}$

47. $\sqrt{4x^2-20x+25}$ **48.** $\sqrt{9x^2+12x+4}$

9.2 IRRATIONAL NUMBERS AND REAL NUMBERS

⬤ IRRATIONAL NUMBERS

Recall that all rational numbers can be named by fractional notation

$\dfrac{a}{b}$,   where a and b are integers and $b \neq 0$.

Rational numbers can be named in other ways, such as with decimal notation, but they are all nameable with fractional notation. Suppose we try to find a rational number for $\sqrt{2}$. We look for a number a/b for which $a/b \cdot a/b = 2$. We can find rational numbers whose squares are quite close to 2:

$$\left(\frac{14}{10}\right)^2 = (1.4)^2 = 1.96,$$

$$\left(\frac{141}{100}\right)^2 = (1.41)^2 = 1.9881,$$

$$\left(\frac{1414}{1000}\right)^2 = (1.414)^2 = 1.999396,$$

$$\left(\frac{14142}{10000}\right)^2 = (1.4142)^2 = 1.99996164,$$

but we can never find one whose square is exactly 2. This can be proved but we will not do it here. Since $\sqrt{2}$ is not a rational number, we call it an *irrational* number.

> An *irrational* number is a number that cannot be named by fractional notation a/b, where a and b are integers and $b \neq 0$.

Unless a whole number is a perfect square, its square root is irrational.

Examples   Identify the rational numbers and the irrational numbers.

1. $\sqrt{3}$ is irrational.   3 is not a perfect square.

2. $\sqrt{25}$ is rational.   25 is a perfect square.

3. $\sqrt{35}$ is irrational.   35 is not a perfect square.

There are many irrational numbers. Many are *not* obtained by taking square roots. For example,

π is irrational,

and it is not a square root of any rational number.

DO EXERCISES 1–5.

⬤⬤ DECIMAL NOTATION FOR IRRATIONAL NUMBERS

Decimal notation for a rational number either ends or repeats. For example,

$$\frac{1}{4} = 0.25, \quad \frac{1}{3} = 0.\boxed{3}33\ldots, \quad \frac{5}{11} = 0.\boxed{45}4545\ldots.$$

ends   3 repeats   45 repeats

1. Define an irrational number.

Identify the rational numbers and the irrational numbers.

2. $\sqrt{5}$

3. $\sqrt{36}$

4. $\sqrt{32}$

5. $\sqrt{101}$

Identify the rational numbers and the irrational numbers.

6. $\dfrac{-3}{5}$

7. $\dfrac{95}{37}$

8. 6.12

9. 0.353535 . . . (Numeral repeats)

10. 3.01001000100001 . . .
(Numeral does not repeat)

Find the number halfway between each pair of numbers.

11. $\dfrac{3}{64}$ and $\dfrac{4}{64}$

12. -0.678 and -0.6782

13. 5.698 and 5.6999

14. Describe the set of real numbers.

Use Table 1 to approximate these square roots.

15. $\sqrt{7}$

16. $\sqrt{72}$

Decimal notation for an irrational number never ends and it does not repeat. The number π is an example:

$$\pi = 3.1415926535.\ldots$$

Decimal notation for π never ends and never repeats. Note that 3.14 and 22/7 are both rational approximations for π. Decimal notation for 22/7 is 3.142857142857. . . . It repeats. Here are some other examples of irrational numbers:

2.818118111811118111118 . . .   Numeral does not repeat

0.03503550355503555550 . . .   Numeral does not repeat

DO EXERCISES 6–10.

●●● REAL NUMBERS

The rational numbers are very close together. Yet no matter how close together two rational numbers are, we can find infinitely many rational numbers between them. By averaging, we can find the number *halfway* between.

Example 4 Find the number halfway between $\frac{1}{32}$ and $\frac{2}{32}$.

We average: $\dfrac{\frac{1}{32} + \frac{2}{32}}{2} = \dfrac{\frac{3}{32}}{2} = \dfrac{3}{64}$.   Add and divide by 2.

Example 5 Find the number halfway between -0.32 and -0.323.

$$\dfrac{(-0.32) + (-0.323)}{2} = \dfrac{-0.643}{2} = -0.3215$$

DO EXERCISES 11–13.

Although the rational numbers seem to fill up the number line, they do not. There are many points on the line for which there is no rational number. For these points, there are *irrational* numbers.

> The *real numbers* **consist of the rational numbers and the irrational numbers. There is a real number for each point on a number line.**

DO EXERCISE 14.

●● APPROXIMATING IRRATIONAL NUMBERS

Most of the time we use rational numbers to approximate irrational numbers. Table 1 on the inside back cover contains rational approximations for square roots; for example,

$$\sqrt{10} \approx 3.162.$$ Rounded to three decimal places

The symbol $\approx$ means "is approximately equal to."

DO EXERCISES 15 AND 16.

NAME	CLASS	ANSWERS

ANSWERS
1.
2.
3.
4.
5.
6.
7.
8.
9.
10.
11.
12.
13.
14.
15.
16.
17.
18.
19.
20.
21.
22.
23.
24.

EXERCISE SET 9.2

● Identify the rational numbers and the irrational numbers.

1. $\sqrt{2}$ **2.** $\sqrt{6}$ **3.** $-\sqrt{8}$ **4.** $-\sqrt{10}$

5. $\sqrt{49}$ **6.** $\sqrt{100}$ **7.** $\sqrt{98}$ **8.** $-\sqrt{12}$

●● Identify the rational numbers and the irrational numbers.

9. $-\dfrac{2}{3}$ **10.** $\dfrac{136}{51}$ **11.** 23   **12.** 4.23

13. 0.424242 . . .   (Numeral repeats)   **14.** 0.1565656 . . .   (Numeral repeats)

15. 4.28228222822228 . . .   (Numeral does not repeat)

16. 7.76776777677776 . . .   (Numeral does not repeat)

17. -1 **18.** 0

19. -45.6919119111911119 . . .   (Numeral does not repeat)

20. -63.03003000300003 . . .   (Numeral does not repeat)

●●● Find the number halfway between each pair of numbers.

21. $\dfrac{2}{7}, \dfrac{5}{7}$ **22.** $-\dfrac{3}{5}, -\dfrac{4}{5}$ **23.** $\dfrac{2}{3}, \dfrac{7}{10}$ **24.** $\dfrac{7}{8}, \dfrac{9}{10}$

ANSWERS

25. _____

26. _____

27. _____

28. _____

29. _____

30. _____

31. _____

32. _____

33. _____

34. _____

35. _____

36. _____

37. _____

38. _____

39. _____

40. _____

41. _____

25. $6\frac{1}{3}, 6\frac{1}{4}$ **26.** $8\frac{1}{6}, 8\frac{1}{7}$ **27.** $-1.45, -1.452$ **28.** $6.374, 6.375$

29. $0.25, \frac{3}{8}$ **30.** $0.58, \frac{5}{8}$

Use Table 1 to approximate these square roots.

31. $\sqrt{5}$ **32.** $\sqrt{43}$ **33.** $\sqrt{81}$ **34.** $\sqrt{87}$

35. $\sqrt{93}$ **36.** $\sqrt{17}$ **37.** $\sqrt{63}$ **38.** $\sqrt{50}$

Complete.

39. Every _____ number can be named as the quotient of two _____.

40. The real numbers consist of the rational numbers and the _____ numbers.

41. *Parking.* When a parking lot has attendants to park cars, it uses spaces for drivers to pull in and leave cars before they are taken to permanent parking stalls. The number N of such spaces needed is given by the formula

$$N = 2.5\sqrt{A},$$

where A is the average number of arrivals in peak hours. Find the number of spaces needed when the average number of arrivals in peak hours is 25, 36, 49, and 64.

9.3   MULTIPLICATION AND FACTORING

● MULTIPLICATION

To see how we can multiply with radical notation, look at the following examples.

Example 1   Simplify.

a) $\sqrt{9} \cdot \sqrt{4} = 3 \cdot 2 = 6$ This is a product of square roots.

b) $\sqrt{9 \cdot 4} = \sqrt{36} = 6$ This is the square root of a product.

Example 2   Simplify.

a) $\sqrt{4} \cdot \sqrt{25} = 2 \cdot 5 = 10$

b) $\sqrt{4 \cdot 25} = \sqrt{100} = 10$

Example 3

$\sqrt{-9}\sqrt{-4}$ Meaningless (negative radicands)

DO EXERCISE 1.

We can multiply radical expressions by multiplying the radicands* provided the radicands are not negative.

> **For any nonnegative radicands A and B, $\sqrt{A} \cdot \sqrt{B} = \sqrt{A \cdot B}$. (The product of square roots, provided they exist, is the square root of the product of the radicands.)**

Examples   Multiply. Assume that all radicands are nonnegative.

4. $\sqrt{5}\sqrt{7} = \sqrt{5 \cdot 7} = \sqrt{35}$

5. $\sqrt{8}\sqrt{8} = \sqrt{8 \cdot 8} = \sqrt{64} = 8$

6. $\sqrt{\dfrac{2}{3}}\sqrt{\dfrac{4}{5}} = \sqrt{\dfrac{2}{3} \cdot \dfrac{4}{5}} = \sqrt{\dfrac{8}{15}}$

7. $\sqrt{2x}\sqrt{3x-1} = \sqrt{2x(3x-1)} = \sqrt{6x^2 - 2x}$

DO EXERCISES 2–5.

---

*A proof (optional).   We consider a product $\sqrt{A}\sqrt{B}$, where A and B are not negative. We square this product, to show that we get AB, and thus the product is the square root of AB (or $\sqrt{AB}$).

$$(\sqrt{A}\sqrt{B})^2 = (\sqrt{A}\sqrt{B})(\sqrt{A}\sqrt{B})$$
$$= (\sqrt{A}\sqrt{A})(\sqrt{B}\sqrt{B})$$
$$= AB$$

OBJECTIVES

After finishing Section 9.3, you should be able to:

● Multiply with radical notation.

●● Factor radical expressions and where possible simplify.

●●● Approximate square roots not in Table 1 using factoring and the square root table.

1. Simplify.

a) $\sqrt{4} \cdot \sqrt{16}$

b) $\sqrt{4 \cdot 16}$

Multiply. Assume that all radicands are nonnegative.

2. $\sqrt{3}\sqrt{7}$

3. $\sqrt{5}\sqrt{5}$

4. $\sqrt{x}\sqrt{x+1}$

5. $\sqrt{x+1}\sqrt{x-1}$

Simplify by factoring.

6. $\sqrt{32}$

7. $\sqrt{x^2 + 14x + 49}$

8. $\sqrt{25x^2}$

9. $\sqrt{36m^2}$

10. $\sqrt{76}$

11. $\sqrt{x^2 - 8x + 16}$

12. $\sqrt{64t^2}$

13. $\sqrt{100a^2}$

Approximate these square roots.
Round to three decimal places.

14. $\sqrt{275}$

15. $\sqrt{102}$

■ ■ **FACTORING AND SIMPLIFYING**

We know that for nonnegative radicands,

$$\sqrt{A}\sqrt{B} = \sqrt{AB}.$$

To factor radical expressions we can think of this equation in reverse:

$$\sqrt{AB} = \sqrt{A}\sqrt{B}.$$

All equations are reversible.

In some cases we can simplify after factoring. A radical expression is simplified when its radicand has no factors that are perfect squares.

Examples Simplify by factoring.

8. $\sqrt{18} = \sqrt{9 \cdot 2}$ Factoring the radicand: 9 is a perfect square.

  $= \sqrt{9} \cdot \sqrt{2}$ Factoring the radical expression

  $= 3\sqrt{2}$ $3\sqrt{2}$ means $3 \cdot \sqrt{2}$

The radicand has no factors that are perfect squares.

9. $\sqrt{25x} = \sqrt{25} \cdot \sqrt{x} = 5\sqrt{x}$

10. $\sqrt{x^2 - 6x + 9} = |x - 3|$

11. $\sqrt{36x^2} = \sqrt{36}\sqrt{x^2} = 6|x|$

DO EXERCISES 6–13.

●●● **APPROXIMATING SQUARE ROOTS**

Table 1 goes only to 100. We can use it to find approximate square roots for other numbers. We do this by first factoring out the largest perfect square, if there is one. If there is none, we use any factorization we can find that will give smaller factors.

Examples Approximate these square roots.

12. $\sqrt{160} = \sqrt{16 \cdot 10}$ Factoring the radicand (make one factor a perfect square, if you can)

  $= \sqrt{16}\sqrt{10}$ Factoring the radical expression

  $= 4\sqrt{10}$

  $\approx 4(3.162)$ From Table 1, $\sqrt{10} \approx 3.162$

  ≈ 12.648

13. $\sqrt{341} = \sqrt{11 \cdot 31}$ Factoring (There is no perfect square factor.)

  $= \sqrt{11}\sqrt{31}$

  $\approx 3.317 \times 5.568$ Table 1

  ≈ 18.469 Rounded to 3 decimal places

DO EXERCISES 14 AND 15.

NAME	CLASS	ANSWERS

ANSWERS
1.
2.
3.
4.
5.
6.
7.
8.
9.
10.
11.
12.
13.
14.
15.
16.
17.
18.
19.
20.
21.
22.

EXERCISE SET 9.3

● Multiply. Assume all radicands are nonnegative.

1. $\sqrt{2}\sqrt{3}$ **2.** $\sqrt{3}\sqrt{5}$ **3.** $\sqrt{3}\sqrt{3}$

4. $\sqrt{6}\sqrt{6}$ **5.** $\sqrt{7}\sqrt{2}$ **6.** $\sqrt{14}\sqrt{3}$

7. $\sqrt{\dfrac{2}{5}}\sqrt{\dfrac{3}{4}}$ **8.** $\sqrt{\dfrac{4}{5}}\sqrt{\dfrac{15}{8}}$ **9.** $\sqrt{2}\sqrt{x}$

10. $\sqrt{3}\sqrt{t}$ **11.** $\sqrt{x}\sqrt{x-3}$ **12.** $\sqrt{5}\sqrt{2x-1}$

13. $\sqrt{x+2}\sqrt{x+1}$ **14.** $\sqrt{x+4}\sqrt{x-4}$ **15.** $\sqrt{x+y}\sqrt{x-y}$

16. $\sqrt{a-b}\sqrt{a-b}$ **17.** $\sqrt{43}\sqrt{2x}$ **18.** $\sqrt{35}\sqrt{4x}$

●● Simplify by factoring.

19. $\sqrt{12}$ **20.** $\sqrt{8}$ **21.** $\sqrt{75}$ **22.** $\sqrt{50}$

23. _____

24. _____

25. _____

26. _____

27. _____

28. _____

29. _____

30. _____

31. _____

32. _____

33. _____

34. _____

35. _____

36. _____

37. _____

38. _____

39. _____

40. _____

41. _____

42. _____

43. _____

23. $\sqrt{200x}$ **24.** $\sqrt{300x}$ **25.** $\sqrt{16a^2}$ **26.** $\sqrt{64y^2}$

27. $\sqrt{49t^2}$ **28.** $\sqrt{81p^2}$ **29.** $\sqrt{x^3 - 2x^2}$ **30.** $\sqrt{t^3 + t^2}$

31. $\sqrt{4x^2 - 4x + 1}$ **32.** $\sqrt{x^2 - 12x + 36}$ **33.** $\sqrt{9a^2 - 18ab + 9b^2}$ **34.** $\sqrt{100t^2 + 40t + 4}$

••• Approximate these square roots using Table 1. Round to three decimal places.

35. $\sqrt{125}$ **36.** $\sqrt{180}$ **37.** $\sqrt{360}$ **38.** $\sqrt{105}$

39. $\sqrt{300}$ **40.** $\sqrt{143}$ **41.** $\sqrt{122}$ **42.** $\sqrt{2000}$

---

43. *Speed of a skidding car.* How do police determine the speed of a car after an accident? The formula

$$r = 2\sqrt{5L}$$

can be used to approximate the speed r, in mph, of a car that has left a skid mark of length L, in feet. What was the speed of a car that left skid marks of lengths 20 ft? 70 ft? 90 ft?

9.4 MULTIPLYING AND SIMPLIFYING

■• SIMPLIFYING

In many situations expressions under radicals never represent negative numbers. In such cases, absolute value notation is not necessary.

Examples   Simplify by factoring. Assume that all expressions under radicals represent nonnegative numbers.

1. $\sqrt{48} = \sqrt{16 \cdot 3}$ Factoring the radicand
   $\phantom{\sqrt{48}} = \sqrt{16}\sqrt{3}$ Factoring into radicals
   $\phantom{\sqrt{48}} = 4\sqrt{3}$ Taking the principal square root

2. $\sqrt{20t^2} = \sqrt{4 \cdot t^2 \cdot 5}$ Factoring the radicand
   $\phantom{\sqrt{20t^2}} = \sqrt{4} \cdot \sqrt{t^2} \cdot \sqrt{5}$ Factoring into several radicals
   $\phantom{\sqrt{20t^2}} = 2t\sqrt{5}$ Taking square roots

Remember: A radical expression is simplified when the radicand does not contain factors that are perfect squares.

Absolute value notation is not necessary because expressions are not negative.

3. $\sqrt{3x^2 + 6x + 3} = \sqrt{3(x^2 + 2x + 1)}$ Factoring the radicand
   $\phantom{\sqrt{3x^2 + 6x + 3}} = \sqrt{3}\sqrt{x^2 + 2x + 1}$ Factoring into radicals
   $\phantom{\sqrt{3x^2 + 6x + 3}} = \sqrt{3}\sqrt{(x + 1)^2}$
   $\phantom{\sqrt{3x^2 + 6x + 3}} = \sqrt{3}(x + 1)$ Taking square roots

DO EXERCISES 1–3.

■■ SQUARE ROOTS OF POWERS

We can take square roots of powers.

Examples   Simplify by factoring. Assume that all expressions under radicals represent nonnegative numbers.

4. $\sqrt{x^6} = \sqrt{(x^3)^2} = x^3$

Absolute value notation again is not necessary because expressions are not negative.

To take a square root of a power such as x^6, the exponent must be even. We then take half the exponent.

5. $\sqrt{x^{12}} = x^6 \longleftarrow \frac{1}{2} \cdot 12 = 6$

OBJECTIVES

After finishing Section 9.4, you should be able to:

■• Simplify radical expressions by factoring, assuming all expressions under radicals represent nonnegative numbers.

■■ Simplify radical expressions where radicands are powers.

■■■ Multiply and simplify radical expressions.

Simplify by factoring. Assume that all expressions under radicals represent nonnegative numbers.

1. $\sqrt{32}$

2. $\sqrt{50h^2}$

3. $\sqrt{3x^2 - 6x + 3}$

Simplify by factoring.

4. $\sqrt{x^8}$

5. $\sqrt{(x + 2)^{14}}$

6. $\sqrt{x^{15}}$

Multiply and simplify.

7. $\sqrt{3}\sqrt{6}$

8. $\sqrt{2}\sqrt{50}$

Multiply and simplify.

9. $\sqrt{2x^3}\sqrt{8x^3y^4}$

10. $\sqrt{10xy^2}\sqrt{5x^2y^3}$

When odd powers occur, express the power in terms of the nearest even power. Then simplify the even power.

Example 6   Simplify by factoring. Assume that all expressions under radicals represent nonnegative numbers.

$$\begin{aligned}
\sqrt{x^9} &= \sqrt{x^8\,x} \\
&= \sqrt{x^8}\sqrt{x} \\
&= x^4\sqrt{x}
\end{aligned}$$

DO EXERCISES 4–6.

●●●   MULTIPLYING AND SIMPLIFYING

Sometimes we can simplify after multiplying.

Example 7   Multiply and then simplify by factoring.

$$\begin{aligned}
\sqrt{2}\sqrt{14} &= \sqrt{2 \cdot 14} \qquad \text{Multiplying} \\
&= \sqrt{2 \cdot 2 \cdot 7} \qquad \text{Factoring and looking for} \\
&\qquad\qquad\qquad\quad \text{perfect square factors} \\
&\qquad\qquad\qquad\quad \text{Perfect square} \\
&= \sqrt{2 \cdot 2} \cdot \sqrt{7} \\
&= 2\sqrt{7}
\end{aligned}$$

DO EXERCISES 7 AND 8.

Example 8   Multiply and then simplify by factoring. Assume that all expressions under radicals represent nonnegative numbers.

$$\begin{aligned}
\sqrt{3x^2}\sqrt{9x^3} &= \sqrt{3 \cdot 9x^5} \qquad \text{Multiplying} \\
&= \sqrt{3 \cdot 9 \cdot x^4 \cdot x} \qquad \text{Factoring and looking for} \\
&\qquad\qquad\qquad\qquad \text{perfect square factors} \\
&= \sqrt{9 \cdot x^4 \cdot 3 \cdot x} \\
&\qquad\qquad\qquad\qquad \text{Perfect squares} \\
&= \sqrt{9}\sqrt{x^4}\sqrt{3x} \\
&= 3x^2\sqrt{3x}
\end{aligned}$$

DO EXERCISES 9 AND 10.

EXERCISE SET 9.4

■ ● Simplify by factoring. Assume that all expressions under radicals represent non-negative numbers.

1. $\sqrt{24}$

2. $\sqrt{12}$

3. $\sqrt{40}$

4. $\sqrt{200}$

5. $\sqrt{175}$

6. $\sqrt{243}$

7. $\sqrt{48x}$

8. $\sqrt{40m}$

9. $\sqrt{28x^2}$

10. $\sqrt{20x^2}$

11. $\sqrt{8x^2 + 8x + 2}$

12. $\sqrt{36y + 12y^2 + y^3}$

■ ● ● Simplify by factoring. Assume that all expressions under radicals represent non-negative numbers.

13. $\sqrt{t^6}$

14. $\sqrt{x^4}$

15. $\sqrt{x^5}$

16. $\sqrt{t^{19}}$

17. $\sqrt{(y - 2)^8}$

18. $\sqrt{4(x + 5)^{10}}$

19. $\sqrt{36m^3}$

20. $\sqrt{8a^5}$

1. _____

2. _____

3. _____

4. _____

5. _____

6. _____

7. _____

8. _____

9. _____

10. _____

11. _____

12. _____

13. _____

14. _____

15. _____

16. _____

17. _____

18. _____

19. _____

20. _____

ANSWERS

21. _____

22. _____

23. _____

24. _____

25. _____

26. _____

27. _____

28. _____

29. _____

30. _____

31. _____

32. _____

33. _____

34. _____

35. _____

36. _____

21. $\sqrt{448x^6y^3}$

22. $\sqrt{243x^5y^4}$

▪▪▪ Multiply and then simplify by factoring. Assume that all expressions under radicals represent nonnegative numbers.

23. $\sqrt{3}\sqrt{18}$

24. $\sqrt{5}\sqrt{10}$

25. $\sqrt{18}\sqrt{14}$

26. $\sqrt{12}\sqrt{18}$

27. $\sqrt{10}\sqrt{10}$

28. $\sqrt{11}\sqrt{11}$

29. $\sqrt{5b}\sqrt{15b}$

30. $\sqrt{6a}\sqrt{18a}$

31. $\sqrt{ab}\sqrt{ac}$

32. $\sqrt{xy}\sqrt{xz}$

33. $\sqrt{18x^2y^3}\sqrt{6xy^4}$

34. $\sqrt{12x^3y^2}\sqrt{8xy}$

35. $\sqrt{50ab}\sqrt{10a^2b^4}$

36. $\sqrt{18xy}\sqrt{14x^3y^2}$

9.5   FRACTIONAL RADICANDS

⬤ PERFECT-SQUARE RADICANDS

Fractional radicands with a perfect-square numerator and denominator can be simplified by taking the square root of the numerator and denominator separately.

Examples   Simplify.

1. $\sqrt{\dfrac{25}{9}} = \dfrac{5}{3}$ because $\dfrac{5}{3} \cdot \dfrac{5}{3} = \dfrac{25}{9}$.

2. $\sqrt{\dfrac{1}{16}} = \dfrac{1}{4}$ because $\dfrac{1}{4} \cdot \dfrac{1}{4} = \dfrac{1}{16}$.

Sometimes a fractional expression can be simplified to one that has a perfect-square numerator and denominator.

Example 3   Simplify: $\sqrt{\dfrac{18}{50}}$.

$$\sqrt{\frac{18}{50}} = \sqrt{\frac{9 \cdot 2}{25 \cdot 2}} = \sqrt{\frac{9}{25} \cdot \boxed{\frac{2}{2}}}$$

$$= \sqrt{\frac{9}{25} \cdot \boxed{1}} = \sqrt{\frac{9}{25}} = \frac{3}{5}$$ because $\dfrac{3}{5} \cdot \dfrac{3}{5} = \dfrac{9}{25}$.

Example 4   Simplify: $\sqrt{\dfrac{2560}{2890}}$.

$$\sqrt{\frac{2560}{2890}} = \sqrt{\frac{256 \cdot 10}{289 \cdot 10}} = \sqrt{\frac{256}{289} \cdot \boxed{\frac{10}{10}}}$$

$$= \sqrt{\frac{256}{289} \cdot \boxed{1}} = \sqrt{\frac{256}{289}} = \frac{16}{17}$$ because $\dfrac{16}{17} \cdot \dfrac{16}{17} = \dfrac{256}{289}$.

DO EXERCISES 1–5.

⬤⬤ RATIONALIZING DENOMINATORS

When neither the numerator nor denominator is a perfect square, we can simplify to an expression that has a whole-number radicand. A simplified expression is then in the form $a\sqrt{b}$, where b is a whole number with no perfect-square factors.

Example 5   Simplify: $\sqrt{\dfrac{2}{3}}$.

We multiply by 1, choosing 3/3 for 1. This makes the denominator a perfect square.

$$\sqrt{\frac{2}{3}} = \sqrt{\frac{2}{3} \cdot \boxed{\frac{3}{3}}} \qquad \text{Multiplying by 1}$$

$$= \sqrt{\frac{6}{9}}$$

$$= \sqrt{\frac{1}{9} \cdot 6} \qquad \text{Factoring to get a perfect-square factor}$$

$$= \sqrt{\frac{1}{9}}\sqrt{6} = \frac{1}{3}\sqrt{6}$$

Simplify.

1. $\sqrt{\dfrac{16}{9}}$

2. $\sqrt{\dfrac{1}{25}}$

3. $\sqrt{\dfrac{1}{9}}$

4. $\sqrt{\dfrac{18}{32}}$

5. $\sqrt{\dfrac{2250}{2560}}$

Rationalize the denominator.

6. $\sqrt{\dfrac{3}{5}}$

7. $\sqrt{\dfrac{5}{8}}$

(*Hint:* Multiply the radicand by $\frac{2}{2}$.)

Approximate to three decimal places.

8. $\sqrt{\dfrac{2}{7}}$

9. $\sqrt{\dfrac{5}{8}}$

We can always multiply by 1 to make a denominator a perfect square. This is called *rationalizing the denominator*.

Example 6   Rationalize the denominator: $\sqrt{\dfrac{5}{12}}$.

$$\sqrt{\frac{5}{12}} = \sqrt{\frac{5}{12} \cdot \frac{3}{3}} = \sqrt{\frac{15}{36}} = \sqrt{\frac{1}{36} \cdot 15} = \sqrt{\frac{1}{36}} \sqrt{15} = \frac{1}{6} \sqrt{15}$$

In Example 6, we did not multiply by 12/12. This would have given us a correct answer, but not the simplest. We chose 3/3 to make the denominator the smallest multiple of 12 that is a perfect square.

DO EXERCISES 6 AND 7.

◖◗◗ APPROXIMATING SQUARE ROOTS OF FRACTIONS

Now we can use the square root table to find square roots of fractions.

Example 7   Approximate $\sqrt{\dfrac{3}{5}}$ to three decimal places.

a) $\sqrt{\dfrac{3}{5}} = \sqrt{\dfrac{3}{5} \cdot \dfrac{5}{5}} = \sqrt{\dfrac{15}{25}} = \sqrt{\dfrac{1}{25} \cdot 15} = \sqrt{\dfrac{1}{25}} \sqrt{15} = \dfrac{1}{5} \sqrt{15}$

b) From Table 1, $\sqrt{15} \approx 3.873$. We divide by 5.

$$
\begin{array}{r}
0.7746 \\
5\overline{)3.8730} \\
\underline{35} \\
37 \\
\underline{35} \\
23 \\
\underline{20} \\
30 \\
\end{array}
$$

Then $\sqrt{\dfrac{3}{5}} \approx 0.775$ to three decimal places.

Example 8   Approximate $\sqrt{\dfrac{7}{18}}$ to three decimal places.

a) $\sqrt{\dfrac{7}{18}} = \sqrt{\dfrac{7}{18} \cdot \dfrac{2}{2}} = \sqrt{\dfrac{14}{36}} = \sqrt{\dfrac{1}{36} \cdot 14} = \sqrt{\dfrac{1}{36}} \cdot \sqrt{14} = \dfrac{1}{6} \sqrt{14}$

b) From Table 1, $\sqrt{14} \approx 3.742$.

We divide by 6 and get

$$\sqrt{\frac{7}{18}} \approx 0.624. \quad \text{Rounded to three decimal places}$$

DO EXERCISES 8 AND 9.

NAME _____ CLASS _____

EXERCISE SET 9.5

◼● Simplify. Assume that all expressions under radicals represent nonnegative numbers.

1. $\sqrt{\dfrac{9}{49}}$

2. $\sqrt{\dfrac{16}{25}}$

3. $\sqrt{\dfrac{1}{36}}$

4. $\sqrt{\dfrac{1}{4}}$

5. $-\sqrt{\dfrac{16}{81}}$

6. $-\sqrt{\dfrac{25}{49}}$

7. $\sqrt{\dfrac{64}{289}}$

8. $\sqrt{\dfrac{81}{361}}$

9. $\sqrt{\dfrac{1690}{1960}}$

10. $\sqrt{\dfrac{1440}{6250}}$

11. $\sqrt{\dfrac{36}{a^2}}$

12. $\sqrt{\dfrac{25}{x^2}}$

13. $\sqrt{\dfrac{9a^2}{625}}$

14. $\sqrt{\dfrac{x^2y^2}{256}}$

◼●● Rationalize the denominator. Assume that all expressions under radicals represent positive numbers.

15. $\sqrt{\dfrac{2}{5}}$

16. $\sqrt{\dfrac{2}{7}}$

17. $\sqrt{\dfrac{3}{8}}$

18. $\sqrt{\dfrac{7}{8}}$

19. $\sqrt{\dfrac{1}{2}}$

20. $\sqrt{\dfrac{1}{3}}$

21. $\sqrt{\dfrac{3}{x}}$

22. $\sqrt{\dfrac{a}{b}}$

ANSWERS

1. _____

2. _____

3. _____

4. _____

5. _____

6. _____

7. _____

8. _____

9. _____

10. _____

11. _____

12. _____

13. _____

14. _____

15. _____

16. _____

17. _____

18. _____

10. _____

20. _____

21. _____

22. _____

ANSWERS

23. _____

24. _____

25. _____

26. _____

27. _____

28. _____

29. _____

30. _____

31. _____

●●● Approximate to three decimal places.

23. $\sqrt{\dfrac{3}{7}}$ **24.** $\sqrt{\dfrac{3}{2}}$ **25.** $\sqrt{\dfrac{1}{3}}$ **26.** $\sqrt{\dfrac{1}{5}}$

27. $\sqrt{\dfrac{7}{20}}$ **28.** $\sqrt{\dfrac{3}{20}}$ **29.** $\sqrt{\dfrac{12}{5}}$ **30.** $\sqrt{\dfrac{8}{3}}$

---

31. *Pendulums.* The *period* T of a pendulum is the time it takes to make a move from one side to the other and back. A formula for the period is

$$T = 2\pi\sqrt{\dfrac{L}{32}},$$

where T is in seconds and L is in feet. Find the periods of pendulums of lengths 2 ft, 8 ft, 64 ft, and 100 ft. Use 3.14 for π.

9.6 ADDITION AND SUBTRACTION

■ We can add any two real numbers. The sum of 5 and $\sqrt{2}$ can be expressed as

$$5 + \sqrt{2}.$$

We cannot simplify this unless we use rational approximations. When we have *like radicals*, however, a sum can be simplified using the distributive laws and collecting like terms. *Like radicals* have the same radicands.

Example 1   Add $3\sqrt{5}$ and $4\sqrt{5}$ and simplify by collecting like radical terms, if possible.

$$3\sqrt{5} + 4\sqrt{5} = (3 + 4)\sqrt{5} \qquad \text{Using the distributive law to factor out } \sqrt{5}$$

$$= 7\sqrt{5}$$

To simplify like this, the radical expressions must be the same. Sometimes we can make them the same.

Examples   Add or subtract. Simplify, if possible, by collecting like radical terms, if possible.

2. $\sqrt{2} - \sqrt{8} = \sqrt{2} - \sqrt{4 \cdot 2}$ Factoring 8

$$= \sqrt{2} - \sqrt{4}\sqrt{2}$$

$$= \sqrt{2} - 2\sqrt{2}$$

$$= 1\sqrt{2} - 2\sqrt{2}$$

$$= (1 - 2)\sqrt{2} \qquad \text{Using the distributive law to factor out the common factor } \sqrt{2}$$

$$= -1 \cdot \sqrt{2}$$

$$= -\sqrt{2}$$

3. $\sqrt{x^3 - x^2} + \sqrt{4x - 4}$

$$= \sqrt{x^2(x - 1)} + \sqrt{4(x - 1)} \qquad \text{Factoring radicands}$$

$$= \sqrt{x^2}\sqrt{x - 1} + \sqrt{4}\sqrt{x - 1}$$

$$= x\sqrt{x - 1} + 2\sqrt{x - 1} \qquad \text{Assume that all expressions under radicals are nonnegative.}$$

$$= (x + 2)\sqrt{x - 1} \qquad \text{Using the distributive law to factor out the common factor } \sqrt{x - 1}$$

Warning! Do not make the mistake of thinking that the sum of square roots is the square root of a sum. For example,

$$\sqrt{9} + \sqrt{16} = 3 + 4 = 7$$

but

$$\sqrt{9 + 16} = \sqrt{25} = 5.$$

In general,

$$\sqrt{a} + \sqrt{b} \neq \sqrt{a + b}.$$

After finishing Section 9.6, you should be able to:

■ Add or subtract with radical notation, using the distributive law to simplify.

Add or subtract and simplify by collecting like radical terms, if possible.

1. $3\sqrt{2} + 9\sqrt{2}$

2. $8\sqrt{5} - 3\sqrt{5}$

3. $2\sqrt{10} - 7\sqrt{40}$

4. $\sqrt{24} + \sqrt{54}$

5. $\sqrt{9x + 9} - \sqrt{4x + 4}$
(Assume that all expressions under radicals are nonnegative.)

Add or subtract.

6. $\sqrt{2} + \sqrt{\dfrac{1}{2}}$

Sometimes rationalizing denominators will enable us to factor and then combine expressions.

Example 4   Add: $\sqrt{3} + \sqrt{\dfrac{1}{3}}$.

$$\sqrt{3} + \sqrt{\dfrac{1}{3}} = \sqrt{3} + \sqrt{\dfrac{1}{3} \cdot \boxed{\dfrac{3}{3}}} \qquad \text{Multiplying by 1}$$

$$= \sqrt{3} + \sqrt{\dfrac{3}{9}}$$

$$= \sqrt{3} + \sqrt{\dfrac{1}{9} \cdot 3}$$

$$= \sqrt{3} + \sqrt{\dfrac{1}{9}}\sqrt{3}$$

$$= \boxed{\sqrt{3}} + \dfrac{1}{3}\boxed{\sqrt{3}}$$

$$= \left(1 + \dfrac{1}{3}\right)\boxed{\sqrt{3}} \qquad \text{Factoring and simplifying}$$

$$= \dfrac{4}{3}\sqrt{3}$$

DO EXERCISES 6 AND 7.

7. $\sqrt{\dfrac{5}{3}} - \sqrt{\dfrac{3}{5}}$

SOMETHING EXTRA

AN APPLICATION: WIND CHILL TEMPERATURE

In cold weather one feels colder when there is wind than when there is not. The *wind chill temperature* is what the temperature would have to be with no wind to give the same chilling effect as when there is wind. A formula for finding the wind chill temperature, T_w, is

$$T_w = 91.4 - \dfrac{(10.45 + 6.68\sqrt{v} - 0.447v)(457 - 5T)}{110},$$

where T is the actual temperature as given by a thermometer, in degrees Fahrenheit, and v is the wind speed in mph.

EXERCISES

You will need a calculator. You can get the square roots from Table 1. Find the wind chill temperature in each case. Round to the nearest one degree.

1. $T = 30°F$, $v = 25$ mph   **2.** $T = 20°F$, $v = 20$ mph

3. $T = 20°F$, $v = 40$ mph   **4.** $T = -10°F$, $v = 30$ mph

NAME   CLASS   ANSWERS

EXERCISE SET 9.6

▆   Add or subtract. Simplify by collecting like radical terms, if possible. Assume that all expressions under radicals represent nonnegative numbers.

1. $3\sqrt{2} + 4\sqrt{2}$

2. $7\sqrt{5} + 3\sqrt{5}$

3. $6\sqrt{a} - 14\sqrt{a}$

4. $10\sqrt{x} - 13\sqrt{x}$

5. $3\sqrt{12} + 2\sqrt{3}$

6. $5\sqrt{8} + 15\sqrt{2}$

7. $\sqrt{27} - 2\sqrt{3}$

8. $\sqrt{45} - \sqrt{20}$

9. $\sqrt{72} + \sqrt{98}$

10. $\sqrt{45} + \sqrt{80}$

11. $3\sqrt{18} - 2\sqrt{32} - 5\sqrt{50}$

12. $2\sqrt{12} + 4\sqrt{27} - 5\sqrt{48}$

1. _____

2. _____

3. _____

4. _____

5. _____

6. _____

7. _____

8. _____

9. _____

10. _____

11. _____

12. _____

13. $\sqrt{4x} + \sqrt{81x^3}$

14. $\sqrt{27} - \sqrt{12x^2}$

13. _____

14. _____

15. $\sqrt{8x + 8} + \sqrt{2x + 2}$

16. $\sqrt{x^5 - x^2} + \sqrt{9x^3 - 9}$

15. _____

16. _____

17. $3x\sqrt{y^3 x} - x\sqrt{yx^3} + y\sqrt{y^3 x}$

18. $4a\sqrt{a^2 b} + a\sqrt{a^2 b^3} - 5\sqrt{b^3}$

17. _____

18. _____

19. $\sqrt{3} - \sqrt{\dfrac{1}{3}}$

20. $\sqrt{2} - \sqrt{\dfrac{1}{2}}$

21. $5\sqrt{2} + 3\sqrt{\dfrac{1}{2}}$

19. _____

20. _____

21. _____

22. $\sqrt{\dfrac{2}{3}} - \sqrt{\dfrac{1}{6}}$

23. $\sqrt{\dfrac{1}{12}} - \sqrt{\dfrac{1}{27}}$

24. $\sqrt{\dfrac{2}{3}} + \sqrt{\dfrac{3}{2}}$

22. _____

23. _____

24. _____

25. Can you find any pairs of numbers a and b, for which
$\sqrt{a} + \sqrt{b} = \sqrt{a + b}$?
If so, name them.

25. _____

9.7 DIVISION

■•■ DIVISION

To multiply with radical notation, we multiply radicands. To divide, we divide the radicands.

> **For any nonnegative radicands A and B,**
> $$\frac{\sqrt{A}}{\sqrt{B}} = \sqrt{\frac{A}{B}}.$$
> **(The quotient of square roots, provided they exist, is the square root of the quotient of the radicands.)**

Examples Divide and simplify. Assume that all expressions under radicals represent positive numbers.

1. $\dfrac{\sqrt{27}}{\sqrt{3}} = \sqrt{\dfrac{27}{3}} = \sqrt{9} = 3$

2. $\dfrac{\sqrt{7}}{\sqrt{14}} = \sqrt{\dfrac{7}{14}} = \sqrt{\dfrac{1}{2}} = \sqrt{\dfrac{1}{2} \cdot \dfrac{2}{2}} = \sqrt{\dfrac{2}{4}} = \sqrt{\dfrac{1}{4} \cdot 2} = \sqrt{\dfrac{1}{4}} \cdot \sqrt{2} = \dfrac{1}{2}\sqrt{2}$

3. $\dfrac{\sqrt{30a^3}}{\sqrt{6a^2}} = \sqrt{\dfrac{30a^3}{6a^2}} = \sqrt{5a}$

DO EXERCISES 1–3.

■•■• RATIONALIZING DENOMINATORS

Expressions with radicals are considered simpler if there are no radicals in denominators. We can simplify by multiplying by 1, but this time we do it a bit differently.

Example 4 Rationalize the denominator.

$$\frac{\sqrt{2}}{\sqrt{3}} = \frac{\sqrt{2}}{\sqrt{3}} \cdot \boxed{\frac{\sqrt{3}}{\sqrt{3}}} = \frac{\sqrt{2} \cdot \sqrt{3}}{\sqrt{3} \cdot \sqrt{3}} = \frac{\sqrt{6}}{3}, \quad \text{or} \quad \frac{1}{3}\sqrt{6}$$

Example 5 Rationalize the denominator. Assume that all expressions under radicals represent positive numbers.

$$\frac{\sqrt{49a^5}}{\sqrt{12}} = \frac{\sqrt{49a^5}}{\sqrt{12}} \cdot \boxed{\frac{\sqrt{3}}{\sqrt{3}}} = \frac{\sqrt{49a^5}\sqrt{3}}{\sqrt{12}\sqrt{3}} = \frac{\sqrt{49a^5}\sqrt{3}}{\sqrt{36}}$$

$$= \frac{\sqrt{49a^4 \cdot 3a}}{\sqrt{36}} = \frac{7a^2\sqrt{3a}}{6}$$

DO EXERCISES 4 AND 5.

OBJECTIVES

After finishing Section 9.7, you should be able to:

■•■ Divide with radical notation.

■•■• Rationalize denominators.

Divide and simplify. Assume that all expressions under radicals represent positive numbers.

1. $\dfrac{\sqrt{5}}{\sqrt{45}}$

2. $\dfrac{\sqrt{2}}{\sqrt{6}}$

3. $\dfrac{\sqrt{42x^5}}{\sqrt{7x^2}}$

Rationalize the denominator. Assume that all expressions under radicals represent positive numbers.

4. $\dfrac{\sqrt{5}}{\sqrt{7}}$

5. $\dfrac{\sqrt{x}}{\sqrt{y}}$

ANSWERS ON PAGE A–31

SOMETHING EXTRA

**CALCULATOR CORNER: FINDING SQUARE ROOTS
ON A CALCULATOR**

Some calculators have a square root key. If yours does not, you can use Table 1 if the number occurs there, or you can use the following method.

> 1. **Make a guess.**
> 2. **Square the guess on a calculator.**
> 3. **Adjust the guess.**

Example   Approximate $\sqrt{27}$ to three decimal places using a calculator.

First, we make a guess. Let's guess 4. Using a calculator we find 4^2, which is 16. Since $16 < 27$, we know that the square root is at *least* 4.

Next we try 5. We find 5^2, which is 25. This is also less than 27, so the square root is at least 5.

Next we try 6: $6^2 = 36$, which is greater than 27, so we know that $\sqrt{27}$ is between 5 and 6. Thus, $\sqrt{27}$ is 5 plus some fraction: $\sqrt{27} = 5\ldots$.

Now we find the tenths digit by trying 5.1, 5.2, 5.3, and so on.

$(5.1)^2 = 26.01$ This is less than 27.

$(5.2)^2 = 27.04$ This is greater than 27.

The tenths digit is 1, so $\sqrt{27} = 5.1\ldots$.

Now we find the hundredths digit by trying 5.11, 5.12, 5.13, and so on.

$(5.11)^2 = 26.1121$

$(5.12)^2 = 26.2144$

$\quad\vdots\qquad\qquad\vdots$ All of these are less than 27, so the hundredths

$(5.18)^2 = 26.8324$ digit is 9.

The hundredths digit is 9, so $\sqrt{27} = 5.19\ldots$.

Now we find the thousandths digit by trying 5.191, 5.192, and so on.

$(5.196)^2 = 26.998416$ This is less than 27.

$(5.197)^2 = 27.008809$ This is greater than 27.

The thousandths digit is 6, so $\sqrt{27} \approx 5.196\ldots$.

Now we find the ten-thousandths digit by trying 5.1961, 5.1962, and so on.

$(5.1961)^2 = 26.99945521$ This is less than 27.

$(5.1962)^2 = 27.00049444$ This is greater than 27.

$\sqrt{27} \approx 5.1961\ldots$

How far do we continue? It depends on the desired accuracy. We decide to stop here and round back to three decimal places. Thus we get

$\sqrt{27} \approx 5.196$

as a final answer. Check this in Table 1.

Tables can be made this way, but there are other methods.

EXERCISES

Approximate these square roots to three decimal places using a calculator.

1. $\sqrt{17}$ 2. $\sqrt{80}$

3. $\sqrt{110}$ 4. $\sqrt{69}$

5. $\sqrt{200}$ 6. $\sqrt{10.5}$

7. $\sqrt{890}$ 8. $\sqrt{265.78}$

9. $\sqrt{2}$ 10. $\sqrt{0.2344}$

11. $\sqrt{\pi}$ 12. $\sqrt{2 + \sqrt{3}}$

NAME

CLASS

EXERCISE SET 9.7

● Divide and simplify. Assume that all expressions under radicals represent positive numbers.

1. $\dfrac{\sqrt{18}}{\sqrt{2}}$

2. $\dfrac{\sqrt{20}}{\sqrt{5}}$

3. $\dfrac{\sqrt{60}}{\sqrt{15}}$

4. $\dfrac{\sqrt{108}}{\sqrt{3}}$

5. $\dfrac{\sqrt{75}}{\sqrt{15}}$

6. $\dfrac{\sqrt{18}}{\sqrt{3}}$

7. $\dfrac{\sqrt{12}}{\sqrt{75}}$

8. $\dfrac{\sqrt{18}}{\sqrt{32}}$

9. $\dfrac{\sqrt{8x}}{\sqrt{2x}}$

10. $\dfrac{\sqrt{18b}}{\sqrt{2b}}$

11. $\dfrac{\sqrt{63y^3}}{\sqrt{7y}}$

12. $\dfrac{\sqrt{48x^3}}{\sqrt{3x}}$

13. $\dfrac{\sqrt{15x^5}}{\sqrt{3x}}$

14. $\dfrac{\sqrt{30a^5}}{\sqrt{5a}}$

15. $\dfrac{\sqrt{3x}}{\sqrt{\dfrac{3x}{4}}}$

16. $\dfrac{\sqrt{5x}}{\sqrt{\dfrac{5x}{9}}}$

ANSWERS

1. _____

2. _____

3. _____

4. _____

5. _____

6. _____

7. _____

8. _____

9. _____

10. _____

11. _____

12. _____

13. _____

14. _____

15. _____

16. _____

ANSWERS

17. _____

18. _____

19. _____

20. _____

21. _____

22. _____

23. _____

24. _____

25. _____

26. _____

27. _____

28. _____

29. _____

30. _____

31. _____

32. _____

● ●   Rationalize the denominator. Assume that all variables under radicals represent positive numbers.

17. $\dfrac{\sqrt{2}}{\sqrt{5}}$

18. $\dfrac{\sqrt{3}}{\sqrt{2}}$

19. $\dfrac{2}{\sqrt{2}}$

20. $\dfrac{3}{\sqrt{3}}$

21. $\dfrac{\sqrt{48}}{\sqrt{32}}$

22. $\dfrac{\sqrt{56}}{\sqrt{40}}$

23. $\dfrac{\sqrt{450}}{\sqrt{18}}$

24. $\dfrac{\sqrt{224}}{\sqrt{14}}$

25. $\dfrac{\sqrt{3}}{\sqrt{x}}$

26. $\dfrac{\sqrt{2}}{\sqrt{y}}$

27. $\dfrac{4y}{\sqrt{3}}$

28. $\dfrac{8x}{\sqrt{5}}$

29. $\dfrac{\sqrt{a^3}}{\sqrt{8}}$

30. $\dfrac{\sqrt{x^3}}{\sqrt{27}}$

31. $\dfrac{\sqrt{16a^4b^6}}{\sqrt{128a^6b^6}}$

32. $\dfrac{\sqrt{45mn^2}}{\sqrt{32m}}$

9.8 RADICALS AND RIGHT TRIANGLES

● RIGHT TRIANGLES

In a right triangle, the longest side is called the *hypotenuse*. The other two sides are called *legs*. We generally use the letters a and b for the lengths of the legs and c for the length of the hypotenuse. They are related as follows.

> *The Pythagorean Property of Right Triangles:* **In any right triangle if a and b are the lengths of the legs and c is the length of the hypotenuse, then**
>
> $$a^2 + b^2 - c^2.$$

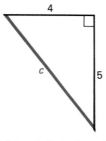

If we know the lengths of any two sides, we can find the length of the third side.

Example 1   Find the length of the hypotenuse of this right triangle.

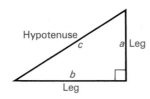

$$4^2 + 5^2 = c^2$$
$$16 + 25 = c^2$$
$$41 = c^2$$
$$c = \sqrt{41}$$
$$c \approx 6.403 \qquad \text{Table 1}$$

Example 2   Find the length of the leg of this right triangle.

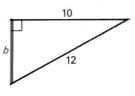

$$10^2 + b^2 = 12^2$$
$$100 + b^2 = 144$$
$$b^2 = 144 - 100$$
$$b^2 = 44$$
$$b = \sqrt{44}$$
$$b \approx 6.633 \qquad \text{Table 1}$$

DO EXERCISES 1 AND 2.

OBJECTIVES

After finishing Section 9.8, you should be able to:

● Given the lengths of any two sides of a right triangle, find the length of the third side.

●● Solve applied problems involving right triangles.

1. Find the length of the hypotenuse in this right triangle.

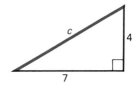

2. Find the length of the leg of this right triangle.

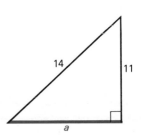

3. Find the length of the leg of this right triangle.

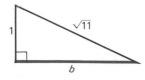

4. Find the length of the leg of this right triangle.

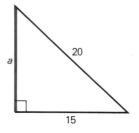

5. How long is a guy wire reaching from the top of a 15-ft pole to a point on the ground 10 ft from the pole?

Example 3   Find the length of the leg of this right triangle.

$$1^2 + b^2 = (\sqrt{7})^2$$
$$1 + b^2 = 7$$
$$b^2 = 7 - 1$$
$$b^2 = 6$$
$$b = \sqrt{6}$$
$$b \approx 2.449$$

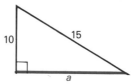

Example 4   Find the length of the leg of this right triangle.

$$a^2 + 10^2 = 15^2$$
$$a^2 + 100 = 225$$
$$a^2 = 225 - 100$$
$$a^2 = 125$$
$$a = \sqrt{125}$$
$$a = \sqrt{25 \cdot 5}$$
$$a = \sqrt{25} \cdot \sqrt{5}$$
$$a = 5\sqrt{5}$$
$$a \approx 5 \times 2.236$$
$$a \approx 11.18$$

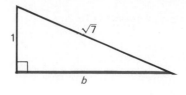

DO EXERCISES 3 AND 4.

● ●   **APPLIED PROBLEMS**

Example 5   A 12-foot ladder is leaning against a building. The bottom of the ladder is 7 ft from the building. How high is the top of the ladder?

a) We first make a drawing. In it we see a right triangle.

b) Now $7^2 + h^2 = 12^2$. We solve this equation.

$$49 + h^2 = 144$$
$$h^2 = 144 - 49$$
$$h^2 = 95$$
$$h = \sqrt{95}$$
$$h \approx 9.747 \text{ ft}$$

DO EXERCISE 5.

NAME   CLASS

EXERCISE SET 9.8

1. _____

▪● Find the length of the third side of each right triangle.

2. _____

1.
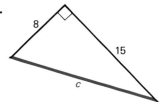

2.
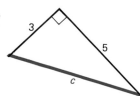

3. _____

4. _____

3.
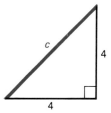

4.
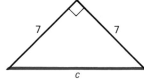

5. _____

6. _____

5.
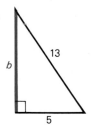

6.
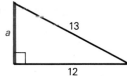

7. _____

8. _____

9. _____

7.
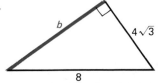

8.
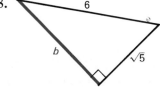

10. _____

11. _____

12. _____

13. _____

In a right triangle, find the length of the side not given.

9. $a = 10, b = 24$ **10.** $a = 5, b = 12$

14. _____

11. $a = 9, c = 15$ **12.** $a = 18, c = 30$

15. _____

13. $b = 1, c = \sqrt{5}$ **14.** $b = 1, c = \sqrt{2}$

16. _____

15. $a = 1, c = \sqrt{3}$ **16.** $a = \sqrt{3}, b = \sqrt{5}$

17. _____

17. $c = 10, b = 5\sqrt{3}$ **18.** $a = 5, b = 5$

18. _____

ANSWERS

■● ● Solve. Don't forget to make drawings.

19. A 10-meter ladder is leaning against a building. The bottom of the ladder is 5 m from the building. How high is the top of the ladder?

19. _____

20. Find the length of a diagonal of a square whose sides are 3 cm long.

21. How long is a guy wire reaching from the top of a 12-ft pole to a point 8 ft from the pole?

20. _____

22. How long must a wire be to reach from the top of a 13-m telephone pole to a point on the ground 9 m from the foot of the pole?

23. A slow-pitch softball diamond is actually a square 65 ft on a side. How far is it from home to second base?

21. _____

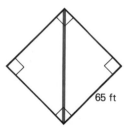

22. _____

24. A baseball diamond is actually a square 90 ft on a side. How far is it from first to third base?

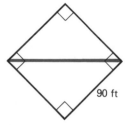

23. _____

24. _____

An *equilateral* triangle is shown at the right.

25. _____

25. Find an expression for its height h in terms of a.

26. _____

26. Find an expression for its area A in terms of a.

9.9   EQUATIONS WITH RADICALS

● SOLVING EQUATIONS WITH RADICALS

To solve equations with radicals, we first convert them to equations without radicals. We do this by squaring both sides of the equation. The following new principle is used.

> *The Principle of Squaring:* **If an equation $a = b$ is true, then the equation $a^2 = b^2$ is true.**

Example 1   Solve: $\sqrt{2x} - 4 = 7$.

$$\sqrt{2x} - 4 = 7$$
$$\sqrt{2x} = 11 \qquad \text{Adding 4, to get the radical alone on one side}$$
$$(\sqrt{2x})^2 = 11^2 \qquad \text{Squaring both sides}$$
$$2x = 121$$
$$x = \frac{121}{2}$$

Check:

$$\begin{array}{c|c} \sqrt{2x} - 4 = 7 & \\ \hline \sqrt{2 \cdot \dfrac{121}{2}} - 4 & 7 \\ \sqrt{121} - 4 & \\ 11 - 4 & \\ 7 & \end{array}$$

It is important to check! When we square both sides of an equation, the new equation may have solutions that the first one does not. For example, the equation

$$x = 1$$

has just one solution, the number 1. When we square both sides we get

$$x^2 = 1,$$

which has two solutions, 1 and $\neg 1$.

DO EXERCISE 1.

Example 2   Solve: $\sqrt{x + 1} = \sqrt{2x - 5}$.

$$(\sqrt{x + 1})^2 = (\sqrt{2x - 5})^2 \qquad \text{Squaring both sides}$$
$$x + 1 = 2x - 5$$
$$x = 6$$

Since 6 checks, it is the solution.

DO EXERCISES 2 AND 3.

OBJECTIVES

After finishing Section 9.9, you should be able to:

● Solve equations with radicals.
●● Solve applied problems involving radical equations.

Solve.

1. $\sqrt{3x} - 5 = 3$

Solve.

2. $\sqrt{3x + 1} = \sqrt{2x + 3}$

3. $\sqrt{x - 2} - 5 = 3$

(*Hint:* Get $\sqrt{x - 2}$ alone on one side.)

4. How far can you see to the horizon through an airplane window at a height of 8000 m?

⊙⊙ APPLICATIONS

How far can you see from a given height? There is a formula for this. At a height of h meters you can see V kilometers to the horizon. These numbers are related as follows:

$$V = 3.5\sqrt{h} \tag{1}$$

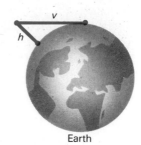

Earth

Example 3   How far to the horizon can you see through an airplane window at a height, or altitude, of 9000 meters?

We substitute 9000 for h in equation (1) and simplify.

$$V = 3.5\sqrt{9000}$$
$$V = 3.5\sqrt{900 \cdot 10}$$
$$V = 3.5 \times 30 \times \sqrt{10}$$
$$V \approx 3.5 \times 30 \times 3.162$$
$$V \approx 332.010 \text{ km}$$

5. How far can a sailor see to the horizon from the top of a 20-m mast?

DO EXERCISES 4 AND 5.

Example 4   A person can see 50.4 kilometers to the horizon from the top of a cliff. How high is the cliff?

We substitute 50.4 for V in equation (1) and solve.

$$50.4 = 3.5\sqrt{h}$$
$$\frac{50.4}{3.5} = \sqrt{h}$$
$$14.4 = \sqrt{h}$$
$$(14.4)^2 = (\sqrt{h})^2$$
$$(14.4)^2 = h$$
$$207.36 \text{ m} = h$$

The cliff is 207.36 meters high.

6. A sailor can see 91 km to the horizon from the top of a mast. How high is the mast?

DO EXERCISE 6.

NAME	CLASS	ANSWERS

EXERCISE SET 9.9

▮● Solve.

1. $\sqrt{x} = 5$ **2.** $\sqrt{x} = 7$ **3.** $\sqrt{x} = 6.2$

4. $\sqrt{x} = 4.3$ **5.** $\sqrt{x + 3} = 20$ **6.** $\sqrt{x + 4} = 11$

7. $\sqrt{2x + 4} = 25$ **8.** $\sqrt{2x + 1} = 13$ **9.** $3 + \sqrt{x - 1} = 5$

10. $4 + \sqrt{y - 3} = 11$ **11.** $6 - 2\sqrt{3n} = 0$ **12.** $8 - 4\sqrt{5n} = 0$

13. $\sqrt{5x - 7} = \sqrt{x + 10}$ **14.** $\sqrt{4x - 5} = \sqrt{x + 9}$ **15.** $\sqrt{x} = -7$

16. $\sqrt{x} = -5$ **17.** $\sqrt{2y + 6} = \sqrt{2y - 5}$ **18.** $2\sqrt{3x - 2} = \sqrt{2x - 3}$

ANSWERS

1. _____

2. _____

3. _____

4. _____

5. _____

6. _____

7. _____

8. _____

9. _____

10. _____

11. _____

12. _____

13. _____

14. _____

15. _____

16. _____

17. _____

18. _____

ANSWERS

■ ■ Solve.

Use $V = 3.5\sqrt{h}$ for Exercises 19–22.

19. How far can you see to the horizon through an airplane window at a height of 9800 m?

19. _____

20. How far can a sailor see to the horizon from the top of a 24-m mast?

20. _____

21. A person can see 371 km to the horizon from an airplane window. How high is the airplane?

21. _____

22. A sailor can see 99.4 km to the horizon from the top of a mast. How high is the mast?

22. _____

The formula $r = 2\sqrt{5L}$ can be used to approximate the speed r, in mph, of a car that has left a skid mark of length L, in feet.

23. How far will a car skid at 50 mph? at 70 mph?

23. _____

24. How far will a car skid at 60 mph? at 100 mph?

24. _____

25. _____

The formula $T = 2\pi\sqrt{L/32}$ can be used to find the period T, in seconds, of a pendulum of length L, in feet.

25. ▦ What is the length of a pendulum that has a period of 1.6 sec? Use 3.14 for π.

26. ▦ What is the length of a pendulum that has a period of 3 sec? Use 3.14 for π.

26. _____

EXTENSION EXERCISES

SECTION 9.1

Simplify.

1. $\sqrt{(39)^2}$ **2.** $(\sqrt{39})^2$ **3.** $\sqrt{3^2 + 4^2}$ **4.** $\sqrt{5^2 - 4^2}$ **5.** $\sqrt{\sqrt{16}}$

SECTION 9.2

6. Find the number one third of the way from $2\frac{3}{4}$ to $4\frac{5}{8}$.   **7.** Find the number one fifth of the way from -1 to $-\frac{4}{3}$.

8. What number is halfway between x and y?   **9.** What number is one fourth of the way between z and w if $z < w$?

SECTION 9.3

Simplify by factoring. Assume variables represent nonnegative numbers.

10. $\sqrt{0.01}$ **11.** $\sqrt{0.25}$ **12.** $\sqrt{x^4}$ **13.** $\sqrt{9a^6}$ **14.** $\sqrt{z^8 w^{10}}$ **15.** $\sqrt{x^{-2}}$

16. Multiply $(x^2 + \sqrt{2}xy + y^2)$ by $(x^2 - \sqrt{2}xy + y^2)$.   **17.** Factor $x^8 + y^8$.

SECTION 9.4

Multiply and simplify by factoring. Assume that all expressions under radicals represent nonnegative numbers.

18. $\sqrt{27(x + 1)}\sqrt{12y(x + 1)^2}$ **19.** $\sqrt{x}\sqrt{2x}\sqrt{10x^5}$

20. $\sqrt{x^{8n}}$ **21.** $\sqrt{0.04x^{4n}}$

SECTION 9.5

Simplify. Assume that all expressions under radicals represent positive numbers.

22. $\sqrt{\dfrac{1}{x^2} - \dfrac{2}{xy} + \dfrac{1}{y^2}}$ **23.** $\sqrt{2 - \dfrac{4}{z^2} + \dfrac{2}{z^4}}$

Rationalize the denominator. Assume that all expressions under radicals represent positive numbers.

24. $\sqrt{\dfrac{1}{3x^5}}$ **25.** $\sqrt{\dfrac{3}{8x}}$ **26.** $\dfrac{3}{\sqrt{6z}}$ **27.** $\sqrt{\dfrac{3y}{2 + y^2}}$

SECTION 9.6

Multiply. Simplify by collecting like radical terms, if possible. Assume that all expressions under radicals represent nonnegative numbers.

28. $(\sqrt{2} + \sqrt{6})(\sqrt{2} - \sqrt{6})$ **29.** $(3\sqrt{5} - 2)(\sqrt{5} + 1)$ **30.** $(\sqrt{5} - 2\sqrt{2})(\sqrt{10} - 1)$

31. $(\sqrt{x} - \sqrt{y})(\sqrt{x} + 5\sqrt{y})$ **32.** $(b^2 - b\sqrt{z} + z)(b + \sqrt{z})$

SECTION 9.7

Perform the indicated operations and simplify. Assume that all expressions under radicals represent positive numbers.

33. $\dfrac{\sqrt{\dfrac{3x}{2}}}{\sqrt{\dfrac{x^3}{5}}}$ **34.** $\sqrt{1 + x^2} + \dfrac{1}{\sqrt{1 + x^2}}$

Observe that $(\sqrt{a} + \sqrt{b})(\sqrt{a} - \sqrt{b}) = (\sqrt{a})^2 - (\sqrt{b})^2 = a - b$. Use this to rationalize denominators in the following exercises.

35. $\dfrac{2}{\sqrt{3} - \sqrt{5}}$

36. $\dfrac{1 - \sqrt{7}}{3 + \sqrt{7}}$

37. $\dfrac{\sqrt{3} - \sqrt{2}}{2\sqrt{3} + \sqrt{2}}$

38. $\dfrac{2 - \sqrt{x + y}}{3 + \sqrt{x + y}}$

SECTION 9.8

39. The length and width of a rectangle are given by consecutive integers. The area of the rectangle is 90 cm². Find the length of the diagonal of the rectangle.

40. Two cars leave a service station at the same time. One car travels east at a speed of 50 mph, and the other travels south at a speed of 60 mph. After one-half hour, how far apart are they?

41. Find the length of the diagonal of a cube with side s.

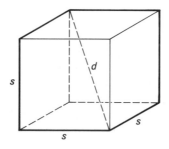

42. Find x.

a)
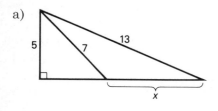

b)
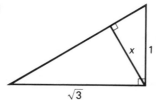

c)
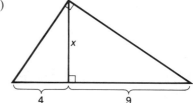

SECTION 9.9

Solve.

43. $x - 1 = \sqrt{x + 5}$

44. $A = \sqrt{1 + \dfrac{a^2}{b^2}}$, for b.

45. $\sqrt{3x + 1} = 1 + \sqrt{x + 4}$

46. $4 + \sqrt{10 - x} = 6 + \sqrt{4 - x}$

(*Hint:* Use the principle of squaring twice.)

NAME _____   SCORE _____

TEST OR REVIEW—CHAPTER 9

If you miss an item, review the indicated section and objective.

[9.1, ■●■]   Find the following.

    1. $\sqrt{36}$ **2.** $-\sqrt{81}$

[9.1, ●●■]   Identify the radicand.

    3. $3x\sqrt{\dfrac{x}{2+x}}$

[9.1, ⚁]   Find the following. Assume that letters can represent any real number.

    4. $\sqrt{m^2}$ **5.** $\sqrt{49t^2}$ **6.** $\sqrt{(x-4)^2}$

[9.2, ■●■]   Identify the rational numbers and the irrational numbers.

    7. $\sqrt{3}$ **8.** $\sqrt{36}$ **9.** $-\sqrt{12}$ **10.** $-\sqrt{4}$

[9.3, ■●■]   Multiply. Assume that radicands are nonnegative.

    11. $\sqrt{3}\sqrt{7}$ **12.** $\sqrt{x-3}\sqrt{x+3}$ **13.** $\sqrt{2x}\sqrt{3y}$

[9.3, ●●■]   Simplify by factoring.

    14. $\sqrt{48}$ **15.** $\sqrt{x^2+16x+64}$

[9.3, ●●●]   Approximate to three decimal places. Use Table 1.

    16. $\sqrt{108}$

[9.4, ●●●]   Multiply and simplify. Assume that all expressions under radicals represent nonnegative numbers.

    17. $\sqrt{5}\sqrt{8}$ **18.** $\sqrt{5x}\sqrt{10xy^2}$

1. _____

2. _____

3. _____

4. _____

5. _____

6. _____

7. _____

8. _____

9. _____

10. _____

11. _____

12. _____

13. _____

14. _____

15. _____

16. _____

17. _____

18. _____

19. _____

20. _____

21. _____

22. _____

23. _____

24. _____

25. _____

26. _____

27. _____

28. _____

29. _____

30. _____

31. _____

32. _____

33. _____

34. _____

35. _____

For the remainder of this test, assume that all expressions under radicals represent positive numbers.

[9.5, ●]   Simplify.

19. $\sqrt{\dfrac{25}{64}}$ **20.** $\sqrt{\dfrac{20}{45}}$ **21.** $\sqrt{\dfrac{49}{t^2}}$

[9.5, ●●]   Rationalize the denominator.

22. $\sqrt{\dfrac{1}{2}}$ **23.** $\sqrt{\dfrac{1}{8}}$ **24.** $\sqrt{\dfrac{5}{y}}$

[9.5, ●●●]   Approximate to three decimal places. Use Table 1.

25. $\sqrt{\dfrac{1}{8}}$ **26.** $\sqrt{\dfrac{11}{20}}$

[9.6, ●]   Add or subtract. Simplify by collecting like radical terms, if possible.

27. $5\sqrt{12} + 3\sqrt{12}$ **28.** $\sqrt{80} - \sqrt{45}$ **29.** $3\sqrt{2} - 5\sqrt{\dfrac{1}{2}}$

[9.7, ●]   Divide and simplify.

30. $\dfrac{\sqrt{27}}{\sqrt{45}}$ **31.** $\dfrac{\sqrt{45x^2y}}{\sqrt{54y}}$

[9.7, ●●]   **32.** Rationalize the denominator: $\dfrac{2}{\sqrt{3}}$.

[9.8, ●]   **33.** Find a.

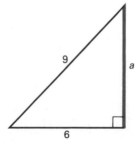

[9.8, ●●]   **34.** Find the length of a diagonal of a square whose sides are 7 m long.

[9.9, ●]   **35.** Solve: $\sqrt{2x - 3} = 7$.

QUADRATIC
EQUATIONS

10

Quadratic equations can be used to solve a problem like this: What must the diameter of a pizza be so that it has the same area as two 10-inch pizzas?

READINESS CHECK: SKILLS FOR CHAPTER 10

1. Multiply: $(3x + 1)^2$.

2. What must be added to $x^2 - 8x$ to make it the square of a binomial?

3. Factor by completing the square:

$x^2 - 8x + 7$.

4. Find $\sqrt{9}$.

Simplify.

5. $\sqrt{20}$

6. $\sqrt{\dfrac{2890}{2560}}$

7. Approximate. Round to the nearest hundredth: $\sqrt{17}$.

8. Rationalize the denominator:

$\sqrt{\dfrac{7}{3}}$.

Solve.

9. $x^2 - 6x = 0$

10. $x^2 - 5x + 6 = 0$

11. Graph: $5x + 3y = 15$.

10.1 INTRODUCTION TO QUADRATIC EQUATIONS

■ ○ **WRITING QUADRATIC EQUATIONS IN STANDARD FORM**

The following are quadratic equations. Each is of second degree.

$$4x^2 + 7x - 5 = 0, \qquad 3x^2 - \frac{1}{2}x = 9, \qquad 5y^2 = -6y$$

> **An equation of the type** $ax^2 + bx + c = 0$, **where** a, b, **and** c **are real number constants and** $a > 0$, **is called the** *standard form of a quadratic equation*.

Examples   Write standard form and determine a, b, and c.

1. $4x^2 + 7x - 5 = 0$ The equation is already in standard form.

$a = 4$; $b = 7$; $c = -5$

2. $3x^2 - 0.5x = 9$
$3x^2 - 0.5x - 9 = 0$ Adding -9. This is standard form.

$a = 3$; $b = -0.5$; $c = -9$

3. $-4y^2 = 5y$
$-4y^2 - 5y = 0$ Adding $-5y$

└── Not positive!

$4y^2 + 5y = 0$ Multiplying by -1. This is standard form.

$a = 4$; $b = 5$; $c = 0$

DO EXERCISES 1–3.

OBJECTIVES

After finishing Section 10.1, you should be able to:

■ ○ Write a quadratic equation in standard form $ax^2 + bx + c = 0$, $a > 0$, **and determine the coefficients** a, b, **and** c.

■ ■ Solve quadratic equations of the type $ax^2 = k$, $a \neq 0$.

Write standard form and determine a, b, and c.

1. $x^2 = 7x$

2. $3 - x^2 = 9x$

3. $3x + 5x^2 = x^2 - 4 + x$

●● SOLVING EQUATIONS OF THE TYPE $ax^2 = k$

When b is 0, we solve for x^2 and take the principal square root.

Example 4   Solve: $5x^2 = 15$.

$$x^2 = 3 \qquad \text{Solving for } x^2; \text{ multiplying by } \tfrac{1}{5}$$

$$|x| = \sqrt{3}$$

Since $|x|$ is either x or $-x$,

$$x = \sqrt{3} \quad \text{or} \quad -x = \sqrt{3}$$
$$x = \sqrt{3} \quad \text{or} \quad x = -\sqrt{3}.$$

Check.   For $\sqrt{3}$:   For $-\sqrt{3}$:

$5x^2 = 15$		$5x^2 = 15$	
$5(\sqrt{3})^2$	15	$5(-\sqrt{3})^2$	15
$5 \cdot 3$		$5 \cdot 3$	
15		15	

The solutions are $\sqrt{3}$ and $-\sqrt{3}$.

DO EXERCISE 4.

Example 5   Solve: $\dfrac{1}{3}x^2 = 0$.

$$x^2 = 0 \qquad \text{Multiplying by 3}$$
$$|x| = \sqrt{0} \qquad \text{Taking the principal square root}$$
$$|x| = 0$$

The only number with absolute value 0 is 0. It checks, so it is the solution.

DO EXERCISE 5.

Example 6   Solve: $-3x^2 + 7 = 0$.

$$-3x^2 = -7 \qquad \text{Adding } -7$$

$$x^2 = \frac{-7}{-3} \qquad \text{Multiplying by } -\frac{1}{3}$$

$$x^2 = \frac{7}{3}$$

$$|x| = \sqrt{\frac{7}{3}} \qquad \text{Taking the square root}$$

$$x = \sqrt{\frac{7}{3}} \quad \text{or} \quad x = -\sqrt{\frac{7}{3}}$$

$$x = \sqrt{\frac{7}{3} \cdot \frac{3}{3}} \quad \text{or} \quad x = -\sqrt{\frac{7}{3} \cdot \frac{3}{3}} \qquad \text{Rationalizing the denominators}$$

$$x = \frac{\sqrt{21}}{3} \quad \text{or} \quad x = -\frac{\sqrt{21}}{3}$$

Solve.

4. $4x^2 = 20$

Solve.

5. $2x^2 = 0$

Solve.

6. $2x^2 - 3 = 0$

Check: For $\dfrac{\sqrt{21}}{3}$:   For $-\dfrac{\sqrt{21}}{3}$:

$$
\begin{array}{c|c}
-3x^2 + 7 = 0 & \\
\hline
-3\left(\dfrac{\sqrt{21}}{3}\right)^2 + 7 & 0 \\
-3 \cdot \dfrac{21}{9} + 7 & \\
-7 + 7 & \\
0 &
\end{array}
\qquad
\begin{array}{c|c}
-3x^2 + 7 = 0 & \\
\hline
-3\left(-\dfrac{\sqrt{21}}{3}\right)^2 + 7 & 0 \\
-3 \cdot \dfrac{21}{9} + 7 & \\
-7 + 7 & \\
0 &
\end{array}
$$

The solutions are $\dfrac{\sqrt{21}}{3}$ and $-\dfrac{\sqrt{21}}{3}$.

DO EXERCISES 6 AND 7.

SOMETHING EXTRA

CALCULATOR CORNER

The length d of a diagonal of a rectangular solid is given by

$$d = \sqrt{a^2 + b^2 + c^2},$$

where a, b, and c are the lengths of the sides.

7. $4x^2 - 9 = 0$

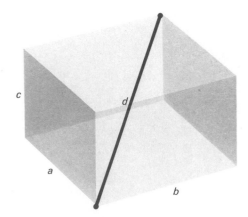

EXERCISES

1. Find the length of a diagonal of a rectangular solid whose sides have lengths $a = 2$, $b = 4$, and $c = 5$.

2. Will a baton of length 57 cm fit inside a briefcase that is 12.5 cm by 45 cm by 31.5 cm?

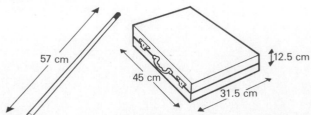

NAME   CLASS

EXERCISE SET 10.1

▪● Write standard form and determine a, b, and c.

1. $x^2 = 3x + 2$ **2.** $2x^2 = 3$

3. $7x^2 = 4x - 3$ **4.** $5 = -2x^2 + 3x$

5. $2x - 1 = 3x^2 + 7$ **6.** $4x^2 - 3x + 2 = 3x^2 + 7x - 5$

●● Solve.

7. $x^2 = 4$ **8.** $x^2 = 1$ **9.** $x^2 = 49$ **10.** $x^2 = 16$

11. $x^2 = 7$ **12.** $x^2 = 11$ **13.** $3x^2 = 30$ **14.** $5x^2 = 35$

1. _____

2. _____

3. _____

4. _____

5. _____

6. _____

7. _____

8. _____

9. _____

10. _____

11. _____

12. _____

13. _____

14. _____

ANSWERS

15. $3x^2 = 24$ **16.** $5x^2 = 60$ **17.** $4x^2 - 25 = 0$ **18.** $9x^2 - 4 = 0$

15. _____

16. _____

17. _____

18. _____

19. $3x^2 - 49 = 0$ **20.** $5y^2 - 16 = 0$ **21.** $4y^2 - 3 = 9$ **22.** $5x^2 - 100 = 0$

19. _____

20. _____

21. _____

22. _____

23. *Falling object*. The distance S, in feet, traveled by a body falling freely from rest in t seconds is given by

$$S = 16t^2.$$

This formula is an approximation since it does not account for air resistance. How far will an object fall in 1 sec, 2 sec, 5 sec, and 20 sec?

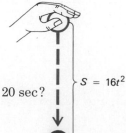

$S = 16t^2$

23. _____

24. _____

24. ▦ Solve: $4.82x^2 = 12{,}000.$

10.2   SOLVING BY FACTORING

●   EQUATIONS OF THE TYPE $ax^2 + bx = 0$

When c is 0 (and $b \neq 0$), we can factor and use the principle of zero products.

Example 1   Solve: $7x^2 + 2x = 0$.

$$x(7x + 2) = 0 \qquad \text{Factoring}$$

$$x = 0 \quad \text{or} \quad 7x + 2 = 0 \qquad \text{Principle of zero products}$$

$$x = 0 \quad \text{or} \qquad 7x = -2$$

$$x = 0 \quad \text{or} \qquad x = -\tfrac{2}{7}$$

Check: For 0:   For $-\dfrac{2}{7}$:

$$\begin{array}{c|c} 7x^2 + 2x = 0 \\ \hline 7 \cdot 0^2 + 2 \cdot 0 & 0 \\ 0 & \end{array}$$

$$\begin{array}{c|c} 7x^2 + 2x = 0 \\ \hline 7\left(-\dfrac{2}{7}\right)^2 + 2\left(-\dfrac{2}{7}\right) & 0 \\ 7\left(\dfrac{4}{49}\right) - \dfrac{4}{7} & \\ \dfrac{4}{7} - \dfrac{4}{7} & \\ 0 & \end{array}$$

The solutions are 0 and $-\tfrac{2}{7}$.

When we use the principle of zero products, we need not check except to detect errors in solving.

Example 2   Solve: $20x^2 - 15x = 0$.

$$5x(4x - 3) = 0$$

$$5x = 0 \quad \text{or} \quad 4x - 3 = 0$$

$$x = 0 \quad \text{or} \qquad 4x = 3$$

$$x = 0 \quad \text{or} \qquad x = \tfrac{3}{4}$$

The solutions are 0 and $\tfrac{3}{4}$.

A quadratic equation of this type will always have 0 as one solution and a nonzero number as the other solution.

DO EXERCISES 1 AND 2.

●●   EQUATIONS OF THE TYPE $ax^2 + bx + c = 0$

When neither b nor c is 0, we can sometimes solve by factoring.

Example 3   Solve: $5x^2 - 8x + 3 = 0$.

$$(5x - 3)(x - 1) = 0 \qquad \text{Factoring}$$

$$5x - 3 = 0 \quad \text{or} \quad x - 1 = 0$$

$$5x = 3 \quad \text{or} \qquad x = 1$$

$$x = \tfrac{3}{5} \quad \text{or} \qquad x = 1$$

The solutions are $\tfrac{3}{5}$ and 1.

OBJECTIVES

After finishing Section 10.2, you should be able to:

● Solve equations of the type $ax^2 + bx = 0$, $a \neq 0$, $b \neq 0$, by factoring.

●● Solve quadratic equations of the type $ax^2 + bx + c = 0$, $a \neq 0$, $b \neq 0$, $c \neq 0$, by factoring.

●●● Solve applied problems involving quadratic equations.

Solve.

1. $3x^2 + 5x = 0$

2. $10x^2 - 6x = 0$

Solve.

3. $3x^2 + x - 2 = 0$

4. $(x - 1)(x + 1) = 5(x - 1)$

5. Use $d = \dfrac{n^2 - 3n}{2}$.

a) A heptagon has 7 sides. How many diagonals does it have?

b) A polygon has 44 diagonals. How many sides does it have?

Example 4   Solve: $(y - 3)(y - 2) = 6(y - 3)$.

We write standard form and then try to factor.

$$y^2 - 5y + 6 = 6y - 18$$
$$y^2 - 11y + 24 = 0 \qquad \text{Standard form}$$
$$(y - 8)(y - 3) = 0$$
$$y - 8 = 0 \quad \text{or} \quad y - 3 = 0$$
$$y = 8 \quad \text{or} \qquad y = 3$$

The solutions are 8 and 3.

DO EXERCISES 3 AND 4.

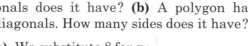 **SOLVING PROBLEMS**

Example 5   The number of diagonals, d, of a polygon of n sides is given by

$$d = \frac{n^2 - 3n}{2}.$$

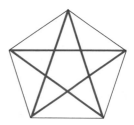

(a) An octagon has 8 sides. How many diagonals does it have? **(b)** A polygon has 27 diagonals. How many sides does it have?

a) We substitute 8 for n:

$$d = \frac{n^2 - 3n}{2}$$

$$d = \frac{8^2 - 3 \cdot 8}{2} \qquad \text{Substituting 8 for } n$$

$$d = \frac{64 - 24}{2} = 20.$$

An octagon has 20 diagonals.

b) We substitute 27 for d and solve for n:

$$\frac{n^2 - 3n}{2} = d$$

$$\frac{n^2 - 3n}{2} = 27 \qquad \text{Substituting 27 for } d$$

$$n^2 - 3n = 54 \qquad \text{Multiplying by 2 to clear of fractions}$$
$$n^2 - 3n - 54 = 0$$
$$(n - 9)(n + 6) = 0$$
$$n - 9 = 0 \quad \text{or} \quad n + 6 = 0$$
$$n = 9 \quad \text{or} \qquad n = -6.$$

Since the number of sides cannot be negative, -6 cannot be a solution. But 9 checks, so the polygon has 9 sides (it is a nonagon).

DO EXERCISE 5.

NAME

CLASS

EXERCISE SET 10.2

■ ● Solve.

1. $x^2 + 7x = 0$

2. $x^2 - 5x = 0$

3. $3x^2 + 2x = 0$

4. $5x^2 + 2x = 0$

5. $4x^2 + 4x = 0$

6. $10x^2 - 30x = 0$

7. $55x^2 - 11x = 0$

8. $14x^2 - 3x = 0$

● ● Solve.

9. $x^2 - 16x + 48 = 0$

10. $x^2 + 7x + 6 = 0$

11. $x^2 + 4x - 21 = 0$

12. $x^2 \quad 9x \mid 14 = 0$

13. $x^2 + 10x + 25 = 0$

14. $x^2 - 2x + 1 = 0$

15. $2x^2 - 13x + 15 = 0$

16. $3a^2 \quad 10a \quad 8 - 0$

1. _____

2. _____

3. _____

4. _____

5. _____

6. _____

7. _____

8. _____

9. _____

10. _____

11. _____

12. _____

13. _____

14. _____

15. _____

16. _____

17. _____

18. _____

19. _____

20. _____

21. _____

22. _____

23. _____

24. _____

25. _____

26. _____

27. _____

28. _____

17. $3x^2 - 7x = 20$

18. $2x^2 + 12x = -10$

19. $t(t - 5) = 14$

20. $3y^2 + 8y = 12y + 15$

21. $t(9 + t) = 4(2t + 5)$

22. $(2x - 3)(x + 1) = 4(2x - 3)$

●●● Solve.

Use $d = \dfrac{n^2 - 3n}{2}$ for Exercises 23–26.

23. A hexagon has 6 sides. How many diagonals does it have?

24. A decagon has 10 sides. How many diagonals does it have?

25. A polygon has 14 diagonals. How many sides does it have?

26. A polygon has 9 diagonals. How many sides does it have?

27. Solve for x.

$ax^2 + bx = 0$

28. ▦ Solve.

$0.0025x^2 + 70{,}400x = 0$

10.3   SQUARES OF BINOMIALS

●   SOLVING EQUATIONS OF THE TYPE $(x + k)^2 = d$

In equations of the type $(x + k)^2 = d$, we have the square of a binomial equal to a constant.

Example 1   Solve: $(x - 5)^2 = 9$.

$$|x - 5| = \sqrt{9} \qquad \text{Taking principal square root}$$
$$|x - 5| = 3$$
$$x - 5 = 3 \quad \text{or} \quad x - 5 = -3$$
$$x = 8 \quad \text{or} \qquad x = 2$$

The solutions are 8 and 2.

Example 2   Solve: $(x + 2)^2 = 7$.

$$|x + 2| = \sqrt{7}$$
$$x + 2 = \sqrt{7} \qquad \text{or} \quad x + 2 = -\sqrt{7}$$
$$x = -2 + \sqrt{7} \quad \text{or} \qquad x = -2 - \sqrt{7}$$

The solutions are $-2 + \sqrt{7}$ and $-2 - \sqrt{7}$, or simply $-2 \pm \sqrt{7}$ (read "-2 plus or minus $\sqrt{7}$").

DO EXERCISES 1–3.

● ●   COMPLETING THE SQUARE

We can make $x^2 + 10x$ the square of a binomial by adding the proper number to it.* This is called *completing the square*.

$$x^2 + 10x$$
$$\downarrow$$
$$\frac{10}{2} = 5 \qquad \text{Taking half the } x\text{-coefficient}$$
$$\downarrow$$
$$5^2 = 25 \qquad \text{Squaring}$$
$$\downarrow$$
$$x^2 + 10x + 25 \qquad \text{Adding}$$

The trinomial $x^2 + 10x + 25$ is the square of $x + 5$.

Examples   Complete the square.

3. $x^2 - 12x$

$$\left(\frac{-12}{2}\right)^2 = (-6)^2 - 36 \qquad \begin{array}{l}\text{Taking half the } x\text{-coefficient and} \\ \text{squaring}\end{array}$$
$$\downarrow$$
$$x^2 - 12x + 36$$

The trinomial $x^2 - 12x + 36$ is the square of $x - 6$.

* See Section 5.6.

OBJECTIVES

After finishing Section 10.3, you should be able to:

●   Solve equations of the type $(x + k)^2 = d$.

● ●   Complete the square for a binomial $x^2 + bx$.

● ● ●   Use quadratic equations to solve certain interest problems.

Solve.

1. $(x - 3)^2 = 16$

2. $(x + 3)^2 = 10$

3. $(x - 1)^2 = 5$

Complete the square.

4. $x^2 - 8x$

5. $x^2 + 10x$

6. $x^2 + 7x$

7. $x^2 - 3x$

Solve.

8. $1000 invested at 14% compounded annually for 2 years will grow to what amount?

4. $x^2 \underbrace{- 5x}$

$$\left(\frac{-5}{2}\right)^2 = \frac{25}{4}$$

$$x^2 - 5x + \frac{25}{4}$$

The trinomial $x^2 - 5x + \frac{25}{4}$ is the square of $x - \frac{5}{2}$.

DO EXERCISES 4–7.

●●●   APPLICATIONS: INTEREST PROBLEMS

If you put money in a savings account, the bank will pay you interest. At the end of a year, the bank will start paying you interest on both the original amount and the interest. This is called *compounding interest annually.*

> **An amount of money P is invested at interest rate r. In t years it will grow to the amount A given by**
> $$A = P(1 + r)^t.$$

Example 5 $1000 invested at 16% for 2 years compounded annually will grow to what amount?

$$A = P(1 + r)^t$$
$$A = 1000(1 + 0.16)^2 \qquad \text{Substituting into the formula}$$
$$A = 1000(1.16)^2$$
$$A = 1000(1.3456)$$
$$A = 1345.60 \qquad \text{Computing}$$

The amount is $1345.60.

DO EXERCISE 8.

Example 6 $2560 is invested at interest rate r compounded annually. In 2 years it grows to $2890. What is the interest rate?

We substitute 2560 for P, 2890 for A, and 2 for t in the formula, and solve for r.

$$A = P(1 + r)^t$$
$$2890 = 2560(1 + r)^2$$
$$\frac{2890}{2560} = (1 + r)^2$$
$$\frac{289}{256} = (1 + r)^2$$
$$\sqrt{\frac{289}{256}} = |1 + r| \qquad \text{Taking the square root}$$
$$\frac{17}{16} = |1 + r|$$

Chapter 10

10.3 Squares of Binomials **467**

We then have

$$\frac{17}{16} = 1 + r \quad \text{or} \quad -\frac{17}{16} = 1 + r$$

$$-\frac{16}{16} + \frac{17}{16} = r \quad \text{or} \quad -\frac{16}{16} - \frac{17}{16} = r$$

$$\frac{1}{16} = r \quad \text{or} \quad -\frac{33}{16} = r$$

Since the interest rate cannot be negative,

$$0.0625 = r \qquad \left(\frac{1}{16} = 0.0625\right)$$
or $\quad 6.25\% = r.$

The interest rate must be 6.25% for $2560 to grow to $2890 in 2 years.

DO EXERCISE 9.

Example 7 For $2000 to double itself in 2 years, what would the interest rate have to be?

We substitute 2000 for P, 4000 for A, and 2 for t in the formula.

$$A = P(1 + r)^t$$

$$4000 = 2000(1 + r)^2$$

$$\frac{4000}{2000} = (1 + r)^2$$

$$2 = (1 + r)^2$$

$$\sqrt{2} = |1 + r| \qquad \text{Taking the square root}$$

$$\sqrt{2} = 1 + r \quad \text{or} \quad -\sqrt{2} = 1 + r$$

$$-1 + \sqrt{2} = r \quad \text{or} \quad -1 - \sqrt{2} = r$$

Since the interest rate cannot be negative,

$$-1 + (1.414) \approx r \qquad \text{Using Table 1}$$

$$0.414 \approx r$$

$$41.4\% \approx r$$

The interest rate would have to be 41.4% for the $2000 to double itself. Such a rate would be hard to get.

DO EXERCISE 10.

9. Suppose $2560 is invested at interest rate r compounded annually, and grows to $3240 in 2 years. What is the interest rate?

10. Suppose $1000 is invested at interest rate r compounded annually, and grows to $3000 (it triples) in 2 years. What is the interest rate?

SOMETHING EXTRA

CALCULATOR CORNER: COMPOUND INTEREST

We have considered the formula

$$A = P(1 + r)^t$$

for interest compounded annually. If interest is compounded more than once a year, say quarterly, we can find a formula like the one above as follows:

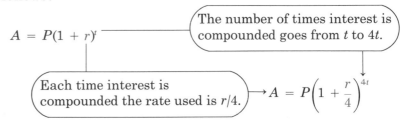

$$A = P(1 + r)^t$$

The number of times interest is compounded goes from t to $4t$.

Each time interest is compounded the rate used is $r/4$. $\rightarrow A = P\left(1 + \dfrac{r}{4}\right)^{4t}$

In general:

> **If principal P is invested at interest rate r, compounded n times a year, in t years it will grow to the amount A given by**
>
> $$A = P\left(1 + \frac{r}{n}\right)^{nt}.$$

Example   Suppose $1000 is invested at 16%, compounded quarterly. How much is in the account at the end of 2 years?

Substituting 1000 for P, 0.16 for r, 4 for n, and 2 for t, we get

$$A = 1000\left(1 + \frac{0.16}{4}\right)^{4 \cdot 2} = 1000(1.04)^8$$

$$\approx 1000(1.368569) \approx \$1368.57.$$

EXERCISES

Use your calculator. A power key $\boxed{a^b}$ will be needed for many of the exercises. Use 365 days for one year.

1. Suppose $1000 is invested at 16%, compounded semiannually $(n = 2)$. How much is in the account at the end of 2 years?

2. Suppose $1000 is invested at 12%, compounded quarterly. How much is in the account at the end of 3 years?

3. Suppose $1000 is invested at 16%. How much is in the account at the end of 1 year, if interest is compounded (a) annually? (b) semi-annually? (c) quarterly? (d) daily? (e) hourly?

4. Suppose $1 is invested at the interest rate of 100%, even though it would be hard to get such a rate. How much is in the account at the end of 1 year, if interest is compounded (a) annually? (b) semi-annually? (c) quarterly? (d) daily? (e) hourly?

NAME   CLASS   ANSWERS

EXERCISE SET 10.3

■ •   Solve.

1. $(x + 2)^2 = 25$ **2.** $(x - 2)^2 = 49$ **3.** $(x + 1)^2 = 6$

4. $(x + 3)^2 = 21$ **5.** $(x - 3)^2 = 6$ **6.** $(x + 13)^2 = 8$

● ●   Complete the square.

7. $x^2 - 2x$ **8.** $x^2 - 4x$ **9.** $x^2 + 18x$ **10.** $x^2 + 22x$

11. $x^2 - x$ **12.** $x^2 + x$ **13.** $x^2 + 5x$ **14.** $x^2 - 9x$

● ● ●   Solve. Use $A = P(1 + r)^t$ for Exercises 15–25. What is the interest rate?

15. $1000 grows to $1210 in 2 years.   **16.** $1000 grows to $1440 in 2 years.

1. _____

2. _____

3. _____

4. _____

5. _____

6. _____

7. _____

8. _____

9. _____

10. _____

11. _____

12. _____

13. _____

14. _____

15. _____

16. _____

17. $2560 grows to $3610 in 2 years.   **18.** $4000 grows to $4410 in 2 years.

17. _____

18. _____

19. $6250 grows to $7290 in 2 years.   **20.** $6250 grows to $6760 in 2 years.

19. _____

20. _____

21. $2500 grows to $3600 in 2 years.   **22.** $1600 grows to $2500 in 2 years.

21. _____

22. _____

23. _____

23. ▦ $1000 is invested at interest rate r. In 2 years it grows to $1267.88. What is the interest rate?

24. _____

24. ▦ $4000 is invested at interest rate r. In 2 years it grows to $5267.03. What is the interest rate?

25. ▦ In two years you want to have $3000. How much do you need to invest now if you can get an interest rate of 15.75 % compounded annually?

25. _____

10.4   COMPLETING THE SQUARE

● ▮ If we can write a quadratic equation in the form $(x + k)^2 = d$, we can solve it by taking the principal square root.

Example 1   Solve: $x^2 + 6x + 8 = 0$.

$$x^2 + 6x \qquad = -8 \qquad \text{Adding } -8$$

We take half of 6 and square it, to get 9. Then we add 9 on *both* sides of the equation. This makes the left side the square of a binomial. That is, it *completes the square*.

$$x^2 + 6x + \boxed{9} = -8 + \boxed{9}$$

$$(x + 3)^2 = 1$$

$$|x + 3| = \sqrt{1} \qquad \text{Taking the square root}$$

$$|x + 3| = 1$$

$$x + 3 = 1 \quad \text{or} \quad x + 3 = -1$$

$$x = -2 \quad \text{or} \qquad x = -4$$

The solutions are -2 and -4.

This method of solving is called *completing the square*.

Example 2   Solve $x^2 - 4x - 7 = 0$ by completing the square.

$$x^2 - 4x \qquad = 7 \qquad \text{Adding } 7$$

$$x^2 - 4x + \boxed{4} = 7 + \boxed{4} \qquad \text{Adding } 4: \left(\frac{-4}{2}\right)^2 = (-2)^2 = 4$$

$$(x - 2)^2 = 11$$

$$|x - 2| = \sqrt{11}$$

$$x - 2 = \sqrt{11} \qquad \text{or} \quad x - 2 = -\sqrt{11}$$

$$x = 2 + \sqrt{11} \quad \text{or} \qquad x = 2 - \sqrt{11}$$

The solutions are $2 \pm \sqrt{11}$.

DO EXERCISES 1–3.

Example 3   Solve $x^2 + 3x - 10 = 0$ by completing the square.

$$x^2 + 3x - 10 = 0$$

$$x^2 + 3x \qquad = 10$$

$$x^2 + 3x + \boxed{\frac{9}{4}} = 10 + \boxed{\frac{9}{4}} \qquad \text{Adding } \frac{9}{4}: \left(\frac{3}{2}\right)^2 = \frac{9}{4}$$

$$\left(x + \frac{3}{2}\right)^2 = \frac{40}{4} + \frac{9}{4}$$

$$\left(x + \frac{3}{2}\right)^2 = \frac{49}{4}$$

$$\left|x + \frac{3}{2}\right| = \sqrt{\frac{49}{4}} = \frac{7}{2}$$

OBJECTIVE

After finishing Section 10.4, you should be able to:

● ▮ Solve quadratic equations by completing the square.

Solve by completing the square.

1. $x^2 + 8x + 12 = 0$

2. $x^2 - 10x + 22 = 0$

3. $x^2 + 6x - 1 = 0$

Solve by completing the square.

4. $x^2 - 3x - 10 = 0$

We then have

$$x + \frac{3}{2} = \frac{7}{2} \quad \text{or} \quad x + \frac{3}{2} = -\frac{7}{2}$$

$$x = \frac{4}{2} \quad \text{or} \qquad x = -\frac{10}{2}$$

$$x = 2 \quad \text{or} \qquad x = -5.$$

The solutions are 2 and -5.

DO EXERCISES 4 AND 5.

When the coefficient of x^2 is not 1, we can make it 1.

5. $x^2 + 5x - 14 = 0$

Example 4   Solve $2x^2 - 3x - 1 = 0$ by completing the square.

$$x^2 - \frac{3}{2}x - \frac{1}{2} \quad = 0 \qquad \text{Multiplying by } \frac{1}{2} \text{ to make the } x^2\text{-coefficient 1}$$

$$x^2 - \frac{3}{2}x \qquad\quad = \frac{1}{2}$$

$$x^2 - \frac{3}{2}x + \boxed{\frac{9}{16}} = \frac{1}{2} + \boxed{\frac{9}{16}} \qquad \text{Adding } \frac{9}{16}: \left[\frac{1}{2}\left(-\frac{3}{2}\right)\right]^2 = \left[-\frac{3}{4}\right]^2 = \frac{9}{16}$$

Solve by completing the square.

6. $2x^2 + 3x - 3 = 0$

$$\left(x - \frac{3}{4}\right)^2 = \frac{8}{16} + \frac{9}{16}$$

$$\left(x - \frac{3}{4}\right)^2 = \frac{17}{16}$$

$$\left|x - \frac{3}{4}\right| = \sqrt{\frac{17}{16}}$$

$$\left|x - \frac{3}{4}\right| = \frac{\sqrt{17}}{4}$$

$$x - \frac{3}{4} = \frac{\sqrt{17}}{4} \qquad \text{or} \quad x - \frac{3}{4} = -\frac{\sqrt{17}}{4}$$

$$x = \frac{3}{4} + \frac{\sqrt{17}}{4} \quad \text{or} \qquad x = \frac{3}{4} - \frac{\sqrt{17}}{4}$$

7. $3x^2 - 2x - 3 = 0$

The solutions are $\dfrac{3 \pm \sqrt{17}}{4}$.

DO EXERCISES 6 AND 7.

EXERCISE SET 10.4

■●■ Solve by completing the square. Show your work.

1. $x^2 - 6x - 16 = 0$

2. $x^2 + 8x + 15 = 0$

3. $x^2 + 22x + 21 = 0$

4. $x^2 + 14x - 15 = 0$

5. $x^2 - 2x - 5 = 0$

6. $x^2 - 4x - 11 = 0$

7. $x^2 - 18x + 74 = 0$

8. $x^2 - 22x + 102 = 0$

9. $x^2 + 7x - 18 = 0$

1. _____

2. _____

3. _____

4. _____

5. _____

6. _____

7. _____

8. _____

9. _____

ANSWERS

10. $x^2 + 5x - 6 = 0$

11. $x^2 + x - 6 = 0$

12. $x^2 + 10x - 4 = 0$

10. _____

11. _____

13. $x^2 - 7x - 2 = 0$

14. $x^2 + 3x - 28 = 0$

15. $x^2 + \dfrac{3}{2}x - \dfrac{1}{2} = 0$

12. _____

13. _____

14. _____

16. $2x^2 + 3x - 17 = 0$

17. $3x^2 + 4x - 1 = 0$

18. $2x^2 - 9x - 5 = 0$

15. _____

16. _____

17. _____

19. $4x^2 + 12x - 7 = 0$

20. $9x^2 - 6x - 9 = 0$

18. _____

19. _____

20. _____

10.5   THE QUADRATIC FORMULA

◼•◼   SOLVING BY USING THE QUADRATIC FORMULA

Each time you solve by completing the square, you do about the same thing. In situations like this in mathematics, when we do about the same kind of computation many times, we look for a formula so we can speed up our work. Consider any quadratic equation in standard form:

$$ax^2 + bx + c = 0, \qquad a > 0.$$

Let's solve by completing the square.

$$x^2 + \frac{b}{a}x + \frac{c}{a} = 0 \qquad \text{Multiplying by } \frac{1}{a}$$

$$x^2 + \frac{b}{a}x \qquad = -\frac{c}{a} \qquad \text{Adding } -\frac{c}{a}$$

Half of $\dfrac{b}{a}$ is $\dfrac{b}{2a}$. The square is $\dfrac{b^2}{4a^2}$. We add $\dfrac{b^2}{4a^2}$ on both sides.

$$x^2 + \frac{b}{a}x + \frac{b^2}{4a^2} = -\frac{c}{a} + \frac{b^2}{4a^2}$$

$$\left(x + \frac{b}{2a}\right)^2 = -\frac{4ac}{4a^2} + \frac{b^2}{4a^2}$$

$$\left(x + \frac{b}{2a}\right)^2 = \frac{b^2 - 4ac}{4a^2}$$

$$\left|x + \frac{b}{2a}\right| = \sqrt{\frac{b^2 - 4ac}{4a^2}} \qquad \text{Taking the square root}$$

$$x + \frac{b}{2a} = \sqrt{\frac{b^2 - 4ac}{4a^2}} \quad \text{or} \quad x + \frac{b}{2a} = -\sqrt{\frac{b^2 - 4ac}{4a^2}}$$

Since $a > 0$, $|a| = a$. Then

$$x + \frac{b}{2a} = \frac{\sqrt{b^2 - 4ac}}{2a} \quad \text{or} \quad x + \frac{b}{2a} = -\frac{\sqrt{b^2 - 4ac}}{2a}.$$

Thus,

$$x + \frac{b}{2a} = \pm \frac{\sqrt{b^2 - 4ac}}{2a},$$

so

$$x = -\frac{b}{2a} + \frac{\sqrt{b^2 - 4ac}}{2a} \quad \text{or} \quad x = -\frac{b}{2a} - \frac{\sqrt{b^2 - 4ac}}{2a}.$$

The solutions are given by the following.

The Quadratic Formula: $x = \dfrac{-b \pm \sqrt{b^2 - 4ac}}{2a}.$

OBJECTIVES

After finishing Section 10.5, you should be able to:

◼•◼ Solve quadratic equations using the quadratic formula.

◼•◼ Find approximate solutions using a square root table.

Solve using the quadratic formula.

1. $2x^2 = 4 - 7x$

Example 1   Solve $5x^2 - 8x = -3$ using the quadratic formula.

First find standard form and determine a, b, and c.

$$5x^2 - 8x + 3 = 0$$
$$a = 5, b = -8, c = 3$$

Then use the quadratic formula:

$$x = \frac{-b \pm \sqrt{b^2 - 4ac}}{2a}$$

$$x = \frac{-(-8) \pm \sqrt{(-8)^2 - 4 \cdot 5 \cdot 3}}{2 \cdot 5}$$

$$x = \frac{8 \pm \sqrt{64 - 60}}{10}$$

$$x = \frac{8 \pm \sqrt{4}}{10}$$

$$x = \frac{8 \pm 2}{10}$$

$$x = \frac{8 + 2}{10} \quad \text{or} \quad x = \frac{8 - 2}{10}$$

$$x = \frac{10}{10} \quad \text{or} \quad x = \frac{6}{10}$$

$$x = 1 \quad \text{or} \quad x = \frac{3}{5}.$$

The solutions are 1 and $\frac{3}{5}$.

It turns out that the equation in Example 1 could have been solved by factoring. Actually, factoring would have been easier.

> **To solve a quadratic equation:**
>
> 1. **Try factoring.**
> 2. **If it is not possible to factor or if factoring seems difficult, use the quadratic formula.**
>
> **The solutions of a quadratic equation can always be found using the quadratic formula. They cannot** always **be found by factoring.**

DO EXERCISE 1.

> **When $b^2 - 4ac \geqslant 0$, the equation has solutions. When $b^2 - 4ac < 0$, the equation has no real-number solutions. The expression $b^2 - 4ac$ is called the** *discriminant.*

When using the quadratic formula, it is wise to compute the discriminant first. If it is negative, there are no solutions.

Example 2   Solve $3x^2 = 7 - 2x$ using the quadratic formula.

Find standard form and determine a, b, and c.

$$3x^2 + 2x - 7 = 0$$
$$a = 3, b = 2, c = -7$$

We compute the discriminant :

$$b^2 - 4ac = 2^2 - 4 \cdot 3 \cdot (-7)$$
$$= 4 + 84$$
$$= 88.$$

This is positive, so there are solutions. They are given by

$$x = \frac{-2 \pm \sqrt{88}}{6} \qquad \text{Substituting into the quadratic formula}$$

$$x = \frac{-2 \pm \sqrt{4 \cdot 22}}{6}$$

$$x = \frac{-2 \pm 2\sqrt{22}}{6}$$

$$x = \frac{2(-1 \pm \sqrt{22})}{2 \cdot 3} \qquad \text{Factoring out 2 in the numerator and denominator}$$

$$x = \frac{-1 \pm \sqrt{22}}{3}.$$

The solutions are $\dfrac{-1 + \sqrt{22}}{3}$ and $\dfrac{-1 - \sqrt{22}}{3}$.

DO EXERCISE 2.

● ●   APPROXIMATING SOLUTIONS

A square root table can be used to approximate solutions.

Example 3   Approximate the solutions to the equation in Example 2.

From Table 1, we see that

$$\sqrt{22} \approx 4.690$$

$$\frac{-1 + \sqrt{22}}{3} \approx \frac{-1 + 4.690}{3} \qquad\qquad \frac{-1 - \sqrt{22}}{3} \approx \frac{-1 - 4.690}{3}$$

$$\approx \frac{3.69}{3} \qquad\qquad\qquad\qquad \approx \frac{-5.69}{3}$$

$$\approx 1.2 \text{ to the} \qquad\qquad\qquad \approx -1.9 \text{ to the}$$
$$\qquad \text{nearest tenth;} \qquad\qquad\qquad\qquad \text{nearest tenth.}$$

DO EXERCISE 3.

Solve using the quadratic formula.

2. $5x^2 - 8x = 3$

3. Approximate the solutions to the equation in Exercise 2 above. Round to the nearest tenth.

ANSWERS ON PAGE A–33

SOMETHING EXTRA

AN APPLICATION: HANDLING DIMENSION SYMBOLS (PART III)

Changes of unit can be achieved by substitutions. The table of measures on the inside front cover may be helpful.

Example 1   Change to inches: 25 yd.

$$25 \text{ yd} = 25 \cdot 1 \text{ yd}$$
$$= 25 \cdot 3 \text{ ft} \qquad \text{Substituting 3 ft for 1 yd}$$
$$= 25 \cdot 3 \cdot 1 \text{ ft}$$
$$= 25 \cdot 3 \cdot 12 \text{ in.} \qquad \text{Substituting 12 in. for 1 ft}$$
$$= 900 \text{ in.}$$

Example 2   Change to meters: 4 km.

$$4 \text{ km} = 4 \cdot 1 \text{ km}$$
$$= 4 \cdot 1000 \text{ m} \qquad \text{Substituting 1000 m for 1 km}$$
$$= 4000 \text{ m}$$

The notion of "multiplying by 1" can also be used to change units.

Example 3   Change to yd: 7.2 in.

$$7.2 \text{ in.} = 7.2 \text{ in.} \cdot \boxed{\frac{1 \text{ ft}}{12 \text{ in.}}} \cdot \boxed{\frac{1 \text{ yd}}{3 \text{ ft}}} \qquad \text{Both of these are equal to 1.}$$

$$= \frac{7.2}{12 \cdot 3} \cdot \boxed{\frac{\text{in.}}{\text{in.}}} \cdot \boxed{\frac{\text{ft}}{\text{ft}}} \cdot \text{yd} = 0.2 \text{ yd}$$

Example 4   Change to mm: 55 cm.

$$55 \text{ cm} = 55 \text{ cm} \cdot \boxed{\frac{1 \text{ m}}{100 \text{ cm}}} \cdot \boxed{\frac{1000 \text{ mm}}{1 \text{ m}}} = \frac{55 \cdot 1000}{100} \cdot \boxed{\frac{\text{cm}}{\text{cm}}} \cdot \boxed{\frac{\text{m}}{\text{m}}} \cdot \text{mm}$$

$$= 550 \text{ mm}$$

Example 5   Change to $\frac{\text{m}}{\text{sec}}$: $95 \frac{\text{km}}{\text{hr}}$.

$$95 \frac{\text{km}}{\text{hr}} = 95 \frac{\text{km}}{\text{hr}} \cdot \boxed{\frac{1000 \text{ m}}{1 \text{ km}}} \cdot \boxed{\frac{1 \text{ hr}}{60 \text{ min}}} \cdot \boxed{\frac{1 \text{ min}}{60 \text{ sec}}}$$

$$= \frac{95 \cdot 1000}{60 \cdot 60} \cdot \frac{\text{km}}{\text{km}} \cdot \frac{\text{hr}}{\text{hr}} \cdot \frac{\text{min}}{\text{min}} \cdot \frac{\text{m}}{\text{sec}} = 26.4 \frac{\text{m}}{\text{sec}}$$

Below is a shortcut for the procedure in Example 5. It can also be used in Examples 3 and 4.

$$95 \frac{\text{km}}{\text{hr}} = \overset{19}{\cancel{95}} \frac{\text{km}}{\text{hr}} \cdot \frac{1000 \text{ m}}{1 \text{ km}} \cdot \frac{1 \text{ hr}}{60 \text{ min}} \cdot \frac{1 \text{ min}}{60 \text{ sec}} = 26.4 \frac{\text{m}}{\text{sec}}$$

EXERCISES

Perform the following changes of unit, using substitution or multiplying by 1.

1. 72 in., change to ft

2. 17 hr, change to min

3. 2 days, change to sec

4. 360 sec, change to hr

5. $60 \frac{\text{kg}}{\text{m}}$, change to $\frac{\text{g}}{\text{cm}}$

6. $44 \frac{\text{ft}}{\text{sec}}$, change to $\frac{\text{mi}}{\text{hr}}$

7. 216 m^2, change to cm^2

8. $60 \frac{\text{lb}}{\text{ft}^3}$, change to $\frac{\text{ton}}{\text{yd}^3}$

9. $\frac{\$36}{\text{day}}$, change to $\frac{\cancel{c}}{\text{hr}}$

10. 1440 person-hr, change to person-days

11. $186,000 \frac{\text{mi}}{\text{sec}}$ (speed of light), change to $\frac{\text{mi}}{\text{yr}}$. Let 365 days = 1 yr.

12. $1100 \frac{\text{ft}}{\text{sec}}$ (speed of sound), change to $\frac{\text{mi}}{\text{yr}}$. Let 365 days = 1 yr.

NAME   CLASS   ANSWERS

EXERCISE SET 10.5

■● Solve. Try factoring first. If factoring is not possible or is difficult, use the quadratic formula.

1. $x^2 - 4x = 21$ **2.** $x^2 + 7x = 18$ **3.** $x^2 = 6x - 9$

4. $x^2 = 8x - 16$ **5.** $3y^2 - 2y - 8 = 0$ **6.** $4y^2 + 12y = 7$

7. $x^2 - 9 = 0$ **8.** $x^2 - 4 = 0$ **9.** $y^2 - 10y + 26 = 4$

10. $x^2 + 4x + 4 = 7$ **11.** $x^2 - 2x = 2$ **12.** $x^2 + 6x = 1$

1. _____

2. _____

3. _____

4. _____

5. _____

6. _____

7. _____

8. _____

9. _____

10. _____

11. _____

12. _____

13. $4y^2 + 3y + 2 = 0$

14. $2t^2 + 6t + 5 = 0$

15. $3p^2 + 2p = 3$

13. _____

14. _____

15. _____

16. $3z^2 - 2z = 2$

17. $(y + 4)(y + 3) = 15$

18. $x^2 + (x + 2)^2 = 7$

16. _____

17. _____

■ ■ Use Table 1 to approximate the solutions to the nearest tenth.

18. _____

19. $x^2 - 4x - 7 = 0$

20. $x^2 = 5$

21. $y^2 - 6y - 1 = 0$

19. _____

20. _____

21. _____

22. $4x^2 + 4x = 1$

23. $3x^2 + 4x - 2 = 0$

24. $2y^2 + 2y - 3 = 0$

22. _____

23. _____

24. _____

10.6   FRACTIONAL AND RADICAL EQUATIONS

■● FRACTIONAL EQUATIONS

We can solve some fractional equations by first deriving a quadratic equation.

Example 1   Solve: $\dfrac{3}{x-1} + \dfrac{5}{x+1} = 2$.

The LCM is $(x-1)(x+1)$. We multiply by this.

$$(x-1)(x+1) \cdot \left(\frac{3}{x-1} + \frac{5}{x+1} \right) = 2 \cdot (x-1)(x+1)$$

We use the distributive law on the left.

$$(x-1)(x+1) \cdot \frac{3}{x-1} + (x-1)(x+1) \cdot \frac{5}{x+1} = 2(x-1)(x+1)$$

$$3(x+1) + 5(x-1) = 2(x-1)(x+1)$$
$$3x + 3 + 5x - 5 = 2(x^2 - 1)$$
$$8x - 2 = 2x^2 - 2$$
$$-2x^2 + 8x = 0$$
$$2x^2 - 8x = 0 \quad \text{Multiplying by } -1$$
$$2x(x-4) = 0 \quad \text{Factoring}$$
$$2x = 0 \quad \text{or} \quad x - 4 = 0$$
$$x = 0 \quad \text{or} \qquad x = 4$$

Check: For 0:

$$\begin{array}{c} \dfrac{3}{x-1} + \dfrac{5}{x+1} = 2 \\[2mm] \hline \dfrac{3}{0-1} + \dfrac{5}{0+1} \ \bigg|\ 2 \\[3mm] \dfrac{3}{-1} + \dfrac{5}{1} \\[3mm] -3 + 5 \\[2mm] 2 \end{array}$$

For 4:

$$\begin{array}{c} \dfrac{3}{x-1} + \dfrac{5}{x+1} = 2 \\[2mm] \hline \dfrac{3}{4-1} + \dfrac{5}{4+1} \ \bigg|\ 2 \\[3mm] \dfrac{3}{3} + \dfrac{5}{5} \\[3mm] 1 + 1 \\[2mm] 2 \end{array}$$

Both numbers check. The solutions are 0 and 4.

DO EXERCISE 1.

OBJECTIVES

After finishing Section 10.6, you should be able to:

■● Solve certain fractional equations by first deriving a quadratic equation.

■●● Solve certain radical equations by first using the principle of squaring to derive a quadratic equation.

Solve.

1. $\dfrac{20}{x+5} - \dfrac{1}{x-4} = 1$

Solve.

2. $\sqrt{x + 2} = 4 - x$

Solve.

3. $\sqrt{30 - 3x} + 4 = x$

■● ●■   **RADICAL EQUATIONS**

We can solve some radical equations by first using the principle of squaring to find a quadratic equation. When we do this we must be sure to check.

Example 2   Solve: $x - 5 = \sqrt{x + 7}$.

$$(x - 5)^2 = (\sqrt{x + 7})^2 \qquad \text{Principle of squaring}$$
$$x^2 - 10x + 25 = x + 7$$
$$x^2 - 11x + 18 = 0$$
$$(x - 9)(x - 2) = 0$$
$$x - 9 = 0 \quad \text{or} \quad x - 2 = 0$$
$$x = 9 \quad \text{or} \qquad x = 2$$

Check: For 9:   For 2:

$x - 5 = \sqrt{x + 7}$		$x - 5 = \sqrt{x + 7}$	
$9 - 5$	$\sqrt{9 + 7}$	$2 - 5$	$\sqrt{2 + 7}$
4	4	-3	3

The number 9 checks, but 2 does not. Thus the solution is 9.

DO EXERCISE 2.

Example 3   Solve: $\sqrt{27 - 3x} + 3 = x$.

$$\sqrt{27 - 3x} = x - 3 \qquad \text{Adding } -3 \text{ to get the radical alone on one side}$$
$$(\sqrt{27 - 3x})^2 = (x - 3)^2 \qquad \text{Principle of squaring}$$
$$27 - 3x = x^2 - 6x + 9$$
$$0 = x^2 - 3x - 18 \qquad \text{We can have 0 on the left.}$$
$$0 = (x - 6)(x + 3) \qquad \text{Factoring}$$
$$x - 6 = 0 \quad \text{or} \quad x + 3 = 0$$
$$x = 6 \quad \text{or} \qquad x = -3$$

Check: For 6:   For -3:

$\sqrt{27 - 3x} + 3 = x$		$\sqrt{27 - 3x} + 3 = x$	
$\sqrt{27 - 3 \cdot 6} + 3$	6	$\sqrt{27 - 3 \cdot (-3)} + 3$	-3
$\sqrt{9} + 3$		$\sqrt{27 + 9} + 3$	
$3 + 3$		$\sqrt{36} + 3$	
6		$6 + 3$	
		9	

There is only one solution, 6.

DO EXERCISE 3.

EXERCISE SET 10.6

● Solve.

1. $\dfrac{8}{x + 2} + \dfrac{8}{x - 2} = 3$

2. $\dfrac{24}{x - 2} + \dfrac{24}{x + 2} = 5$

3. $\dfrac{1}{x} + \dfrac{1}{x + 6} = \dfrac{1}{4}$

4. $\dfrac{1}{x} + \dfrac{1}{x + 9} = \dfrac{1}{20}$

5. $1 + \dfrac{12}{x^2 - 4} = \dfrac{3}{x - 2}$

6. $\dfrac{5}{t - 3} - \dfrac{30}{t^2 - 9} = 1$

7. $\dfrac{r}{r - 1} + \dfrac{2}{r^2 - 1} = \dfrac{8}{r + 1}$

8. $\dfrac{x + 2}{x^2 - 2} = \dfrac{2}{3 - x}$

9. $\dfrac{4 - x}{x - 4} + \dfrac{x + 3}{x - 3} = 0$

10. $\dfrac{y + 2}{y} = \dfrac{1}{y + 2}$

11. $\dfrac{x^2}{x - 4} - \dfrac{7}{x - 4} = 0$

12. $\dfrac{x^2}{x + 3} - \dfrac{5}{x + 3} = 0$

13. $x + 2 = \dfrac{3}{x + 2}$

14. $x - 3 = \dfrac{5}{x - 3}$

ANSWERS

15. _____

16. _____

17. _____

18. _____

19. _____

20. _____

21. _____

22. _____

23. _____

24. _____

25. _____

26. _____

27. _____

28. _____

29. _____

30. _____

15. $\dfrac{1}{x} + \dfrac{1}{x + 6} = \dfrac{1}{5}$

16. $\dfrac{1}{x} + \dfrac{1}{x + 1} = \dfrac{1}{3}$

● ● Solve.

17. $x - 7 = \sqrt{x - 5}$

18. $\sqrt{x + 7} = x - 5$

19. $\sqrt{x + 18} = x - 2$

20. $x - 9 = \sqrt{x - 3}$

21. $2\sqrt{x - 1} = x - 1$

22. $x + 4 = 4\sqrt{x + 1}$

23. $\sqrt{5x + 21} = x + 3$

24. $\sqrt{27 - 3x} = x - 3$

25. $x = 1 + 6\sqrt{x - 9}$

26. $\sqrt{2x - 1} + 2 = x$

27. $\sqrt{x^2 + 6} - x + 3 = 0$

28. $\sqrt{x^2 + 5} - x + 2 = 0$

29. $\sqrt{(p + 6)(p + 1)} - 2 = p + 1$

30. $\sqrt{(4x + 5)(x + 4)} = 2x + 5$

10.7 FORMULAS

■•■ To solve a formula for a given letter, we try to get the letter alone on one side.

Example 1 Solve for h: $V = 3.5\sqrt{h}$ (the distance to the horizon).

$V^2 = (3.5)^2(\sqrt{h})^2$ Squaring both sides

$V^2 = 12.25h$

$\dfrac{V^2}{12.25} = h$ Multiplying by $\dfrac{1}{12.25}$ to get h alone

DO EXERCISE 1.

Example 2 Solve for g: $T = 2\pi\sqrt{\dfrac{L}{g}}$ (the period of a pendulum).

$T^2 = (2\pi)^2\left(\sqrt{\dfrac{L}{g}}\right)^2$ Squaring both sides

$T^2 = 4\pi^2\dfrac{L}{g}$

$T^2 = \dfrac{4\pi^2 L}{g}$

$gT^2 = 4\pi^2 L$ Multiplying by g, to clear of fractions

$g = \dfrac{4\pi^2 L}{T^2}$ Multiplying by $\dfrac{1}{T^2}$, to get g alone

DO EXERCISES 2 AND 3.

In most formulas the letters represent nonnegative numbers, so you don't need to use absolute values when taking square roots.

Example 3 (*Torricelli's theorem*) In hydrodynamics the speed v of a liquid leaving a tank from an orifice is related to the height h of the top of the water above the orifice by the formula

$h = \dfrac{v^2}{2g}.$

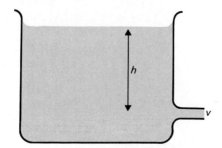

Solve for v:

$2gh = v^2$ Multiplying by $2g$, to clear of fractions

$\sqrt{2gh} = v.$ Taking the square root. We assume v is negative.

DO EXERCISE 4.

1. Solve for L: $r = 2\sqrt{5L}$.

   (A formula for the speed of a skidding car)

2. Solve for L: $T = 2\pi\sqrt{\dfrac{L}{g}}$.

3. Solve for m: $c = \sqrt{\dfrac{E}{m}}$.

4. Solve for r: $A = \pi r^2$.

   (The area of a circle)

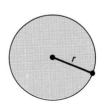

5. Solve for d: $C = P(d - 1)^2$.

6. Solve for n: $N = n^2 - n$.

7. Solve for t: $h = vt + 8t^2$.

Example 4 Solve for r:

$$A = P(1 + r)^2 \quad \text{(a compound interest formula)}.$$

$$\frac{A}{P} = (1 + r)^2 \qquad \text{Multiplying by } \frac{1}{P}$$

$$\sqrt{\frac{A}{P}} = 1 + r \qquad \text{Taking the square root}$$

$$-1 + \sqrt{\frac{A}{P}} = r \qquad \text{Adding } -1, \text{ to get } r \text{ alone}$$

DO EXERCISE 5.

Sometimes you need to use the quadratic formula to solve for a given letter.

Example 5 Solve for n:

$$d = \frac{n^2 - 3n}{2} \quad \text{(the number of diagonals of a polygon)}.$$

$$n^2 - 3n = 2d \qquad \text{Multiplying by 2, to clear of fractions}$$
$$n^2 - 3n - 2d = 0 \qquad \text{Finding standard form}$$
$$a = 1, \quad b = -3, \quad c = -2d \qquad \text{All letters are considered constants except } n.$$

$$n = \frac{-b \pm \sqrt{b^2 - 4ac}}{2a}$$

$$n = \frac{-(-3) \pm \sqrt{(-3)^2 - 4 \cdot 1 \cdot (-2d)}}{2 \cdot 1} \qquad \text{Substituting into the quadratic formula}$$

$$n = \frac{3 \pm \sqrt{9 + 8d}}{2}$$

DO EXERCISE 6.

Example 6 Solve for t: $S = gt + 16t^2$.

$$16t^2 + gt - S = 0 \qquad \text{Finding standard form}$$
$$a = 16, \quad b = g, \quad c = -S$$

$$t = \frac{-b \pm \sqrt{b^2 - 4ac}}{2a}$$

$$t = \frac{-g \pm \sqrt{g^2 - 4 \cdot 16 \, (-S)}}{2 \cdot 16} \qquad \text{Substituting into the quadratic formula}$$

$$t = \frac{-g \pm \sqrt{g^2 + 64S}}{32}$$

DO EXERCISE 7.

NAME   CLASS   ANSWERS

EXERCISE SET 10.7

▪◦ Solve for the indicated letter.

1. $N = 2.5\sqrt{A}$; A 2. $T = 2\pi\sqrt{\dfrac{L}{32}}$; L 3. $Q = \sqrt{\dfrac{aT}{c}}$; T

4. $v = \sqrt{\dfrac{2gE}{m}}$; E 5. $E = mc^2$; c 6. $S = 4\pi r^2$; r

7. $Q = ad^2 - cd$; d 8. $P = kA^2 + mA$; A 9. $c^2 = a^2 + b^2$; a

10. $c = \sqrt{a^2 + b^2}$; b 11. $S = \dfrac{1}{2}gt^2$; t 12. $V = \pi r^2 h$; r

1. _____

2. _____

3. _____

4. _____

5. _____

6. _____

7. _____

8. _____

9. _____

10. _____

11. _____

12. _____

13. _____

14. _____

15. _____

16. _____

17. a) _____

b) _____

18. a) _____

b) _____

13. $A = \pi r^2 + 2\pi rh$; r

14. $A = 2\pi r^2 + 2\pi rh$; r

15. $A = \dfrac{\pi r^2 S}{360}$; r

16. $H = \dfrac{D^2 N}{2.5}$; D

17. The circumference C of a circle is given by $C = 2\pi r$.

   a) Solve $C = 2\pi r$ for r.

   b) Express the area $A = \pi r^2$ in terms of the circumference C.

18. The distance S, in feet, traveled by a body falling from rest in t seconds is given by $S = 16t^2$.

   a) Solve for t.

   b) ▦ The Sears Tower in Chicago is 1451 ft tall. How long would it take an object to drop from the top?

10.8 APPLIED PROBLEMS

● ● We now use quadratic equations to solve more applied problems.

Example 1   A picture frame measures 20 cm by 14 cm. 160 square centimeters of picture shows. Find the width of the frame.

We first make a drawing. Let $x =$ the width of the frame. To translate, we recall that area is length $\times$ width. Thus,

$$A = lw = 160$$
$$(20 - 2x)(14 - 2x) = 160.$$

We solve.

$$280 - 68x + 4x^2 = 160$$
$$4x^2 - 68x + 120 = 0$$
$$x^2 - 17x + 30 = 0 \qquad \text{Multiplying by } \tfrac{1}{4}$$
$$(x - 15)(x - 2) = 0 \qquad \text{Factoring}$$
$$x - 15 = 0 \quad \text{or} \quad x - 2 = 0 \qquad \text{Principle of zero products}$$
$$x = 15 \quad \text{or} \qquad x = 2$$

We check in the original problem. 15 is not a solution because when $x = 15$, $20 - 2x = -10$, and the length of the picture cannot be negative. When $x = 2$, $20 - 2x = 16$. This is the length. When $x = 2$, $14 - 2x = 10$. This is the width. The area is 16×10, or 160. This checks, so the width of the frame is 2 cm.

DO EXERCISE 1.

Example 2   The hypotenuse of a right triangle is 6 m long. One leg is 1 m longer than the other. Find the lengths of the legs. Round to the nearest tenth.

We first make a drawing. Let $x =$ the length of one leg. Then $x + 1$ is the length of the other leg. To translate we use the Pythagorean property.

$$x^2 + (x + 1)^2 = 6^2$$
$$x^2 + x^2 + 2x + 1 = 36$$
$$2x^2 + 2x - 35 = 0$$

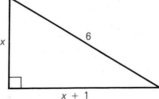

Since we cannot factor, we use the quadratic formula.

$$a = 2, \quad b = 2, \quad c = -35$$

$$x = \frac{-b \pm \sqrt{b^2 - 4ac}}{2a} = \frac{-2 \pm \sqrt{2^2 - 4 \cdot 2(-35)}}{2 \cdot 2}$$

$$= \frac{-2 \pm \sqrt{4 + 280}}{4} = \frac{-2 \pm \sqrt{284}}{4}$$

$$= \frac{-2 \pm \sqrt{4 \cdot 71}}{4} = \frac{-2 \pm 2 \cdot \sqrt{71}}{2 \cdot 2} = \frac{-1 \pm \sqrt{71}}{2}$$

1. A rectangular garden is 80 m by 60 m. Part of the garden is torn up to install a strip of lawn around the garden. The new area of the garden is 800 m². How wide is the strip of lawn?

ANSWER ON PAGE A–34

2. The hypotenuse of a right triangle is 4 cm long. One leg is 1 cm longer than the other. Find the lengths of the legs. Round to the nearest tenth.

The square root table on the inside back cover gives an approximation: $\sqrt{71} \approx 8.426$.

$$\frac{-1 + \sqrt{71}}{2} \approx 3.7, \qquad \frac{-1 - \sqrt{71}}{2} \approx -4.7$$

Since the length of a leg cannot be negative, -4.7 does not check; 3.7 does: $3.7^2 + 4.7^2 = 13.69 + 22.09 = 35.78$ and $\sqrt{35.78} \approx 5.98 \approx 6$. Thus one leg is about 3.7 m long and the other is about 4.7 m long.

DO EXERCISE 2.

Example 3   The current in a stream moves at a speed of 2 km/h. A boat travels 24 km upstream and 24 km downstream in a total time of 5 hours. What is the speed of the boat in still water?

First make a drawing. The distances are the same. Let r represent the speed of the boat in still water. Then, when traveling upstream the speed of the boat is $r - 2$. When traveling downstream, the speed of the boat is $r + 2$. We let t_1 represent the time it takes the boat to go upstream, and t_2 the time it takes to go downstream. We summarize in a table.

Upstream
$r - 2$
t_1 hours   24 km

Downstream
$r + 2$
t_2 hours   24 km

	d	r	t
Upstream	24	$r - 2$	t_1
Downstream	24	$r + 2$	t_2

3. The speed of a boat in still water is 12 km/h. The boat travels 45 km upstream and 45 km downstream in a total time of 8 hours. What is the speed of the stream? (*Hint:* Let s = the speed of the stream. Then $12 - s$ is the speed upstream and $12 + s$ is the speed downstream.)

Since $d = rt$, we know that $t = d/r$. Then using the rows of the table, we have $t_1 = 24/(r - 2)$ and $t_2 = 24/(r + 2)$. Since the total time is 5 hours, $t_1 + t_2 = 5$, and we have

$$\frac{24}{r - 2} + \frac{24}{r + 2} = 5.$$

The translation is complete. Now we solve. The LCM $= (r - 2)(r + 2)$.

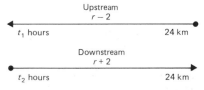

$$(r - 2)(r + 2) \cdot \frac{24}{r - 2} + (r - 2)(r + 2) \cdot \frac{24}{r + 2} = (r^2 - 4)5$$

$$24(r + 2) + 24(r - 2) = 5r^2 - 20$$
$$24r + 48 + 24r - 48 = 5r^2 - 20$$
$$-5r^2 + 48r + 20 = 0$$
$$5r^2 - 48r - 20 = 0 \qquad \text{Multiplying by } -1$$
$$(5r + 2)(r - 10) = 0 \qquad \text{Factoring}$$

$5r + 2 = 0 \quad$ or $\quad r - 10 = 0 \qquad$ Principle of zero products
$\qquad 5r = -2 \quad$ or $\qquad r = 10$
$\qquad r = -\frac{2}{5} \quad$ or $\qquad r = 10$

Since speed cannot be negative, $-\frac{2}{5}$ cannot be a solution. But 10 checks, so the speed of the boat in still water is 10 km/h.

DO EXERCISE 3.

NAME CLASS   ANSWERS

EXERCISE SET 10.8

█ ● █ Solve.

1. A picture frame measures 20 cm by 12 cm. There are 84 cm² of picture showing. Find the width of the frame.

1. _____

2. A picture frame measures 18 cm by 14 cm. There are 192 cm² of picture showing. Find the width of the frame.

2. _____

3. The hypotenuse of a right triangle is 25 ft long. One leg is 17 ft longer than the other. Find the lengths of the legs.

3. _____

4. The hypotenuse of a right triangle is 26 yd long. One leg is 14 yd longer than the other. Find the lengths of the legs.

4. _____

5. The length of a rectangle is 2 cm greater than the width. The area is 80 cm². Find the length and width.

5. _____

6. The length of a rectangle is 3 m greater than the width. The area is 70 m². Find the length and width.

6. _____

7. The width of a rectangle is 4 cm less than the length. The area is 320 cm². Find the length and width.

7. _____

8. The width of a rectangle is 3 cm less than the length. The area is 340 cm². Find the length and width.

8. _____

9. The length of a rectangle is twice the width. The area is 50 m². Find the length and width.

9. _____

10. The length of a rectangle is twice the width. The area is 32 cm². Find the length and width.

10. _____

Give approximate answers for Exercises 11–16. Round to the nearest tenth.

11. The hypotenuse of a right triangle is 8 m long. One leg is 2 m longer than the other. Find the lengths of the legs.

11. _____

12. The hypotenuse of a right triangle is 5 cm long. One leg is 2 cm longer than the other. Find the lengths of the legs.

12. _____

13. The length of a rectangle is 2 in. greater than the width. The area is 20 in^2. Find the length and width.

13. _____

14. The length of a rectangle is 3 ft greater than the width. The area is 15 ft^2. Find the length and width.

14. _____

15. The length of a rectangle is twice the width. The area is 10 m^2. Find the length and width.

15. _____

16. The length of a rectangle is twice the width. The area is 20 cm². Find the length and width.

16. _____

17. The current in a stream moves at a speed of 3 km/h. A boat travels 40 km upstream and 40 km downstream in a total time of 14 hours. What is the speed of the boat in still water?

17. _____

18. The current in a stream moves at a speed of 3 km/h. A boat travels 45 km upstream and 45 km downstream in a total time of 8 hours. What is the speed of the boat in still water?

18. _____

19. The current in a stream moves at a speed of 4 mph. A boat travels 4 mi upstream and 12 mi downstream in a total time of 2 hours. What is the speed of the boat in still water?

19. _____

20. The current in a stream moves at a speed of 4 mph. A boat travels 5 mi upstream and 13 mi downstream in a total time of 2 hours. What is the speed of the boat in still water?

20. _____

21. The speed of a boat in still water is 10 km/h. The boat travels 12 km upstream and 28 km downstream in a total time of 4 hours. What is the speed of the stream?

22. The speed of a boat in still water is 8 km/h. The boat travels 60 km upstream and 60 km downstream in a total time of 16 hours. What is the speed of the stream?

21. _____

23. An airplane flies 738 mi against the wind and 1062 mi with the wind in a total time of 9 hours. The speed of the airplane in still air is 200 mph. What is the speed of the wind?

22. _____

23. _____

24. An airplane flies 520 km against the wind and 680 km with the wind in a total time of 4 hours. The speed of the airplane in still air is 300 km/h. What is the speed of the wind?

24. _____

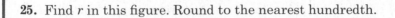

25. Find r in this figure. Round to the nearest hundredth.

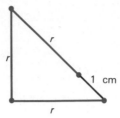

25. _____

26. A 20-ft pole is struck by lightning, and, while not completely broken, falls over and touches the ground 10 ft from the bottom of the pole. How high up the pole did the lightning strike?

26. _____

27. What should the diameter d of a pizza be so that it has the same area as two 10-inch pizzas? Do you get more to eat with a 13-inch pizza or two 10-inch pizzas?

28. ▦ In this figure, the area of the shaded region is 24 cm². Find r if $R = 6$ cm. Round to the nearest hundredth.

27. _____

28. _____

10.9 GRAPHS OF QUADRATIC EQUATIONS

● GRAPHING QUADRATIC EQUATIONS, $y = ax^2 + bx + c$

Graphs of quadratic equations, $y = ax^2 + bx + c$ (where $a \neq 0$), are always cup-shaped. They all have a *line of symmetry* like the dashed line shown in the figures below. If you fold on this line, the two halves will match exactly. The arrows show that the curve goes on forever.

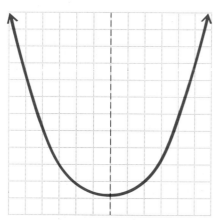

Thin parabola

These curves are called *parabolas*. Some parabolas are thin and others are flat, but they all have the same general shape.

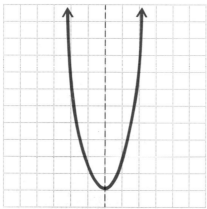

Thin parabola

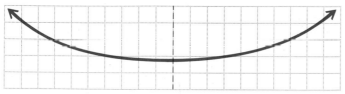

Flat parabola

To graph a quadratic equation, choose some numbers for x and compute the corresponding values of y.

OBJECTIVES

After finishing Section 10.9, you should be able to:

● Without graphing, tell whether the graph of an equation of the type $y = ax^2 + bx + c$ opens upward or downward. Then graph the equation.

●● Approximate the solutions of $0 = ax^2 + bx + c$ by graphing.

Example 1   Graph $y = x^2 + 2x - 3$.

x	y
1	0
0	-3
-1	-4
-2	-3
-3	0
-4	5
2	5

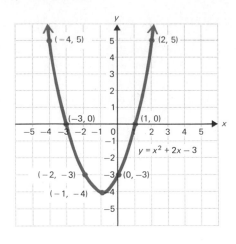

Example 2   Graph $y = -2x^2 + 3$.

x	y
0	3
1	1
-1	1
2	-5
-2	-5

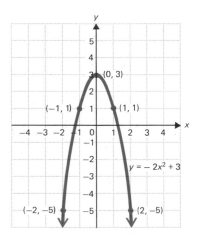

The graph in Example 1 opens upward and the coefficient of x^2 is 1, which is positive. The graph in Example 2 opens downward and the coefficient of x^2 is -2, which is negative.

Graphs of quadratic equations $y = ax^2 + bx + c$ are all parabolas. They are *smooth* cup-shaped symmetric curves, with no sharp points or kinks in them.

The graph of $y = ax^2 + bx + c$ opens upward if $a > 0$. It opens downward if $a < 0$.

In drawing parabolas, be sure to plot enough points to see the general shape of each graph.

If your graphs look like any of the following, they are incorrect.

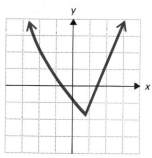

a) Sharp point is wrong.

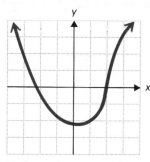

b) Outward nonsymmetric curve is wrong.

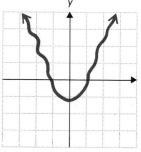

c) Kinks are wrong.

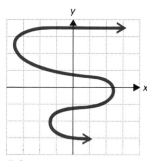

d) S-shaped curve is wrong.

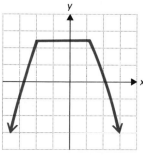

e) Flat nose is wrong.

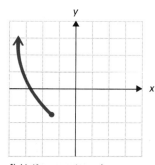

f) Half a cup-shaped curve is wrong.

DO EXERCISES 1 AND 2.

●●   **APPROXIMATING SOLUTIONS OF $0 = ax^2 + bx + c$**

Graphing can be used to approximate the solutions of

$$0 = ax^2 + bx + c.$$

We graph the equation $y = ax^2 + bx + c$. If the graph crosses the x-axis, the points of crossing will give us solutions.

1. a) Without graphing tell whether the graph of

   $$y = x^2 + 6x + 9$$

   opens upward or downward.

   b) Graph the equation.

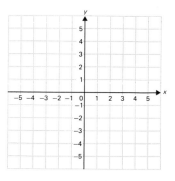

2. a) Without graphing tell whether the graph of

   $$y = -3x^2 + 6x$$

   opens upward or downward.

   b) Graph the equation.

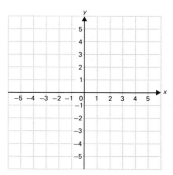

ANSWERS ON PAGE A–34

Approximate the solutions by graphing.

3. $x^2 - 4x + 4 = 0$

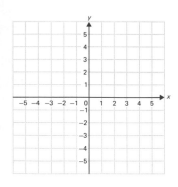

4. $-2x^2 - 4x + 1 = 0$

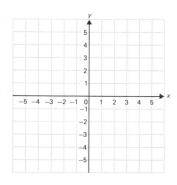

Example 3 Approximate the solutions of

$$-2x^2 + 3 = 0$$

by graphing.

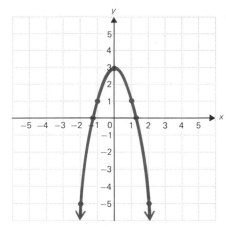

The graph was found in Example 2. The graph crosses the x-axis at about $(-1.2, 0)$ and $(1.2, 0)$. So the solutions are about -1.2 and 1.2.

DO EXERCISES 3 AND 4.

SOMETHING EXTRA

CALCULATOR CORNER: FINDING THE MEAN AND STANDARD DEVIATION

Statistics is used in many fields. Chances are good that you will take a statistics course later in your college program. The notions of *mean* $\overline{X}$ and *standard deviation s* are used in statistics. We show here how to find them using a calculator.

Example Consider the data 15, 16, 22, 43, 35.

$$\overline{X} = mean = \text{the arithmetic average}$$

$$= \frac{15 + 16 + 22 + 43 + 35}{5} = 26.2$$

To find the standard deviation we first need the average of the squares, $\overline{S}$, given by

$$\overline{S} = \frac{15^2 + 16^2 + 22^2 + 43^2 + 35^2}{5} = 807.8.$$

Then the standard deviation is the square root of $\overline{S}$ minus the square of the mean:

$$s = \sqrt{\overline{S} - (\overline{X})^2} = \sqrt{807.8 - (26.2)^2} = \sqrt{121.36} \approx 11.02.$$

EXERCISES

Find the mean and standard deviation of each set of data.

1. 16, 17, 23, 44, 36   **2.** 1, 4, 9, 9, 16, 16, 25, 25, 25, 25, 49, 49, 49, 64, 81

NAME _____   CLASS _____

EXERCISE SET 10.9

🔲 Without graphing, tell whether the graph of the equation opens upward or downward. Then graph the equation.

1. $y = x^2$

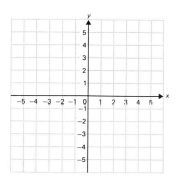

2. $y = 2x^2$

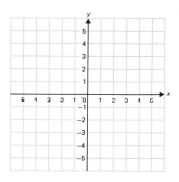

3. $y = -1 \cdot x^2$

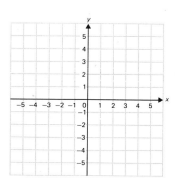

4. $y = x^2 - 1$

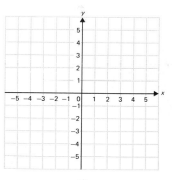

5. $y = -x^2 + 2x$

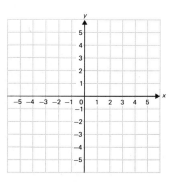

6. $y = x^2 + x - 6$

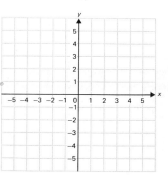

7. $y = 8 - x - x^2$

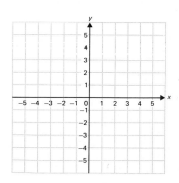

8. $y = x^2 + 2x + 1$

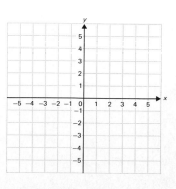

ANSWERS

15. _____

16. _____

17. _____

18. _____

19. _____

20. _____

21. _____

22. _____

9. $y = x^2 - 2x + 1$

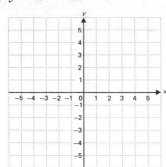

10. $y = -\dfrac{1}{2}x^2$

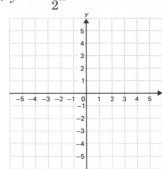

11. $y = -x^2 + 2x + 3$

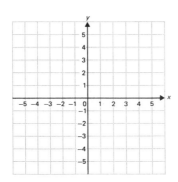

12. $y = -x^2 - 2x + 3$

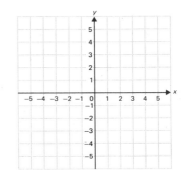

13. $y = -2x^2 - 4x + 1$

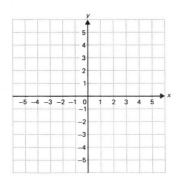

14. $y = 2x^2 + 4x - 1$

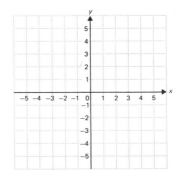

⚫⚫ Approximate the solutions by graphing. Use graph paper.

15. $x^2 - 5 = 0$

16. $x^2 + 2x = 0$

17. $8 - x - x^2 = 0$

18. $x^2 + 10x + 25 = 0$

19. $x^2 - 8x + 16 = 0$

20. $x^2 + 3 = 0$

21. $x^2 + 8 = 0$

22. $2x^2 + 4x - 1 = 0$

EXTENSION EXERCISES

SECTION 10.1

Solve for x.

1. $\dfrac{1}{4}x^2 + \dfrac{1}{6} = \dfrac{2}{3}x^2$ **2.** $\dfrac{4}{x^2 - 3} = \dfrac{6}{x^2}$ **3.** $3ax^2 - 9b = 3b^2$ **4.** $x^2 + 9a^2 = 9 + ax^2$

SECTION 10.2

Solve.

5. $x^2 + \sqrt{3}x = 0$ **6.** $\sqrt{5}x^2 - x = 0$ **7.** $\sqrt{7}x^2 + \sqrt{3}x = 0$

8. $\dfrac{5}{y + 4} - \dfrac{3}{y - 2} = 4$ **9.** $\dfrac{2z + 11}{2z + 8} = \dfrac{3z - 1}{z - 1}$

10. $y^4 - 4y^2 + 4 = 0$ (*Hint:* Let $x = y^2$. Then $x^2 = y^4$. Substitute to obtain a quadratic equation in x and solve. Remember to solve for y after finding x.)

11. $z - 10\sqrt{z} + 9 = 0$ (Let $x = \sqrt{z}$.)

SECTION 10.3

Solve (for x or y).

12. $\dfrac{x - 1}{9} = \dfrac{1}{x - 1}$ **13.** $(y - b)^2 = 4b^2$ **14.** $2(3x + 1)^2 = 8$ **15.** $5(5x - 2)^2 - 7 = 13$

Complete the square.

16. $x^2 - ax$ **17.** $x^2 + (2b - 4)x$

SECTION 10.4

Solve for x by completing the square.

18. $x^2 - ax - 6a^2 = 0$ **19.** $x^2 + 4bx + 2b = 0$ **20.** $x^2 - x - c^2 - c = 0$

21. $3x^2 - bx + 1 = 0$ **22.** $ax^2 + 4x + 3 = 0$ **23.** $4x^2 - 4x + c = 0$

SECTION 10.5

Solve for x. Try factoring first. If factoring is not possible or is difficult, use the quadratic formula.

24. $ax^2 + 2x = 3$ **25.** $2bx^2 - 5x + 3b = 0$ **26.** $4x^2 - 4cx + c^2 - 3d^2 = 0$

SECTION 10.6

27. Solve.

a) $\dfrac{4}{x} - \dfrac{5}{2x + 3} = 2$ b) $\dfrac{7}{1 + x} - 1 = \dfrac{5x}{x^2 + 3x + 2}$ c) $\dfrac{x}{x + 1} = 4 + \dfrac{1}{3x^2 - 3}$

28. Solve using two methods.

$$x + 1 + 3\sqrt{x + 1} - 4 = 0$$

a) First, use the principle of squaring. (Remember to check.)
b) Second, let $y = \sqrt{x + 1}$. (Then $y^2 = x + 1$.) Substitute to obtain a quadratic equation in y and solve. (Remember to solve for x after finding y.)

SECTION 10.7

29. Solve for x.

$$3ax^2 - x - 3ax + 1 = 0$$

30. Solve for x in terms of y.

$$6y^2 - 8xy = 9y - 3x - 2x^2$$

SECTION 10.8

31. A ladder 10 ft long leans against a wall. The bottom of the ladder is 6 ft from the wall. How much would the lower end of the ladder have to be pulled away so that the top end would be pulled down the same amount?

32. Trains A and B leave the same city at right angles at the same time. Train B travels 5 mph faster than train A. After 2 hr they are 50 mi apart. Find the speed of each train.

33. Find the side of a square whose diagonal is 3 cm longer than a side.

34. Two pipes are connected to the same tank. When working together they can fill the tank in 2 hr. The larger pipe, working alone, can fill the tank in 3 hr less than the smaller one. How long would the smaller one take, working alone, to fill the tank?

SECTION 10.9

35. Graph the equation $y = x^2 - x - 6$. Use your graph to approximate the solutions to the following equations.
a) $x^2 - x - 6 = 2$ (*Hint:* Graph $y = 2$ on the same set of axes as your graph of $y = x^2 - x - 6$.)
b) $x^2 - x - 6 = -3$

NAME   SCORE   ANSWERS

TEST OR REVIEW—CHAPTER 10

If you miss an item, review the indicated section and objective.

Solve.

[10.1, ●●]   **1.** $8x^2 = 24$

                                                                 1. _____

[10.2, ●]   **2.** $5x^2 - 7x = 0$

                                                                  2. _____

[10.2, ●●]   **3.** $3y^2 + 5y = 2$

                                                                   3. _____

[10.3, ●]   **4.** $(x + 8)^2 = 13$

                                                                   4. _____

[10.4, ●]   Solve by completing the square. Show your work.

               5. $x^2 - 2x - 10 = 0$

                                                                 5. _____

[10.5, ●]   Solve using the quadratic formula.

               6. $x^2 + 6x - 9 = 0$

                                                                 6. _____

[10.5, ●●]   Use Table 1 to approximate solutions to the nearest tenth.

               7. $x^2 + 6x - 9 = 0$

                                                                 7. _____

ANSWERS

Solve.

[10.6, ■] 8. $\dfrac{1}{4-r} - \dfrac{1}{4+r} = \dfrac{1}{3}$

8. _____

[10.6, ■ ■] 9. $\sqrt{x+5} = x-1$

9. _____

[10.7, ■] 10. Solve for T: $V = 5\sqrt{\dfrac{aT}{L}}$.

10. _____

[10.3, ■■■] 11. $1000 is invested at interest rate r, compounded annually. In 2 years it grows to $1690. What is the interest rate?

11. _____

[10.8, ■] 12. The hypotenuse of a right triangle is 5 m long. One leg is 3 m longer than the other. Find the lengths of the legs. Round to the nearest tenth.

12. _____

[10.8, ■] 13. The current in a stream moves at a speed of 2 km/h. A boat travels 56 km upstream and 64 km downstream in a total time of 4 hours. What is the speed of the boat in still water?

13. _____

[10.9, ■] 14. a) Without graphing, tell whether the graph of $y = x^2 - 4x - 2$ opens upward or downward.
b) Graph the equation.

14. a) _____

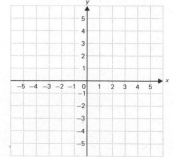

b) See graph.

INEQUALITIES
AND SETS

11

When sizes, such as diameters of gears, are compared, inequalities and set notation are useful.

1. Graph $\dfrac{5}{3}$ on a number line.

Use the proper symbol $<$, $>$, or $=$.

2. $\dfrac{5}{2}$ $\dfrac{8}{2}$ **3.** $\dfrac{4}{9}$ $\dfrac{5}{11}$

Find the absolute value of each integer.

4. 9   **5.** 0   **6.** -8

Solve and check.

7. $x + 8 = -2$ **8.** $y - 5 = 1$

9. $-\dfrac{3}{4}x = \dfrac{1}{8}$ **10.** $-2y - 3 = 11$

11. Graph: $2x - 3y = 6$.

11.1 USING THE ADDITION PRINCIPLE

◦ THE ADDITION PRINCIPLE

In Section 1.5 we defined the symbols $>$ (greater than) and $<$ (less than). The symbol $\geqslant$ means *greater than or equal to*. The symbol $\leqslant$ means *less than or equal to*. For example, $3 \leqslant 4$ and $3 \leqslant 3$ are both true, but $-3 \leqslant -4$ and $0 \geqslant 2$ are both false. An *inequality* is a number sentence with $>$, $<$, $\geqslant$, or $\leqslant$ for its verb—for example,

$$-4 < 5, \quad x < 2, \quad 2x + 5 \geqslant 0, \quad \text{and} \quad -3y + 7 \leqslant -8.$$

The letter x in $x < 2$ is called a *variable*. Some replacements for x make the inequality true. Some make it false. When we find all the numbers that make an inequality true, we have *solved* the inequality.

DO EXERCISE 1.

> **The replacements that make an inequality true are called its** *solutions.* **To** *solve* **an inequality means to find all of its solutions.**

Consider the true inequality

$$3 < 7.$$

Add $\boxed{2}$ on both sides and we get another true inequality:

$$3 + \boxed{2} < 7 + \boxed{2} \quad \text{or} \quad 5 < 9.$$

After finishing Section 11.1, you should be able to:

◦ **Solve inequalities using the addition principle.**

1. Consider the inequality

$$x < 2.$$

a) Find three replacements that make the inequality true.
b) Find three replacements that make the inequality false.

Similarly, if we add -3 to both numbers we get another true inequality:

$$3 + (\ -3\ ) < 7 + (\ -3\ ) \quad \text{or} \quad 0 < 4.$$

> *The Addition Principle for Inequalities:* **If any number is added on both sides of a true inequality, we get another true inequality.**

The addition principle holds whether the inequality contains $<$, $>$, $\leqslant$, or $\geqslant$.

Let's see how we use the addition principle to solve inequalities.

Example 1   Solve: $x + 2 > 8$.

We use the addition principle, adding -2:

$$x + 2 + \boxed{-2} > 8 + \boxed{-2}$$
$$x > 6.$$

By using the addition principle we get an inequality for which we can determine the solution easily.

Any number greater than 6 makes the last sentence true, hence is a solution of that sentence. Any such number is also a solution of the original sentence. Thus we have it solved.

We cannot check all the solutions of an inequality* by substitution, as we can check solutions of equations. There are too many of them. However, we don't really need to check. Let us see why. Consider the first and last inequalities

$$x + 2 > 8 \quad \text{and} \quad x > 6.$$

Any number that makes the first one true must make the last one true. We know this by the addition principle. Now the question is, will any number that makes the last one true also be a solution of the first one? Let us use the addition principle again, adding 2:

$$x > 6$$
$$x + \boxed{2} > 6 + \boxed{2}$$
$$x + 2 > 8.$$

Now we know that any number that makes $x > 6$ true also makes $x + 2 > 8$ true. Therefore, the sentences $x > 6$ and $x + 2 > 8$ have the same solutions. Any time we use the addition principle a similar thing happens. Thus whenever we use the principle with inequalities the first and last sentences will have the same solutions.

---

* A partial check could be done by substituting a number greater than 6, say 7, into the original inequality:

$$\begin{array}{c|c} x + 2 > 8 \\ \hline 7 + 2 & 8 \\ 9 & \end{array}$$

Since $9 > 8$ is true, 7 is a solution.

Solve. Write set notation for your answer.

2. $x + 3 > 5$

3. $x - 5 < 8$

4. $5x + 1 < 4x - 2$

Solve.

5. $x + \dfrac{2}{3} \leqslant \dfrac{4}{5}$

6. $5y + 2 \leqslant -1 + 4y$

Example 2   Solve: $3x + 1 < 2x - 3$.

$$3x + 1 < 2x - 3.$$

$$3x + 1 \boxed{-1} < 2x - 3 \boxed{-1} \qquad \text{Adding } -1$$

$$3x < 2x - 4 \qquad \text{Simplifying}$$

$$3x \boxed{-2x} < 2x - 4 \boxed{-2x} \qquad \text{Adding } -2x$$

$$x < -4 \qquad \text{Simplifying}$$

Any number less than -4 is a solution. The following are some of the solutions:

$$-5, \qquad -6, \qquad -4.1, \qquad -2045, \qquad -18\pi, \qquad -\sqrt{30}.$$

To describe all the solutions we use the set notation

$$\{x | x < -4\},$$

which is read:

The set of all x such that x is less than -4.

DO EXERCISES 2–4.

Example 3   Solve: $x + \dfrac{1}{3} \geqslant \dfrac{5}{4}$.

$$x + \dfrac{1}{3} \geqslant \dfrac{5}{4}$$

$$x + \dfrac{1}{3} \boxed{-\dfrac{1}{3}} \geqslant \dfrac{5}{4} \boxed{-\dfrac{1}{3}} \qquad \text{Adding } -\dfrac{1}{3}$$

$$x \geqslant \dfrac{5}{4} \cdot \boxed{\dfrac{3}{3}} - \dfrac{1}{3} \cdot \boxed{\dfrac{4}{4}} \qquad \begin{array}{l}\text{Finding a common} \\ \text{denominator}\end{array}$$

$$x \geqslant \dfrac{15}{12} - \dfrac{4}{12}$$

$$x \geqslant \dfrac{11}{12}$$

Any number greater than or equal to $\frac{11}{12}$ is a solution. We say that the *solution set* is

$$\left\{ x | x \geqslant \dfrac{11}{12} \right\}.$$

Here is a handy way to help read and write set notation:

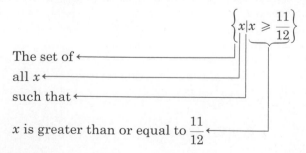

The set of ←

all x ←

such that ←

x is greater than or equal to $\dfrac{11}{12}$ ←

DO EXERCISES 5 AND 6.

NAME   CLASS   ANSWERS

EXERCISE SET 11.1

■ Solve using the addition principle. Write set notation for answers.

1. $x + 7 > 2$ **2.** $x + 6 > 3$ **3.** $y + 5 > 8$ **4.** $y + 7 > 9$

5. $x + 8 \leqslant -10$ **6.** $a + 12 < 6$ **7.** $x - 7 \leqslant 9$ **8.** $x - 6 > 2$

9. $y - 7 > -12$ **10.** $2x + 3 > x + 5$ **11.** $2x + 4 > x + 7$

12. $3x + 9 \leqslant 2x + 6$ **13.** $3x - 6 \geqslant 2x + 7$ **14.** $3x - 9 \geqslant 2x + 11$

15. $5x - 6 < 4x - 2$ **16.** $6x - 8 < 5x - 9$ **17.** $-7 + c > 7$

1. _____

2. _____

3. _____

4.

5. _____

6. _____

7. _____

8. _____

9. _____

10. _____

11. _____

12. _____

13. _____

14. _____

15. _____

16. _____

17. _____

ANSWERS

18. $-9 + b > 9$

19. $y + \dfrac{1}{4} \leqslant \dfrac{1}{2}$

20. $y + \dfrac{1}{3} \leqslant \dfrac{5}{6}$

18. _____

19. _____

20. _____

21. $x - \dfrac{1}{3} > \dfrac{1}{4}$

22. $x - \dfrac{1}{8} > \dfrac{1}{2}$

23. $-14x + 21 > 21 - 15x$

21. _____

22. _____

24. $-10x + 15 > 18 - 11x$

25. $3(r + 2) < 2r + 4$

26. $4(r + 5) \geqslant 3r + 7$

23. _____

24. _____

25. _____

27. $0.8x + 5 \geqslant 6 - 0.2x$

28. $0.7x + 6 \leqslant 7 - 0.3x$

26. _____

27. _____

28. _____

29. A student is taking an introductory algebra course in which four tests are to be given. To get a B, a total of 320 points is needed. The student got scores of 81, 76, and 73 on the first three tests. Determine (in terms of an inequality) those scores that will allow the student to get a B.

29. _____

30. _____

30. If $2x - 5 \geqslant 8$, is $2x - 3 \geqslant 9$?

31. _____

31. ▦ Solve: $17x + 9{,}479{,}756 \leqslant 16x - 8{,}579{,}243$.

11.2 USING THE MULTIPLICATION PRINCIPLE

▌•▐ THE MULTIPLICATION PRINCIPLE

Consider the true inequality

$3 < 7.$

Multiply both numbers by $\boxed{2}$ and we get another true inequality:

$3 \cdot \boxed{2} < 7 \cdot \boxed{2}$

or

$6 < 14.$

Multiply both numbers by $\boxed{-3}$ and we get the false inequality

$3 \cdot (\boxed{-3}) < 7 \cdot (\boxed{-3})$

or

$-9 < -21.$ False

However, if we reverse the inequality symbol we get a true inequality:

$-9 > -21.$ True

> *The Multiplication Principle for Inequalities:* **If we multiply on both sides of a true inequality by a positive number, we get another true inequality. If we multiply by a negative number and the inequality symbol is reversed, we get another true inequality.**

The multiplication principle also holds for inequalities containing $\geqslant$ or $\leqslant$.

Example 1   Solve: $4x < 28.$

$$\frac{1}{4} \cdot 4x < \frac{1}{4} \cdot 28 \qquad \text{Multiplying by } \frac{1}{4}$$

The symbol stays the same.

$$x < 7 \qquad \text{Simplifying}$$

The solution set is $\{x | x < 7\}$.

DO EXERCISES 1 AND 2.

Example 2   Solve: $-2y < 18.$

$$-\frac{1}{2}(-2y) > -\frac{1}{2} \cdot 18 \qquad \text{Multiplying by } -\frac{1}{2}$$

The symbol has to be reversed!

$$y > -9 \qquad \text{Simplifying}$$

The solution set is $\{y | y > -9\}$.

DO EXERCISES 3 AND 4.

OBJECTIVES

After finishing Section 11.2, you should be able to:

▌•▐ Solve inequalities using the multiplication principle.

▌••▐ Solve inequalities using the addition and multiplication principles together.

Solve.

1. $8x < 64$

2. $5y \geqslant 160$

Solve.

3. $-4x \leqslant 24$

4. $-5y > 13$

Solve.

5. $7 - 4x < 8$

Solve.

6. $13x + 5 \leqslant 12x + 4$

Solve.

7. $24 - 7y \geqslant 11y - 14$

● ●   USING THE PRINCIPLES TOGETHER

We use the addition and multiplication principles together in solving inequalities in much the same way as in solving equations. We generally use the addition principle first.

Example 3   Solve: $6 - 5y > 7$.

$$\boxed{-6} + 6 - 5y > \boxed{-6} + 7 \qquad \text{Adding } -6$$

$$-5y > 1 \qquad \text{Simplifying}$$

$$\boxed{-\frac{1}{5}} \cdot (-5y) < \boxed{-\frac{1}{5}} \cdot 1 \qquad \text{Multiplying by } -\frac{1}{5}$$

The symbol has to be reversed!

$$y < -\frac{1}{5}$$

The solution set is $\{y | y < -\frac{1}{5}\}$.

DO EXERCISE 5.

Example 4   Solve: $5x + 9 \leqslant 4x + 3$.

$$5x + 9 \boxed{-9} \leqslant 4x + 3 \boxed{-9} \qquad \text{Adding } -9$$

$$5x \leqslant 4x - 6 \qquad \text{Simplifying}$$

$$5x \boxed{-4x} \leqslant 4x - 6 \boxed{-4x} \qquad \text{Adding } -4x$$

$$x \leqslant -6 \qquad \text{Simplifying}$$

The solution set is $\{x | x \leqslant -6\}$.

DO EXERCISE 6.

Example 5   Solve: $17 - 5y < 8y - 5$.

$$17 - 5y \boxed{-17} < 8y - 5 \boxed{-17} \qquad \text{Adding } -17$$

$$-5y < 8y - 22 \qquad \text{Simplifying}$$

$$\boxed{-8y} - 5y < \boxed{-8y} + 8y - 22 \qquad \text{Adding } -8y$$

$$-13y < -22 \qquad \text{Simplifying}$$

$$\boxed{-\frac{1}{13}} \cdot (-13y) > \boxed{-\frac{1}{13}} \cdot (-22) \qquad \text{Multiplying by } -\frac{1}{13}$$

$$y > \frac{22}{13}$$

The solution set is $\{y | y > \frac{22}{13}\}$.

DO EXERCISE 7.

NAME _____   CLASS _____

EXERCISE SET 11.2

● Solve using the multiplication principle.

1. $5x < 35$ **2.** $8x \geqslant 32$ **3.** $9y \leqslant 81$ **4.** $10x > 240$

5. $7x < 13$ **6.** $8y < 17$ **7.** $12x > -36$ **8.** $16x < -64$

9. $5y \geqslant -2$ **10.** $7x > -4$ **11.** $-2x \leqslant 12$ **12.** $-3y \leqslant 15$

13. $-4y \geqslant -16$ **14.** $-7x < -21$ **15.** $-3x < -17$ **16.** $-5y > -23$

17. $-2y > \dfrac{1}{7}$ **18.** $-4x \leqslant \dfrac{1}{9}$ **19.** $-\dfrac{6}{5} \leqslant -4x$ **20.** $-\dfrac{7}{8} > -56t$

1. _____
2. _____
3. _____
4. _____
5. _____
6. _____
7. _____
8. _____
9. _____
10. _____
11. _____
12. _____
13. _____
14. _____
15. _____
16. _____
17. _____
18. _____
19. _____
20. _____

■ ● ■ Solve using the addition and multiplication principles.

21. $2x + 5 < 3$ **22.** $3x - 4 > 5$

21. _____

22. _____

23. $-3x + 7 \geqslant -2$ **24.** $-4x - 3 \leqslant -5$

23. _____

24. _____

25. $6t - 8 \leqslant 4t + 1$ **26.** $5b + 7 \geqslant b - 1$

25. _____

26. _____

27. $3 - 6c > 15$ **28.** $5 - 3y < -4$

27. _____

28. _____

29. $15x - 21 \geqslant 8x + 7$ **30.** $21 - 15x < -8x - 7$

29. _____

30. _____

31. Badger Rent-a-Car rents compact cars for \$23.95 plus 20¢ per mile. Thirsty Rent-a-Car rents compact cars for \$22.95 plus 22¢ per mile. For what mileages would the cost of renting a compact be cheaper at Badger?

31. _____

32. Badger Rent-a-Car rents an intermediate-size car at a daily rate of \$24.95 plus \$0.20 per mile. A businessperson is not to exceed a daily car rental budget of \$96. Determine (in terms of an inequality) those mileages that will allow the businessperson to stay within the budget.

32. _____

11.3 GRAPHS OF INEQUALITIES

● INEQUALITIES IN ONE VARIABLE

We graph inequalities in one variable on a number line.

Example 1 Graph $x < 2$.

The solutions of $x < 2$ are those numbers less than 2. They are shown on the graph by shading all points to the left of 2.

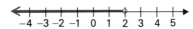

Example 2 Graph $y \geqslant -3$.

The solutions of $y \geqslant -3$ are shown by shading the point for -3 and all points to the right of -3.

DO EXERCISES 1 AND 2.

Example 3 Graph $3x + 2 < 5x - 1$.

First solve:
$$3x + 2 < 5x - 1$$
$$2 < 2x - 1 \qquad \text{Adding } -3x$$
$$3 < 2x \qquad \text{Adding } 1$$
$$\frac{3}{2} < x \qquad \text{Multiplying by } \frac{1}{2}$$

DO EXERCISES 3 AND 4.

●● GRAPHING INEQUALITIES WITH ABSOLUTE VALUE

Example 4 Graph $|x| < 3$.

The absolute value of a number is its distance from 0 on a number line. For the absolute value of a number to be less than 3 it must be between 3 and -3. Therefore, we shade the points between these two numbers.

DO EXERCISES 5 AND 6.

Graph on a number line.

1. $x < 4$

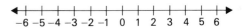

2. $y \geqslant -5$

Graph on a number line.

3. $x + 2 > 1$

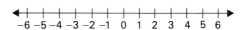

4. $4x + 6 < 7x - 3$

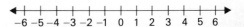

Graph on a number line.

5. $|x| < 5$

6. $|x| \leqslant 4$

Graph on a number line.

7. $|x| \geqslant 3$

Example 5   Graph $|x| \geqslant 2$.

For the absolute value of a number to be greater than or equal to 2 its distance from 0 must be 2 or more. Thus the number must be 2 or greater, or it must be less than or equal to -2. Therefore, we shade the point for 2 and all points to its right. We also shade the point for -2 and all points to its left.

DO EXERCISES 7 AND 8.

●●● INEQUALITIES IN TWO VARIABLES

The solutions of inequalities in two variables are ordered pairs.

8. $|y| > 5$

Example 6   Determine whether $(-3, 2)$ is a solution of $5x - 4y < 13$.

We use alphabetical order of variables. We replace x by -3 and y by 2.

$$\begin{array}{c|c} 5x - 4y < 13 & \\ \hline 5(-3) - 4 \cdot 2 & 13 \\ -15 - 8 & \\ -23 & \end{array}$$

Since $-23 < 13$ is true, $(-3, 2)$ is a solution.

DO EXERCISE 9.

Example 7   Graph $y > x$.

We first graph the line $y = x$ for comparison. Every solution of $y = x$ is an ordered pair such as $(3, 3)$. The first and second coordinates are the same. The graph of $y = x$ is shown to the left below.

9. Determine whether $(4, 3)$ is a solution of $3x - 2y < 1$.

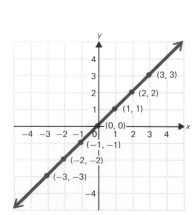

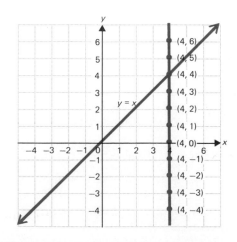

Now look at the graph to the right above. We consider a vertical line and ordered pairs on it. For all points above $y = x$, the second coordinate is greater than the first, $y > x$. For all points below the line, $y < x$. The same thing happens for any vertical line. Then for all points above $y = x$, the ordered pairs are solutions. We shade the half-plane above $y = x$. This is the graph of $y > x$. Points on $y = x$ are not in the graph, so we draw it dashed.

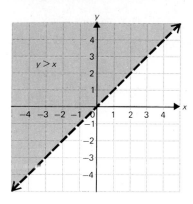

DO EXERCISE 10.

Example 8 Graph $y \leqslant x - 1$.

First sketch the line $y = x - 1$. Points on the line $y = x - 1$ are also in the graph of $y \leqslant x - 1$, so we draw the line solid. For points above the line, $y > x - 1$. These points are not in the graph. For points below the line, $y < x - 1$. These are in the graph, so we shade the lower half-plane.

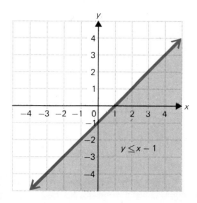

DO EXERCISE 11.

Example 9 Graph $6x - 2y < 10$.

We first solve for y, getting y on the left.

$$-2y < -6x + 10 \qquad \text{Adding } -6x$$

$$y > -\frac{1}{2}(-6x + 10) \qquad \text{Multiplying by } -\frac{1}{2}$$

       └─────────── Here the symbol has to be reversed.

$$y > 3x - 5$$

Now we graph the line $y = 3x - 5$. This line forms the boundary of the solutions of the inequality. In this case points on the line are not solutions of the inequality.

10. Graph $y < x$.

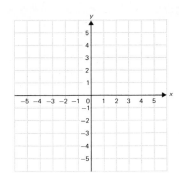

11. Graph $y \geqslant x + 2$.

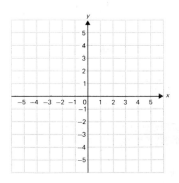

12. Graph $2x + 4y < 8$.

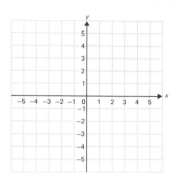

13. Graph $3x - 5y < 15$.

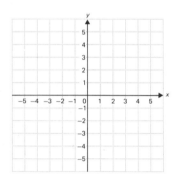

14. Graph $2x + 3y \geqslant 12$.

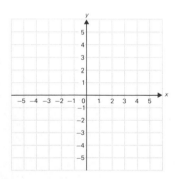

We shade the half-plane above the line.

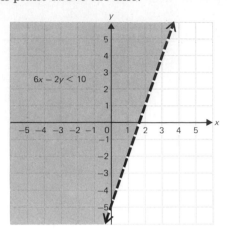

DO EXERCISE 12.

A *linear inequality* is one that we can get from a linear equation by changing the equals symbol to an inequality symbol. Every linear equation has a graph that is a straight line. The graph of a linear inequality is a half-plane, sometimes including the line along the edge. In the following example we give a different method of graphing. We graph the line using intercepts. Then we determine which side to shade by substituting a point from either half-plane.

Example 10   Graph $2x + 3y < 6$.

a) First graph the line $2x + 3y = 6$. The intercepts are $(3, 0)$ and $(0, 2)$. We use a dashed line for the graph since we have $<$.

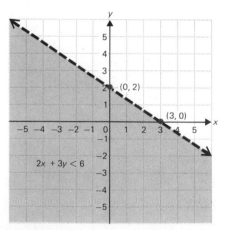

b) Pick a point that does not belong to the line. Substitute to determine whether this point is a solution. The origin $(0, 0)$ is usually an easy one to use: $2 \cdot 0 + 3 \cdot 0 < 6$ is true, so the origin is a solution. This means we shade the lower half-plane. Had the substitution given us a false inequality we would have shaded the other half-plane.

If the line goes through the origin, then we must test some other point not on the line. The point $(1, 1)$ is often a good one to try.

DO EXERCISES 13 AND 14.

NAME   CLASS

EXERCISE SET 11.3

⬤ Graph on a number line.

1. $x < 5$

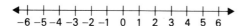

2. $x < 3$

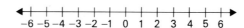

3. $y \geqslant -4$

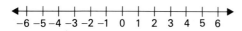

4. $x \geqslant -7$

5. $t - 3 \leqslant -7$

6. $x - 4 \leqslant -8$

7. $2x + 6 < 14$

8. $4x - 8 \geqslant 12$

9. $4y + 9 > 11y - 12$

10. $6x + 11 \leqslant 14x + 7$

⬤⬤ Graph on a number line.

11. $|x| < 2$

12. $|t| \leqslant 1$

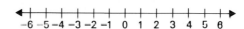

13. $|a| \geqslant 4$

14. $|y| > 5$

●●● Graph on a plane.

15. $y > x - 2$

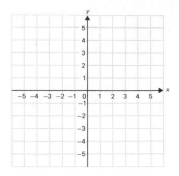

16. $y \leqslant x - 3$

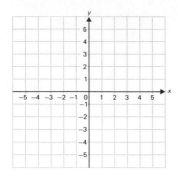

17. $6x - 2y \leqslant 12$

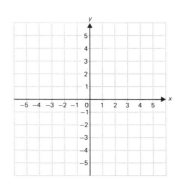

18. $2x + 3y < 12$

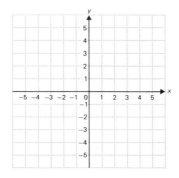

19. $3x - 5y \geqslant 15$

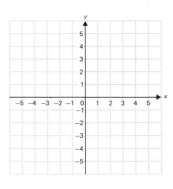

20. $5x + 2y > 10$

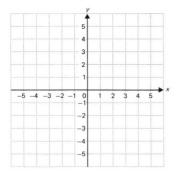

21. $y - 2x < 4$

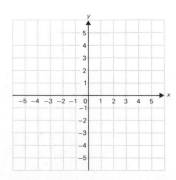

22. $2x - y \leqslant 4$

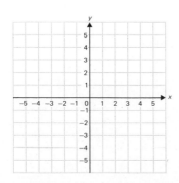

11.4 SETS

• NAMING SETS

We have discussed notation like the following:

$$\{x | x < 6\},$$

where we were considering the real numbers. If we were just considering whole numbers, the above set could also be named, or symbolized,

$$\{0, 1, 2, 3, 4, 5\}.$$

This way of naming a set is known as the *roster* method. In words, it is

The set of whole numbers 0 through 5.

DO EXERCISES 1 AND 2.

• • MEMBERSHIP

The symbol $\in$ means *is a member of* or *belongs to*. Thus

$$x \in A$$

means

x is a member of A

or

x belongs to A.

Examples   Determine whether true or false.

1. $1 \in \{1, 2, 3\}$ True
2. $1 \in \{2, 3\}$ False
3. $4 \in \{x | x \text{ is an even whole number}\}$ True
4. $5 \in \{x | x \text{ is an even whole number}\}$ False

Set membership can be illustrated with a diagram as in the figure below.

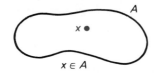

DO EXERCISES 3–6.

• • • INTERSECTIONS

The *intersection* of two sets A and B is the set of members common to both sets and is indicated by the symbol

$$A \cap B.$$

Thus,

$$\{0, 1, 3, 5, 25\} \cap \{2, 3, 4, 5, 6, 7, 9\}$$

represents the set

$$\{3, 5\}.$$

OBJECTIVES

After finishing Section 11.4, you should be able to:

▪ Name sets using the roster method.

▪▪ Determine whether a given object is a member of a set.

▪▪▪ Find the intersection of two sets.

▪▪:▪▪ Find the union of two sets.

Name each set using the roster method.

1. The set of whole numbers 2 through 10.

2. The set of even whole numbers between 30 and 40.

Determine whether true or false.

3. $4 \in \{a, b, 3, 4, 5\}$

4. $c \in \{a, b, 3, 4, 5\}$

5. $\dfrac{2}{3} \in \{x | x \text{ is a rational number}\}$

6. Heads $\in$ The set of outcomes of flipping a penny

Find.

7. $\{a, 1, 0, t, 6, 9, 14\} \cap \{2, 1, s, a, 9\}$

8. $\{1, 2\} \cap \{2, 3\}$

Find.

9. $\{1, 2\} \cap \{3, 4\}$

10. $A \cap B$, where

$A = \{x | x$ is a rational number$\}$

and

$B = \{x | x$ is an irrational number$\}$

Find.

11. $\{0, 1, 2, 3, 4\} \cup \{2, 3, 4, 5, 6, 7\}$

12. $A \cup B$, where

$A = \{x | x$ is an even integer$\}$

and

$B = \{x | x$ is an odd integer$\}$

Set intersection is illustrated by the diagram below.

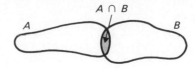

The solution of the system of equations

$$x + 2y = 7,$$
$$x - y = 4$$

is the ordered pair (5, 1). It is the intersection of the graphs of the two lines.

DO EXERCISES 7 AND 8.

The set without members is known as the *empty set*, and is often named $\emptyset$. Each of the following is a description of the empty set:

The set of all six-eyed algebra teachers;

$\{2, 3\} \cap \{5, 6, 7\}$;

$\{x | x$ is an even whole number$\} \cap \{x | x$ is an odd whole number$\}$.

The system of equations

$$x + 2y = 7,$$
$$x + 2y = 9$$

has no solution. The lines are parallel. Their intersection is empty.

DO EXERCISES 9 AND 10.

⦂⦂ UNIONS

Two sets A and B may be combined to form a new set. It contains the members of A as well as those of B. The new set is called the *union* of A and B, and is represented by the symbol

$$A \cup B.$$

Thus,

$$\{0, 5, 7, 13, 27\} \cup \{0, 2, 3, 4, 5\}$$

represents the set

$$\{0, 2, 3, 4, 5, 7, 13, 27\}.$$

Set union is illustrated by the diagram below.

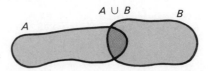

The solution set of the equation $(x - 3)(x + 2) = 0$ is $\{3, -2\}$. This set is the union of the solution sets of the equations $x - 3 = 0$ and $x + 2 = 0$, which are $\{3\}$ and $\{-2\}$.

DO EXERCISES 11 AND 12.

NAME   CLASS   ANSWERS

EXERCISE SET 11.4

▪◦   Name each set using the roster method.

1. The set of whole numbers 3 through 8

2. The set of whole numbers 101 through 107

3. The set of odd numbers between 40 and 50

4. The set of multiples of 5 between 10 and 40

5. $\{x|x \text{ is a square root of } 3\}$ 6. $\{x|x \text{ is the cube of } 0.2\}$

▪◦◦   Determine whether true or false.

7. $2 \in \{x|x \text{ is an odd number}\}$ 8. $7 \in \{x|x \text{ is an odd number}\}$

9. Elton John $\in$ The set of all rock stars   10. Apple $\in$ The set of all fruit

11. $-3 \in \{-4, -3, 0, 1\}$ 12. $0 \in \{-4, -3, 0, 1\}$

1. _____

2. _____

3. _____

4. _____

5. _____

6. _____

7. _____

8. _____

9. _____

10. _____

11. _____

12. _____

ANSWERS

●●● Find each intersection.

13. $\{a, b, c, d, e\} \cap \{c, d, e, f, g\}$

14. $\{a, e, i, o, u\} \cap \{q, u, i, c, k\}$

13. _____

14. _____

15. $\{1, 2, 5, 10\} \cap \{0, 1, 7, 10\}$

16. $\{0, 1, 7, 10\} \cap \{0, 1, 2, 5\}$

15. _____

16. _____

17. $\{1, 2, 5, 10\} \cap \{3, 4, 7, 8\}$

18. $\{a, e, i, o, u\} \cap \{m, n, f, g, h\}$

17. _____

18. _____

⦂⦂ Find each union.

19. $\{a, e, i, o, u\} \cup \{q, u, i, c, k\}$

20. $\{a, b, c, d, e\} \cup \{c, d, e, f, g\}$

19. _____

20. _____

21. $\{0, 1, 7, 10\} \cup \{0, 1, 2, 5\}$

22. $\{1, 2, 5, 10\} \cup \{0, 1, 7, 10\}$

21. _____

22. _____

23. $\{a, e, i, o, u\} \cup \{m, n, f, g, h\}$

24. $\{1, 2, 5, 10\} \cup \{a, b\}$

23. _____

24. _____

EXTENSION EXERCISES

SECTION 11.1

1. Find the set of sensible replacements for each expression. Write set notation for your answers.

   a) $\sqrt{x - 1}$

   b) $\sqrt{x^2 + 4}$

   c) $\sqrt{x^2 - 4}$

   d) $\sqrt{x^2 + x - 2}$

2. Find the numbers of c for which each quadratic equation has real number solutions.

   a) $2x^2 + 5x + c = 0$

   b) $x^2 + cx - 3 = 0$

SECTION 11.2

3. The height of a triangle is 25 cm. What lengths of the base will make the area greater than or equal to 40 cm²?

4. The length of a rectangle is 24 cm. What widths will make the perimeter more than 84 m?

5. A salesperson can choose to be paid in one of two ways.

   Plan A: A salary of $600 per month, plus a commission of 4% of gross sales.

   Plan B: A salary of $800 per month, plus a commission of 6% of gross sales over $10,000.

   For what gross sales is Plan A better than Plan B, assuming that gross sales are always more than $10,000?

6. A mason can be paid in one of two ways.

   Plan A: $300 plus $3.00 per hour.
   Plan B: $8.50 per hour.

   If a job takes n hours, for what values of n is Plan B better for the mason?

SECTION 11.3

Graph the following pair of inequalities using the same set of axes. Then determine whether each point satisfies *both* inequalities.

$$x + y \leq 1$$
$$x - y < 1$$

7. $(0, 0)$

8. $(1, 1)$

9. $(1, 0)$

10. $(0, 1)$

11. $(-1, -1)$

12. $(1, -3)$

13. $(5, 2)$

14. $(-4, 2)$

SECTION 11.4

15. a) Find the union of the set of integers and the set of whole numbers.
    c) Find the union of the rational numbers and the irrational numbers.
    b) Find the intersection of the odd integers and the even integers.
    d) Find the intersection of the positive integers and the even rational numbers.

16. For a set A, find the following.
    a) $A \cup \emptyset$
    b) $A \cup A$
    c) $A \cap A$
    d) $A \cap \emptyset$

17. A set is *closed* under an operation if, when the operation is performed on its members, the result is in the set. For example, the set of real numbers is closed under the operation of addition since the sum of two real numbers is a real number.

a) Is the set of even numbers closed under addition?

b) Is the set of odd numbers closed under addition?

c) Is the set {0, 1} closed under addition?

d) Is the set {0, 1} closed under multiplication?

e) Is the set of real numbers closed under multiplication?

f) Is the set of integers closed under division?

NAME   SCORE   ANSWERS

TEST OR REVIEW—CHAPTER 11

If you miss an item, review the indicated section and objective.

[11.1, ]   Solve using the addition principle.

   1. $y + 9 \geqslant 3$ **2.** $3x + 5 < 2x - 6$

2. _____

[11.2, ]   Solve using the multiplication principle.

   3. $9x \geqslant 63$ **4.** $-3y < -21$

3. _____

4. _____

[11.2, ]   Solve using the addition and multiplication principles.

   5. $2 + 6y > 14$ **6.** $4 - 8x < 13 + 3x$

5. _____

[11.3, ]   Graph on a number line.

   7. $y \leqslant 6$ **8.** $6x - 3 < x + 2$

6. _____

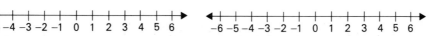

7. See graph.

[11.3, ]   Graph on a number line.

   9. $|x| \leqslant 2$ **10.** $|x| > 1$

8. See graph.

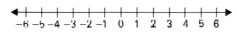 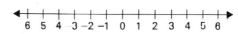

9. See graph.

10. See graph.

ANSWERS

[11.3, 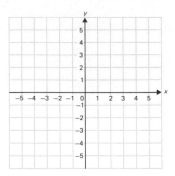]   Graph these inequalities on a plane.

11. $y < x + 3$

11. See graph.

12. $2x + 3y \leqslant 6$

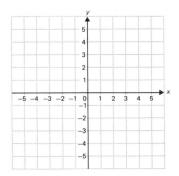

12. See graph.

[11.4, 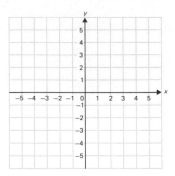]   Find this intersection.

13. $\{8, 9, 10, 5, 7, 0\} \cap \{2, 0, 7, 8, 11\}$

13.

[11.4, 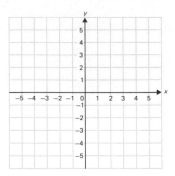]   Find this union.

14. $\{8, 9, 10, 5, 7, 0\} \cup \{2, 0, 7, 8, 11\}$

14.

NAME _____   SCORE _____

FINAL EXAMINATION

This is a practice test. Be sure to review several days before you take it. You can use the chapter tests as a guide.

When you finish this test, check the answers in the back of the book. If you miss an item, go back and review. The bracketed numbers in the answers give the appropriate page numbers.

Chapter 1

1. What is the meaning of x^3?

2. Write fractional notation for 13.6.

3. Write decimal notation for $\dfrac{17}{25}$.

4. Divide and simplify: $\dfrac{5}{6} \div \dfrac{25}{24}$.

Chapter 2

5. Simplify: $|-7|$.

6. Add: $-6 + 12 + (-4) + 7$.

7. Subtract: $2.8 - (-12.2)$.

8. Divide and simplify: $-\dfrac{3}{8} \div \dfrac{5}{2}$.

9. Multiply: $(-9) \cdot 7$.

10. Rename, using a negative exponent:

$\dfrac{1}{3^4}$.

11. Multiply and simplify: $x^{-6} \cdot x^2$

12. Divide and simplify: $\dfrac{y^3}{y^{-4}}$.

1. _____

2. _____

3. _____

4. _____

5. _____

6. _____

7. _____

8. _____

9. _____

10. _____

11. _____

12. _____

Chapter 3

13. Solve: $3x = -24$.

13. _____

14. Solve: $3x + 7 = 2x - 5$.

14. _____

15. Solve: $3(y - 1) - 2(y + 2) = 0$.

15. _____

16. Solve for b: $A = \dfrac{4b}{t}$.

16. _____

17. The sum of two consecutive integers is 49. What are the integers?

17. _____

Chapter 4

18. Collect like terms and arrange in descending order.

$$2x - 3 + 5x^3 - 2x^3 + 7x^3 + x$$

18. _____

19. Add: $(4x^3 + 3x^2 - 5) + (3x^3 - 5x^2 + 4x - 12)$

19. _____

20. Subtract: $(6x^2 - 4x + 1) - (-2x^2 + 7)$.

20. _____

21. Multiply: $-2x^2(4x^2 - 3x + 1)$.

21. _____

22. Multiply: $(2x - 3)(3x^2 - 4x + 2)$.

22. _____

23. Multiply: $(t - 5)(t + 5)$.

23. _____

24. Multiply: $(3m - 2)^2$.

24. _____

ANSWERS

Chapter 5

Factor completely.

25. $8x^2 - 4x$ **26.** $25x^2 - 4$ **27.** $6x^2 - 5x - 6$ **28.** $x^2 - 8x + 16$

29. $x(x - 8) - 5(x - 8)$ **30.** Solve: $x^2 - 8x + 15 = 0$.

31. The square of a number plus the number itself is 20. Find the number.

32. Factor by completing the square. Show your work.

$x^2 - 8x + 12$

Chapter 6

33. Graph: $y = \dfrac{1}{3}x - 2$.

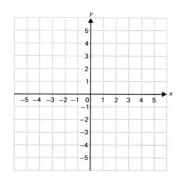

34. Graph: $2x + 3y = -6$.

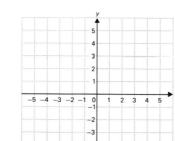

35. Graph: $y = -3$.

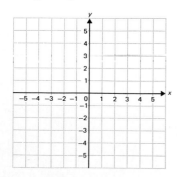

36. Solve by the substitution method.

$$y - x = 1$$
$$y = 3 - x$$

25. _____

26. _____

27. _____

28. _____

29. _____

30. _____

31. _____

32. _____

33. See graph. _____

34. See graph. _____

35. See graph. _____

36. _____

Solve by the addition method.

37. $x + y = 17$
$\quad\;\; x - y = \;\; 7$

38. $4x - 3y = 3$
$\quad\;\; 3x - 2y = 4$

37. _____

39. The sum of two numbers is 24. Three times the first number minus the second number is 20. Find the numbers.

38. _____

Chapter 7

40. Collect like terms.

$\quad 5x^2yz^3 - 3xy^2 + 2x^2yz^3 + 5x^2y^2 - 2xy^2$

39. _____

40. _____

41. Arrange in descending powers of x.

$\quad x^2 + 3ax^3 + 5a^4x^4 + 2a$

41. _____

42. Add.

$\quad (x^3y - 5x^2y^2 + xy^3 + 2) + (x^3y - 2x^2y^2 + 4)$

42. _____

43. Subtract.

$\quad (15x^2y^3 + 10xy^2 + 5) - (5xy^2 - x^2y^2 - 2)$

43. _____

Multiply.

44. _____

44. $(x^2 - 2y)(x^2 + 2y)$

45. $(3x + 4y^2)^2$

45. _____

Factor.

46. $25a^2b^2 - 1$ **47.** $(x - y)^2 - 25$

46. _____

48. $9x^2 + 30xy + 25y^2$ **49.** $15x^2 + 14xy - 8y^2$

47. _____

48. _____

50. $2ac - 6ab - 3db + dc$ **51.** $16x^4 - y^8$

49. _____

50. _____

52. $3a^4 + 6a^2 - 72$

51. _____

Chapter 8

52. _____

53. Multiply and simplify.   **54.** Divide and simplify.

$$\frac{4}{2x - 6} \cdot \frac{x - 3}{x + 3}$$ $$\frac{3a^4}{a^2 - 1} \div \frac{2a^3}{a^2 - 2a + 1}$$

53. _____

54. _____

55. Add: $\dfrac{3}{3x - 1} + \dfrac{4}{5x}$.   **56.** Subtract: $\dfrac{2}{x^2 - 16} - \dfrac{x - 3}{x^2 - 9x + 20}$

55. _____

56. _____

57. Solve: $\dfrac{8}{x} - \dfrac{3}{x + 5} = \dfrac{5}{3}$.   **58.** Solve for m: $\dfrac{1}{t} = \dfrac{1}{m} - \dfrac{1}{n}$.

57. _____

58. _____

ANSWERS

59. _____

60. _____

61. _____

62. _____

63. _____

64. _____

65. _____

66. _____

67. _____

68. _____

69. _____

70. _____

Chapter 9

Simplify.

59. $\sqrt{50}$

60. $\dfrac{\sqrt{72}}{\sqrt{45}}$

61. Add and simplify: $4\sqrt{12} + 2\sqrt{12}$.

62. Solve: $3 - x = \sqrt{x^2 - 3}$.

Chapter 10

63. Solve: $x^2 - x - 6 = 0$.

64. Solve: $x^2 + 3x = 5$.

65. Graph: $y = x^2 + 2x + 1$.

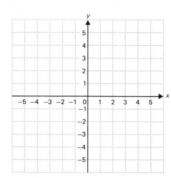

66. The length of a rectangle is 7 m more than the width. The length of a diagonal is 13 m. Find the length.

Chapter 11

67. Solve: $5 - 9x \leqslant 19 + 5x$.

68. Graph: $4x - 3y > 12$.

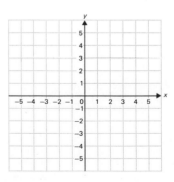

69. Find this intersection.

$\{a, b, c, d, e\} \cap \{g, h, c, d, f\}$

70. Find this union.

$\{a, b, c, d, e\} \cup \{g, h, c, d, f\}$

ANSWERS

CHAPTER 1

MARGIN EXERCISES, SECTION 1.1, pp. 2–6

Answers may vary in Exercises 1 through 7.

1. $\frac{4}{6}, \frac{10}{15}, \frac{6}{9}$ **2.** $\frac{5}{10}, \frac{2}{4}, \frac{3}{6}$ **3.** $\frac{6}{10}, \frac{9}{15}, \frac{12}{20}$ **4.** $\frac{2}{2}, \frac{5}{5}, \frac{14}{14}$ **5.** $\frac{12}{3}, \frac{4}{1}, \frac{20}{5}$ **6.** $\frac{8}{10}, \frac{12}{15}, \frac{24}{30}$ **7.** $\frac{16}{14}, \frac{40}{35}, \frac{24}{21}$ **8.** $\frac{2}{3}$ **9.** $\frac{19}{9}$

10. $\frac{8}{7}$ **11.** $\frac{1}{2}$ **12.** 4 **13.** $\frac{5}{2}$ **14.** $\frac{35}{16}$ **15.** $\frac{23}{15}$ **16.** $\frac{7}{12}$ **17.** $\frac{2}{15}$ **18.** $\frac{7}{36}$

EXERCISE SET 1.1, pp. 7–8

1. $\frac{8}{6}$, etc. **3.** $\frac{12}{22}$, etc. **5.** $\frac{4}{22}$, etc. **7.** $\frac{25}{5}, \frac{20}{4}$, etc. **9.** $\frac{4}{3}$ **11.** $\frac{1}{2}$ **13.** 2 **15.** $\frac{10}{3}$ **17.** $\frac{1}{8}$ **19.** $\frac{51}{8}$ **21.** 1 **23.** $\frac{7}{6}$

25. $\frac{5}{6}$ **27.** $\frac{1}{2}$ **29.** $\frac{5}{18}$ **31.** $\frac{31}{60}$

MARGIN EXERCISES, SECTION 1.2, pp. 9–12

1. $\frac{162}{100}$ **2.** $\frac{35,431}{1000}$ **3.** 0.875 **4.** 0.8 0.$\overline{81}$ **6.** 2.$\overline{5}$ **7.** 10^4 **8.** $5 \cdot 5 \cdot 5 \cdot 5$ **9.** $x \cdot x \cdot x \cdot x \cdot x$ **10.** 4 **11.** 1 **12.** n^5

13. yyy **14.** 140 **15.** 8

EXERCISE SET 1.2, pp. 13–14

1. $\frac{291}{10}$ **3.** $\frac{467}{100}$ **5.** $\frac{362}{100}$ **7.** $\frac{18,789}{1000}$ **9.** 0.25 **11.** 0.6 **13.** 0.$\overline{2}$ **15.** 0.125 **17.** 0.$\overline{45}$ **19.** 0.08$\overline{3}$ **21.** $5 \cdot 5$

23. $m \cdot m \cdot m$ **25.** xxx **27.** $yyyy$ **29.** 1 **31.** p **33.** M **35.** 1 **37.** 100 **39.** 100,000 **41.** 0.0256 **43.** 320

45. Answers may vary; for a nine-digit readout, 6^{12} is too large, so 6^n is too large where n is 12 or larger.

MARGIN EXERCISES, SECTION 1.3, pp. 15–18

1. 22 **2.** 30 **3.** 26 **4.** 30 **5.** 27 **6.** 27 **7.** 27 **8.** 1820 **9.** 1820 **10.** 23 **11.** 25 **12.** Comm. law, mult.

13. Assoc. law, add. **14.** Assoc. law, add. **15.** Assoc. law, mult. **16.** (a) 28; (b) 28 **17.** (a) 77; (b) 77 **18.** 23

19. 15 **20.** (a) 240; (b) 240 **21.** $4(x + y)$ **22.** $5(a + b)$ **23.** $7(p + q + r)$ **24.** $5(x + y)$; 35 **25.** $7(x + y)$; 49

EXERCISE SET 1.3, pp. 19–20

1. 22 **3.** 89 **5.** 41 **7.** 59 **9.** Comm. law, add. **11.** Assoc. law, add. **13.** 52 **15.** $9(x + y)$ **17.** $\frac{1}{2}(a + b)$

19. $1.5(x + z)$ **21.** $4(x + y + z)$ **23.** $\frac{4}{7}(a + b + c + d)$ **25.** $9(x + y)$; 135 **27.** $10(x + y)$; 150 **29.** $5(a + b)$; 45

31. $20(a + b)$; 180 **33.** $(20 + 5) \cdot 8 = 20 \cdot 8 + 5 \cdot 8 = 160 + 40 = 200$; only Part I is given.

35. $(10 + 2) \cdot 39 = 10 \cdot 39 + 2 \cdot 39 = 390 + 78 = 468$

MARGIN EXERCISES, SECTION 1.4, pp. 21–22

1. $5y + 15$ **2.** $4x + 8y + 20$ **3.** $8m + 24n + 32p$ **4.** $5(x + 2)$ **5.** $3(4 + x)$ **6.** $3(2x + 4 + 3y)$ **7.** $5(x + 2y + 5)$

8. $Q(1 + ab)$ **9.** $8y$ **10.** $5x$ **11.** $1.03x$ **12.** $18p + 9q$ **13.** $11x + 8y$ **14.** $16y$ **15.** $7s + 13w$ **16.** $9x + 10y$

17. $6a + 1.07b$

EXERCISE SET 1.4, pp. 23–24

1. $3x + 3$ **3.** $4 + 4y$ **5.** $36t + 27z$ **7.** $7x + 28 + 42y$ **9.** $15x + 45 + 35y$ **11.** $2(x + 2)$ **13.** $6(x + 4)$

15. $3(3x + y)$ **17.** $7(2x + 3y)$ **19.** $5(1 + 2x + 3y)$ **21.** $8(a + 2b + 8)$ **23.** $3(x + 6y + 5z)$ **25.** $5x + 12$ **27.** $11a$

29. $8x + 9z$ **31.** $43a + 150c$ **33.** $1.09x + 1.2t$ **35.** $24u + 5t$ **37.** $50 + 6t + 8y$ **39.** $b + \dfrac{4}{3}$ **41.** $\dfrac{13}{4}y$

MARGIN EXERCISES, SECTION 1.5, pp. 25–28

1. $\dfrac{4}{7}$ **2.** 1 **3.** $\dfrac{2}{3}$ **4.** 1 **5.** $\dfrac{11}{4}$ **6.** $\dfrac{7}{15}$ **7.** $\dfrac{1}{5}$ **8.** 3 **9.** $\dfrac{21}{20}$ **10.** $\dfrac{5}{6}$ **11.** $\dfrac{45}{28}$ **12.** $\dfrac{8}{21}$ **13.** $\dfrac{12}{35}$ **14.** $\dfrac{14}{45}$

15.   **16.**

17. True   **18.** True   **19.** False   **20.** $<$ **21.** $=$ **22.** $>$ **23.** $>$ **24.** $<$ **25.** $>$ **26.** $>$

EXERCISE SET 1.5, pp. 29–30

1. $\dfrac{4}{3}$ **3.** 8 **5.** 1 **7.** $\dfrac{35}{18}$ **9.** $\dfrac{10}{3}$ **11.** $\dfrac{1}{2}$ **13.** $\dfrac{5}{36}$ **15.** 500 **17.** $\dfrac{3}{40}$ **19.**

21.   **23.** $=$ **25.** $>$ **27.** $>$ **29.** $=$ **31.** $<$ **33.** 3.2 **35.** $<$

MARGIN EXERCISES, SECTION 1.6, pp. 31–34

Answers may vary in Exercises 1 through 3.

1. $2 + 3 = 5; 8 - 5 = 3; 4 + 5.8 = 9.8$ **2.** $2 + 3 = 6; 8 - 5 = 4; 4 + 5.8 = 8.8$ **3.** $x + 2 = 5; y - 1 = 8; 10 - y = 3$

4. $4, 8, 10$; any three numbers other than 7   **5.** 7 **6.** 6 **7.** 4 **8.** 5 **9.** 8 **10.** 273 **11.** 4.87 **12.** $\dfrac{9}{8}$ **13.** 7

14. $\dfrac{9}{2}$, or 4.5 **15.** 5.6 **16.** Yes   **17.** No   **18.** Yes   **19.** No   **20.** No   **21.** No

EXERCISE SET 1.6, pp. 35–36

1. 2 **3.** 9 **5.** 5 **7.** 20 **9.** 7 **11.** 5 **13.** 19 **15.** 5.66 **17.** 2818 **19.** $\dfrac{5}{12}$ **21.** $\dfrac{1}{6}$ **23.** 4 **25.** $\dfrac{5}{4}$, or 1.25
27. 0.24 **29.** 3.1 **31.** 8.5 **33.** $\dfrac{140}{3}$ **35.** $470{,}188$

MARGIN EXERCISES, SECTION 1.7, p. 37–38

1. $x + 37 = 73; 36$ **2.** $\dfrac{2}{3}x = 44; 66$ **3.** $145 = p + 132; 13$ million   **4.** $1.16x = 14{,}500; \$12{,}500$

EXERCISE SET 1.7, pp. 39–42

1. 72 **3.** 17 **5.** 9 **7.** $78, 114 = 4A$ **9.** $35 = \dfrac{2}{5}t$ **11.** $78.3 = 13.5 + m$ **13.** $640 = 1.6s$ **15.** $1175 = 1.8c$

17. $19{,}528.5 \text{ km}^2$ **19.** $87\dfrac{1}{2}$ words per min   **21.** $64.8°\text{C}$ **23.** 400 kilowatt hours   **25.** $\$652.78$ **27.** Salary now

MARGIN EXERCISES, SECTION 1.8, pp. 43–46

1. 0.462 **2.** 1 **3.** $\dfrac{67}{100}$ **4.** $\dfrac{456}{1000}$ **5.** $\dfrac{1}{400}$ **6.** 677% **7.** 99.44% **8.** 25% **9.** 37.5% **10.** $66.6\overline{6}\%$, or $66\dfrac{2}{3}\%$

11. 11.04 **12.** 10 **13.** 32% **14.** 25% **15.** 225 **16.** 50 **17.** $x = 19\% \cdot 586{,}400; 111{,}416$ sq mi

18. $x + 7\%x = 8988; \$8400$ **19.** $x + 12\% x = 20{,}608; \$18{,}400$

1. 0.76   **3.** 0.547   **5.** $\dfrac{20}{100}$ **7.** $\dfrac{786}{1000}$ **9.** 454%   **11.** 99.8%   **13.** 12.5%   **15.** 68%   **17.** $x = 65\% \cdot 840$; 546

19. $24\% \cdot x = 20.4$; 85   **21.** $x\% \cdot 80 = 100$; 125%   **23.** $76 = x\% \cdot 88$; approx. 86.36%   **25.** $208 = 26\% \cdot x$; \$800

27. $x = 5\% \cdot 428.86$; \$21.44; \$450.30   **29.** $x + 9\%x = 8502$; \$7800   **31.** $(8\% - 7.4\%) \cdot 9600 = x$; \$57.60

EXERCISE SET 1.8A, pp. 51–52

1. 0.38   **3.** 0.654   **5.** 0.0824   **7.** 0.00012   **9.** 0.0073   **11.** 1.25   **13.** $\dfrac{30}{100}$ **15.** $\dfrac{135}{1000}$ **17.** $\dfrac{32}{1000}$ **19.** $\dfrac{120}{100}$ **21.** $\dfrac{35}{10,000}$

23. $\dfrac{42}{100,000}$ **25.** 62%   **27.** 62.3%   **29.** 720%   **31.** 200%   **33.** 7.2%   **35.** 0.57%   **37.** 17%   **39.** 70%   **41.** 35%

43. 50%   **45.** 60%   **47.** $33\dfrac{1}{3}\%$ **49.** 95   **51.** 25%   **53.** 64

MARGIN EXERCISES, SECTION 1.9, pp. 53–58

1. 21 m^2; 20 m   **2.** 25 m^2; 20 m   **3.** 60 cm^2 **4.** 38 m^2 **5.** 35 cm^2 **6.** 64^0 **7.** 200 m^3 **8.** 18 m   **9.** 56.52 cm

10. 254.34 ft^2

SOMETHING EXTRA—CALCULATOR CORNER: ESTIMATING π, p. 58

1. 3.1415929   **2.** 3.1416   **3.** 3.3436735   **4.** 3.1578644

EXERCISE SET 1.9, pp. 59–62

1. 1100 cm^2; 138 cm   **3.** 225 ft^2; 60 ft   **5.** 120 cm^2 **7.** 200 in^2 **9.** 280 m^2 **11.** 74°   **13.** 36°   **15.** 384 m^3

17. 4.8 m; 15.072 m   **19.** $\dfrac{3}{2}$ in.; $\dfrac{33}{7}$ in.   **21.** 18.0864 m^2 **23.** 31,400 ft^2 **25.** 12.3 cm   **27.** 7.2 yd

EXTENSION EXERCISES, pp. 63–64

1. $=$ **3.** $\neq$ **5.** $\neq$ **7.** b, c, and e **9.** No. For example, $(12 \div 6) \div 2 = 1$, but $12 \div (6 \div 2) = 4$.   **11.** $2\left(\dfrac{1}{3}x + \dfrac{1}{7}y + 1\right)$

13. $5(x + 3y + 5)$ **15.** Two are $\dfrac{13}{20}$ and $\dfrac{7}{10}$. There are many others.

17. (a) $\dfrac{7}{13}, \dfrac{7}{11}, \dfrac{7}{9}, \dfrac{7}{8}, \dfrac{7}{5}, \dfrac{7}{3}$; (b) $\dfrac{1}{15}, \dfrac{2}{15}, \dfrac{4}{15}, \dfrac{7}{15}, \dfrac{8}{15}, \dfrac{11}{15}$; (c) $\dfrac{1}{2}, \dfrac{2}{3}, \dfrac{9}{13}, \dfrac{8}{11}, \dfrac{7}{8}, \dfrac{6}{5}$ **19.** No   **21.** No   **23.** Yes   **25.** 5

27. 24   **29.** 43   **31.** (a) 68%; (b) \$75   **33.** 1600   **35.** 3 cm   **37.** (a) 26 in.; (b) 18.28 cm

TEST OR REVIEW—CHAPTER 1, pp. 65–66

1. $\dfrac{2}{3}$ **2.** $\dfrac{23}{15}$ **3.** $\dfrac{9}{32}$ **4.** $y \cdot y \cdot y \cdot y$ **5.** $\dfrac{569}{100}$ **6.** 0.375   **7.** 62   **8.** Assoc. law, mult.   **9.** $5(x + y)$; 50

10. $6(3x + y)$ **11.** $4(9x + 4 + y)$ **12.** $40a + 24b + 16$ **13.** $40y + 10a$ **14.** $10a + 1.23b$ **15.** $\dfrac{4}{3}$ **16.** $<$ **17.** 8.4

18. 6   **19.** $\dfrac{3}{5}x = 18$; 30   **20.** 0.458   **21.** 56%   **22.** $6\dfrac{2}{3}\%$ **23.** 64 ft^2 **24.** 26 m^2 **25.** 314 yd^2

CHAPTER 2

READINESS CHECK, p. 68

1. [1.1,■■■] $\frac{3}{5}$ **2.** [1.2, ■■■] $w \cdot w \cdot w \cdot w$ **3.** [1.3, ■■■] Distributive   **4.** [1.4, ■■■] $3(x + 3 + 4y)$

5. [1.4, ■■] $21z + 7y + 14$ **6.** [1.5, ■■] $\frac{28}{3}$

MARGIN EXERCISES, SECTION 2.1, pp. 68–71

1. -1 **2.** 2   **3.** 0   **4.** 4   **5.** 1   **6.** -2 **7.** -5 **8.** 4   **9.** 13   **10.** -28 **11.** 0   **12.** >   **13.** >   **14.** <   **15.** >
16. <   **17.** <   **18.** 8   **19.** 10   **20.** 29   **21.** 6   **22.** 0

SOMETHING EXTRA—CALCULATOR CORNER, p. 72

1. 1, 9, 36, 100; 225, 441   **2.** 2^n: 2, 4, 8, 16, 32, 64, 128; 2^{-n}: 0.5, 0.25, 0.125, 0.0625, 0.0313, 0.0156, 0.0078

EXERCISE SET 2.1, pp. 73–74

1. 6   **3.** -6 **5.** -7 **7.** 1   **9.** 12   **11.** -70 **13.** 0   **15.** -34 **17.** 1   **19.** -7 **21.** 14   **23.** 0   **25.** >   **27.** <
29. <   **31.** <   **33.** >   **35.** <   **37.** 3   **39.** 10   **41.** 0   **43.** 24   **45.** 53   **47.** 8   **49.** 6   **51.** 6   **53.** -5
55. 10   **57.** $-x$

MARGIN EXERCISES, SECTION 2.2, pp. 75–76

1. -3 **2.** -8 **3.** 4   **4.** 0   **5.** $4 + (-5) = -1$ **6.** $-2 + (-4) = -6$ **7.** $-3 + 8 = 5$ **8.** -11 **9.** -12 **10.** -34
11. -22 **12.** 2   **13.** -4 **14.** -2 **15.** 3   **16.** 0   **17.** 0   **18.** 0   **19.** 0   **20.** -12

EXERCISE SET 2.2, pp. 77–78

1. -7 **3.** -4 **5.** 0   **7.** -13 **9.** -16 **11.** -11 **13.** 0   **15.** 0   **17.** 8   **19.** -8 **21.** -25 **23.** -17 **25.** 0
27. 0   **29.** 3   **31.** -9 **33.** -5 **35.** 7   **37.** -3 **39.** 0   **41.** -5 **43.** -21 **45.** 2   **47.** -26 **49.** -22 **51.** 32
53. 0   **55.** 45   **57.** -198 **59.** 52   **61.** 30,937

MARGIN EXERCISES, SECTION 2.3, pp. 79–80

1. 4   **2.** -6 **3.** Negative two minus six; -8 **4.** Negative two minus negative three; 1
5. Negative seven minus negative three; -4 **6.** Negative four minus four; -8 **7.** Negative six minus two; -8
8. (a) -2; (b) -2 **9.** (a) -11; (b) -11 **10.** (a) 4; (b) 4   **11.** (a) -2; (b) -2 **12.** -6 **13.** 7   **14.** -11 **15.** 1   **16.** -4

EXERCISE SET 2.3, pp. 81–82

1. -4 **3.** -7 **5.** -6 **7.** 0   **9.** -4 **11.** -7 **13.** -6 **15.** 0   **17.** 0   **19.** 14   **21.** 11   **23.** -14 **25.** 5
27. -7 **29.** -1 **31.** 18   **33.** -5 **35.** -3 **37.** -21 **39.** 5   **41.** -8 **43.** 12   **45.** -23 **47.** -68 **49.** -73
51. 116   **53.** 0   **55.** $-309,882$

MARGIN EXERCISES, SECTION 2.4, pp. 83–84

1. 20, 10, 0, -10, -20, -30 **2.** -18 **3.** -100 **4.** -80 **5.** -10, 0, 10, 20, 30   **6.** 12   **7.** 32   **8.** 35   **9.** -2
10. 5   **11.** -3 **12.** 8   **13.** -6

EXERCISE SET 2.4, pp. 85–86

1. -16 **3.** -42 **5.** -24 **7.** -72 **9.** 16 **11.** 42 **13.** 24 **15.** 72 **17.** -120 **19.** 1000 **21.** 90

23. 200 **25.** -6 **27.** -13 **29.** -2 **31.** 4 **33.** -8 **35.** 2 **37.** -12 **39.** -8

41. $-9, -98, -987, -9876; -98{,}765$

MARGIN EXERCISES, SECTION 2.5, pp. 87–88

1. -8.7 **2.** $\dfrac{8}{9}$ **3.** 7.74 **4.** $\dfrac{10}{3}$ **5.** 8.32 **6.** $-\dfrac{5}{4}$ **7.** 12 **8.** $\dfrac{5}{6}$ **9.** -17.2 **10.** 4.1 **11.** $\dfrac{8}{3}$ **12.** 3.5 **13.** -6.2

14. $-\dfrac{2}{9}$ **15.** $-\dfrac{19}{20}$ **16.** 510.8 ml   **17.** -9 **18.** $-\dfrac{1}{5}$ **19.** -7.5 **20.** $-\dfrac{1}{24}$

EXERCISE SET 2.5, pp. 89 90

1. 4.7 **3.** $-\dfrac{7}{2}$ **5.** 7 **7.** 26.9 **9.** $\dfrac{1}{3}$ **11.** $-\dfrac{7}{6}$ **13.** 9.3 **15.** -90.3 **17.** -12.4 **19.** $\dfrac{9}{10}$ **21.** 34.8 **23.** -567

25. 19.2 **27.** $\dfrac{2}{3}$ **29.** 89.3 **31.** $\dfrac{14}{3}$ **33.** -1.8 **35.** -8.1 **37.** $-\dfrac{1}{5}$ **39.** $-\dfrac{8}{7}$ **41.** $-\dfrac{3}{8}$ **43.** $-\dfrac{29}{35}$ **45.** $-\dfrac{11}{15}$

47. -6.3 **49.** -4.3 **51.** -4 **53.** $-\dfrac{1}{4}$ **55.** $\dfrac{1}{12}$ **57.** $-\dfrac{17}{12}$ **59.** $\dfrac{1}{8}$ **61.** 19.9 **63.** -9

65. -0.01 **67.** 67.85

MARGIN EXERCISES, SECTION 2.6, pp. 91–94

1. -30 **2.** $-\dfrac{10}{27}$ **3.** $-\dfrac{7}{10}$ **4.** -30.033 **5.** 64 **6.** $\dfrac{20}{63}$ **7.** $\dfrac{2}{3}$ **8.** 13.455 **9.** -30 **10.** -30.75 **11.** $-\dfrac{5}{3}$ **12.** 120

13. $\dfrac{3}{2}$ **14.** $-\dfrac{4}{5}$ **15.** $-\dfrac{1}{3}$ **16.** -5 **17.** $-\dfrac{2}{3}, \dfrac{3}{2}; \dfrac{5}{4}, -\dfrac{4}{5}$; 0, does not exist; $-1, 1; 4.5, -\dfrac{1}{4.5}$ or $-\dfrac{10}{45}$ **18.** $-\dfrac{20}{21}$ **19.** $-\dfrac{12}{5}$

20. $\dfrac{16}{7}$ **21.** -7 **22.** $\dfrac{19}{20}$ **23.** $\dfrac{8}{5}$ **24.** $-\dfrac{10}{3}$ **25.** $-\dfrac{5}{6}$ **26.** $-\dfrac{5}{6}, \dfrac{-5}{6}$ **27.** $\dfrac{-8}{7}, \dfrac{8}{-7}$ **28.** $\dfrac{-10}{3}, -\dfrac{10}{3}$

EXERCISE SET 2.6, pp. 95–96

1. -72 **3.** -12.4 **5.** 24 **7.** 21.7 **9.** $-\dfrac{2}{5}$ **11.** $\dfrac{1}{12}$ **13.** -17.01 **15.** $-\dfrac{5}{12}$ **17.** -90 **19.** 420 **21.** $\dfrac{2}{7}$ **23.** -60

25. 150 **27.** $-\dfrac{1}{5}$ **29.** 4 **31.** $-\dfrac{5}{7}$ **33.** $-\dfrac{11}{4}$ **35.** $-\dfrac{9}{8}$ **37.** $\dfrac{5}{3}$ **39.** $\dfrac{9}{14}$ **41.** $\dfrac{9}{64}$ **43.** -1 **45.** -2 **47.** $\dfrac{11}{13}$

49. $-\dfrac{23}{14}$ **51.** 1 **53.** -8

MARGIN EXERCISES, SECTION 2.7, pp. 97–98

1. (a) 8; (b) 8 **2.** (a) -4; (b) -4 **3.** (a) -25; (b) -25 **4.** (a) -20; (b) -20 **5.** $5x, -4y, 3$ **6.** $-4y, -2x, 3z$

7. $3x - 15$ **8.** $5x - 5y + 20$ **9.** $-2x + 6$ **10.** $bx - 2by + 4bz$ **11.** $4(x - 2)$ **12.** $3(x - 2y + 3)$ **13.** $b(x + y - z)$

14. $3x$ **15.** $6x$ **16.** $0.59x$ **17.** $3x + 3y$ **18.** $-4x - 5y - 7$

EXERCISE SET 2.7, pp. 99–100

1. 12 **3.** -8 **5.** $8x, -1.4y$ **7.** $-5x, 3y, -14z$ **9.** $8(x - 3)$ **11.** $4(8 - x)$ **13.** $2(4x + 5y - 11)$ **15.** $a(x - 7)$

17. $a(x + y - z)$ **19.** $7x - 14$ **21.** $-7y + 14$ **23.** $-21 + 3t$ **25.** $-4x - 12y$ **27.** $-14x - 28y + 21$ **29.** $8x$

31. $16y$ **33.** $-11x$ **35.** $0.17x$ **37.** $4x + 2y$ **39.** $7x + y - 11$ **41.** $0.8x + 0.5y$ **43.** $\dfrac{3}{5}x + \dfrac{3}{5}y$ **45.** $1.08P$

MARGIN EXERCISES, SECTION 2.8, pp. 101–102

1. $-x - 2$ **2.** $-5x + 2y + 8$ **3.** $-6 + t$ **4.** $-x + y$ **5.** $4a - 3t + 10$ **6.** $-18 + m + 2n - 4z$

7. $2x - 9$ **8.** $3y + 2$ **9.** 6   **10.** 4   **11.** $5x - y - 8$

EXERCISE SET 2.8, pp. 103–104

1. $-2x - 7$ **3.** $-5x + 8$ **5.** $-4a + 3b - 7c$ **7.** $-6x - 8y - 5$ **9.** $-3x - 5y + 6$ **11.** $8x + 6y + 43$ **13.** $5x - 3$

15. $7a - 9$ **17.** $5x - 6$ **19.** $-5x - 2y$ **21.** $y - 5z$ **23.** -16 **25.** 56   **27.** 19   **29.** 22   **31.** $12x - 2$

33. $16x - 4$ **35.** $3x + 30$ **37.** $9x - 18$ **39.** $-4x - 64$

MARGIN EXERCISES, SECTION 2.9, pp. 105–108

1. $\dfrac{1}{4^3}$, or $\dfrac{1}{4 \cdot 4 \cdot 4}$, or $\dfrac{1}{64}$ **2.** $\dfrac{1}{5^2}$, or $\dfrac{1}{5 \cdot 5}$, or $\dfrac{1}{25}$ **3.** $\dfrac{1}{2^4}$, or $\dfrac{1}{2 \cdot 2 \cdot 2 \cdot 2}$, or $\dfrac{1}{16}$ **4.** 3^{-2} **5.** 5^{-4} **6.** 7^{-3} **7.** $\dfrac{1}{5^3}$ **8.** $\dfrac{1}{7^5}$

9. $\dfrac{1}{10^4}$ **10.** 3^8 **11.** 5^2 **12.** 6^{-7} **13.** x^{-4} **14.** y^{-2} **15.** x^{-8} **16.** 4^3 **17.** 7^{-5} **18.** a^7 **19.** b **20.** x^4

21. x^7 **22.** 3^{20} **23.** x^{-12} **24.** y^{15} **25.** x^{-32} **26.** $16x^{20}y^{-12}$ **27.** $25x^{10}y^{-12}z^{-6}$ **28.** $27y^{-6}x^{-15}z^{24}$ **29.** \$3121.79

EXERCISE SET 2.9, pp. 109–110

1. $\dfrac{1}{3^2}$, or $\dfrac{1}{3 \cdot 3}$, or $\dfrac{1}{9}$ **3.** $\dfrac{1}{10^4}$, or $\dfrac{1}{10 \cdot 10 \cdot 10 \cdot 10}$, or $\dfrac{1}{10,000}$ **5.** 4^{-3} **7.** x^{-3} **9.** a^{-4} **11.** p^{-n} **13.** $\dfrac{1}{7^3}$ **15.** $\dfrac{1}{a^3}$ **17.** $\dfrac{1}{y^4}$

19. $\dfrac{1}{z^n}$ **21.** 2^7 **23.** 3^3 **25.** x^{-1} **27.** x^7 **29.** x^{-13} **31.** 1   **33.** 7^3 **35.** x^2 **37.** x^9 **39.** z^{-4} **41.** x^3 **43.** 1

45. 2^6 **47.** 5^{-6} **49.** x^{12} **51.** $x^{-12}y^{-15}$ **53.** $x^{24}y^8$ **55.** $9x^6y^{-16}z^{-6}$ **57.** \$2508.80   **59.** \$22,318.40

MARGIN EXERCISES, SECTION 2.10, pp. 111–113

1. 4.6×10^{11} **2.** 9.3×10^7 **3.** 1.235×10^{-8} **4.** 1.7×10^{-24} **5.** 3.14×10^{-4} **6.** 2.18×10^8 **7.** 789,300,000,000

8. 0.0000567   **9.** 5.6×10^{-15} **10.** 7.462×10^{-13} **11.** 2.0×10^3 **12.** 5.5×10^2

SOMETHING EXTRA—CALCULATOR CORNER: NUMBER PATTERNS, p. 114

1. 1; 121; 12,321; 1,234,321; 123,454,321   **2.** 24; 2904; 295,704; 29,623,704; 2,962,903,704   **3.** 48; 408; 4008; 40,008; 400,008

4. 81; 9801; 998,001; 99,980,001; 9,999,800,001

EXERCISE SET 2.10, pp. 115–116

1. 7.8×10^{10} **3.** 9.07×10^{17} **5.** 3.74×10^{-6} **7.** 1.8×10^{-8} **9.** 10^7 **11.** 10^{-9} **13.** 784,000,000

15. 0.0000000008764   **17.** 100,000,000   **19.** 0.0001   **21.** 6×10^9 **23.** 3.38×10^4 **25.** 8.1477×10^{-13} **27.** 2.5×10^{13}

29. 5.0×10^{-4} **31.** 3.0×10^{-21} **33.** 2.478125×10^{-1}

EXTENSION EXERCISES, pp. 117–118

1. True   **3.** Not true for $x \geqslant 0$ **5.** Not true for $-1 \leqslant x \leqslant 1$ **7.** Not true for $x < 0$ and $y > 0$ or vice versa

9. When $x < 0$ **11.** 1   **13.** 0   **15.** -9 **17.** 16   **19.** -32 **21.** -2

23. (a) $-\dfrac{13}{2}$ and $-\dfrac{33}{5}$ are two. There are many others. (b) 3.01 and 3.11 are two. There are many others.

25. $-\dfrac{5}{6}, -\dfrac{5}{12}, -\dfrac{1}{3}, -\dfrac{1}{4}, -\dfrac{1}{16}, 0$ **27.** -168 **29.** 16   **31.** -1 **33.** (a) Positive; (b) Negative   **35.** $\dfrac{a + b}{-a}$

37. $\frac{1}{2}h(a + b)$ **39.** $-15y - 81$ **41.** $-2x - 1$ **43.** False   **45.** False   **47.** False   **49.** True

51. (a) No; (b) No; (c) No; (d) No; (e) Yes; (f) No   **53.** (a) 1.6×10^2; (b) 2.5×10^{-11}

TEST OR REVIEW—CHAPTER 2, pp. 119–120

1. 9   **2.** $\frac{3}{4}$ **3.** $\frac{5}{6}$ **4.** -17 **5.** 5.8   **6.** $-\frac{5}{21}$ **7.** -7 **8.** 19.9   **9.** $-\frac{5}{4}$ **10.** 40   **11.** $-\frac{5}{8}$ **12.** -360 **13.** -2

14. $\frac{27}{49}$ **15.** $7(x - 4)$ **16.** $-15x + 6$ **17.** $-2a + 4b$ **18.** $13y - 8$ **19.** $5x - 6y + 30$ **20.** 5^{-5} **21.** $\frac{1}{6^4}$ **22.** y^{-4}

23. a^8 **24.** x^{-12} **25.** \$1442.90   **26.** 2.3×10^{-5}

CHAPTER 3

READINESS CHECK, p. 122

1. [2.2, ●] -11 **2.** [2.3, ●●] 16   **3.** [2.6, ●●] $-\frac{5}{12}$ **4.** [2.6, ●●●] $\frac{1}{3}$ **5.** [2.7, ●●●] $3x - 15$ **6.** [2.7, ●●●●] $-7x$

7. [2.8, ●●] $3w + 5$

MARGIN EXERCISES, SECTION 3.1, p. 123

1. -5 **2.** 13.2   **3.** -2 **4.** 10.8

SOMETHING EXTRA—CALCULATOR CORNER: NUMBER PATTERNS, p. 124

1. 9; 1089; 110,889; 11,108,889; 1,111,088,889   **2.** 54; 6534; 665,334; 66,653,334; 6,666,533,334   **3.** 111; 1221; 12,321; 123,321; 1,233,321   **4.** 111; 222; 333; 444; 555

EXERCISE SET 3.1, pp. 125–126

1. 4   **3.** -20 **5.** $\frac{1}{3}$ **7.** $\frac{41}{24}$ **9.** 4   **11.** $\frac{1}{2}$ **13.** -5.1 **15.** 10.9   **17.** $-\frac{7}{4}$ **19.** -4 **21.** 16   **23.** -5 **25.** $1\frac{5}{6}$

27. $6\frac{2}{3}$ **29.** $-7\frac{6}{7}$ **31.** 342.246

MARGIN EXERCISES, SECTION 3.2, pp. 127–128

1. 5   **2.** -12 **3.** $-\frac{7}{4}$ **4.** 7800   **5.** -3

SOMETHING EXTRA—CALCULATOR CORNER, p. 128

AEDG of length 893

EXERCISE SET 3.2, pp. 129–130

1. 6   **3.** -4 **5.** -6 **7.** 63   **9.** $\frac{3}{5}$ **11.** -20 **13.** 36   **15.** $\frac{3}{2}$ **17.** 10   **19.** -2 **21.** 8   **23.** 13.38

25. $\frac{9}{2}$ **27.** 15.9

MARGIN EXERCISES, SECTION 3.3, pp. 131–134

1. 5   **2.** 4   **3.** 4   **4.** 39   **5.** $-\dfrac{3}{2}$ **6.** -4.3 **7.** -3 **8.** 800   **9.** 1   **10.** 2   **11.** 2   **12.** $\dfrac{17}{2}$ **13.** $\dfrac{8}{3}$ **14.** -4.3

EXERCISE SET 3.3, pp. 135–138

1. 5   **3.** 10   **5.** -8 **7.** 6   **9.** 5   **11.** -20 **13.** 6   **15.** 8400   **17.** 7   **19.** 7   **21.** 3   **23.** 5   **25.** 2   **27.** 10

29. 4   **31.** 8   **33.** $-\dfrac{3}{5}$ **35.** -4 **37.** $\dfrac{10}{7}$ **39.** 4.4233464

MARGIN EXERCISES, SECTION 3.4, p. 139

1. 2   **2.** 3   **3.** -2 **4.** $-\dfrac{1}{2}$

SOMETHING EXTRA—HANDLING DIMENSION SYMBOLS (PART I), p. 140

1. $5\,\dfrac{\text{mi}}{\text{hr}}$ **2.** $34\,\dfrac{\text{km}}{\text{hr}}$ **3.** $2.2\,\dfrac{\text{m}}{\text{sec}}$ **4.** $19\,\dfrac{\text{ft}}{\text{min}}$ **5.** 52 ft   **6.** 102 sec   **7.** 31 m   **8.** 32 hr   **9.** $\dfrac{23}{20}$ lb   **10.** 12 km

11. 22 g   **12.** $105\,\dfrac{\text{m}}{\text{sec}}$

EXERCISE SET 3.4, pp. 141–144

1. 6   **3.** 2   **5.** 6   **7.** 8   **9.** 4   **11.** 1   **13.** 17   **15.** -8 **17.** $-\dfrac{5}{3}$ **19.** -3 **21.** 2   **23.** 5   **25.** 434.08657

MARGIN EXERCISES, SECTION 3.5, pp. 145–146

1. 3, -4 **2.** 7, 3   **3.** $-\dfrac{1}{4}, \dfrac{2}{3}$ **4.** $0, \dfrac{17}{3}$

SOMETHING EXTRA—CALCULATOR CORNER: NUMBER PATTERNS, p. 146

1. 1, 4, 9, 16, 25, 36; 49, 64   **2.** $\dfrac{8\cdot 9}{2}$ or 36, $\dfrac{9\cdot 10}{2}$ or 45; $\dfrac{n(n+1)}{2}$

EXERCISE SET 3.5, pp. 147–148

1. $-8, -6$ **3.** $3, -5$ **5.** 12, 11   **7.** 0, 13   **9.** $0, -21$ **11.** $-\dfrac{5}{2}, -4$ **13.** $\dfrac{1}{3}, -2$ **15.** $0, \dfrac{2}{3}$ **17.** 0, 18   **19.** $\dfrac{1}{9}, \dfrac{1}{10}$

21. 2, 6   **23.** 0, 2   **25.** $\dfrac{1}{3}, 20$ **27.** $-10, \dfrac{10}{3}$

29. In (a), 4 checks so it is a solution; 3 is not a solution since it does not check. In (b) neither number checks, so neither is a solution. In both (a) and (b) the principle of zero products does not apply because the right side is 8 instead of 0.

MARGIN EXERCISES, SECTION 3.6, pp. 149–152

1. 3 ft, 5 ft   **2.** 5   **3.** $5700   **4.** 18, 20   **5.** 206.2 miles   **6.** Length is 20 m; width is 10 m

7. 30°, 60°, 90°   **8.** $11,500

EXERCISE SET 3.6, pp. 153–158

1. 19   **3.** -10 **5.** 40   **7.** 20 m; 40 m; 120 m   **9.** 37; 39   **11.** 56; 58   **13.** 35; 36; 37   **15.** 61; 63; 65

17. Length is 90 m; width is 65 m   **19.** Length is 49 m; width is 27 m   **21.** 22.5°   **23.** $4400   **25.** $16   **27.** 450.5 mi

29. 28°; 84°; 68°   **31.** Approx. 3.92 billion   **33.** 1984   **35.** $6400

MARGIN EXERCISES, SECTION 3.7, pp. 159–160

1. $I = \dfrac{E}{R}$ **2.** $\dfrac{C}{\pi} = D$ **3.** $4A - a - b - d = c$ **4.** $I = \dfrac{9R}{A}$

EXERCISE SET 3.7, pp. 161–162

1. $b = \dfrac{A}{h}$ **3.** $r = \dfrac{d}{t}$ **5.** $P = \dfrac{I}{rt}$ **7.** $a = \dfrac{F}{m}$ **9.** $w = \dfrac{1}{2}(P - 2l)$ **11.** $r^2 = \dfrac{A}{\pi}$ **13.** $b = \dfrac{2A}{h}$ **15.** $m = \dfrac{E}{c^2}$

17. $3A - a - c = b$ **19.** $t = \dfrac{3k}{v}$ **21.** $b = \dfrac{2A - ah}{h}$ **23.** $D^2 = \dfrac{2.5H}{N}$ **25.** $S = \dfrac{360A}{\pi r^2}$ **27.** $t = \dfrac{R - 3.85}{-0.0075}$

EXTENSION EXERCISES, pp. 163–164

1. $a + 4$ **3.** $1 - c - a$ **5.** $\dfrac{b}{3c}$ **7.** $\dfrac{4b}{a}$ **9.** $x = -\dfrac{1}{3}, y = -5$ **11.** 2   **13.** -4 **15.** $1, -2$ **17.** $1, -\dfrac{7}{2}, \dfrac{5}{3}$ **19.** 76

21. $7\dfrac{1}{2}\%$ **23.** A quadruples   **25.** An increase in a

TEST OR REVIEW—CHAPTER 3, pp. 165–166

1. 6.1   **2.** 1   **3.** 12   **4.** -7 **5.** 4   **6.** -5 **7.** 6   **8.** -6 **9.** $-9, 6$ **10.** $0, \dfrac{7}{4}$ **11.** 9   **12.** 57, 59

13. Width is 11 cm; length is 17 cm   **14.** $102°; 34°; 44°$ **15.** $s = \dfrac{P}{4}$ **16.** $h = \dfrac{3V}{B}$

CHAPTER 4

READINESS CHECK, p. 168

1. [1.4, ▪] $3s + 3t + 24$ **2.** [2.7, ●●●] $-7x - 28$ **3.** [2.7, ⠿] $5x$ **4.** [2.2, ●] -10 **5.** [2.1, ●] 6

6. [2.3, ●●] 12   **7.** [2.4, ●] 30   **8.** [2.7, ●] $6x, -5y, 9z$ **9.** [2.8, ●] $-4x + 7y - 2$ **10.** [2.8, ●●] $-4y - 2$

11. [2.9, ●●] x^{10}

MARGIN EXERCISES, SECTION 4.1, pp. 168–170

1. $x^2 + 2x - 8; x^3 + x^2 - 2x - 5, x - 7$, answers may vary   **2.** -19 **3.** -104 **4.** -13 **5.** 8   **6.** 28¢ per mile

7. $-9x^3 + (-4x^5)$ **8.** $-2x^3 + 3x^7 + (-7x)$ **9.** $3x^2, 6x, \dfrac{1}{2}$ **10.** $-4y^5, 7y^2, -3y, -2$ **11.** $4x^3$ and $-x^3$

12. $4t^4$ and $-7t^4$; $-9t^3$ and $10t^3$ **13.** $8x^2$ **14.** $2x^3 + 7$ **15.** $-\dfrac{1}{4}x^5 + 2x^2$ **16.** $-4x^3$ **17.** $5x^3$ **18.** $25 - 3x^5$

19. $6x$ **20.** $4x^3 + 4$ **21.** $-\dfrac{1}{4}x^3 + 4x^2 + 7$ **22.** $3x^2 + x^3 + 9$

EXERCISE SET 4.1, pp. 171–172

1. -18 **3.** 19   **5.** -12 **7.** Approx. 449   **9.** 2   **11.** 4   **13.** 11   **15.** $2, -3x, x^2$ **17.** $6x^2$ and $-3x^2$

19. $2x^4$ and $-3x^4, 5x$ and $-7x$ **21.** $-3x$ **23.** $-8x$ **25.** $11x^3 + 4$ **27.** $x^3 - x$ **29.** $4b^5$ **31.** $\dfrac{3}{4}x^5 - 2x - 42$

33. x^4 **35.** $\dfrac{15}{16}x^3 - \dfrac{7}{6}x^2$

MARGIN EXERCISES, SECTION 4.2, pp. 173–174

1. $6x^7 + 3x^5 - 2x^4 + 4x^3 + 5x^2 + x$ **2.** $7x^5 - 5x^4 + 2x^3 + 4x^2 - 3$ **3.** $14t^7 - 10t^5 + 7t^2 - 14$ **4.** $-2x^2 - 3x + 2$

5. $10x^4 - 8x - \dfrac{1}{2}$ **6.** $4, 2, 1, 0; 4$ **7.** $5, 6, 1, -1, 4$ **8.** x **9.** x^3, x^2, x, x^0 **10.** x^2, x **11.** x^3 **12.** Monomial

13. None of these   **14.** Binomial   **15.** Trinomial

EXERCISE SET 4.2, pp. 175–176

1. $x^5 + 6x^3 + 2x^2 + x + 1$ **3.** $15x^9 + 7x^8 + 5x^3 - x^2 + x$ **5.** $-5y^8 + y^7 + 9y^6 + 8y^3 - 7y^2$ **7.** $x^6 + x^4$

9. $13x^3 - 9x + 8$ **11.** $-5x^2 + 9x$ **13.** $12x^4 - 2x + \dfrac{1}{4}$ **15.** $3, 2, 1, 0; 3$ **17.** $2, 1, 6, 4; 6$ **19.** $-3, 6$ **21.** $6, 7, -8, -2$

23. x^2, x **25.** x^3, x^2, x^0 **27.** None missing   **29.** Trinomial   **31.** None of these   **33.** Binomial   **35.** Monomial

37. $x^2 - 16$

MARGIN EXERCISES, SECTION 4.3, pp. 177–178

1. $x^2 + 7x + 3$ **2.** $-4x^5 + 7x^4 + 3x^3 + 2x^2 + 4$ **3.** $24x^4 + 5x^3 + x^2 + 1$ **4.** $2x^3 + \dfrac{10}{3}$ **5.** $2x^2 - 3x - 1$

6. $8x^3 - 2x^2 - 8x + \dfrac{5}{2}$ **7.** $-8x^4 + 4x^3 + 12x^2 + 5x - 8$ **8.** $-x^3 + x^2 + 3x + 3$ **9.** 224   **10.** $\dfrac{7}{2}x^2$ **11.** 224

EXERCISE SET 4.3, pp. 179–182

1. $-x + 5$ **3.** $x^2 - 5x - 1$ **5.** $3x^5 + 13x^2 + 6x - 3$ **7.** $-4x^4 + 6x^3 + 6x^2 + 2x + 4$ **9.** $12x^2 + 6$

11. $5x^4 - 2x^3 - 7x^2 - 5x$ **13.** $9x^8 + 8x^7 - 3x^4 + 2x^2 - 2x + 5$ **15.** $-\dfrac{1}{2}x^4 + \dfrac{2}{3}x^3 + x^2$

17. $0.01x^5 + x^4 - 0.2x^3 + 0.2x + 0.06$ **19.** $-3x^4 + 3x^2 + 4x$ **21.** $3x^5 - 3x^4 - 3x^3 + x^2 + 3x$ **23.** $5x^3 - 9x^2 + 4x - 7$

25. $\dfrac{1}{4}x^4 - \dfrac{1}{4}x^3 + \dfrac{3}{2}x^2 + 6\dfrac{3}{4}x + \dfrac{1}{4}$ **27.** $-x^4 + 3x^3 + 2x + 1$ **29.** $x^4 + 4x^2 + 12x - 1$ **31.** $x^5 - 6x^4 + 4x^3 - x^2 + 1$

33. $7x^4 - 2x^3 + 7x^2 + 4x + 9$ **35.** $3x^5 + x^4 + 10x^3 + x^2 + 3x - 6$ **37.** $1.05x^4 + 0.36x^3 + 14.22x^2 + x + 0.97$

39. (a) $5x^2 + 4x$; (b) $57, 352$ **41.** $48.544x^6 - 0.795x^5 + 890x$

MARGIN EXERCISES, SECTION 4.4, pp. 183–184

1. $-12x^4 + 3x^2 - 4x$ **2.** $4x^4 - 3x^2 + 4x$ **3.** $13x^6 - 2x^4 + 3x^2 - x + \dfrac{5}{13}$ **4.** $7y^3 - 2y^2 + y - 3$ **5.** $-4x^3 + 6x - 3$

6. $-5x^4 - 3x^2 - 7x + 5$ **7.** $-14x^{10} + \dfrac{1}{2}x^5 - 5x^3 + x^2 - 3x$ **8.** $2x^3 + 2x + 8$ **9.** $x^2 - 6x - 2$

10. $-8x^4 - 5x^3 + 8x^2 - 1$ **11.** $x^3 - x^2 - \dfrac{4}{3}x - 0.9$ **12.** $2x^3 + 5x^2 - 2x - 5$ **13.** $-x^5 - 2x^3 + 3x^2 - 2x + 2$

EXERCISE SET 4.4, pp. 185–186

1. $5x$ **3.** $x^2 - 10x + 2$ **5.** $-12x^4 + 3x^3 - 3$ **7.** $-3x + 7$ **9.** $-4x^2 + 3x - 2$ **11.** $4x^4 + 6x^2 - \dfrac{3}{4}x + 8$

13. $2x^2 + 14$ **15.** $-2x^5 - 6x^4 + x + 2$ **17.** $9x^2 + 9x - 8$ **19.** $\dfrac{3}{4}x^3 - \dfrac{1}{2}x$ **21.** $0.06x^3 - 0.05x^2 + 0.01x + 1$

23. $3x + 6$ **25.** $4x^3 - 3x^2 + x + 1$ **27.** $11x^4 + 12x^3 - 9x^2 - 8x - 9$ **29.** $-4x^5 + 9x^4 + 6x^2 + 16x + 6$
31. $x^4 - x^3 + x^2 - x$ **33.** $569.607x^3 - 15.168x$

MARGIN EXERCISES, SECTION 4.5, pp. 187–189

1. $-15x$ **2.** $-x^2$ **3.** x^2 **4.** $-x^5$ **5.** $12x^7$ **6.** $-8x^{11}$ **7.** $7y^5$ **8.** 0 **9.** $8x^2 + 16x$ **10.** $-15x^3 + 6x^2$
11. $x^2 + 13x + 40$ **12.** $x^2 + x - 20$ **13.** $5x^2 - 17x - 12$ **14.** $6x^2 - 19x + 15$ **15.** $x^4 + 3x^3 + x^2 + 15x - 20$
16. $6x^5 - 20x^3 + 15x^2 + 14x - 35$ **17.** $3x^3 + 13x^2 - 6x + 20$ **18.** $20x^4 - 16x^3 + 32x^2 - 32x - 16$
19. $6x^4 - x^3 - 18x^2 - x + 10$

SOMETHING EXTRA—EXPANDED NOTATION AND POLYNOMIALS, p. 190

1. $8 \cdot 10^3 + 7 \cdot 10^2 + 6 \cdot 10 + 2; 8x^3 + 7x^2 + 6x + 2$ **2.** $7 \cdot 10^2 + 8 \cdot 10 + 6; 7x^2 + 8x + 6$
3. $1 \cdot 10^4 + 6 \cdot 10^3 + 4 \cdot 10^2 + 3 \cdot 10 + 2; x^4 + 6x^3 + 4x^2 + 3x + 2$ **4.** $7 \cdot 10^3 + 6 \cdot 10 + 3; 7x^3 + 6x + 3$

EXERCISE SET 4.5, pp. 191–192

1. $-12x$ **3.** $42x^2$ **5.** $30x$ **7.** $-2x^3$ **9.** x^6 **11.** $6x^6$ **13.** 0 **15.** $-0.02x^{10}$ **17.** $8x^2 - 12x$ **19.** $-6x^4 - 6x^3$
21. $4x^6 - 24x^5$ **23.** $-x^2 - 4x + 12$ **25.** $2x^2 - 15x + 25$ **27.** $9x^2 - 25$ **29.** $2x^2 + \frac{5}{2}x - \frac{3}{4}$ **31.** $x^3 + 7x^2 + 7x + 1$
33. $-10x^3 - 19x^2 - x + 3$ **35.** $3x^4 - 6x^3 - 7x^2 + 18x - 6$ **37.** $6t^4 + t^3 - 16t^2 - 7t + 4$ **39.** $x^4 - 1$

MARGIN EXERCISES, SECTION 4.6, pp. 193–194

1. $8x^3 - 12x^2 + 16x$ **2.** $10y^6 + 8y^5 - 10y^4$ **3.** $x^2 + 7x + 12$ **4.** $x^2 - 2x - 15$ **5.** $2x^2 + 9x + 4$
6. $2x^3 - 4x^2 - 3x + 6$ **7.** $12x^5 + 6x^2 + 10x^3 + 5$ **8.** $y^6 - 49$ **9.** $-2x^7 + x^5 + x^3$ **10.** $x^2 \quad \dfrac{16}{25}$
11. $x^5 + 0.5x^3 - 0.5x^2 - 0.25$ **12.** $8 + 2x^2 - 15x^4$ **13.** $30x^5 - 3x^4 - 6x^3$

EXERCISE SET 4.6, pp. 195–196

1. $4x^2 + 4x$ **3.** $-3x^2 + 3x$ **5.** $x^5 + x^2$ **7.** $6x^3 - 18x^2 + 3x$ **9.** $x^3 + 3x + x^2 + 3$ **11.** $x^4 + x^3 + 2x + 2$
13. $x^2 - x - 6$ **15.** $9x^2 + 15x + 6$ **17.** $5x^2 + 4x - 12$ **19.** $9x^2 - 1$ **21.** $4x^2 - 6x + 2$ **23.** $x^2 - \dfrac{1}{16}$
25. $x^2 - 0.01$ **27.** $2x^3 + 2x^2 + 6x + 6$ **29.** $-2x^2 - 11x + 6$ **31.** $x^2 + 14x + 49$ **33.** $1 - x - 6x^2$
35. $x^5 - x^2 + 3x^3 - 3$ **37.** $x^3 - x^2 - 2x + 2$ **39.** $3x^6 - 6x^2 - 2x^4 + 4$ **41.** $6x^7 + 18x^5 + 4x^2 + 12$
43. $8x^6 + 65x^3 + 8$ **45.** $4x^3 - 12x^2 + 3x - 9$ **47.** $4x^6 + 4x^5 + x^4 + x^3$

MARGIN EXERCISES, SECTION 4.7, pp. 197 198

1. $x^2 - 25$ **2.** $4x^2 - 9$ **3.** $x^2 - 4$ **4.** $x^2 - 49$ **5.** $9t^2 - 25$ **6.** $4x^6 - 1$ **7.** $x^2 + 16x + 64$ **8.** $x^2 - 10x + 25$
9. $x^2 + 4x + 4$ **10.** $a^2 - 8a + 16$ **11.** $4x^2 + 20x + 25$ **12.** $16x^4 - 24x^3 + 9x^2$ **13.** $y^2 + 18y + 81$
14. $9x^4 - 30x^2 + 25$ **15.** $x^2 + 11x + 30$ **16.** $t^2 - 16$ **17.** $-8x^5 + 20x^4 + 40x^2$ **18.** $81x^4 + 18x^2 + 1$
19. $4a^2 + 6a - 40$ **20.** $25x^2 + 5x + \dfrac{1}{4}$ **21.** $4x^2 - 2x + \dfrac{1}{4}$

SOMETHING EXTRA—FACTORS AND SUMS, p. 200

63	36	72	140	96	48	168	110	90	432	63
7	18	36	14	12	6	21	11	9	24	3
9	2	2	10	8	8	8	10	10	18	21
16	20	38	24	20	14	29	21	19	42	24

EXERCISE SET 4.7, pp. 201–202

1. $x^2 - 16$ **3.** $4x^2 - 1$ **5.** $25m^2 - 4$ **7.** $4x^4 - 9$ **9.** $9x^8 - 16$ **11.** $x^{12} - x^4$ **13.** $x^8 - 9x^2$ **15.** $x^{24} - 9$

17. $4x^{16} - 9$ **19.** $x^2 + 4x + 4$ **21.** $9x^4 + 6x^2 + 1$ **23.** $x^2 - x + \dfrac{1}{4}$ **25.** $9 + 6x + x^2$ **27.** $x^4 + 2x^2 + 1$

29. $4 - 12x^4 + 9x^8$ **31.** $25 + 60x^2 + 36x^4$ **33.** $9 - 12x^3 + 4x^6$ **35.** $4x^3 + 24x^2 - 12x$ **37.** $4x^4 - 2x^2 + \dfrac{1}{4}$

39. $-1 + 9p^2$ **41.** $15t^5 - 3t^4 + 3t^3$ **43.** $36x^8 + 48x^4 + 16$ **45.** $12x^3 + 8x^2 + 15x + 10$ **47.** $64 - 96x^4 + 36x^8$

49. $4567.0564x^2 + 435.891x + 10.400625$

EXTENSION EXERCISES, pp. 203–204

1. $-6a^2b - 2b^2$ **3.** $x^3 - 2x^2 - 6x + 3$ **5.** (a) Compare $(ax + b) + (cx + d)$ and $(cx + d) + (ax + b)$; $(ax + b) + (cx + d) = ax + b + cx + d = (a + c)x + (b + d)$ (collecting like terms); $(cx + d) + (ax + b) = cx + d + ax + b = (c + a)x + (d + b) = (a + c)x + (b + d)$ (commutative property of addition of real numbers). Since $(ax + b) + (cx + d) = (cx + d) + (ax + b)$, addition of binomials is commutative. (b) Compare $(ax^2 + bx + c) + (dx^2 + ex + f)$ and $(dx^2 + ex + f) + (ax^2 + bx + c)$ as in part (a).   **7.** No, it is not associative. Compare $(5x + 2) - [(2x - 4) - (-x + 1)]$ and $[(5x + 2) - (2x - 4)] - (-x + 1)$.

9. $x^3 - 7x + 6$ **11.** $x^3 - 5x^2 + 8x - 4$ **13.** $8 \text{ ft} \times 16 \text{ ft}$ **15.** -7 **17.** $16x^4 - 1$ **19.** $5a^2 + 12a - 9$ **21.** $\dfrac{1}{17}$

TEST OR REVIEW—CHAPTER 4, pp. 205–206

1. $-5x^2 + 2x + 1$ **2.** $\dfrac{1}{5}a^3 - \dfrac{58}{5}a$ **3.** $2x^5 + 3x^4 - 6x^3 - 8x^2 + 2x - 11$ **4.** $3x^5 - 9x^4 + 3x^3 + 2$ **5.** $8x^2 - 4x - 6$

6. $x^5 - 3x^3 - 2x^2 + 8$ **7.** $6a^3 + 5a^2 - 8a - 3$ **8.** $-18y^4 + 30y^3 + 12y^2$ **9.** $x^2 - 81$ **10.** $x^2 - 18x + 81$

11. $2x^2 - 11x - 21$ **12.** $49x^2 - 14x + 1$ **13.** $6p^5 + 9p^3 - 10p^2 - 15$ **14.** $4x^6 - 49$ **15.** $9x^2 + 12x + 4$

16. $15x^7 - 40x^6 + 50x^5 + 10x^4$

CHAPTER 5

READINESS CHECK, p. 208

1. [1.4, ▨] $4(y + 7 + 3z)$ **2.** [2.7, ▨] $8(x - 4)$ **3.** [3.3, ▨] 2 **4.** [3.5, ▨] $-\dfrac{3}{4}, 7$ **5.** [4.5, ▨] $-12x^{13}$

6. [4.6, ▨] $16x^3 - 48x^2 + 8x$ **7.** [4.6, ▨] $x^2 + 2x - 24$ **8.** [4.6, ▨] $28w^2 + 17w - 6$ **9.** [4.7, ▨] $t^2 - 18t + 81$

10. [4.7, ▨] $25x^2 + 30x + 9$ **11.** [4.7, ▨] $p^2 - 16$

MARGIN EXERCISES, SECTION 5.1, pp. 208–210

1. (a) $12x^2$; (b) $3x \cdot 4x$, $2x \cdot 6x$, answers may vary   **2.** (a) $16x^3$; (b) $(2x)(8x^2)$, $(4x)(4x^2)$, answers may vary

3. $8x \cdot x^3$; $4x^2 \cdot 2x^2$; $2x^3 \cdot 4x$, answers may vary   **4.** $7x \cdot 3x$, $(-7x)(-3x)$, $(21x)(x)$, answers may vary

5. $6x^4 \cdot x$, $(-2x^3)(-3x^2)$, $(3x^3)(2x^2)$, answers may vary   **6.** (a) $3x + 6$; (b) $3(x + 2)$

7. (a) $2x^3 + 10x^2 + 8x$; (b) $2x(x^2 + 5x + 4)$ **8.** $x(x + 3)$ **9.** $x^2(3x^4 - 5x + 2)$ **10.** $3x^2(3x^2 - 5x + 1)$

11. $\frac{1}{4}(3x^3 + 5x^2 + 7x + 1)$ **12.** $7x^3(5x^4 - 7x^3 + 2x^2 - 9)$ **13.** $(x + 2)(x + 5)$ **14.** $(x + 3)(x - 4)$

15. $(2x - 3)(x + 4)$ **16.** $(4x - 3)(4x + 5)$

EXERCISE SET 5.1, pp. 211–212

1. $6x^2 \cdot x$, $3x^2 \cdot 2x$, $(-3x^2)(-2x)$, answers may vary   **3.** $(-9x^4) \cdot x$, $(-3x^2)(3x^3)$, $(-3x)(3x^4)$, answers may vary

5. $(8x^2)(3x^2)$, $(-8x^2)(-3x^2)$, $(4x^3)(6x)$, answers may vary   **7.** $x(x - 4)$ **9.** $x^2(x + 6)$ **11.** $8x^2(x^2 - 3)$

13. $17x(x^4 + 2x^2 + 3)$ **15.** $5(2x^3 + 5x^2 + 3x - 4)$ **17.** $\frac{x^3}{3}(5x^3 + 4x^2 + x + 1)$ **19.** $(y + 1)(y + 4)$ **21.** $(x + 1)(x + 4)$

23. $(2x + 3)(3x + 2)$ **25.** $(x - 4)(3x - 4)$ **27.** $(5x + 3)(7x - 8)$ **29.** $(2x - 3)(2x + 3)$ **31.** $(2x^2 + 5)(x^2 + 3)$

MARGIN EXERCISES, SECTION 5.2, pp. 213–217

1. $(x + 3)(x + 4)$ **2.** $(x - 7)(x - 5)$ **3.** $(x - 2)(x + 1)$ **4.** $(2x + 1)(3x + 2)$ **5.** $3(2x + 3)(x + 1)$

6. $2(x + 3)(x - 1)$ **7.** $2(2x + 3)(x - 1)$ **8.** $(2x - 1)(3x - 1)$ **9.** $(2x + 5)(x - 3)$ **10.** $(4x + 1)(3x - 5)$

11. $(3x - 4)(x - 5)$ **12.** $2(5x - 4)(2x - 3)$

SOMETHING EXTRA—CALCULATOR CORNER: NESTED MULTIPLICATION, p. 218

1. $x(x(x(x(x + 3) - 1) + 1) - 1) + 9$; 6432.0122; $170,616.6106$; $-5,584,092$

2. $x(x(x(5x - 17) + 2) - 1) + 11$; 1476.2115; $43,101.696$; $1,607,136$

3. $x(x(x(2x - 3) + 5) - 2) + 18$; 1279.3152; $22,235.0912$; $598,892$

4. $x(x(x(x(-2x + 4) + 8) - 4) - 3) + 24$; -4120.9615; $-208,075.1491$; $13,892,691$

EXERCISE SET 5.2, pp. 219–220

1. $(x + 3)(x + 5)$ **3.** $(x - 1)(x - 1)$ **5.** $(x + 3)(x + 4)$ **7.** $(x + 5)(x - 3)$ **9.** $(y + 8)(y + 1)$ **11.** $(x^2 + 2)(x^2 + 3)$

13. $(x - 4)(x + 7)$ **15.** $(4 + x)(4 + x)$ **17.** $(a - 11)(a - 1)$ **19.** $\left(x - \frac{1}{5}\right)\left(x - \frac{1}{5}\right)$ **21.** $(y - 0.4)(y + 0.2)$

23. $(3x + 1)(x + 1)$ **25.** $4(3x - 2)(x + 3)$ **27.** $(2x + 1)(x - 1)$ **29.** $(3x - 2)(3x + 8)$ **31.** $5(3x + 1)(x - 2)$

33. $(3x + 4)(4x + 5)$ **35.** $(7x - 1)(2x + 3)$ **37.** $(3x^2 + 2)(3x^2 + 4)$ **39.** $(3x - 7)(3x - 7)$ **41.** $2x(3x + 5)(x - 1)$

43. Not factorable

MARGIN EXERCISES, SECTION 5.3, pp. 221–223

1. (a), (b), (d), (f)   **2.** $(x + 1)^2$ **3.** $(x - 1)^2$ **4.** $(x + 2)^2$ **5.** $(5x - 7)^2$ **6.** $(4x - 7)^2$ **7.** (a), (e), (f)   **8.** $(x - 3)(x + 3)$

9. $(x - 8)(x + 8)$ **10.** $5(1 - 2x^3)(1 + 2x^3)$

SOMETHING EXTRA—CALCUALTOR CORNER: A NUMBER PATTERN, p. 224

1. $1, 1, 1, 1, 1, 1$ **2.** $1, 1, 1$ **3.** $x^2 - (x + 1)(x - 1) = 1$; $x^2 - [x^2 - 1] = x^2 - x^2 + 1 = 1$

EXERCISE SET 5.3, pp. 225–226

1. Yes   **3.** No   **5.** $(x + 3)^2$ **7.** $2(x - 1)^2$ **9.** $x(x - 9)^2$ **11.** $(y - 6)^2$ **13.** $(8 - y)^2$ **15.** $3(2x + 3)^2$ **17.** $(x^3 - 8)^2$

19. $(7 - 3x)^2$ **21.** $5(y^2 + 1)^2$ **23.** No   **25.** Yes   **27.** $(x - 2)(x + 2)$ **29.** $2(2x - 7)(2x + 7)$

31. $(2x + 5)(2x - 5)$ **33.** $x(4 - 9x)(4 + 9x)$ **35.** $x^2(4 - 5x)(4 + 5x)$ **37.** $(7x^2 + 9)(7x^2 - 9)$ **39.** $a^2(a^5 + 2)(a^5 - 2)$

41. $(11a^4 + 10)(11a^4 - 10)$ **43.** $(x^2 + 1)(x - 1)(x + 1)$

45. Not true; $(12 - 5)^2 = 7^2 = 49$, $(12 + 5)(12 - 5) = 17 \cdot 7 = 119$

MARGIN EXERCISES, SECTION 5.4, pp. 227–228

1. $3(m^2 + 1)(m - 1)(m + 1)$ **2.** $(x^3 + 4)^2$ **3.** $2x^2(x + 1)(x + 3)$ **4.** $(3x^2 - 2)(x + 4)$ **5.** $8x(x - 5)(x + 5)$

EXERCISE SET 5.4, pp. 229–230

1. $2(x - 8)(x + 8)$ **3.** $(a - 5)^2$ **5.** $(2x - 3)(x - 4)$ **7.** $x(x + 12)^2$ **9.** $(x + 3)(x + 2)$ **11.** $6(2x + 3)(2x - 3)$

13. $4x(5x + 9)(x - 2)$ **15.** Not factorable   **17.** $(x^2 + 7)(x^2 - 3)$ **19.** $x^3(x - 7)^2$ **21.** $2(2 - x)(5 + x)$

23. Not factorable   **25.** $4(x^2 + 4)(x + 2)(x - 2)$ **27.** $(1 + y^4)(1 + y^2)(1 + y)(1 - y)$ **29.** $x^3(x^2 - 4x - 3)$

31. $\left(6a - \dfrac{5}{4}\right)^2$

MARGIN EXERCISES, SECTION 5.5, pp. 231–234

1. $-2, 3$ **2.** $7, -4$ **3.** 3 **4.** $0, 4$ **5.** $4, -4$ **6.** $5, -5$ **7.** $7, 8$ **8.** $-4, 5$ **9.** Length is 5 cm; width is 3 cm

10. (a) 342; (b) 9   **11.** 21 and 22; -22 and -21

EXERCISE SET 5.5, pp. 235–240

1. $-1, -5$ **3.** $0, 5$ **5.** -3 **7.** $\dfrac{5}{3}, -1$ **9.** $4, -\dfrac{5}{3}$ **11.** $\dfrac{2}{3}, -\dfrac{1}{4}$ **13.** $0, \dfrac{3}{5}$ **15.** $-\dfrac{3}{4}, 1$

17. 13 and 14, or -13 and -14 **19.** 13 and 15; -13 and -15 **21.** 5   **23.** 6 m   **25.** 506   **27.** 12   **29.** 780   **31.** 20

MARGIN EXERCISES, SECTION 5.6, pp. 241–242

1. $(x + 5)(x - 3)$ **2.** $(x - 9)(x + 7)$ **3.** $(x + 7)(x + 1)$ **4.** $(x - 2)(x + 18)$ **5.** $x^2 + 12x + 36$ **6.** $x^2 - 6x + 9$

7. $x^2 + 14x + 49$ **8.** $x^2 - 22x + 121$ **9.** $(x + 3)(x - 1)$ **10.** $(x + 4)(x + 2)$ **11.** $(x - 10)(x - 12)$

EXERCISE SET 5.6, pp. 243–244

1. $(x + 13)(x - 1)$ **3.** $(x + 3)(x + 11)$ **5.** $(x - 8)(x - 10)$ **7.** $(x^2 + 8)(x^2 + 12)$ **9.** $3(x^3 + 9)(x^3 - 7)$

11. $(a^2 - 2)(a + 2)(a - 2)$ **13.** $x^2 + 4x + 4$ **15.** $x^2 - 14x + 49$ **17.** $a^2 + 20d + 100$ **19.** $(x + 4)(x + 2)$

21. $(x - 6)(x - 2)$ **23.** $2(x + 8)(x - 16)$ **25.** $(a - 19)^2$ **27.** $10(b - 12)(b + 7)$ **29.** $(x + 18)(x - 44)$ **31.** 1437.774724

EXTENSION EXERCISES, pp. 245–246

1. $3x(x - 6)(x + 1)$ **3.** (a) No; (b) Yes; (c) No; (d) Yes; (e) No; (f) Yes   **5.** $[2(x - 3) + 3][3(x - 3) - 2] = (2x - 3)(3x - 11)$

7. $2x^2(4 - \pi)$ **9.** $[(a + 1) + 3][(a + 1) - 3] = (a + 4)(a - 2)$ **11.** $[4 + (2x + 1)][4 - (2x + 1)] = (5 + 2x)(3 - 2x)$

13. 9   **15.** 9   **17.** $(x + 3)(x - 3)(x^2 + 2)$ **19.** $(x + 2)(x - 2)(x - 1)$ **21.** $(y - 1)^3$ **23.** $x^2 - 2x - 3 = 0$

25. $x^3 - 2x^2 - 5x + 6 = 0$ **27.** 15 cm × 30 cm

TEST OR REVIEW—CHAPTER 5, pp. 247–248

1. $4x^3(x + 5)$ **2.** $(x - 5)(x + 3)$ **3.** $(x - 5)(x - 3)$ **4.** $x(x - 6)(x + 5)$ **5.** $3(x + 8)(x - 2)$ **6.** $(10x + 1)(x + 10)$

7. $(x - 3)^2$ **8.** $3(2x + 5)^2$ **9.** $(3a - 2)(3a + 2)$ **10.** $2(x - 5)(x + 5)$ **11.** $(4t^2 + 1)(2t + 1)(2t - 1)$ **12.** $-7, 5$

13. 16 and 18, or -16 and -18 **14.** $(x - 9)(x + 1)$

CHAPTER 6

READINESS CHECK, p. 250

1. $[3.1, \ \blacksquare\bullet\blacksquare \ ] \ -\dfrac{13}{6}$ **2.** $[3.3, \ \blacksquare\bullet\blacksquare \ ] \ 10$ **3.** $[3.3, \ \blacksquare\bullet\bullet\blacksquare \ ] \ -5$ **4.** $[4.1, \ \blacksquare\bullet\blacksquare \ ] \ -11$ **5.** $[4.1, \ \blacksquare\bullet\blacksquare \ ] \ 52$

MARGIN EXERCISES, SECTION 6.1, pp. 250–251

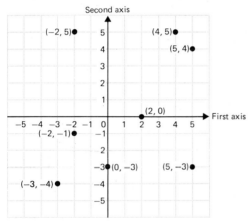

9. Both are negative numbers   **10.** First positive; second negaitve

11. I   **12.** III   **13.** IV   **14.** II   **15.** $B(-3, 5)$; $C(-4, -3)$; $D(2, -4)$;

$E(1, 5)$; $F(-2, 0)$; $G(0, 3)$

SOMETHING EXTRA—AN APPLICATION: COORDINATES, p. 252

1. Latitude 32.5° north, longitude 64.5° west   **2.** Latitude 27° north, longitude 81° west

EXERCISE SET 6.1, pp. 253–254

1.

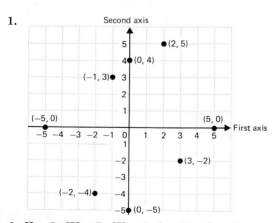

3. II   **5.** IV   **7.** III   **9.** I   **11.** Negative; negative

13. $A(3, 3)$; $B(0, -4)$; $C(-5, 0)$; $D(-1, -1)$; $E(2, 0)$

15.

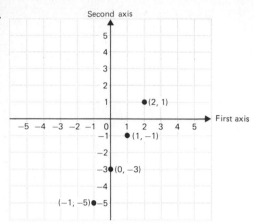

MARGIN EXERCISES, SECTION 6.2, pp. 255–259

1. No   **2.** Yes

3.

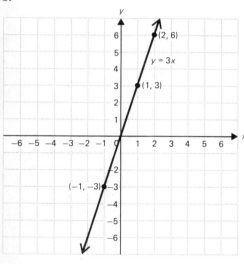

4.

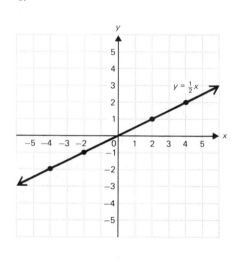

5.

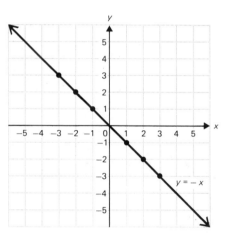

6.

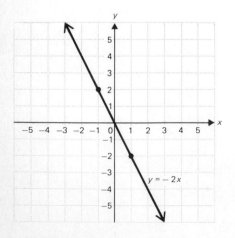

7.

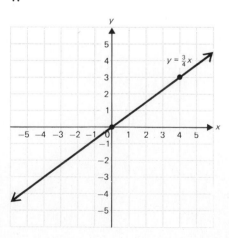

8.

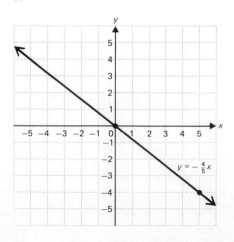

9. $y = x + 3$ looks like $y = x$ moved *up* 3 units.

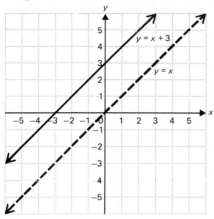

10. $y = x - 1$ looks like $y = x$ moved *down* 1 unit.

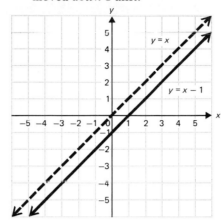

11. $y = 2x + 3$ looks like $y = 2x$ moved *up* 3 units.

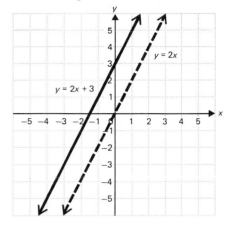

12.

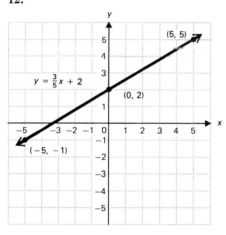

13.

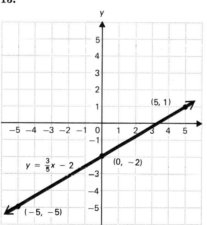

14.

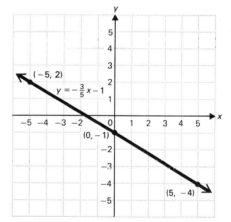

15.

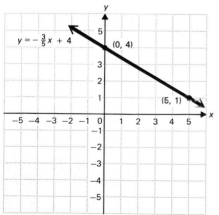

SOMETHING EXTRA—AN APPLICATION, p. 260

1. 57.848%   **2.** 62.6%   **3.** 66.56%   **4.** 70.52%

EXERCISE SET 6.2, pp. 261–262

1. Yes   **3.** No   **5.** No

7.

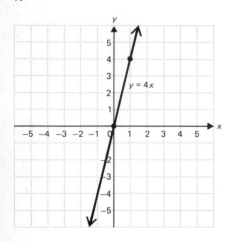

9.

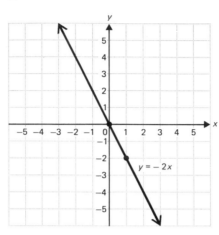

11.

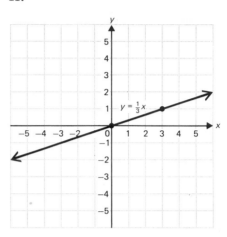

13.

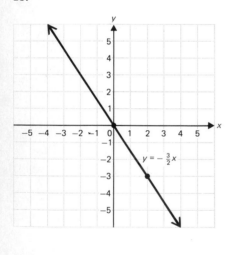

15.

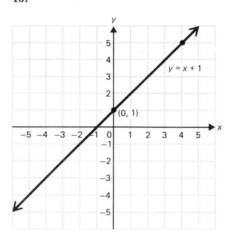

17.

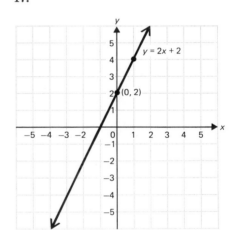

19.

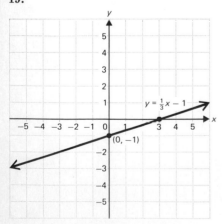

21.

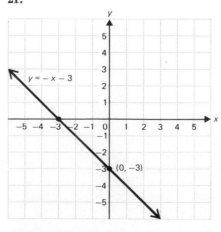

23.

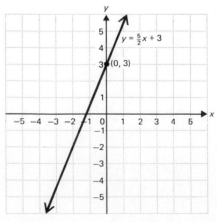

25.

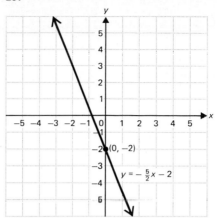

MARGIN EXERCISES, SECTION 6.3, pp. 263–265

1. (a) (4, 0); (b) (0, 3)

2. x-intercept is (3, 0); y-intercept is (0, 2)

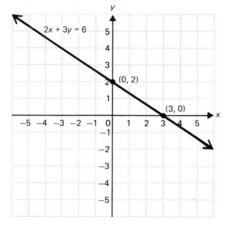

3. x-intercept is $(-3, 0)$; y-intercept is (0, 4)

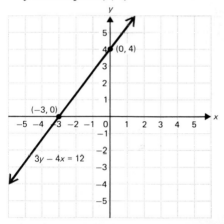

4.

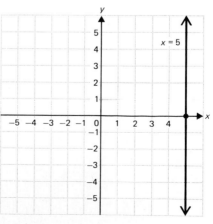

5.

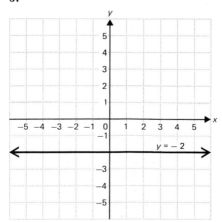

6.

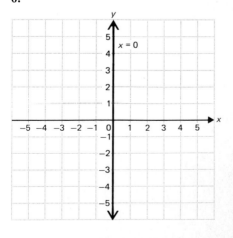

7.

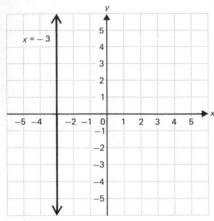

SOMETHING EXTRA—DEPRECIATION: THE STRAIGHT-LINE METHOD, p. 266

1. $3250, $2762.50, $2275, $1787.50, $1300

EXERCISE SET 6.3, pp. 267–268

1. x-intercept is $(3, 0)$;
y-intercept is $(0, -5)$

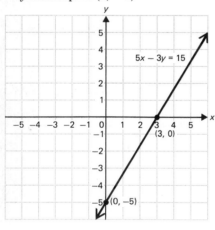

3. x-intercept is $(2, 0)$;
y-intercept is $(0, 4)$

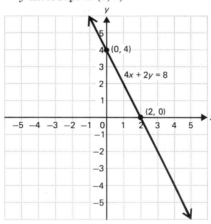

5. x-intercept is $(1, 0)$;
y-intercept is $(0, -1)$

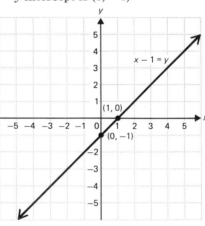

7. x-intercept is $\left(\frac{1}{2}, 0\right)$;
y-intercept is $(0, -1)$

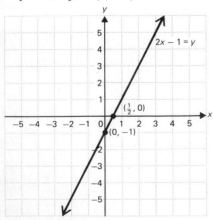

9. x-intercept is $(3, 0)$;
y-intercept is $(0, -4)$

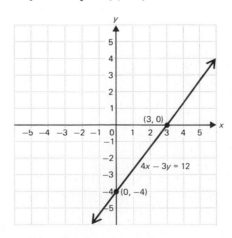

11. x-intercept is $\left(\frac{6}{7}, 0\right)$;
y-intercept is $(0, 3)$

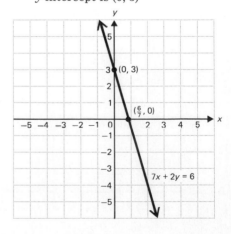

13. x-intercept is $(-1, 0)$;
   y-intercept is $(0, -4)$

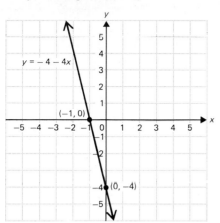

15.

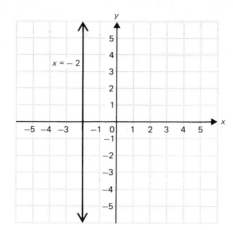

17.

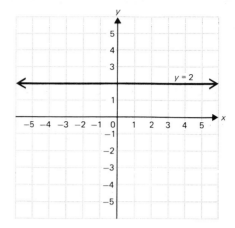

19.

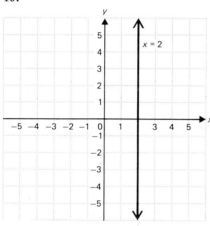

21.

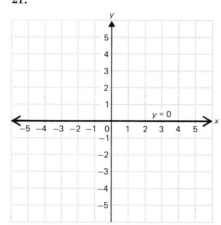

23.

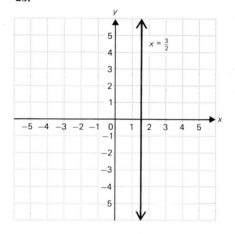

MARGIN EXERCISES, SECTION 6.4, pp. 269–270

1. $x + y = 115$, $x - y = 21$, where x is one number and y is the other

2. $31.95 + 0.33m = c$, $34.95 + 0.29m = c$, where m = mileage and c = cost

3. $2l + 2w = 76$, $l = w + 17$, where l = length and w = width

EXERCISE SET 6.4, pp. 271–272

1. $x + y = 58$, $x - y = 16$, where x = one number and y = the other number

3. $2l + 2w = 400$, $w = l - 40$, where l = length and w = width

5. $23.95 + 0.20m = c$, $24.95 + 0.27m = c$, where m = mileage and c = cost

7. $x - y = 16$, $3x = 7y$, where x = the larger number and y = the smaller

9. $x + y = 180$, $y = 3x + 8$, where x and y are the angles   **11.** $x + y = 90$, $x - y = 34$, where x and y are the angles

13. $x + y = 820$, $y = x + 140$, where x = number of hectares of Riesling and y = hectares of Chardonnay

MARGIN EXERCISES, SECTION 6.5, pp. 273–274

1. Yes   **2.** No   **3.** $(2, -3)$ **4.** No solution; lines are parallel.

EXERCISE SET 6.5, pp. 275–276

1. Yes   **3.** No   **5.** Yes   **7.** $(-12, 11)$ **9.** $(4, 3)$ **11.** No solution

MARGIN EXERCISES, SECTION 6.6, pp. 277–278

1. $(3, 2)$ **2.** $(3, -1)$ **3.** $\left(\dfrac{24}{5}, -\dfrac{8}{5}\right)$

EXERCISE SET 6.6, pp. 279–280

1. $(1, 3)$ **3.** $(1, 2)$ **5.** $(4, 3)$ **7.** $(-2, 1)$ **9.** $(-1, -3)$ **11.** $\left(\dfrac{17}{3}, \dfrac{16}{3}\right)$ **13.** $\left(\dfrac{25}{8}, -\dfrac{11}{4}\right)$ **15.** $(-3, 0)$ **17.** $(6, 3)$

19. $(4.3821792, 4.3281211)$

MARGIN EXERCISES, SECTION 6.7, pp. 281–284

1. $(3, 2)$ **2.** $(1, -1)$ **3.** $(1, 4)$ **4.** $(1, 1)$ **5.** $(1, -1)$ **6.** $\left(\dfrac{17}{13}, -\dfrac{7}{13}\right)$ **7.** No solution   **8.** $(1, -1)$

EXERCISE SET 6.7, pp. 285–286

1. $(9, 1)$ **3.** $(5, 3)$ **5.** $\left(3, -\dfrac{1}{2}\right)$ **7.** $\left(-1, \dfrac{1}{5}\right)$ **9.** $(-3, -5)$ **11.** No solution   **13.** $(2, -2)$ **15.** $(1, -1)$ **17.** $(-2, 3)$

19. $(8, 6)$ **21.** $(-1.8088509, 3.3817573)$

MARGIN EXERCISES, SECTION 6.8, pp. 287–288

1. 75 miles   **2.** A is 47; B is 21   **3.** Length is 27.5 cm; width is 10.5 cm

EXERCISE SET 6.8, pp. 289–292

1. 60 miles   **3.** Sammy is 44; his daughter is 22   **5.** 28 and 12   **7.** 43° and 137°   **9.** 62° and 28°

11. 480 hectares Chardonnay; 340 hectares Riesling   **13.** Length is 120 m; width is 80 m

15. 23 pheasants; 12 rabbits

MARGIN EXERCISES, SECTION 6.9, pp. 293–296

1. 168 km   **2.** 275 km/h   **3.** 324 mi   **4.** 3 hr

EXERCISE SET 6.9, pp. 297–300

1. 2 hr   **3.** 4.5 hr   **5.** $7\frac{1}{2}$ hr after the first train leaves   **7.** 14 km/h   **9.** 384 km   **11.** 330 km/h   **13.** 15 mi

15. 317.02702 km/h

MARGIN EXERCISES, SECTION 6.10, pp. 301–304

1. 7 quarters, 13 dimes   **2.** 125 adults, 41 children   **3.** $22\frac{1}{2}$ L of 50%, $7\frac{1}{2}$ L of 70%

4. 30 lb of A, 20 lb of B

EXERCISE SET 6.10, pp. 305–308

1. 70 dimes; 33 quarters   **3.** 300 nickels; 100 dimes   **5.** 203 adults; 226 children   **7.** 130 adults; 70 students

9. 40 g of A; 60 g of B **11.** 43.75 L   **13.** 80 L of 30%; 120 L of 50%

15. 6 kg of cashews; 4 kg of pecans

EXTENSION EXERCISES, pp. 309–310

1. I or IV   **3.** I or III   **5.** $(-1, -5)$ **7.**

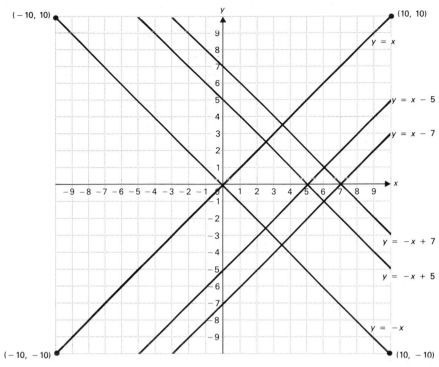

9. $-\dfrac{3}{2}$ **11.** Let x = Patrick's age, y = his father's age. $x = 0.2y$, $x + 20 = 0.52(y + 20)$

13. Let b = base, h = height. $\dfrac{1}{3}(b + 2) = h - 1$, $\dfrac{1}{2}(b + 2)(h - 1) = 24$ **15.** (a) No; (b) Yes   **17.** $(7, -1)$ **19.** $(2, -1)$

21. 6 yrs, 30 yrs   **23.** $b = 10$ ft, $h = 5$ ft   **25.** After $\dfrac{4}{3}$ hr   **27.** $12,500 at 12%, $14,500 at 13%   **29.** 75

31. 10 at $20 per day, 5 at $25 per day

TEST OR REVIEW—CHAPTER 6, pp. 311–312

1.

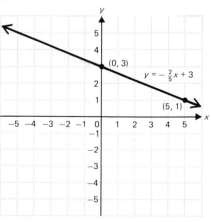

2.

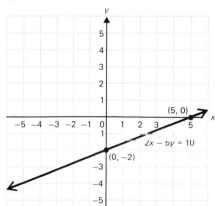

3.

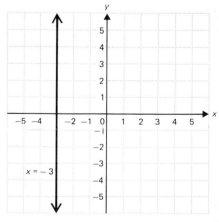

4. $(0, 5)$ **5.** $(1, 4)$ **6.** $(-2, 4)$ **7.** 5 and -3 **8.** 135 km/h   **9.** 297 orchestra seats; 211 balcony seats   **10.** 40 L of each

CHAPTER 7

READINESS CHECK, p. 314

1. [4.1, ▪] -15 **2.** [4.1, ▪▪] $3x^4 + 3x^3$ **3.** [4.3, ▪] $x^2 + 3x - 5$ **4.** [4.4, ▪▪▪] $7x^2 - 7x + 9$

5. [4.5, ▪▪▪] $3x^3 - 11x^2 - 22x + 10$ **6.** [4.6, ▪] $x^2 - x - 6$ **7.** [4.7, ▪] $9t^2 - 1$ **8.** [4.7, ▪▪] $p^2 + 8p + 16$

9. [5.1, ▪▪▪] $(x + 2)(x - 7)$ **10.** [5.2, ▪] $(3x + 1)(x - 2)$ **11.** [5.3, ▪▪] $(3t - 1)^2$ **12.** [5.3, ▪▪] $x^3(x + 4)(x - 4)$

13. [3.4, ▪] 3 **14.** [3.7, ▪] $r = \dfrac{d}{t}$

MARGIN EXERCISES, SECTION 7.1, pp. 314–316

1. -7940 **2.** -176 **3.** 32 **4.** $-3, 3, -2, 1, 2$ **5.** $3, 7, 1, 1, 0; 7$ **6.** $2x^2y + 3xy$ **7.** $5pq + 4$

8. $5xy^4 - 3xy^3 - 7xy^2 + 3xy$ **9.** $-2 + 5xy^2z + 2x^2yz + 5x^3yz^2$

EXERCISE SET 7.1, pp. 317–318

1. -1 **3.** -7 **5.** \$12,597.12   **7.** 44.4624 in^2 **9.** Coefficients: $1, -2, 3, -5$; degrees: $4, 2, 2, 0$; 4

11. Coefficients: $17, -3, -7$; degrees: $5, 5, 0$; 5 **13.** $-a - 2b$ **15.** $3x^2y - 2xy^2 + x^2$ **17.** $8u^2v - 5uv^2$

19. $-8au + 10av$ **21.** a^3 **23.** $-y^3 - xy^2 + 5x^2y$ **25.** $-y^3 - xy^2 + x^3$ **27.** $4n^2 - 3mn^3 + 5m^2n - m^4n$

29. $3y^2 + 2xy + x^2$ **31.** $-11uv^2 + 7u^2v + 5uv$

MARGIN EXERCISES, SECTION 7.2, pp. 319–320

1. $-4x^3 + 2x^2 - 4x + 2$ **2.** $14x^3y + 7x^2y - 3xy - 2y$ **3.** $-5p^2q^4 + 2p^2q^2 + 3p^2q + 6pq^2 + 3q + 5$

4. $-8s^4t + 6s^3t^2 + 2s^2t^3 - s^2t^2$ **5.** $-9p^4q + 10p^3q^2 - 4p^2q^3 - 9q^4$ **6.** $x^5y^5 + 2x^4y^2 + 3x^3y^3 + 6x^2$

7. $p^5q - 4p^3q^3 + 3pq^3 + 6q^4$

EXERCISE SET 7.2, pp. 321–322

1. $x^2 - 4xy + 3y^2$ **3.** $3r + 7$ **5.** $-x^2 - 8xy - y^2$ **7.** $2ab$ **9.** $-2a + 10b - 5c + 8d$ **11.** $6z^2 + 7zu - 3u^2$

13. $a^4b^2 - 7a^2b + 10$ **15.** $a^4 + a^3 - a^2y - ay + a + y - 1$ **17.** $a^6 - b^2c^2$ **19.** $y^6x + y^4x + y^4 + 2y^2 + 1$

21. $r^3 + r^2y + rs^2 + r^2s + rsy + s^3$ **23.** $5x^2y^2 + xy - 4$ **25.** $5y^2z^4 + 4yz^2 - 3$ **27.** $a^4 - b^4$

MARGIN EXERCISES, SECTION 7.3, pp. 323–324

1. $3x^3y + 6x^2y^3 + 2x^3 + 4x^2y^2$ **2.** $2x^2 - 11xy + 15y^2$ **3.** $16x^2 + 40xy + 25y^2$ **4.** $9x^4 - 12x^3y^2 + 4x^2y^4$

5. $4x^2y^4 - 9x^2$ **6.** $16y^2 - 9x^2y^4$ **7.** $9y^2 + 24y + 16 - 9x^2$ **8.** $4a^2 - 25b^2 - 10bc - c^2$

EXERCISE SET 7.3, pp. 325–326

1. $12x^2y^2 + 2xy - 2$ **3.** $12 - c^2d^2 - c^4d^4$ **5.** $m^3 + m^2n - mn^2 - n^3$ **7.** $x^9y^9 - x^6y^6 + x^5y^5 - x^2y^2$

9. $x^2 + 2xh + h^2$ **11.** $r^6t^4 - 8r^3t^2 + 16$ **13.** $p^8 + 2m^2n^2p^4 + m^4n^4$ **15.** $4a^6 - 2a^3b^3 + \dfrac{1}{4}b^6$

17. $3a^3 - 12a^2b + 12ab^2$ **19.** $4a^2 - b^2$ **21.** $c^4 - d^2$ **23.** $a^2b^2 - c^2d^4$ **25.** $x^2 + 2xy + y^2 - 9$

27. $x^2 - y^2 - 2yz - z^2$ **29.** $a^2 - b^2 - 2bc - c^2$ **31.** $-2ab + 2b^2$ **33.** $4ab$ **35.** $4y - 4$ **37.** $a^2 - 2ab + b^2 - b + a$

39. $A^3 + 3A^2B + 3AB^2 + B^3$

MARGIN EXERCISES, SECTION 7.4, pp. 327–328

1. $x^2y(x^2y + 2x + 3)$ **2.** $2p^4q^2(5p^2 - 2pq + q^2)$ **3.** $(a - b)(2x + 5 + y^2)$ **4.** $(a + b)(x^2 + y)$

5. $(x^2 + y^2)^2$ **6.** $-(2x - 3y)^2$ **7.** $(xy + 4)(xy + 1)$ **8.** $2(x^2y^3 + 5)(x^2y^3 - 2)$

9. $t^3(t - 2m)(t + m)$

EXERCISE SET 7.4, pp. 329–330

1. $12n^2(1 + 2n)$ **3.** $9xy(xy - 4)$ **5.** $2\pi r(h + r)$ **7.** $(a + b)(2x + 1)$ **9.** $(x - 1 - y)(x + 1)$ **11.** $(n + p)(n + 2)$

13. $(2x + z)(x - 2)$ **15.** $(x - y)^2$ **17.** $(3c + d)^2$ **19.** $(7m^2 - 8n)^2$ **21.** $(y^2 + 5z^2)^2$ **23.** $\left(\dfrac{1}{2}a + \dfrac{1}{3}b\right)^2$

25. $(a + b)(a - 2b)$ **27.** $(m + 20n)(m - 18n)$ **29.** $(mn - 8)(mn + 4)$ **31.** $a^3(ab + 5)(ab - 2)$ **33.** $a^3(a - b)(a + 5b)$

35. $(x^3 - y)(x^3 + 2y)$

MARGIN EXERCISES, SECTION 7.5, pp. 331–332

1. $(3a + 4x^2)(3a - 4x^2)$ **2.** $(5xy^2 + 2a)(5xy^2 - 2a)$ **3.** $5(1 - xy^3)(1 + xy^3)$ **4.** $y^2(4 + 9x^2)(2 - 3x)(2 + 3x)$

5. $(m^2 + 5t^2)^2$ **6.** $5(2p + 3q)(p - 4q)$ **7.** $a(3 + 5a)(a - 2b)$

EXERCISE SET 7.5, pp. 333–334

1. $(x - y)(x + y)$ **3.** $(ab - 3)(ab + 3)$ **5.** $(3x^2y - b)(3x^2y + b)$ **7.** $3(x + 4y)(x - 4y)$ **9.** $(8z - 5cd)(8z + 5cd)$

11. $7(p^2 + q^2)(p - q)(p + q)$ **13.** $(9a^2 + b^2)(3a - b)(3a + b)$ **15.** $2m^2(3m + 1)^2$ **17.** $(x + 3)(y - 2)(y + 2)$

19. $p(p + t)(p - 2t)$ **21.** $-(a + 3b)(a - 2b)$ **23.** $-(k - 18l)^2$ **25.** $ab(b^2 - b - 1)$

27. $(3b + a)(b - 6a)$ **29.** $(a + b)(a - b)(c + 2)(c - 2)$ **31.** $-(p - 10t)^2$ **33.** $2a(a - b)$

35. $(s^2 + t^2)(k + 3)(k - 3)$ **37.** $h(2x + h)$

MARGIN EXERCISES, SECTION 7.6, pp. 335–336

1. $y = \dfrac{3 + b}{a}$ **2.** $x = \dfrac{6}{b}$ **3.** $y = \dfrac{29 - 3t}{8}$

EXERCISE SET 7.6, pp. 337–338

1. $x = \dfrac{2a - 12}{3}$ **3.** $y = \dfrac{3b - 1}{2}$ **5.** $y = \dfrac{9}{2c}$ **7.** $x = \dfrac{7b}{4}$ **9.** $z = 2$ **11.** $x = \dfrac{a^2 - b^2}{a - 3}$ **13.** $x = \dfrac{m + 1}{m^2}$

15. $x = \dfrac{c}{rt}$ **17.** $x = \dfrac{c}{2a - b}$ **19.** $x = \dfrac{5m}{2}$

MARGIN EXERCISES, SECTION 7.7, p. 339

1. $M = \dfrac{fd^2}{km}$ **2.** $a^2 = \dfrac{6V - \pi h^3}{3\pi h}$

SOMETHING EXTRA—APPLICATIONS, p. 340

1. $D = \dfrac{1}{2}L - \dfrac{1}{4}C_1 - \dfrac{1}{4}C_2$

EXERCISE SET 7.7, pp. 341–342

1. $r = \dfrac{S}{2\pi h}$ **3.** $b = \dfrac{2A}{h}$ **5.** $n = \dfrac{S + 360}{180}$ **7.** $b = \dfrac{3V - kB - 4kM}{k}$ **9.** $r = \dfrac{S - a}{S - l}$ **11.** $h = \dfrac{2A}{b_1 + b_2}$ **13.** $a = \dfrac{v^2 pL}{r}$

15. $b_1 = \dfrac{2A - hb_2}{h}$ **17.** $E = \dfrac{180A}{\pi r^2}$ **19.** $M = \dfrac{-V - hB - hc}{4h}$ **21.** $L = \dfrac{ay}{v^2 p}$ **23.** $p = \dfrac{ar}{v^2 L}$ **25.** $n = \dfrac{a}{c(1 + b)}$

27. $F = \dfrac{9C + 160}{5}$ **29.** $q = \dfrac{mf + t}{m}$

EXTENSION EXERCISES, pp. 343–344

1. $5x^2 y^4$ **3.** $2\pi ab - \pi b^2$ **5.** $a^2 - 4b^2$ **7.** (a) 8; (b) 9; (c) 5 negative, 6 positive; (d) $220 x^9 y^3$; (e) 8   **9.** $[(y + 4) + x]^2$

11. $2(x + 2y)(x - 2y)(x^2 + 3y^2)$ **13.** $(5 + 3x + y)(5 - 3x - y)$ **15.** $(y + x - 4)(y - x + 6)$ **17.** $\dfrac{c + d}{2}$

19. $\dfrac{a^2 - 2ab + b^2}{2b + a}$ **21.** $2c(d + b)$ **23.** $\dfrac{abc}{12}$ **25.** V is multiplied by 8   **27.** $0, -3$

TEST OR REVIEW—CHAPTER 7, pp. 345–346

1. $4x^2 yz^3 + x^2 y^2$ **2.** $a^3 + 3a^2 b + 3ab^2 + b^3$ **3.** $2a^3 b - 10a^2 b^2 - ab^3 + 11$ **4.** $4x + 3y + 5z$

5. $2y^4 z^3 - y^3 z^2 - 4y^2 z^2 - 4yz + 3$ **6.** $a^3 b^3 + 2a^3 b - 3a^2 b^2 - 6a^2$ **7.** $a^4 + 4a^2 b + 4b^2$ **8.** $c^4 - d^2$

9. $pq(p^3 - p^2 - 1)$ **10.** $(2u + v)(w - 3x)$ **11.** $(4xy + 1)(4xy - 1)$ **12.** $(4c^2 + d^2)(2c - d)(2c + d)$

13. $c^2(c + x)(c - x)$ **14.** $(3p + 7q)^2$ **15.** $(5x - y)(7x - 3y)$ **16.** $x = -\dfrac{3}{4a}$ **17.** $P = \dfrac{A}{1 + rt}$

CHAPTER 8

READINESS CHECK, p. 348

1. [1.1, ●●●] $\dfrac{1}{16}$ **2.** [1.1, ●●●] 1   **3.** [1.5, ●] $\dfrac{5}{2}$ **4.** [1.5, ●] $\dfrac{3}{2}$ **5.** [1.1, ●●●] $\dfrac{9}{10}$ **6.** [1.1, ●●●] $\dfrac{37}{45}$

7. [4.4, ●●●] $9x + 12$ **8.** [4.7, ●] $x^2 - 4$ **9.** [5.2, ●] $(x + 2)(x + 1)$ **10.** [5.3, ●] $(x - 3)^2$

11. [5.1, ●] $2x(3x + 2)$ **12.** [5.3, ●●] $(5t - 2)(5t + 2)$ **13.** [5.2, ●] $(2x - 1)(x - 1)$ **14.** [5.4, ●] $x^3(x + 2)(x - 1)$

15. [3.3, ●●] $\dfrac{12}{5}$ **16.** [5.5, ●] 2,3

MARGIN EXERCISES, SECTION 8.1, pp. 349–352

1. $\dfrac{(x + 3)(x + 2)}{5(x + 4)}$ **2.** $\dfrac{-3 \cdot 4}{(2x + 1)(2x - 1)}$ **3.** $\dfrac{x(2x + 1)}{x(3x - 2)}$ **4.** $\dfrac{(x + 1)(x + 2)}{(x - 2)(x + 2)}$ **5.** $\dfrac{-1(x - 8)}{-1(x - y)}$ **6.** 5   **7.** $\dfrac{x}{3}$ **8.** $\dfrac{2x + 1}{3x + 2}$

9. $\dfrac{x + 1}{2x + 1}$ **10.** $x + 2$ **11.** $\dfrac{y + 2}{4}$ **12.** $\dfrac{a - 2}{a - 3}$ **13.** $\dfrac{x - 5}{2}$

EXERCISE SET 8.1, pp. 353–354

1. $\dfrac{(x - 2)(x - 2)}{(x - 5)(x + 5)}$ **3.** $\dfrac{(c - 3d)(c + 3d)}{(c + d)(c - d)}$ **5.** $\dfrac{(2a - 1)(3a - 1)}{(2a - 1)(3a + 2)}$ **7.** $\dfrac{3x + 2}{3x - 2}$ **9.** $\dfrac{a + b}{a - b}$ **11.** $\dfrac{t - 5}{t - 4}$ **13.** $\dfrac{x + 2}{2(x - 4)}$

15. $\dfrac{a - 3}{a - 4}$ **17.** $\dfrac{6}{x - 3}$ **19.** $\dfrac{a^2 + 1}{a + 1}$ **21.** $\dfrac{t}{t + 2}$ **23.** $\dfrac{12a}{a - 2}$ **25.** $\dfrac{1}{a}$ **27.** $\dfrac{21,375}{x^2 - 119,094.01}$

MARGIN EXERCISES, SECTION 8.2, pp. 355–356

1. $\dfrac{2}{7}$ **2.** $\dfrac{2x^3 - 1}{x^2 + 5}$ **3.** $\dfrac{1}{x - 5}$ **4.** $x^2 - 3$ **5.** $\dfrac{6}{35}$ **6.** $\dfrac{(x - 3)(x - 2)}{(x + 5)(x + 5)}$ **7.** $\dfrac{x - 3}{x + 2}$ **8.** $\dfrac{(x - 3)(x - 2)}{x + 2}$ **9.** $\dfrac{y + 1}{y - 1}$

EXERCISE SET 8.2, pp. 357–358

1. $\dfrac{x}{4}$ **3.** $\dfrac{1}{x^2 - y^2}$ **5.** $\dfrac{x^2 - 4x + 7}{x^2 + 2x - 5}$ **7.** $\dfrac{3}{10}$ **9.** $\dfrac{1}{4}$ **11.** $\dfrac{y^2}{x}$ **13.** $\dfrac{(a + 2)(a + 3)}{(a - 3)(a - 1)}$ **15.** $\dfrac{(x - 1)^2}{x}$ **17.** $\dfrac{1}{2}$ **19.** $\dfrac{3}{2}$

21. $\dfrac{(x + y)^2}{x^2 + y}$ **23.** $\dfrac{x + 3}{x - 5}$ **25.** $\dfrac{1}{(c - 5)^2}$ **27.** $\dfrac{t + 5}{t - 5}$

MARGIN EXERCISES, SECTION 8.3, pp. 359–360

1. $\dfrac{7}{9}$ **2.** $\dfrac{3 + x}{x - 2}$ **3.** $\dfrac{6x + 4}{x - 1}$ **4.** $\dfrac{x - 5}{4}$ **5.** $\dfrac{x - 1}{x - 3}$ **6.** $\dfrac{4}{11}$ **7.** $\dfrac{x^2 + 2x + 1}{2x + 1}$ **8.** $\dfrac{3x - 1}{3}$ **9.** $\dfrac{4x - 3}{x - 2}$

EXERCISE SET 8.3, pp. 361–364

1. 1   **3.** $\dfrac{6}{3 + x}$ **5.** $\dfrac{2x + 3}{x - 5}$ **7.** $\dfrac{1}{4}$ **9.** $-\dfrac{1}{t}$ **11.** $\dfrac{-x + 7}{x - 6}$ **13.** $y + 3$ **15.** $\dfrac{2b - 14}{b^2 - 16}$ **17.** $-\dfrac{1}{y + z}$ **19.** $\dfrac{5x + 2}{x - 5}$

21. -1 **23.** $\dfrac{-x^2 + 9x - 14}{(x - 3)(x + 3)}$ **25.** $\dfrac{1}{2}$ **27.** 1   **29.** $\dfrac{4}{x - 1}$ **31.** $\dfrac{8}{3}$ **33.** $\dfrac{13}{a}$ **35.** $\dfrac{4x - 5}{4}$ **37.** $\dfrac{x - 2}{x - 7}$ **39.** $\dfrac{2x - 16}{x^2 - 16}$

41. $\dfrac{2x - 4}{x - 9}$ **43.** $\dfrac{-9}{2x - 3}$ **45.** $\dfrac{18x + 5}{x - 1}$ **47.** 0   **49.** $\dfrac{20}{2y - 1}$

MARGIN EXERCISES, SECTION 8.4, pp. 365–366

1. 144   **2.** 12   **3.** 10   **4.** 120   **5.** $\dfrac{35}{144}$ **6.** $\dfrac{1}{4}$ **7.** $\dfrac{11}{10}$ **8.** $\dfrac{9}{40}$ **9.** $60x^3y^2$ **10.** $(y + 1)^2(y + 4)$ **11.** $7(t^2 + 16)(t - 2)$

12. $3x(x + 1)^2(x - 1)$, or $3x(x + 1)^2(1 - x)$

EXERCISE SET 8.4, pp. 367–368

1. 108   **3.** 72   **5.** 126   **7.** 360   **9.** 420   **11.** $\dfrac{59}{300}$ **13.** $\dfrac{71}{120}$ **15.** $\dfrac{23}{180}$ **17.** $8a^2b^2$ **19.** c^3d^2

21. $8(x - 1)$, or $8(1 - x)$ **23.** $(a + 1)(a - 1)^2$ **25.** $(3k + 2)(3k - 2)$ or $(3k + 2)(2 - 3k)$ **27.** $18x^3(x - 2)^2(x + 1)$

29. Every 60 years

MARGIN EXERCISES, SECTION 8.5, pp. 369–370

1. $\dfrac{x^2 + 6x - 8}{(x + 2)(x - 2)}$ **2.** $\dfrac{4x^2 - x + 3}{x(x - 1)(x + 1)^2}$ **3.** $\dfrac{8x + 88}{(x + 16)(x + 1)(x + 8)}$ **4.** $\dfrac{2a^2 - 5a + 15}{(a + 2)(a - 9)}$ **5.** $\dfrac{x - 3}{(x + 1)(x + 3)}$

EXERCISE SET 8.5, pp. 371–372

1. $\dfrac{2x + 5}{x^2}$ **3.** $\dfrac{x^2 + 4xy + y^2}{x^2y^2}$ **5.** $\dfrac{4x}{(x - 1)(x + 1)}$ **7.** $\dfrac{x^2 + 6x}{(x + 4)(x - 4)}$ **9.** $\dfrac{3x - 1}{(x - 1)^2}$ **11.** $\dfrac{x^2 + 5x + 1}{(x + 1)^2(x + 4)}$

13. $\dfrac{2x^2 - 4x + 34}{(x - 5)(x + 3)}$ **15.** $\dfrac{3a + 2}{(a + 1)(a - 1)}$ **17.** $\dfrac{2x + 6y}{(x + y)(x - y)}$ **19.** $\dfrac{3x^2 + 19x - 20}{(x + 3)(x - 2)^2}$

MARGIN EXERCISES, SECTION 8.6, p. 373

1. $\dfrac{-x - 7}{15x}$ **2.** $\dfrac{6x^2 - 2x - 2}{3x(x + 1)}$

SOMETHING EXTRA—AN APPLICATION: HANDLING DIMENSION SYMBOLS (PART II), p. 374

1. 4 yd   **2.** 96 oz   **3.** 27 km   **4.** 60 m   **5.** 3 g   **6.** 18 mi   **7.** 2.347 km   **8.** 550 mm   **9.** $0.7 \text{ kg}^2/\text{m}^2$

10. $720 \text{ lb-mi}^2/\text{ft-hr}^2$ **11.** 14 m-kg/sec^2

8

EXERCISE SET 8.6, pp. 375–376

1. $\dfrac{-(x + 4)}{6}$ **3.** $\dfrac{7z - 12}{12z}$ **5.** $\dfrac{4x^2 - 13xt + 9t^2}{3x^2t^2}$ **7.** $\dfrac{2x - 40}{(x + 5)(x - 5)}$ **9.** $\dfrac{3 - 5t}{2t(t - 1)}$ **11.** $\dfrac{2s - st - s^2}{(t + s)(t - s)}$ **13.** $\dfrac{2}{y(y - 1)}$

15. $\dfrac{z - 3}{2z - 1}$ **17.** $\dfrac{1 - 3x}{(2x - 3)(x + 1)}$ **19.** $\dfrac{1}{2c - 1}$

MARGIN EXERCISES, SECTION 8.7, pp. 377–378

1. $\dfrac{20}{21}$ **2.** $\dfrac{2(6 + x)}{5}$ **3.** $\dfrac{7x^2}{3(2 - x^2)}$ **4.** $\dfrac{x}{x - 1}$

EXERCISE SET 8.7, pp. 379–380

1. $\dfrac{25}{4}$ **3.** $\dfrac{1}{3}$ **5.** $\dfrac{1 + 3x}{1 - 5x}$ **7.** -6 **9.** $\dfrac{5}{3y^2}$ **11.** 8 **13.** $-\dfrac{1}{a}$ **15.** $\dfrac{x + y}{x}$ **17.** $\dfrac{x - 2}{x - 3}$

MARGIN EXERCISES, SECTION 8.8, pp. 381–382

1. $x^2 + 3x + 2$ **2.** $2x^2 + x - \dfrac{2}{3}$ **3.** $4x^2 - \dfrac{3}{2}x + \dfrac{1}{2}$ **4.** $2x^2 - 3x + 5$ **5.** $x - 2$ **6.** $x + 4$

7. $x + 4, \text{R } -2$; or $x + 4 + \dfrac{-2}{x + 3}$ **8.** $x^2 + x + 1$

EXERCISE SET 8.8, pp. 383–384

1. $1 - 2u - u^4$ **3.** $5t^2 + 8t - 2$ **5.** $-4x^4 + 4x^2 + 1$ **7.** $1 - 2x^2y + 3x^4y^5$ **9.** $x - 5 + \dfrac{-50}{x - 5}$; or $x - 5, \text{R } -50$

11. $x + 2$ **13.** $x - 2 + \dfrac{-2}{x + 6}$; or $x - 2, \text{R } -2$ **15.** $x^4 + x^3 + x^2 + x + 1$ **17.** $t^2 + 1$ **19.** $x^3 - 6$

MARGIN EXERCISES, SECTION 8.9, pp. 385–388

1. $\dfrac{33}{2}$ **2.** 3 **3.** $\dfrac{3}{2}$ **4.** $-\dfrac{1}{8}$ **5.** 1 **6.** 2 **7.** 4

EXERCISE SET 8.9, pp. 389–392

1. $\dfrac{12}{5}$ **3.** -2 **5.** 3 **7.** 10 **9.** 5 **11.** 3 **13.** $\dfrac{17}{2}$ **15.** No solution   **17.** $-4, -1$ **19.** -1 **21.** 3 **23.** 12

25. $\dfrac{5}{3}$ **27.** $\dfrac{2}{9}$ **29.** $\dfrac{1}{2}$ **31.** No solution

MARGIN EXERCISES, SECTION 8.10, pp. 393–396

1. -3 **2.** 40 km/h, 50 km/h   **3.** $\dfrac{24}{7}$, or $3\dfrac{3}{7}$ hr   **4.** $p = \dfrac{n}{2 - m}$ **5.** $f = \dfrac{pq}{p + q}$

EXERCISE SET 8.10, pp. 397–400

1. $\dfrac{20}{9}$, or $2\dfrac{2}{9}$ **3.** 20 and 15 **5.** 30 km/h, 70 km/h   **7.** 20 mph   **9.** Passenger: 80 km/h, freight: 66 km/h   **11.** $2\dfrac{2}{9}$ hr

13. $5\dfrac{1}{7}$ hr   **15.** $p = \dfrac{qf}{q - f}$ **17.** $A = P(1 + r)$ **19.** $R = \dfrac{r_1 r_2}{r_1 + r_2}$ **21.** $D = \dfrac{BC}{A}$ **23.** $h_2 = \dfrac{p(h_1 - q)}{q}$

MARGIN EXERCISES, SECTION 8.11, pp. 401–403

1. 58 km/L   **2.** 0.280 **3.** 124 km/h   **4.** 2.4 fish/yd^2 **5.** 3.45 **6.** 42 **7.** 2074 **8.** $\$0.20; \$0.008\dfrac{1}{3}$ **9.** $10\dfrac{2}{3}$

SOMETHING EXTRA—CALCULATOR CORNER: PRICE EARNINGS RATIO, p. 404

1. 6.0   **2.** 11.9   **3.** 4.7   **4.** 9.4

EXERCISE SET 8.11, pp. 405–408

1. 9   **3.** 2.3 km/h   **5.** $581\frac{9}{11}$ **7.** 702 km   **9.** 1.92 g   **11.** 287   **13.** (a) 1.92 tons; (b) 14.4 kg   **15.** 2500

17. $1500   **19.** 32 kg

EXTENSION EXERCISES, pp. 409–410

1. $\dfrac{t^2 - 1}{t^2 - 9}$ **3.** $\dfrac{t - 2}{t - 1}$ **5.** (a) Approaches 0; (b) Approaches 2; (c) Approaches 0; (d) Approaches 0; (e) Approaches 0

7. $\dfrac{x(x^2 + 1)}{3(x + y - 1)}$ **9.** $\dfrac{3(y + 2)^2(y + 5)}{(y - 4)(y - 2)}$ **11.** $\dfrac{12(x + 3)}{(x + 4)(x - 3)}$ **13.** 24 min   **15.** $\dfrac{(2z - 3)(z + 6)}{(z + 2)(z - 2)}$

17. $\dfrac{11z^4 - 22z^2 + 6}{(z^2 - 2)(z^2 + 2)(2z^2 - 3)}$ **19.** $\dfrac{x^2 + xy - x^3 + x^2y - y^2x + y^3}{(x^2 + y^2)(x + y)^2(x - y)}$ **21.** $\dfrac{5x + 3}{3x + 2}$

23. $\dfrac{(x - 1)(3x - 2)}{5x - 3}$ **25.** $5x^5 + 5x^4 - 8x^2 - 8x + 2$, R $(10x - 5)$ **27.** $y^3 - ay^2 + a^2y - a^3$, R $(a^2 + a^4)$ **29.** $-0.0\overline{3}$

31. $-\dfrac{1}{6}$ **33.** 6 hr, 12 hr   **35.** (a) $T = \dfrac{FP}{u + FE}$; (b) $n = \dfrac{1 - a + d}{d}$

37. If $p = kq$, then $q = \dfrac{1}{k}p$. Since k is a constant, so is $\dfrac{1}{k}$, and q varies directly as p.

TEST OR REVIEW—CHAPTER 8, pp. 411–412

1. $\dfrac{7x + 3}{x - 3}$ **2.** $\dfrac{a - 6}{5}$ **3.** $\dfrac{2x^2 - 2x}{x + 1}$ **4.** $\dfrac{3}{x - 2}$ **5.** -1 **6.** $\dfrac{2x + 3}{x - 2}$ **7.** $\dfrac{-x^2 + x + 26}{(x - 5)(x + 5)(x + 1)}$ **8.** 2

9. $2x^2 - 8x + 25 + \dfrac{-79}{x + 3}$; or $2x^2 - 8x + 25$, R -79 **10.** 1   **11.** 30 km/h, 20 km/h   **12.** $r = \dfrac{Re}{E - e}$

12. $r = \dfrac{Re}{E - e}$ **13.** 100

CHAPTER 9

READINESS CHECK, p. 414

1. [1.2, ▪▪▪] $5 \cdot 5$ **2.** [2.4, ▪] 25   **3.** [1.1, ▪▪▪] $\dfrac{16}{9}$ **4.** [1.1, ▪▪] $\dfrac{9}{25}$ **5.** [1.1, ▪▪] 3   **6.** [1.2, ▪▪] 0.875

7. [1.2, ▪▪] $0.45\overline{45}$ **8.** [2.1, ▪▪▪] 8   **9.** [2.9, ▪▪] x^6 **10.** [5.1, ▪▪] $x^2(x - 1)$ **11.** [5.3, ▪▪] $(x + 1)^2$ **12.** [3.3, ▪▪] 6

MARGIN EXERCISES, SECTION 9.1, pp. 414–416

1. 6, -6 **2.** 8, -8 **3.** 15, -15 **4.** 4   **5.** 7   **6.** 10   **7.** -4 **8.** -7 **9.** -13 **10.** $45 + x$ **11.** $\dfrac{x}{x + 2}$

12. Yes   **13.** No   **14.** Yes   **15.** No   **16.** Yes   **17.** No   **18.** $|xy|$ **19.** $|xy|$ **20.** $|x - 1|$ **21.** $|x + 4|$ **22.** $5|x|$

EXERCISE SET 9.1, pp. 417–418

1. 1, -1 **3.** 4, -4 **5.** 10, -10 **7.** 13, -13 **9.** 2   **11.** -3 **13.** -8 **15.** 15   **17.** 19   **19.** 18   **21.** $a - 4$

23. $t^2 + 1$ **25.** $\dfrac{3}{x + 2}$ **27.** Yes   **29.** No   **31.** Yes   **33.** No   **35.** $|x|$ **37.** $2|a|$ **39.** 5   **41.** $3|b|$ **43.** $|x - 7|$

45. $|x + 1|$ **47.** $|2x - 5|$

MARGIN EXERCISES, SECTION 9.2, pp. 419–420

1. An irrational number is a number that cannot be named by fractional notation $\frac{a}{b}$, where a and b are integers and $b \neq 0$.

2. Irrational   **3.** Rational   **4.** Irrational   **5.** Irrational   **6.** Rational   **7.** Rational   **8.** Rational   **9.** Rational

10. Irrational   **11.** $\frac{7}{128}$ **12.** -0.6781 **13.** 5.69895

14. The set of real numbers consists of the rational and the irrational numbers.   **15.** 2.646 **16.** 8.485

EXERCISE SET 9.2, pp. 421–422

1. Irrational   **3.** Irrational   **5.** Rational   **7.** Irrational   **9.** Rational   **11.** Rational   **13.** Rational   **15.** Irrational

17. Rational   **19.** Irrational   **21.** $\frac{1}{2}$ **23.** $\frac{41}{60}$ **25.** $6\frac{7}{24}$ **27.** -1.451 **29.** 0.3125 **31.** 2.236 **33.** 9 **35.** 9.644

37. 7.937 **39.** Rational; integers   **41.** $12.5, 15, 17.5, 20$

MARGIN EXERCISES, SECTION 9.3, pp. 423–424

1. (a) 8; (b) 8 **2.** $\sqrt{21}$ **3.** 5 **4.** $\sqrt{x^2 + x}$ **5.** $\sqrt{x^2 - 1}$ **6.** $4\sqrt{2}$ **7.** $|x + 7|$ **8.** $5|x|$ **9.** $6|m|$ **10.** $2\sqrt{19}$

11. $|x - 4|$ **12.** $8|t|$ **13.** $10|a|$ **14.** 16.585 **15.** 10.097

EXERCISE SET 9.3, pp. 425–426

1. $\sqrt{6}$ **3.** 3 **5.** $\sqrt{14}$ **7.** $\sqrt{\frac{3}{10}}$ **9.** $\sqrt{2x}$ **11.** $\sqrt{x^2 - 3x}$ **13.** $\sqrt{x^2 + 3x + 2}$ **15.** $\sqrt{x^2 - y^2}$ **17.** $\sqrt{86x}$ **19.** $2\sqrt{3}$

21. $5\sqrt{3}$ **23.** $10\sqrt{2x}$ **25.** $4|a|$ **27.** $7|t|$ **29.** $|x|\sqrt{x - 2}$ **31.** $|2x - 1|$ **33.** $3|a - b|$ **35.** 11.180 **37.** 18.972

39. 17.320 **41.** 11.043 **43.** 20 mph, 37.42 mph, 42.43 mph

MARGIN EXERCISES, SECTION 9.4, pp. 427–428

1. $4\sqrt{2}$ **2.** $5h\sqrt{2}$ **3.** $\sqrt{3}(x - 1)$ **4.** x^4 **5.** $(x + 2)^7$ **6.** $x^7\sqrt{x}$ **7.** $3\sqrt{2}$ **8.** 10 **9.** $4x^3y^2$ **10.** $5xy^2\sqrt{2xy}$

EXERCISE SET 9.4, pp. 429–430

1. $2\sqrt{6}$ **3.** $2\sqrt{10}$ **5.** $5\sqrt{7}$ **7.** $4\sqrt{3x}$ **9.** $2x\sqrt{7}$ **11.** $\sqrt{2}(2x + 1)$ **13.** t^3 **15.** $x^2 \cdot \sqrt{x}$ **17.** $(y - 2)^4$ **19.** $6m\sqrt{m}$

21. $8x^3y\sqrt{7y}$ **23.** $3\sqrt{6}$ **25.** $6\sqrt{7}$ **27.** 10 **29.** $5b\sqrt{3}$ **31.** $a\sqrt{bc}$ **33.** $6xy^3\sqrt{3xy}$ **35.** $10ab^2\sqrt{5ab}$

MARGIN EXERCISES, SECTION 9.5, pp. 431–432

1. $\frac{4}{3}$ **2.** $\frac{1}{5}$ **3.** $\frac{1}{3}$ **4.** $\frac{3}{4}$ **5.** $\frac{15}{16}$ **6.** $\frac{1}{5}\sqrt{15}$ **7.** $\frac{1}{4}\sqrt{10}$ **8.** 0.535 **9.** 0.791

EXERCISE SET 9.5, pp. 433–434

1. $\frac{3}{7}$ **3.** $\frac{1}{6}$ **5.** $-\frac{4}{9}$ **7.** $\frac{8}{17}$ **9.** $\frac{13}{14}$ **11.** $\frac{6}{a}$ **13.** $\frac{3a}{25}$ **15.** $\frac{1}{5}\sqrt{10}$ **17.** $\frac{1}{4}\sqrt{6}$ **19.** $\frac{1}{2}\sqrt{2}$ **21.** $\frac{1}{x}\sqrt{3x}$ **23.** 0.655

25. 0.577 **27.** 0.592 **29.** 1.549 **31.** 1.57 sec, 3.14 sec, 8.880 sec, 11.101 sec

MARGIN EXERCISES, SECTION 9.6, pp. 435–436

1. $12\sqrt{2}$ **2.** $5\sqrt{5}$ **3.** $-12\sqrt{10}$ **4.** $5\sqrt{6}$ **5.** $\sqrt{x + 1}$ **6.** $\frac{3}{2}\sqrt{2}$ **7.** $\frac{2}{15}\sqrt{15}$

SOMETHING EXTRA—AN APPLICATION: WIND CHILL TEMPERATURE, p. 436

1. $0°$ **2.** $-10°$ **3.** $-22°$ **4.** $-64°$

EXERCISE SET 9.6, pp. 437–438

1. $7\sqrt{2}$ **3.** $-8\sqrt{a}$ **5.** $8\sqrt{3}$ **7.** $\sqrt{3}$ **9.** $13\sqrt{2}$ **11.** $-24\sqrt{2}$ **13.** $(2+9x)\sqrt{x}$ **15.** $3\sqrt{2x+2}$

17. $(3xy - x^2 + y^2)\sqrt{xy}$ **19.** $\dfrac{2}{3}\sqrt{3}$ **21.** $\dfrac{13}{2}\sqrt{2}$ **23.** $\dfrac{1}{18}\sqrt{3}$

25. Any pairs of numbers a, b such that $a = 0$, $b \geqslant 0$; or $a \geqslant 0$, $b = 0$.

MARGIN EXERCISES, SECTION 9.7, p. 439

1. $\dfrac{1}{3}$ **2.** $\dfrac{\sqrt{3}}{3}$ **3.** $x\sqrt{6x}$ **4.** $\dfrac{\sqrt{35}}{7}$ **5.** $\dfrac{\sqrt{xy}}{y}$

SOMETHING EXTRA—CALCULATOR CORNER: FINDING SQUARE ROOTS ON A CALCULATOR, p. 440

1. 4.123   **2.** 8.944   **3.** 10.488   **4.** 8.307   **5.** 14.142   **6.** 3.240   **7.** 29.833   **8.** 16.303   **9.** 1.414   **10.** 0.484

11. 1.772   **12.** 1.932

EXERCISE SET 9.7, pp. 441–442

1. 3   **3.** 2   **5.** $\sqrt{5}$ **7.** $\dfrac{2}{5}$ **9.** 2   **11.** $3y$ **13.** $x^2\sqrt{5}$ **15.** 2   **17.** $\dfrac{\sqrt{10}}{5}$ **19.** $\sqrt{2}$ **21.** $\dfrac{\sqrt{6}}{2}$ **23.** 5   **25.** $\dfrac{\sqrt{3x}}{x}$

27. $\dfrac{4y\sqrt{3}}{3}$ **29.** $\dfrac{a\sqrt{2a}}{4}$ **31.** $\dfrac{\sqrt{2}}{4a}$

MARGIN EXERCISES, SECTION 9.8, pp. 443–444

1. $c = \sqrt{65} \approx 8.062$ **2.** $a = \sqrt{75} \approx 8.660$ **3.** $b = \sqrt{10} \approx 3.162$ **4.** $a = \sqrt{175} = 5\sqrt{7} \approx 13.23$ **5.** $5\sqrt{13} \approx 18.03$

EXERCISE SET 9.8, pp. 445–446

1. $c = 17$ **3.** $c = \sqrt{32} \approx 5.657$ **5.** $b = 12$ **7.** $b = 4$ **9.** $c = 26$ **11.** $b = 12$ **13.** $a = 2$ **15.** $b = \sqrt{2} \approx 1.414$

17. $a = 5$ **19.** $\sqrt{75} \approx 8.660$ m   **21.** $\sqrt{208} = 4\sqrt{13} \approx 14.424$ ft   **23.** $65\sqrt{2} \approx 91.92$ ft   **25.** $h = \dfrac{a}{2}\sqrt{3}$

MARGIN EXERCISES, SECTION 9.9, pp. 431–432

1. $\dfrac{64}{3}$ **2.** 2   **3.** 66   **4.** Approx. 313.040 km   **5.** Approx. 15.652 km   **6.** 676 m

EXERCISE SET 9.9, pp. 433–434

1. 25   **3.** 38.44   **5.** 397   **7.** $\dfrac{621}{2}$ **9.** 5   **11.** 3   **13.** $\dfrac{17}{4}$ **15.** No solution   **17.** No solution   **19.** 346.43 km, approx.

21. 11,236 m   **23.** 125 ft, 245 ft   **25.** 2.0772 sec, approx.

EXTENSION EXERCISES, pp. 451–452

1. 39   **3.** 5   **5.** 2   **7.** $-1\dfrac{1}{15}$ **9.** $\dfrac{3z+w}{4}$ **11.** 0.5   **13.** $3a^3$ **15.** x^{-1} **17.** $(x^4 + \sqrt{2}x^2y^2 + y^4)(x^4 - \sqrt{2}x^2y^2 + y^4)$

19. $2x^3\sqrt{5x}$ **21.** $0.2x^{2n}$ **23.** $\dfrac{(z^2-1)\sqrt{2}}{z^4}$ **25.** $\dfrac{\sqrt{6x}}{4x}$ **27.** $\dfrac{\sqrt{3y(2+y^2)}}{2+y^2}$ **29.** $13 + \sqrt{5}$ **31.** $x + 4\sqrt{xy} - 5y$ **33.** $\dfrac{\sqrt{30}}{2x}$

35. $-\sqrt{3} - \sqrt{5}$ **37.** $\dfrac{8 - 3\sqrt{6}}{10}$ **39.** ≈ 13.5 cm   **41.** $s\sqrt{3}$ **43.** 4   **45.** 5

TEST OR REVIEW—CHAPTER 9, pp. 453–454

1. 6   **2.** -9 **3.** $\dfrac{x}{2 + x}$ **4.** $|m|$ **5.** $7|t|$ **6.** $|x - 4|$ **7.** Irrational   **8.** Rational   **9.** Irrational   **10.** Rational

11. $\sqrt{21}$ **12.** $\sqrt{x^2 - 9}$ **13.** $\sqrt{6xy}$ **14.** $4\sqrt{3}$ **15.** $|x + 8|$ **16.** 10.392   **17.** $2\sqrt{10}$ **18.** $5xy\sqrt{2}$ **19.** $\dfrac{5}{8}$ **20.** $\dfrac{2}{3}$

21. $\dfrac{7}{t}$ **22.** $\dfrac{1}{2}\sqrt{2}$ **23.** $\dfrac{1}{4}\sqrt{2}$ **24.** $\dfrac{1}{y}\sqrt{5y}$ **25.** 0.354   **26.** 0.742   **27.** $16\sqrt{3}$ **28.** $\sqrt{5}$ **29.** $\dfrac{1}{2}\sqrt{2}$ **30.** $\dfrac{\sqrt{15}}{5}$ **31.** $\dfrac{x\sqrt{30}}{6}$

32. $\dfrac{2\sqrt{3}}{3}$ **33.** $a = \sqrt{45} \approx 6.708$ **34.** $\sqrt{98} \approx 9.899$ m   **35.** 26

CHAPTER 10

READINESS CHECK, p. 456

1. [4.7, ▣] $9x^2 + 6x + 1$ **2.** [5.6, ▣] 16   **3.** [5.6, ▣] $(x - 7)(x - 1)$ **4.** [9.1, ▣] 3   **5.** [9.3, ▣] $2\sqrt{5}$

6. [9.5, ▣] $\dfrac{17}{16}$ **7.** [9.2, ▣] 4.12   **8.** [9.7, ▣] $\dfrac{\sqrt{21}}{3}$ **9.** [5.5, ▣] 0, 6   **10.** [5.5, ▣] 2, 3

11. [6.3, ▣]

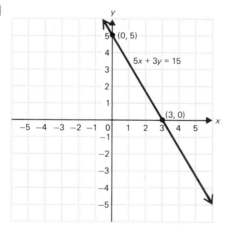

MARGIN EXERCISES, SECTION 10.1, pp. 456–458

1. $x^2 - 7x = 0$; $a = 1, b = -7, c = 0$ **2.** $x^2 + 9x - 3 = 0$; $a = 1, b = 9, c = -3$

3. $4x^2 + 2x + 4 = 0$; $a = 4, b = 2, c = 4$ **4.** $\sqrt{5}, -\sqrt{5}$ **5.** 0   **6.** $\dfrac{\sqrt{6}}{2}, -\dfrac{\sqrt{6}}{2}$ **7.** $\dfrac{3}{2}, -\dfrac{3}{2}$

SOMETHING EXTRA—CALCULATOR CORNER, p. 458

1. $d = \sqrt{45} \approx 6.708$

2. No; the length of the diagonal of the briefcase is 56.334 cm and this is the longest length that will fit in the briefcase.

EXERCISE SET 10.1, pp. 459–460

1. $x^2 - 3x - 2 = 0$; $a = 1, b = -3, c = -2$ **3.** $7x^2 - 4x + 3 = 0$; $a = 7, b = -4, c = 3$

5. $3x^2 - 2x + 8 = 0$; $a = 3, b = -2, c = 8$ **7.** $2, -2$ **9.** $7, -7$ **11.** $\sqrt{7}, -\sqrt{7}$ **13.** $\sqrt{10}, -\sqrt{10}$ **15.** $2\sqrt{2}, -2\sqrt{2}$

17. $\dfrac{5}{2}, -\dfrac{5}{2}$ **19.** $\dfrac{7\sqrt{3}}{3}, -\dfrac{7\sqrt{3}}{3}$ **21.** $\sqrt{3}, -\sqrt{3}$ **23.** 16 ft, 64 ft, 400 ft, 6400 ft

MARGIN EXERCISES, SECTION 10.2, pp. 461–462

1. $0, -\dfrac{5}{3}$ **2.** $0, \dfrac{3}{5}$ **3.** $\dfrac{2}{3}, -1$ **4.** $4, 1$ **5.** (a) 14; (b) 11

EXERCISE SET 10.2, pp. 463–464

1. $0, -7$ **3.** $0, -\dfrac{2}{3}$ **5.** $0, -1$ **7.** $0, \dfrac{1}{5}$ **9.** $4, 12$ **11.** $3, -7$ **13.** -5 **15.** $\dfrac{3}{2}, 5$ **17.** $4, -\dfrac{5}{3}$ **19.** $-2, 7$ **21.** $4, -5$

23. 9 **25.** 7 **27.** $0, -\dfrac{b}{a}$

MARGIN EXERCISES, SECTION 10.3, pp. 465–467

1. $7, -1$ **2.** $-3 \pm \sqrt{10}$ **3.** $1 \pm \sqrt{5}$ **4.** $x^2 - 8x + 16 = (x - 4)^2$ **5.** $x^2 + 10x + 25 = (x + 5)^2$

6. $x^2 + 7x + \dfrac{49}{4} = \left(x + \dfrac{7}{2}\right)^2$ **7.** $x^2 - 3x + \dfrac{9}{4} = \left(x - \dfrac{3}{2}\right)^2$ **8.** $\$1299.60$ **9.** 12.5% **10.** 73.2%

SOMETHING EXTRA—CALCULATOR CORNER: COMPOUND INTEREST, p. 468

1. $\$1360.49$ **2.** $\$1425.76$ **3.** (a) $\$1160.00$; (b) $\$1166.40$; (c) $\$1169.86$; (d) $\$1173.47$; (e) $\$1173.51$

4. (a) $\$2$; (b) $\$2.25$; (c) $\$2.44$; (d) $\$2.7146$; (e) $\$2.7181$

EXERCISE SET 10.3, pp. 469–470

1. $-7, 3$ **3.** $-1 \pm \sqrt{6}$ **5.** $3 \pm \sqrt{6}$ **7.** $x^2 - 2x + 1 = (x - 1)^2$ **9.** $x^2 + 18x + 81 = (x + 9)^2$

11. $x^2 - x + \dfrac{1}{4} = \left(x - \dfrac{1}{2}\right)^2$ **13.** $x^2 + 5x + \dfrac{25}{4} = \left(x + \dfrac{5}{2}\right)^2$ **15.** 10% **17.** 18.75% **19.** 8% **21.** 20% **23.** 12.6%
25. $\$2239.13$

MARGIN EXERCISES, SECTION 10.4, pp. 471–472

1. $-2, -6$ **2.** $5 \pm \sqrt{3}$ **3.** $-3 \pm \sqrt{10}$ **4.** $5, -2$ **5.** $-7, 2$ **6.** $\dfrac{-3 \pm \sqrt{33}}{4}$ **7.** $\dfrac{1 \pm \sqrt{10}}{3}$

EXERCISE SET 10.4, pp. 473–474

1. $-2, 8$ **3.** $-21, -1$ **5.** $1 \pm \sqrt{6}$ **7.** $9 \pm \sqrt{7}$ **9.** $-9, 2$ **11.** $-3, 2$ **13.** $\dfrac{7 \pm \sqrt{57}}{2}$ **15.** $\dfrac{-3 \pm \sqrt{17}}{4}$

17. $\dfrac{-2 \pm \sqrt{7}}{3}$ **19.** $-\dfrac{7}{2}, \dfrac{1}{2}$

MARGIN EXERCISES, SECTION 10.5, pp. 475–477

1. $\dfrac{1}{2}, -4$ **2.** $\dfrac{4 \pm \sqrt{31}}{5}$ **3.** $-0.3, 1.9$

SOMETHING EXTRA—AN APPLICATION: HANDLING DIMENSION SYMBOLS (PART III), p. 478
1. 6 ft   **2.** 1020 min   **3.** 172,800 sec   **4.** 0.1 hr   **5.** 600 g/cm   **6.** 30 mi/hr   **7.** 2,160,000 cm^2 **8.** 0.81 ton/yd^3

9. 150¢/hr   **10.** 60 person-days   **11.** Approx. 5,865,696,000,000 mi/yr   **12.** 6,570,000 mi/yr

EXERCISE SET 10.5, pp. 479–480

1. $-3, 7$ **3.** 3 **5.** $-\dfrac{4}{3}, 2$ **7.** $-3, 3$ **9.** $5 \pm \sqrt{3}$ **11.** $1 \pm \sqrt{3}$ **13.** No real-number solutions   **15.** $\dfrac{-1 \pm \sqrt{10}}{3}$

17. $\dfrac{-7 \pm \sqrt{61}}{2}$ **19.** $-1.3, 5.3$ **21.** $-0.2, 6.2$ **23.** $-1.7, 0.4$

10

MARGIN EXERCISES, SECTION 10.6, pp. 481–482

1. $13, 5$ **2.** 2 **3.** 7

EXERCISE SET 10.6, pp. 483–484

1. $6, -\dfrac{2}{3}$ **3.** $6, -4$ **5.** 1 **7.** $5, 2$ **9.** No solution   **11.** $\sqrt{7}, -\sqrt{7}$ **13.** $-2 \pm \sqrt{3}$ **15.** $2 \pm \sqrt{34}$ **17.** 9 **19.** 7

21. $1, 5$ **23.** 3 **25.** $25, 13$ **27.** No solution   **29.** 3

MARGIN EXERCISES, SECTION 10.7, pp. 485–486

1. $L = \dfrac{r^2}{20}$ **2.** $L = \dfrac{T^2 g}{4\pi^2}$ **3.** $m = \dfrac{E}{c^2}$ **4.** $r = \sqrt{\dfrac{A}{\pi}}$ **5.** $d = \sqrt{\dfrac{C}{P} + 1}$ **6.** $n = \dfrac{1 \pm \sqrt{1 + 4N}}{2}$ **7.** $t = \dfrac{-v \pm \sqrt{v^2 + 32h}}{16}$

EXERCISE SET 10.7, pp. 487–488

1. $A = \dfrac{N^2}{6.25}$ **3.** $T = \dfrac{cQ^2}{a}$ **5.** $c = \sqrt{\dfrac{E}{m}}$ **7.** $d = \dfrac{c \pm \sqrt{c^2 + 4aQ}}{2a}$ **9.** $a = \sqrt{c^2 - b^2}$ **11.** $t = \sqrt{\dfrac{2S}{g}}$

13. $r = \dfrac{-\pi h \pm \sqrt{\pi^2 h^2 + \pi A}}{\pi}$ **15.** $r = 6\sqrt{\dfrac{10A}{\pi S}}$ **17.** (a) $r = \dfrac{C}{2\pi}$; (b) $A = \dfrac{C^2}{4\pi}$

MARGIN EXERCISES, SECTION 10.8, pp. 489–490

1. 20 m   **2.** 2.3 cm; 3.3 cm   **3.** 3 km/h

EXERCISE SET 10.8, pp. 491–496

1. 3 cm   **3.** 7 ft; 24 ft   **5.** Width 8 cm; length 10 cm   **7.** Length 20 cm; width 16 cm   **9.** Width 5 m; length 10 m

11. 4.6 m; 6.6 m   **13.** Width 3.6 in.; length 5.6 in.   **15.** Width 2.2 m; length 4.4 m   **17.** 7 km/h   **19.** 8 mph

21. 4 km/h   **23.** 36 mph   **25.** $1 + \sqrt{2} \approx 2.41$ **27.** $d = 10\sqrt{2} = 14.14$; two 10-inch pizzas

MARGIN EXERCISES, SECTION 10.9, pp. 497–500

1. (a) Upward; (b)

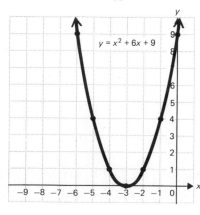

2. (a) Downward; (b)

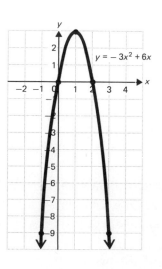

3. 2 **4.** $0.2, -2.2$

SOMETHING EXTRA—CALCULATOR CORNER: FINDING THE MEAN AND STANDARD DEVIATION, p. 500

1. $\bar{X} = 27.2$; $s = 11.02$ **2.** $\bar{X} = 29.8$; $s = 22.70$

EXERCISE SET 10.9, pp. 501–502

1. $y = x^2$

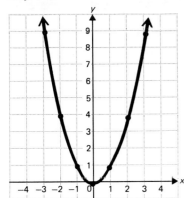

3. $y = -1 \cdot x^2$

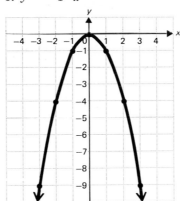

5. $y = -x^2 + 2x$

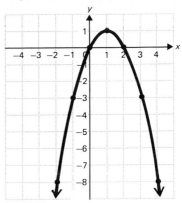

7. $y = 8 - x - x^2$

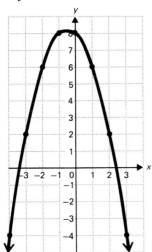

9. $y = x^2 - 2x + 1$

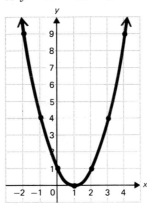

11. $y = -x^2 + 2x + 3$

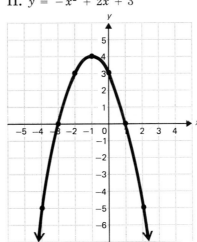

13. $y = -2x^2 - 4x + 1$

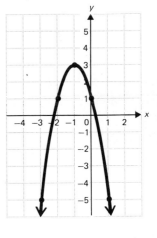

15. $2.2, -2.2$ **17.** $2.4, -3.4$ **19.** 4 **21.** No solution

EXTENSION EXERCISES, pp. 503–504

1. $\pm\dfrac{\sqrt{10}}{5}$ **3.** $\pm\dfrac{\sqrt{a(b^2 + 3b)}}{a}$ **5.** $0, -\sqrt{3}$ **7.** $0, -\dfrac{\sqrt{21}}{7}$ **9.** $-\dfrac{1}{4}, -3$ **11.** $1, 81$ **13.** $-b, 3b$ **15.** $0, \dfrac{4}{5}$

17. $x^2 + (2b - 4)x + (b - 2)^2$ **19.** $-2b \pm \sqrt{4b^2 - 2b}$ **21.** $\dfrac{b \pm \sqrt{b^2 - 12}}{6}$ **23.** $\dfrac{1 \pm \sqrt{1 - c}}{2}$ **25.** $\dfrac{5 \pm \sqrt{25 - 24b^2}}{4b}$

27. (a) $\dfrac{-3 \pm \sqrt{201}}{8}$; (b) $-4, 3$; (c) $\dfrac{-1 \pm 3\sqrt{5}}{6}$ **29.** $\dfrac{1}{3a}, 1$ **31.** 2 ft   **33.** ≈ 7.2 cm   **35.** (a) $\approx 3.4, \approx -2.4$; (b) $\approx 2.3, \approx -1.3$

TEST OR REVIEW—CHAPTER 10, pp. 505–506

1. $\sqrt{3}, -\sqrt{3}$ **2.** $0, \dfrac{7}{5}$ **3.** $\dfrac{1}{3}, -2$ **4.** $-8 \pm \sqrt{13}$ **5.** $1 \pm \sqrt{11}$ **6.** $-3 \pm 3\sqrt{2}$ **7.** $1.2, -7.2$ **8.** $2, -8$ **9.** 4

10. $T = \dfrac{LV^2}{25a}$ **11.** 30%   **12.** 1.7 m; 4.7 m   **13.** 30 km/h   **14.** (a) Upward;
(b) $y = x^2 - 4x - 2$

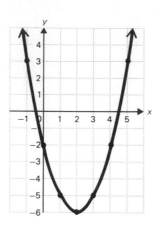

CHAPTER 11

READINESS CHECK, p. 508

1. [1.5, ▨] 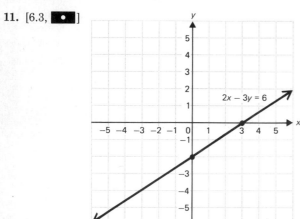   **2.** [1.5, ▨] <   **3.** [1.5, ▨] <   **4.** [2.1, ▨] 9

5. [2.1, ▨] 0   **6.** [2.1, ▨] 8   **7.** [3.1, ▨] -10 **8.** [3.1, ▨] 6   **9.** [3.2, ▨] $-\dfrac{1}{6}$ **10.** [3.3, ▨] -7

11. [6.3, ▨]

$2x - 3y = 6$

MARGIN EXERCISES, SECTION 11.1, pp. 508–510

1. (a) 1, 0, -6; answers may vary; (b) 2, 5, 8; answers may vary   **2.** $\{x|x > 2\}$ **3.** $\{x|x < 13\}$ **4.** $\{x|x < -3\}$
5. $\left\{x|x \leqslant \dfrac{2}{15}\right\}$ **6.** $\{y|y \leqslant -3\}$

EXERCISE SET 11.1, pp. 511–512

1. $\{x|x > -5\}$ **3.** $\{y|y > 3\}$ **5.** $\{x|x \leqslant -18\}$ **7.** $\{x|x \leqslant 16\}$ **9.** $\{y|y > -5\}$ **11.** $\{x|x > 3\}$ **13.** $\{x|x \geqslant 13\}$
15. $\{x|x < 4\}$ **17.** $\{c|c > 14\}$ **19.** $\left\{y|y \leqslant \dfrac{1}{4}\right\}$ **21.** $\left\{x|x > \dfrac{7}{12}\right\}$ **23.** $\{x|x > 0\}$ **25.** $\{r|r < -2\}$ **27.** $\{x|x \geqslant 1\}$
29. $\{S|S \geqslant 90\}$ **31.** $\{x|x \leqslant -18,058,999\}$

MARGIN EXERCISES, SECTION 11.2, pp. 513–514

1. $\{x|x < 8\}$ **2.** $\{y|y \geq 32\}$ **3.** $\{x|x \geq -6\}$ **4.** $\left\{y|y < -\dfrac{13}{5}\right\}$ **5.** $\left\{x|x > -\dfrac{1}{4}\right\}$ **6.** $\{x|x \leq -1\}$ **7.** $\left\{y|y \leq \dfrac{19}{9}\right\}$

EXERCISE SET 11.2, pp. 515–516

1. $\{x|x < 7\}$ **3.** $\{y|y \leq 9\}$ **5.** $\left\{x|x < \dfrac{13}{7}\right\}$ **7.** $\{x|x > -3\}$ **9.** $\left\{y|y \geq -\dfrac{2}{5}\right\}$ **11.** $\{x|x \geq -6\}$ **13.** $\{y|y \leq 4\}$

15. $\left\{x|x > \dfrac{17}{3}\right\}$ **17.** $\left\{y|y < -\dfrac{1}{14}\right\}$ **19.** $\left\{x\Big|\dfrac{3}{10} \geq x\right\}$ **21.** $\{x|x < -1\}$ **23.** $\{x|x \leq 3\}$ **25.** $\left\{t|t \leq \dfrac{9}{2}\right\}$ **27.** $\{c|c < -2\}$

29. $\{x|x \geq 4\}$ **31.** $\{m|m > 50\}$

MARGIN EXERCISES, SECTION 11.3, pp. 517–520

1. 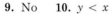   **2.**   **3.**   **4.**

5.    **6.**   **7.**   **8.**

9. No   **10.** $y < x$

11. $y \geq x + 2$

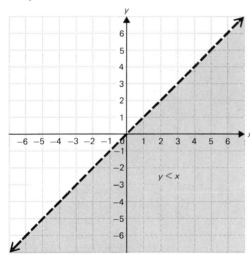

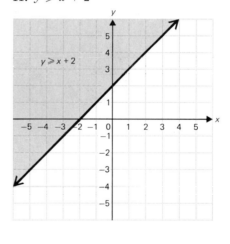

12. $2x + 4y < 8$

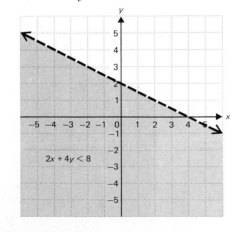

13. $3x - 5y < 15$

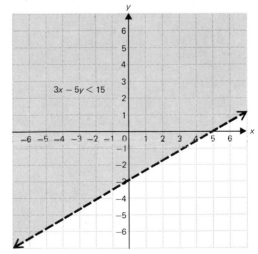

14. $2x + 3y \geqslant 12$

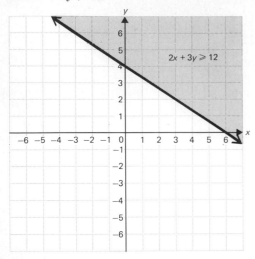

EXERCISE SET 11.3, pp. 521–522

1.

3.

5.

7.

9.

11.

13.

15. $y > x - 2$

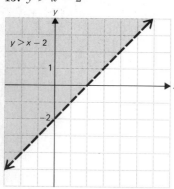

17. $6x - 2y \leqslant 12$

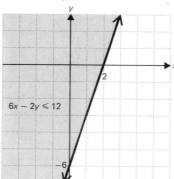

19. $3x - 5y \geqslant 15$

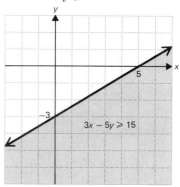

21. $y - 2x < 4$

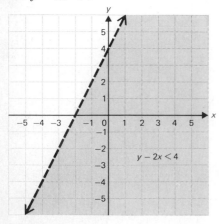

MARGIN EXERCISES, SECTION 11.4, pp. 523–524

1. {2, 3, 4, 5, 6, 7, 8, 9, 10}   **2.** {32, 34, 36, 38}   **3.** True   **4.** False   **5.** True   **6.** True   **7.** {a, 1, 9}   **8.** {2}   **9.** ∅

10. ∅   **11.** {0, 1, 2, 3, 4, 5, 6, 7}   **12.** The set of integers

EXERCISE SET 11.4, pp. 525–526

1. {3, 4, 5, 6, 7, 8}   **3.** {41, 43, 45, 47, 49}   **5.** {$\sqrt{3}$, $-\sqrt{3}$}   **7.** False   **9.** True   **11.** True   **13.** {c, d, e}   **15.** {1, 10}

17. ∅   **19.** {a, e, i, o, u, q, c, k}   **21.** {0, 1, 2, 5, 7, 10}   **23.** {a, e, i, o, u, m, n, f, g, h}

EXTENSION EXERCISES, pp. 527–528

1. (a) $x \geqslant 1$; (b) All real values of x; (c) $x \leqslant -2$ or $x \geqslant 2$; (d) $x \leqslant -2$ or $x \geqslant 1$ **3.** Base $\geqslant \dfrac{16}{5}$ cm

5. Gross sales < $20,000   **7.** Yes   **9.** No   **11.** Yes   **13.** No

15. (a) The set of integers; (b) ∅; (c) The set of real numbers; (d) The positive even integers

17. (a) Yes; (b) No; (c) No; (d) Yes; (e) Yes; (f) No

TEST OF REVIEW—CHAPTER 11, pp. 529–530

1. {y|$y \geqslant -6$}   **2.** {x|$x < -11$}   **3.** {x|$x \geqslant 7$}   **4.** {y|$y > 7$}   **5.** {y|$y > 2$}   **6.** $\left\{ x | x > -\dfrac{9}{11} \right\}$

7. 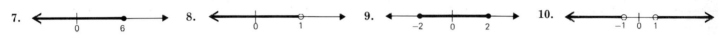   **8.**   **9.**   **10.**

11. $y < x + 3$ **12.** $2x + 3y \leqslant 6$ **13.** {8, 7, 0}

14. {8, 9, 10, 5, 7, 0, 2, 11}

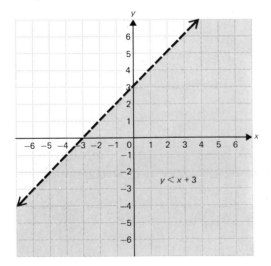

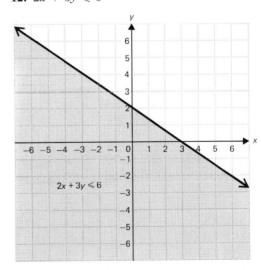

FINAL EXAMINATION, pp. 531–538

1. [10] $x \cdot x \cdot x$ **2.** [9] $\dfrac{136}{10}$ **3.** [9] 0.68   **4.** [26] $\dfrac{4}{5}$ **5.** [71] 7   **6.** [75] 9   **7.** [88] 15   **8.** [93] $-\dfrac{3}{20}$ **9.** [83] -63

10. [105] 3^{-4} **11.** [105] x^{-4} **12.** [106] y^7 **13.** [127] -8 **14.** [131] -12 **15.** [139] 7   **16.** [159] $b = \dfrac{tA}{4}$

17. [150] 24 and 25   **18.** [173] $10x^3 + 3x - 3$ **19.** [177] $7x^3 - 2x^2 + 4x - 17$ **20.** [184] $8x^2 - 4x - 6$

21. [193] $-8x^4 + 6x^3 - 2x^2$ **22.** [188] $6x^3 - 17x^2 + 16x - 6$ **23.** [197] $t^2 - 25$ **24.** [197] $9m^2 - 12m + 4$

25. [209] $4x(2x - 1)$ **26.** [223] $(5x - 2)(5x + 2)$ **27.** [213] $(3x + 2)(2x - 3)$ **28.** [222] $(x - 4)^2$

29. [210] $(x - 5)(x - 8)$ **30.** [231] $3, 5$ **31.** [232] $4, -5$ **32.** [242] $(x - 2)(x - 6)$

33. [258]

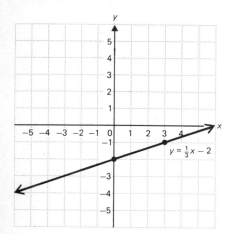

34. [263]

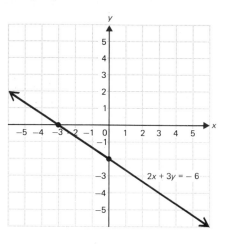

35. [264]

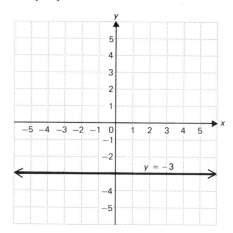

36. [277] $(1, 2)$ **37.** [281] $(12, 5)$ **38.** [283] $(6, 7)$ **39.** [287] 11 and 13   **40.** [316] $7x^2yz^3 + 5x^2y^2 - 5xy^2$

41. [316] $5a^4x^4 + 3ax^3 + x^2 + 2a$ **42.** [319] $2x^3y - 7x^2y^2 + xy^3 + 6$ **43.** [319] $15x^2y^3 + x^2y^2 + 5xy^2 + 7$

44. [320] $x^4 - 4y^2$ **45.** [323] $9x^2 + 24xy^2 + 16y^4$ **46.** [331] $(5ab - 1)(5ab + 1)$ **47.** [331] $(x - y - 5)(x - y + 5)$

48. [327] $(3x + 5y)^2$ **49.** [328] $(5x - 2y)(3x + 4y)$ **50.** [327] $(2a + d)(c - 3b)$ **51.** [331] $(2x + y^2)(2x - y^2)(4x^2 + y^4)$

52. [331] $3(a^2 + 6)(a + 2)(a - 2)$ **53.** [352] $\dfrac{2}{x + 3}$ **54.** [358] $\dfrac{3a(a - 1)}{2(a + 1)}$ **55.** [369] $\dfrac{27x - 4}{5x(3x - 1)}$

56. [373] $\dfrac{-x^2 + x + 2}{(x + 4)(x - 4)(x - 5)}$ **57.** [385] $4, -6$ **58.** [395] $m = \dfrac{tn}{t + n}$ **59.** [424] $5\sqrt{2}$ **60.** [423] $\dfrac{2\sqrt{10}}{5}$ **61.** [435] $12\sqrt{3}$

62. [447] 2   **63.** [461] $3, -2$ **64.** [475] $\dfrac{-3 \pm \sqrt{29}}{2}$

65. [497]

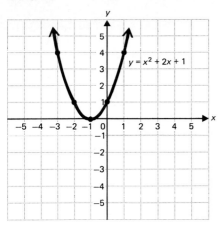

66. [489] 12 m

67. [514] $\{x|x \geq -1\}$

68. [518] $4x - 3y > 12$

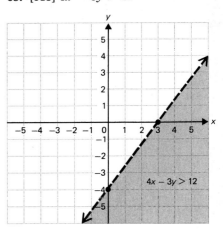

69. [523] $\{c, d\}$　　**70.** [524] $\{a, b, c, d, e, f, g, h\}$

Fractional and Decimal Equivalents

FRACTIONAL NOTATION	DECIMAL NOTATION	PERCENT NOTATION
$\frac{1}{10}$	0.1	10%
$\frac{1}{8}$	0.125	12.5% or $12\frac{1}{2}$%
$\frac{1}{6}$	$0.16\overline{6}$	$16.6\overline{6}$% or $16\frac{2}{3}$%
$\frac{1}{5}$	0.2	20%
$\frac{1}{4}$	0.25	25%
$\frac{3}{10}$	0.3	30%
$\frac{1}{3}$	$0.333\overline{3}$	$33.3\overline{3}$% or $33\frac{1}{3}$%
$\frac{3}{8}$	0.375	37.5% or $37\frac{1}{2}$%
$\frac{2}{5}$	0.4	40%
$\frac{1}{2}$	0.5	50%
$\frac{3}{5}$	0.6	60%
$\frac{5}{8}$	0.625	62.5% or $62\frac{1}{2}$%
$\frac{2}{3}$	$0.666\overline{6}$	$66.6\overline{6}$% or $66\frac{2}{3}$%
$\frac{7}{10}$	0.7	70%
$\frac{3}{4}$	0.75	75%
$\frac{4}{5}$	0.8	80%
$\frac{5}{6}$	$0.83\overline{3}$	$83.3\overline{3}$% or $83\frac{1}{3}$%
$\frac{7}{8}$	0.875	87.5% or $87\frac{1}{2}$%
$\frac{9}{10}$	0.9	90%
$\frac{1}{1}$	1	100%